T0397856

Continuum Mechanics

CONTINUUM MECHANICS

ANDRUS KOPPEL
AND
JAAK OJA
EDITORS

Nova Science Publishers, Inc.
New York

For permission to use material from this book please contact us:
Telephone 631-231-7269; Fax 631-231-8175
Web Site: http://www.novapublishers.com

LIBRARY OF CONGRESS CATALOGING-IN-PUBLICATION DATA

Continuum mechanics / [edited by] Andrus Koppel, Jaak Oja.
p. cm.
Includes index.
ISBN 978-1-60741-585-5 (hardcover)
1. Continuum mechanics. I. Koppel, Andrus. II. Oja, Jaak.
QA808.2.C66 2009
531--dc22

2009027482

Published by Nova Science Publishers, Inc. ✦ New York

CONTENTS

PREFACE

Continuum mechanics is a branch of mechanics that deals with the analysis of the kinematics and mechanical behavior of materials modeled as a continuum (e.g., solids and fluids, liquids and gases). A continuum concept assumes that the substance of the body is distributed throughout — and completely fills — the space it occupies. Differential equations can be employed in solving problems in continuum mechanics. Some of these differential equations are specific to the materials being investigated and are called constitutive equations, while others capture fundamental physical laws, such as the conservation of mass (the continuity equation), the conservation of momentum (the equations of motion and equilibrium), and the conservation of energy (the first law of thermodynamics). This new and important book gathers the latest research from around the globe in this field.

Chapter 1 - The authors study differential geometry very elementarily, but hopefully in a practical way. The authors divide vector spaces into material and spatial spaces since these spaces behave differently in the observer transformation and with objective derivatives (Lie-derivatives). All the vector spaces, which they consider, have a metric tensor thus they are metric vector spaces, and all the finite dimensional manifolds are Riemannian manifolds that are embedded in a Euclidean space. Additionally, the authors may identify a dual vector space by its primary vector space. In classical tensor analysis, this identification is applied, but here they make distinction between primary and dual spaces in the formulation, and the identification can be accomplished later. If the identification of dual and primary vector spaces is done a priori, then push-forward and pull-back operations are not uniquely defined. As an example, the authors will study the rotation manifold and its underlying geometric structure in the terms of differential geometry.

In continuum mechanics, the authors have different manifolds. The placement field of continuum medium takes values in a Hilbert space, where a chart parametrization maps vector-valued functions into vector-valued functions. The placement field needs an infinite number of basis functions in order to present an arbitrary placement field on continuum, yielding infinite-dimensional manifolds.

In Lagrangian mechanics, forces are divided differently into constraint forces and applied forces. In addition, rich mathematical methods like variational calculus and other mechanical principles are included. In terms of differential geometry, Lagrangian mechanics describes a motion on an event manifold with a Lagrangian functional on the tangent bundle of the event manifold. An event manifold is a time-placement manifold which is also a constraint manifold, i.e. the time and placement variables satisfy all the constraints.

The authors think that there is a need for another type of mechanics between Newtonian and Lagrangian mechanics, as it is sometimes done. This mechanics could be named d'Alembertian mechanics where the principle of virtual work is its cornerstone. Here the authors include inertial forces also in the virtual work form. The virtual work may be viewed as a linear form on the tangent field-bundle. This field-bundle is also a tangent bundle of the placement manifold at fixed time. The authors give definitions for the virtual work in the finite-dimensional and infinite-dimensional cases. In addition they give definition for the variation, Lie derivative and Lie variation. The concept of push-forward and pull-back operators is essential for understanding Lie derivatives and variations.

Finally, the authors consider constraint point-manifolds that arise from point-wise holonomic constraint equations. The usual geometric joints of a multibody system like spherical, revolute, cylindrical, universal, helical, prismatic, and sliding joints can be presented via holonomic constraint equations that only depend on displacement at corresponding geometric points. All these constraints generate a smooth point-manifold that can be parametrized. Also, the principle of virtual work and its geometric structure are naturally related with the parametrization of the constraint manifold

Chapter 2 - Over the years many shell finite elements has been developed for applications in aerospace, automotive and shipbuilding industries. Misuse or abuse of some of these shell finite elements in the relatively mature shell finite element technology is not uncommon. Some of these shell finite elements were based on the principles of classical shell theory in which the simplest one is the theory of Love. Others were based on intuitive or heuristic arguments. For reasons of economy, mathematical simplicity and accuracy, lower order flat triangular shell finite elements are popular among designers. This article is concerned with the review, development and application of triangular shell finite elements. Emphasis is on mixed formulation based lower order flat triangular shell finite elements. Finite element representation of shell structures is introduced. Linear analysis of static and dynamic, and nonlinear analysis of static and dynamic problems, are included in this article.

Chapter 3 - This chapter presents an overview of recent results on the electric field control and manipulation of fluids in microfluidic devices. The newer approaches are based on using alternating or a combination of alternating and direct current fields. The alternating field can be locally converted to direct by semiconductor diodes that may be placed at key locations where an electroosmotic force has to be applied to the fluid. Such techniques allow to design and fabricate small micrometer sized pumps and mixers. The latter are important because of the inherent low Reynolds characteristics of the flow in microchannels. The diode mixers are simple to fabricate and can be turned on and off depending on the operational requirements. Combining alternate and direct current fields and diode pumps makes possible the decoupling of the electroosmotic fluid flow from the electrophoretic particle or macromolecular mass flux. This can be exploited for precise analyte focusing, preconcentration and separation.

Chapter 4 - Landslide run-out is a complex phenomenon, much more difficult to simulate by models than flow of fluids. The main complicating aspects concern that landslide material is often heterogeneous and its characteristics may change during the landslide movement due to drainage, hydraulic interaction between fluid and grains, comminution of grains or mixing with surface water or partly or fully liquefied superficial material entrained from the path.

The continuum mechanical theory, treating the heterogeneous and multiphase moving mass as a continuum, has emerged in the last years as a useful tool for describing the evolving geometry and the velocity distribution of a mass flowing down a surface. A hypothetical material, "equivalent fluid", whose rheology is controlled by a small number of parameters is, in fact, introduced to represent the bulk behaviour of a landslide.

After a brief introduction on landslide characteristics and dynamics, new advances in the continuum mechanical description of flow-like landslides are discussed in dedicated sections. Each section deals with one of the main aspects that characterize the physical behaviour of a landslide and presents the simplifying, but nevertheless realistic, assumptions made to streamline their mathematical formulation.

The mathematical formulation is then implemented in a numerical code (RASH3D) to test the capability of each mathematical assumption in allowing the modelling of real phenomenon dynamics. Results of numerical simulations of laboratory tests and real events are discussed in this chapter to this aim.

Chapter 5 - The phenomenon of failure by catastrophic crack propagation in structural materials poses problems of design and analysis in many fields of engineering. Cracks are present to some degree in all structures. They may exist as basic defects in the constituent materials or they may be induced in construction or during service life. The continuum-mechanics can be applied for macro cracks.

Over the past decades the finite element technique has become firmly established as a useful tool for numerical solution of engineering problems. In order to be able to apply the finite element method to the efficient solution of fracture problems, adaptations or further developments must be made. Using the finite element method, a lot of papers deal with the calculation of stress intensity factors for two- and three-dimensional geometries containing cracks of different shapes under various loadings to elastic bodies. In order to increase the accuracy of the results, special singular and transition elements have been used. They are described together with methods for calculating the stress intensity factors from the computed results. These include the displacement substitution method, J-integral and the virtual crack extension technique. Despite of the large number of published finite element stress intensity factor calculations there are not so many papers published on J-integral to elastic-plastic bodies.

At the vicinity of a crack tip the strains are not always small, but they may be large ones, too. In this case the J-integral can also be applied to characterise the cracks in elastic or elastic-plastic bodies.

This chapter describes the computation of the two dimensional J-integral in the case of small and large strains to elastic and elastic-plastic bodies and represents some numerical examples, too.

Chapter 6 - Stiffness degradation for laminated composites such as carbon fiber/epoxy composites is an important physical response to the damage and failure evolution under continuous or cyclic loads. The ability to predict the initial and subsequent evolution process of such damage phenomenon is essential to explore the mechanical properties of laminated composites. This chapter gives a general review on the popular methodologies which deal with the damage initiation, stiffness degradation and final failure strength of composite laminates. These methodologies include the linear/nonlinear stress calculations, the failure criteria for initial microcracking, the stiffness degradation models and solution algorithms in the progressive failure analysis. It should be pointed out that the assumption of constant damage variable which is introduced into the constitutive equations of laminated composites to simulate the stiffness degradation properties is less effective and practical than that of changed damage variable with loads in the framework of continuum damage mechanics (CDM). Also, different damage evolution laws using CDM should be assumed to describe three failure modes: fiber breakage, matrix cracking and interfacial debonding, respectively.

Chapter 7 - The continuum mechanics in terms of the differential forms is proposed. The authors introduce the dual material space-time which consists of the strain space-time and the stress space-time. In this case, there is a one-to-one correspondence between the kinds of the basic equations in the continuum mechanics and the kinds of the basic operators in the differential forms. That is, the kinematic and constitutive equations can be derived by the exterior differential operator and the Hodge star operator, respectively. Other compound equations such as the Navier equation, Laplace (wave) equation and the incompatibility equation can be derived by the combination of the basic operators. This systematic approach allows us to find (i) the anti-exact solution of the Navier equation and (ii) the J-integral in fracture mechanics. The result (ii) means that the continuum mechanics in terms of the differential forms describes a partial aspect of the fracture mechanics. Moreover, the differential form approach allows us to link the deformation field with the non-deformation field such as the electromagnetic field. As an example, the authors take up the piezoelectric and Villari effects and derive the constitutive equations for these effects. These constitutive equations can be interpreted geometrically as the interaction among the geometrical objets of the space-time.

Chapter 8 - In this chapter the authors develop a general framework for the modelling of morphogenesis by introducing a growth process in the structural elements of the cell, which in turn depends on the stress state of the tissue. Some experimental observations suggest this feedback mechanism during embryo development, and only very recently this behaviour has started to be simulated.

The authors here derive the necessary equilibrium equations of a stress controlled growth mechanism in the context of continuum mechanics. In these derivations the authors assume a free energy source which is responsible for the active forces during the elongation process, and a passive hyperelastic response of the material. In addition, they write the necessary conditions that the active elongation law must satisfy in order to be thermodynamically consistent. The authors particularise these equations and conditions for the relevant elements of the cytoskeleton, namely, microfilaments and microtubules. The authors apply the model to simulate the shape changes observed during embryo morphogenesis in truss element. As a salient result, the model reveals that by imposing boundary stress conditions, unbounded elongation would be obtained. Therfore, either prescribed displacements or cross-links between fibres are necessary to reach a homeostatic state.

Chapter 9 - In the nearest-nodes finite element method (NN-FEM), finite elements are mainly used for numerical integration; for each quadrature point, shape functions are constructed from a set of nodes that are the nearest to the quadrature point, nodes from neighbour elements may be involved in the construction. Based on this strategy, there are several techniques available for constructing shape functions. In this paper, the moving local polynomial interpolation method is adopted. Benefiting from the above strategy, NN-FEM has several attractive features. High-order shape functions can be constructed from simplex finite element meshes; Analysis accuracy of NN-FEM is not influenced by element distortion; NN-FEM can deal with extremely large deformation, etc. Furthermore, NN-FEM provides a favourable environment for implementing an adaptive algorithm.

Chapter 10 - In this chapter, an adaptive finite element method is formulated based on the newly developed nearest-nodes finite element method (NN-FEM). In the adaptive NN-FEM, mesh modification is guided by the gradient of strain energy density, i.e. a larger gradient requires a denser mesh and vice versa. A finite element mesh is iteratively modified by a set of operators, including mesh refinement, mesh coarsening and mesh smoothing, to make its density conform with the gradient of strain energy density. The selection of a proper operator for a specific mesh region is determined by a set of criteria that are based on mesh intensity. The iteration loop of mesh modification is stopped when the relative error in the total potential energy is less than a prescribed accuracy. Numerical examples are presented to demonstrate the performance of the proposed adaptive NN-FEM.

Chapter 11 - The natural finite element approach introduced by John Argyris in the early sixties is characterized by the distinction between rigid body motion and deformation, on the one hand, and by the description of the latter in compliance with the element purpose and geometry, on the other hand. For triangular and tetrahedral elements the concept suggests strain and stress measures defined along the sides or the edges respectively as homogeneous normal quantities, free of shear. In the mechanics of continua the corresponding infinitesimal elements represent minimum configurations to define local deformation in two- and three dimensions.

This treatise concerns utilization of the natural approach on the continuum level within a consistent theoretical framework. It is proposed to begin with a reference system of supernumerary coordinates associated with the elementary tetrahedron in the space or with the triangle in the plane. Vectorial quantities are defined, the operations of gradient and divergence are interpreted in this system. The natural deformation rate is deduced from the velocity field, the stress is introduced as work conjugate measure. The condition for local equilibrium is presented in natural quantities as well as the stress definition in association with the resultant forces. The set up of material constitutive relations is exemplified for the elastic solid and for viscous media. Beyond the description of the momentary kinematics as from the velocity field, the appearance of finite deformation is considered basing on displacements. Illustration of the methodology for a plane elastic case terminates the part regarding the mechanics of solids. Extension to fluid motion and to thermal phenomena is appended.

In: Continuum Mechanics
Editors: Andrus Koppel and Jaak Oja, pp.1-52

ISBN: 978-1-60741-585-5

Chapter 1

MANIFOLDS ON CONTINUUM MECHANICS

Jari Mäkinen*

Tampere University of Technology, Department of Mechanics and Design, P.O. Box 589, FIN-33101 Tampere, Finland

ABSTRACT

We study differential geometry very elementarily, but hopefully in a practical way. We divide vector spaces into material and spatial spaces since these spaces behave differently in the observer transformation and with objective derivatives (Lie-derivatives). All the vector spaces, which we consider, have a metric tensor thus they are metric vector spaces, and all the finite dimensional manifolds are Riemannian manifolds that are embedded in a Euclidean space. Additionally, we may identify a dual vector space by its primary vector space. In classical tensor analysis, this identification is applied, but here we make distinction between primary and dual spaces in the formulation, and the identification can be accomplished later. If the identification of dual and primary vector spaces is done a priori, then push-forward and pull-back operations are not uniquely defined. As an example, we will study the rotation manifold and its underlying geometric structure in the terms of differential geometry.

In continuum mechanics, we have different manifolds. The placement field of continuum medium takes values in a Hilbert space, where a chart parametrization maps vector-valued functions into vector-valued functions. The placement field needs an infinite number of basis functions in order to present an arbitrary placement field on continuum, yielding infinite-dimensional manifolds.

In Lagrangian mechanics, forces are divided differently into constraint forces and applied forces. In addition, rich mathematical methods like variational calculus and other mechanical principles are included. In terms of differential geometry, Lagrangian mechanics describes a motion on an event manifold with a Lagrangian functional on the tangent bundle of the event manifold. An event manifold is a time-placement manifold which is also a constraint manifold, i.e. the time and placement variables satisfy all the constraints.

We think that there is a need for another type of mechanics between Newtonian and Lagrangian mechanics, as it is sometimes done. This mechanics could be named

* Academy Research Fellow, E-mail: jari.m.makinen@tut.fi, Fax: +358 3 3115 2107, Phone: +358 50 5366632

d'Alembertian mechanics where the principle of virtual work is its cornerstone. Here we include inertial forces also in the virtual work form. The virtual work may be viewed as a linear form on the tangent field-bundle. This field-bundle is also a tangent bundle of the placement manifold at fixed time. We give definitions for the virtual work in the finite-dimensional and infinite-dimensional cases. In addition we give definition for the variation, Lie derivative and Lie variation. The concept of push-forward and pull-back operators is essential for understanding Lie derivatives and variations.

Finally, we consider constraint point-manifolds that arise from point-wise holonomic constraint equations. The usual geometric joints of a multibody system like spherical, revolute, cylindrical, universal, helical, prismatic, and sliding joints can be presented via holonomic constraint equations that only depend on displacement at corresponding geometric points. All these constraints generate a smooth point-manifold that can be parametrized. Also, the principle of virtual work and its geometric structure are naturally related with the parametrization of the constraint manifold

1. Introduction to Differential Geometry

In this Section, we study differential geometry very elementarily way. Some knowledge on differential geometry is essential comprehending the quantity of finite rotation. In addition, dividing vector spaces into material and spatial spaces is necessary since these spaces behave differently in observer transformation and in objective derivatives (Lie-derivatives).

All the vector spaces, which we consider, have metric tensors thus they are metric vector spaces, and all the finite dimensional manifolds are Riemannian manifolds that are embedded in an Euclidean space. Hence, we can always choose an orthonormal set of basis vectors and we will get rid of those informative (read: terrible) subscript and superscript tensor notation and Christoffel symbols. Additionally, we may identify a dual vector space by its primary vector space. In classical tensor analysis, this identification is applied, but here we make distinction between primary and dual spaces in the formulation, and the identification is accomplished later in the finite element implementation. If the identification of dual and primary vector spaces is done a priori, then push-forward and pull-back operations are not uniquely defined. We also itemize terms a vector space and a linear space where the linear space is considered as a trivial manifold (or a linear manifold, or a flat manifold). Vector spaces usually appear from the tangent spaces of the manifold which are distinct at different points of a nontrivial manifold

1.1 Manifolds and Tensors on Manifolds

In this section, we give the definitions for vector and tensor algebra on topological vector spaces[1], definitions for manifolds, and tensor algebra on manifolds. We recommend consulting, especially, the paper [Stumpf & Hoppe 1997], and the textbooks [Wang & Truesdell 1973] or [Marsden & Hughes 1983] for tensors on manifolds, and textbooks [Arnold 1978] or [Abraham *et al.* 1983] for differentiable manifolds. A reader is assumed to

[1] We consider the topological vector space as a general vector space without explicit knowledge of a metric.

be familiar with classical tensor algebra on Euclidean spaces[2], text books like [Ogden 1984] or [Truesdell 1977] or [Bonet & Wood 1997].

1 Definitions for Covector Space, Dot Product, and Adjoint Operator

The covector space $\mathscr{V}^*$ of the vector space $\mathscr{V}$ is defined by the space of linear maps $\mathscr{V} \to \mathrm{R}$, i.e. $\mathscr{V}^* := \mathscr{L}(\mathscr{V}, \mathrm{R})$. These linear maps are represented by the dot product (duality pairing) defined as

$$\cdot : \mathscr{V}^* \times \mathscr{V} \to \mathrm{R}, \quad (\mathbf{f}, \mathbf{a}) \mapsto \mathbf{f} \cdot \mathbf{a} \in \mathrm{R},$$

which have two properties: bilinearity, i.e. it is linear with respect to each of its two members, and definite, i.e. if $\mathbf{f} \in \mathscr{V}^*$ is fixed and $\mathbf{f} \cdot \mathbf{a} = 0 \ \forall \mathbf{a} \in \mathscr{V}$, then $\mathbf{a} = \mathbf{0}$. Conversely, if $\mathbf{a} \in \mathscr{V}$ is fixed and $\mathbf{f} \cdot \mathbf{a} = 0, \forall \mathbf{f} \in \mathscr{V}^*$, then $\mathbf{f} = \mathbf{0}$. If $\mathbf{f} \cdot \mathbf{a} = 0$, the vector **a** is said to be orthogonal to the covector **f**, and vice versa. Note that a covector space is also a vector space satisfying the vector space properties. Because the vector space and its co-covector space are canonically isomorphic[3], i.e. $\mathscr{V} = \mathscr{V}^{**}$, we have the symmetry property of the dot product: $\mathbf{f} \cdot \mathbf{a} = \mathbf{a} \cdot \mathbf{f}$.

Let $\mathbf{F} \in \mathscr{L}(\mathscr{V}, \mathscr{W})$ be a linear operator from $\mathscr{V} \to \mathscr{W}$. The adjoint operator $\mathbf{F}^* \in \mathscr{L}(\mathscr{W}^*, \mathscr{V}^*)$ is defined with the aid of the dot product as

$$\mathbf{F}^* \mathbf{w} \cdot \mathbf{a} = \mathbf{w} \cdot \mathbf{F} \mathbf{a} \in \mathrm{R} \quad \forall \mathbf{a} \in \mathscr{V}, \ \mathbf{w} \in \mathscr{W}^*,$$

where the first dot product is on the vector space $\mathscr{V}$, and the latter on the vector space $\mathscr{W}$, see Figure 1. ◊□

On notation: we omit $\cdot$-symbol when there is no source of confusion. Then the terms **Fa** and $\mathbf{F} \cdot \mathbf{a}$ are identical. The brackets are used for purpose of dependency, e.g. $\mathbf{F}(\mathbf{x}) \cdot \mathbf{a}$ denotes the linear operator **F(x)** acts (linearly) on **a** where the operator depends on **x**. In this case, the dot symbol may not be omitted. In addition, in the composite mapping of operators, like **FG**, the dot symbol is omitted.

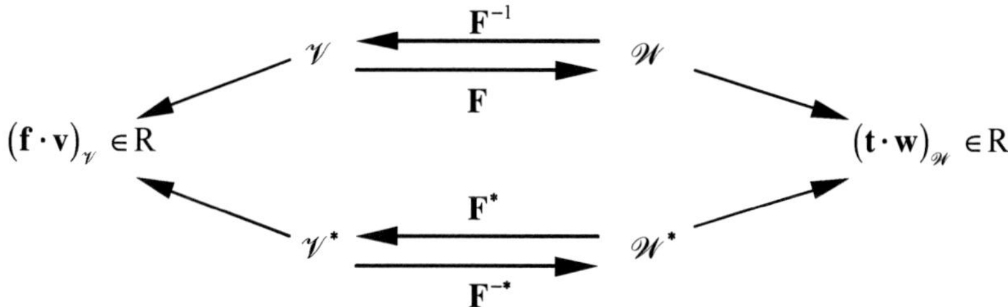

Figure 1. The diagram of domains and ranges for the operator $\mathbf{F} \in \mathscr{L}iso(\mathscr{V}, \mathscr{W})$ and its derivatives.

[2] The Euclidean space is a real, finite-dimensional, linear, inner-product space with an Euclidean metric.

[3] Topological vector spaces are isomorphic, denoted by $\cong$, if there exists a (continuous) linear bijection, called isomorphism, between these spaces. Two vector spaces are isomorphic iff they have the same dimensions. Vector spaces are canonical isomorphic, denoted by $=$, if there exists a natural ('almost trivial') isomorphism.

A covector space is commonly called a dual vector space, and the elements of the covector space are called covectors, or dual vectors, or linear forms (which never make any sense). Additionally, the dot product is also called duality pairing, and an adjoint operator is called a dual operator. We do not make any notational difference between the elements in the vector and covector spaces since we desire to use the notation similar to classical tensor algebra. For example, force quantities like moment and force vectors, and Lagrange multiplies are the elements of covector spaces. We will use the byte 'co-' instead of the word 'dual' because of its simplicity and compactness.

2 Definition for Inverse Operator, and Inverse Adjoint Operator

If the operator $\mathbf{F}$ is a linear bijection (isomorphism), denoted $\mathbf{F} \in \mathscr{L}iso(\mathscr{V},\mathscr{W})$, the inverse operator $\mathbf{F}^{-1} \in \mathscr{L}iso(\mathscr{W},\mathscr{V})$ exists and it is unique. The inverse operator is defined by formulas

$$\mathbf{I} = \mathbf{F}^{-1}\mathbf{F} \quad \text{and} \quad \mathbf{i} = \mathbf{F}\mathbf{F}^{-1},$$

where $\mathbf{I} \in \mathscr{L}iso(\mathscr{V},\mathscr{V})$ is the identity on $\mathscr{V}$, and $\mathbf{i} \in \mathscr{L}iso(\mathscr{W},\mathscr{W})$ is the identity on $\mathscr{W}$. The inverse of the adjoint operator $\mathbf{F}^* \in \mathscr{L}iso(\mathscr{W}^*,\mathscr{V}^*)$ is defined similarly by formulas

$$\mathbf{i}^* = \mathbf{F}^{-*}\mathbf{F}^* \quad \text{and} \quad \mathbf{I}^* = \mathbf{F}^*\mathbf{F}^{-*},$$

where $\mathbf{i}^* \in \mathscr{L}iso(\mathscr{W}^*,\mathscr{W}^*)$ is the identity on $\mathscr{W}^*$, and $\mathbf{I}^* \in \mathscr{L}iso(\mathscr{V}^*,\mathscr{V}^*)$ is the identity on $\mathscr{V}^*$. Note that an inverse adjoint operator is an operator $\mathbf{F}^{-*} \in \mathscr{L}iso(\mathscr{V}^*,\mathscr{W}^*)$, see Figure 1. ◊

3 Definition for Tensor Product and Tensor Space

The tensor product between the vector $\mathbf{a} \in \mathscr{V}$[4] and the covector $\mathbf{f} \in \mathscr{W}^*$ is defined via the dot product by the formula

$$(\mathbf{a} \otimes \mathbf{f}) \cdot \mathbf{w} = (\mathbf{f} \cdot \mathbf{w})\mathbf{a} \in \mathscr{V}, \qquad \forall \mathbf{w} \in \mathscr{W},$$

where the tensor $\mathbf{a} \otimes \mathbf{f}$ belongs to the tensor space produced by $\mathscr{V}$ and $\mathscr{W}^*$, i.e. $\mathbf{a} \otimes \mathbf{f} \in \mathscr{V} \otimes \mathscr{W}^* = \mathscr{L}(\mathscr{W},\mathscr{V})$. The tensor product is a linear mapping for each member separately, i.e. a bilinear operator, because of the bilinearity of the dot product. The tensor is called a two-point tensor if it is defined on two different vector spaces. The general two-point tensor space $\mathscr{T}$ can be denoted by

$$\mathscr{T} := \underbrace{\mathscr{V} \otimes \cdots \otimes \mathscr{V}}_{r} \otimes \underbrace{\mathscr{V}^* \otimes \cdots \otimes \mathscr{V}^*}_{s} \otimes \underbrace{\mathscr{W} \otimes \cdots \otimes \mathscr{W}}_{t} \otimes \underbrace{\mathscr{W}^* \otimes \cdots \otimes \mathscr{W}^*}_{u}$$

that is the space of r-fold on the vector space $\mathscr{V}$, s-fold on the covector space $\mathscr{V}^*$, t-fold on the vector space $\mathscr{W}$, and u-fold on the covector space $\mathscr{W}^*$. This can be shortly denoted by the tensor space $\mathscr{T}(r,s;t,u)$ with the order of $r+s+t+u$. ◊

---

[4] This vector space $\mathscr{V}$ could be a covector space, or more generally, a tensor space

Note that any other permutation of vector spaces is possible, thus e.g. the notation (1,0;0,1) could mean the tensor spaces $\mathscr{V} \otimes \mathscr{W}^*$ or $\mathscr{W}^* \otimes \mathscr{V}$. In the case of one-point tensor spaces, defined on the same vector or covector space, we use a simplified notation: e.g. the tensor space $\mathscr{T}(1,1)$ for the tensor spaces $\mathscr{V} \otimes \mathscr{V}^*$ or $\mathscr{V}^* \otimes \mathscr{V}$ defined on $\mathscr{V}$, or correspondingly for the tensor spaces $\mathscr{W} \otimes \mathscr{W}^*$ or $\mathscr{W}^* \otimes \mathscr{W}$ defined on $\mathscr{W}$.

There are two possible points of view to comprehend a tensor: operational or quantitative. The operational aspect informs 'how it works', and quantitative responds to 'how much is it'. Mathematicians represent the operational point of view and engineers the quantitative point of view. Although we will define the tensor by quantitative, we shall keep in mind its operational aspect: a tensor is a multilinear operator.

4 Definition for Tensors

A tensor is defined an element of a tensor space. Thus after the property of tensor product, the two-point tensor $\mathbf{T}$ of the tensor space $\mathscr{T}(r,s;t,u)$, given in *Def. 3*, is a multilinear mapping

$$\mathbf{T}: \underbrace{\mathscr{V}^* \times \cdots \times \mathscr{V}^*}_{r} \times \underbrace{\mathscr{V} \times \cdots \times \mathscr{V}}_{s} \times \underbrace{\mathscr{W}^* \times \cdots \times \mathscr{W}^*}_{t} \times \underbrace{\mathscr{W} \times \cdots \times \mathscr{W}}_{u} \to \mathrm{R} \quad .$$

The two-point tensor $\mathbf{T}$ is an element of two-point tensor space such that it assigns a tensor for its two-point domain. ◊

The tensor space is a vector space itself by satisfying all vector space properties. Then we may state that the tensors are vectors and the vectors are tensors. However, we consider the first-order tensors as vectors, and the higher-order tensors as tensors. Sometimes the tensors are characterized by their component transformation laws under the change of the basis: the object is a tensor if its components change like tensor components under a coordinate transformation. For example, Christoffel symbols are not tensors. Conversely, the vectors are characterized by direction, magnitude, and, especially, by the parallelogram law: the vector can be added to another vector by the parallelogram law. For example, it is often incorrectly claimed that the finite rotation does not satisfy the parallelogram law, whereupon the finite rotation vector is not a vector quantity. We keep these characterizations rather old-fashioned and they can lead to serious misunderstandings. The vectors and tensors may be characterized by studying if they are elements of corresponding vector and tensor spaces, respectively.

The trace of the second order tensor is usually defined by the contraction of its components. This is a contradiction with the component independence of the tensor, although, the trace is component-independent. We follow the definition of the trace given in [Truesdell 1977; App. II].

5 Definition for Trace and Double-Dot Product

The trace $\mathrm{tr} \in \mathscr{L}(\mathscr{V}^* \times \mathscr{V}, \mathrm{R})$ of the one-point tensor $\mathbf{f} \otimes \mathbf{a} \in \mathscr{V}^* \otimes \mathscr{V}$ is a scalar-valued linear operator defined via the dot product as

$$\mathrm{tr}(\mathbf{f} \otimes \mathbf{a}) := \mathbf{f} \cdot \mathbf{a} \in \mathrm{R} \quad .$$

Also the trace operation for the tensor on $\mathscr{V} \otimes \mathscr{V}^*$ can be applied by noting $\mathscr{V} = \mathscr{V}^{**}$, but it is not defined for two-point tensors. The double-dot product for the tensors $\mathbf{f} \otimes \mathbf{t} \in \mathscr{V}^* \otimes \mathscr{W}^*$ and $\mathbf{v} \otimes \mathbf{w} \in \mathscr{V} \otimes \mathscr{W}$ is defined via the ordinary dot product

$$(\mathbf{f} \otimes \mathbf{t}):(\mathbf{v} \otimes \mathbf{w}) := (\mathbf{f} \cdot \mathbf{v})_{\mathscr{V}} \cdot (\mathbf{t} \cdot \mathbf{w})_{\mathscr{W}} \in \mathrm{R},$$

where the subscripts indicate the vector space of the corresponding dot product. Therefore, the double-dot product is a mapping $\mathscr{L}(\mathscr{V}^* \times \mathscr{W}^* \times \mathscr{V} \times \mathscr{W}, \mathrm{R})$ that is a four-linear operator. ◊

All tensors, which we have considered, have been presented by the tensor product of the vectors, e.g. the tensor $\mathbf{f} \otimes \mathbf{a}$. However, a general tensor can not be expressed directly in that way. We may present a common tensor with basis vectors of tensor space. Let $\{\mathbf{G}_i\}$, with the index $i = 1,2,3$, be an ordered basis for the vector space $\mathscr{V}$ and let $\{\mathbf{g}_i\}$ $(i = 1,2,3)$ be an ordered basis for the vector space $\mathscr{W}$, then we may present a general second-order two-point tensor $\mathbf{T} \in \mathscr{V} \otimes \mathscr{W}$ by the linear combination of the basis vectors, namely (with the conventional summation)

$$\mathbf{T} = T_{ij} \mathbf{G}_i \otimes \mathbf{g}_j \tag{1}$$

where $\mathbf{G}_i \otimes \mathbf{g}_j \in \mathscr{V} \otimes \mathscr{W}$ corresponds the basis vector of the tensor with the coefficient $T_{ij} \in \mathrm{R}$. The coefficient matrix $[T_{ij}] \in \mathrm{R}^{3\times 3}$ is called the component matrix of the tensor $\mathbf{T}$ with respect to the bases $\{\mathbf{G}_i\}$ and $\{\mathbf{g}_i\}$[5]. Higher order tensors are represented a similar way. In order to represent tensors on covector spaces, we have to define the bases for the covector spaces.

6 Definition for Bases of Covector Spaces

Let $\{\mathbf{G}_i\}$ and $\{\mathbf{g}_i\}$ be ordered bases of the vector spaces $\mathscr{V}$ and $\mathscr{W}$, respectively. The bases (dual bases) $\{\mathbf{G}_i^*\}$ and $\{\mathbf{g}_i^*\}$ on the covector spaces $\mathscr{V}^*$ and $\mathscr{W}^*$ are defined by formulas

$$\mathbf{G}_i^* \cdot \mathbf{G}_j = \delta_{ij}, \quad \mathbf{g}_i^* \cdot \mathbf{g}_j = \delta_{ij},$$

where δ_{ij} is the Kronecker's delta symbol. Then, for example, the tensor $\mathbf{T} \in \mathscr{V} \otimes \mathscr{W}^*$ may be represented by $\mathbf{T} = T_{ij} \mathbf{G}_i \otimes \mathbf{g}_j^*$. ◊

We have defined a tensor algebra on a topological vector space. These vector spaces are often induced by a manifold, yielding a tensor algebra on the manifold that we define next.

7 Definition for Manifold

A set $\mathscr{M} \subset \mathrm{E}^n$ is a manifold with dimension *d*, if there exists a bijection[6] $\varphi_i : \mathscr{U}_i \to \mathrm{E}^n$ from an open domain $\mathscr{U}_i \subset \mathrm{E}^d$ in a *d*-dimensional Euclidean parameter space onto some open set in the manifold, $\varphi_i : \mathscr{U}_i \to \varphi_i(\mathscr{U}_i) \subset \mathscr{M}$, such that every point of the manifold is an image

---

[5] The component matrix is an isomorphism between the tensor space $\mathscr{V} \otimes \mathscr{W}$ and the Cartesian space $\mathrm{R}^{3\times 3}$.

[6] a mapping is a bijection if it is injective and surjective, i.e. one-to-one and onto mapping

under a mapping, see Figure 2. A pair $(\mathcal{U}_i, \varphi_i)$ is called a chart or a parametrization chart, and the mapping φ_i is called a chart mapping or a parametrization mapping ◊

8 Definition for Differentiable Manifold

A manifold $\mathcal{M}$ is a differentiable manifold if for every point $a \in \mathcal{M}$ there exist images $\varphi_1(\mathcal{U}_1)$ and $\varphi_2(\mathcal{U}_2)$ where the point $a \in \mathcal{M}$ belongs to, such that the composite mapping $\varphi_2^{-1} \circ \varphi_1$ is a diffeomorphism[7] from $\varphi_1^{-1}(\varphi_1(\mathcal{U}_1) \cap \varphi_2(\mathcal{U}_2))$ onto $\varphi_2^{-1}(\varphi_1(\mathcal{U}_1) \cap \varphi_2(\mathcal{U}_2))$. The composite mapping is called the change of parametrization, see Figure 2. ◊

We note that usually a chart mapping is defined by an inverse mapping from an open set of a manifold into a parameter space. We have defined a chart mapping differently since we could use this terminology when constraint equations are parametrized; also, we note the connection between the finite element method. A vector space, where a manifold is embedded, is called a embedding space; the Euclidean space E^n in Figure 2.

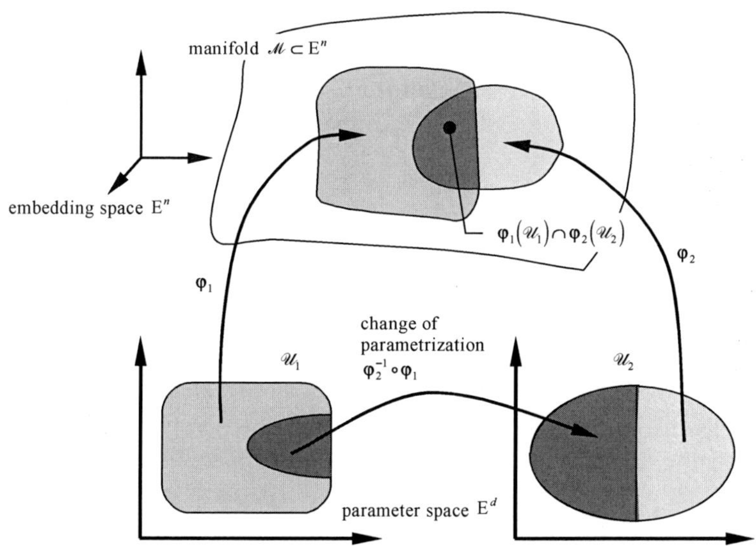

Figure 2. A geometric interpretation for a parametrization of a manifold, when $n = 3$ and $d = 2$.

9 Definition for Tangent Vector and Tangent Space on Manifold

Let $\varphi(t)$ be a parametrized vector-valued curve in the manifold $\mathcal{M}$ through the point $\mathbf{x} \in \mathcal{M}$ such that $\varphi(t = 0) = \mathbf{x}$. The tangent of curve (or equivalent class of curves) $\varphi(t)$ at $t = 0$ on the manifold $\mathcal{M}$ is defined as

$$\dot{\mathbf{x}} = \lim_{t \to 0} \frac{\varphi(t) - \varphi(0)}{t}, \quad \text{where } \varphi(0) = \mathbf{x}, \;\; \varphi(t) \in \mathcal{M}.$$

The tangent vector $\dot{\mathbf{x}}$ belongs to a tangent space of the manifold, namely $\dot{\mathbf{x}} \in T_{\mathbf{x}}\mathcal{M}$, see Figure 3. The tangent (vector) space $T_{\mathbf{x}}\mathcal{M}$ is a set of tangent vectors at $\mathbf{x} \in \mathcal{M}$. ◊

---

[7] a diffeomorphism is a bijection with continuously differentiable mapping and its inverse mapping

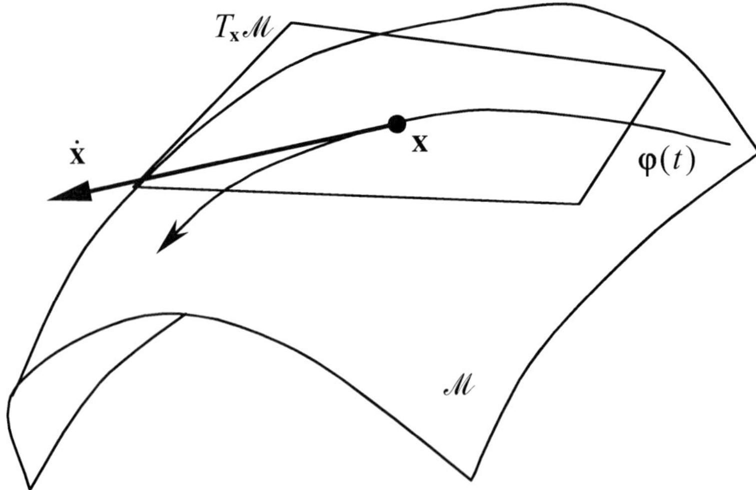

Figure 3. The tangent vector $\dot{\mathbf{x}}$ and its tangent space $T_{\mathbf{x}}\mathcal{M}$ on the manifold $\mathcal{M}$ at the point $\mathbf{x}$.

10 Definition for Tangent Bundle on Manifold

A tangent bundle $T\mathcal{M}$ is defined a union of the tangent spaces on the manifold $\mathcal{M}$ at its every point

$$T\mathcal{M} := \bigcup_{\mathbf{x}\in\mathcal{M}} (\mathbf{x}\,,\, T_{\mathbf{x}}\mathcal{M})\,.$$

The dimension of the tangent bundle is twicc the dimension of the manifold $\mathcal{M}$. Especially, the pair of state vectors, the placement $\mathbf{x}(t)$ and velocity vectors $\mathbf{v}(t)$, belongs to the tangent bundle, $(\mathbf{x},\mathbf{v})(t) \in T\mathcal{M}$. ◊

For a two-point tensor, its domain of points is divided into two separate but not independent regions which are defined in the vector spaces $\mathcal{V}$ and $\mathcal{W}$. It is convenient to choose a material body *B*, containing all material points of body, for one region of the domain and another region which is obtained via a mapping of the material body *B*. The material body *B* is a set of points and its elements are denoted $X, Y, Z, \ldots \in B$.

11 Definition for Current and Initial Reference Placements

Let $\chi_t : B \to E^3$ be a smooth time-dependent embedding of the material body B into the Euclidean space E^3. For each fixed time *t*, the mapping $\chi(t,\cdot)$ is defined as a current placement of the body B along with the current place vector $\mathbf{x}$ of a body-point, namely

$$\mathcal{B} := \chi(t, B), \qquad \mathbf{x} := \chi(t, X), \quad X \in B\,.$$

The initial reference placement $\mathcal{B}_0$ is defined as the special case of the current placement $\mathcal{B}$ by setting $t = 0$, giving

$$\mathcal{B}_0 := \chi(t = 0, B), \qquad \mathbf{X} := \chi(t = 0, X) \quad X \in B\,,$$

where **X** is an initial reference place vector. ◊

Since the initial reference placement $\mathscr{B}_0$ is uneffected in the observation transformation (see e.g. [Ogden 1984; Ch. 2]), we call vectors and tensors defined on the initial reference placement $\mathscr{B}_0$ as material quantities. For example, a reference place vector **X** is called a material place vector, and $\mathscr{B}_0$ the material placement. Sometimes the material description is named as referential or Lagrangian description, and occasionally, some distinction has been accomplished between these phrases.

Contrary to the material placement $\mathscr{B}_0$, the current placement $\mathscr{B}$ and vectors and tensors defined on it are concerned in the observation transformation. Vector and tensors defined on the current placement $\mathscr{B}$ are called spatial quantities, e.g. a current place vector **x** is also named as a spatial place vector, and $\mathscr{B}$ as a spatial placement. A spatial description is sometimes called an Eulerian description.

We will apply the phrases 'material' and 'spatial' for placements, vectors, tensors, fields, spaces and descriptions. A geometric interpretation of the material body *B*, the material placement $\mathscr{B}_0$, and the spatial placement $\mathscr{B}$ is given in Figure 4. Note that placements, likewise place vectors, should be regarded as mappings, not the images of these maps, according to *Def. 11.*

In Figure 4, it is demonstrated that a body-point $X \in B$, which is represented by a vector-valued mapping $\mathbf{X} := \chi_0(X)$, assigns a material vector **A** on the material placement $\mathscr{B}_0$. The material vector belongs to the tangent space of the material placement $\mathscr{B}_0$, namely $T_\mathbf{X}\mathscr{B}_0$, where **X** corresponds a base point[8] of manifold. Correspondingly, the body-point $X \in B$, which is represented by the mapping $\mathbf{x} := \chi(X)$, assigns the spatial vector **a** on the spatial placement $\mathscr{B}$. The spatial vector belongs to the tangent space of the spatial placement $\mathscr{B}$, i.e. $\mathbf{a} \in T_\mathbf{x}\mathscr{B}$, where **x** represents a base point of manifold. Note that the placements $\mathscr{B}_0$ and $\mathscr{B}$ are manifolds.

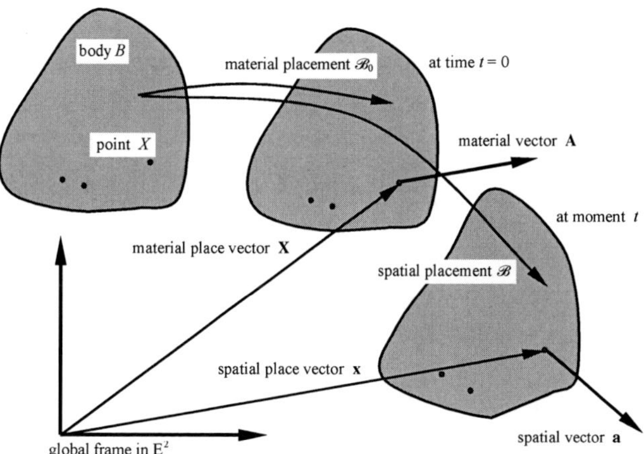

Figure 4. The material body B with the body-point X, the material placement $\mathscr{B}_0$ with the material place vector **X** and the material vector **A**, and the spatial placement $\mathscr{B}$ with the spatial place vector **x** and the spatial vector **a**.

[8] A base point is a point of the manifold where a tangent space is induced.

Now we could set $\mathscr{V} = T_{\mathbf{X}}\mathscr{B}_0$, $\mathscr{W} = T_{\mathbf{x}}\mathscr{B}$ giving, for example, the type of (1,1;1,1) two-point tensor at a body-point $X \in B$ with mappings $\mathbf{X} = \chi_0(X)$ and $\mathbf{x} = \chi(X)$ $\mathbf{T}: T^*_{\mathbf{X}}\mathscr{B}_0 \times T_{\mathbf{X}}\mathscr{B}_0 \times T^*_{\mathbf{x}}\mathscr{B} \times T_{\mathbf{x}}\mathscr{B} \to \mathrm{R}$, where $T^*_{\mathbf{X}}\mathscr{B}_0$ and $T^*_{\mathbf{x}}\mathscr{B}$ are the covector spaces for the vector spaces $T_{\mathbf{X}}\mathscr{B}_0$ and $T_{\mathbf{x}}\mathscr{B}$, respectively. The two-point tensor $\mathbf{T}$ is an element of multilinear operators, denoted as $\mathbf{T} \in \mathscr{L}(T^*_{\mathbf{X}}\mathscr{B}_0 \times T_{\mathbf{X}}\mathscr{B}_0 \times T^*_{\mathbf{x}}\mathscr{B} \times T_{\mathbf{x}}\mathscr{B}, \mathrm{R})$. For the sake of simplicity, we omit body points and mappings when expressing tensors and vectors, and we call the place vectors $\mathbf{X}$ and $\mathbf{x}$ as the material base point and the spatial base point, respectively.

So far we have studied vectors and tensors in vector spaces without knowledge about its metric. A metric of the vector space is a symmetric positive-definite bilinear operator[9] , called a metric tensor. Let pairs $(\mathscr{V}, \mathbf{G})$ and $(\mathscr{W}, \mathbf{g})$ indicate metric vector spaces in the material and spatial representation, with the (material) metric tensor $\mathbf{G} \in \mathscr{L}(\mathscr{V}, \mathscr{V}^*)$ and the (spatial) metric tensor $\mathbf{g} \in \mathscr{L}(\mathscr{W}, \mathscr{W}^*)$. Metric tensors are used for measuring distances and deformation, which is impossible without introducing metric. Since manifolds are embedded in the Euclidean space E^3, we could choose metric tensors as the identity elements. This can be achieved by identifying the metric vector spaces $(\mathscr{V}, \mathbf{G})$ and $(\mathscr{W}, \mathbf{g})$ with the Euclidean vector space E^3. However, this identifying is not accomplished at this moment since it is informative to comprehend the existence of the metric tensor in different operators like deformation and strain tensors.

12 Definition for Inner Product and Transpose Operator

The inner product for a metric vector space $(\mathscr{V}, \mathbf{G})$ is defined by

$$\langle \cdot,\cdot \rangle : \mathscr{V} \times \mathscr{V} \to \mathrm{R}, \ (\mathbf{a},\mathbf{b}) \mapsto \langle \mathbf{a},\mathbf{b} \rangle_{\mathrm{G}} := \mathbf{Ga} \cdot \mathbf{b} \quad (= \mathbf{a}^{\wedge} \cdot \mathbf{b}),$$

where the dot product is defined in *Def.* 1. For simplicity, the covector $\mathbf{Ga}$ is often denoted by $\mathbf{a}^{\wedge}$. The tensor $\mathbf{F} \in \mathscr{L}((\mathscr{V}, \mathbf{G}), (\mathscr{W}, \mathbf{g}))$, its transpose operator $\mathbf{F}^{\mathrm{T}}$ is defined via the inner product

$$\langle \mathbf{F}^{\mathrm{T}}\mathbf{w}, \mathbf{v} \rangle_{\mathrm{G}} = \langle \mathbf{w}, \mathbf{Fv} \rangle_{\mathrm{g}} \quad \forall \mathbf{w} \in \mathscr{W}, \mathbf{v} \in \mathscr{V}.$$

Hence, the transpose operator is a mapping $\mathbf{F}^{\mathrm{T}} \in \mathscr{L}(\mathscr{W}, \mathscr{V})$. After the definition of the inner product, we found a relation between the transpose $\mathbf{F}^{\mathrm{T}}$ and the adjoint operator $\mathbf{F}^*$, yielding $\mathbf{F}^{\mathrm{T}} = \mathbf{G}^{-1}\mathbf{F}^*\mathbf{g}$. Note that the transpose operator depends on metric tensors on contrary to the adjoint operator. ◊

9 a metric is a scalar valued function that induces a linear bijection (isomorphism), called a metric tensor

1.2 Rotation Manifold

A rotation vector is one of the most misunderstood quantities in applied mechanics. In this section, we will demonstrate how a rotation vector and a differentiable manifold are connected. We derive a rotation operator in terms of the rotation vector, see e.g. [Argyris 1982]. At this point, we assume that all vectors live in the Euclidean space E^3. This assumption is not contradictory since vectors in any three-dimensional topological vector space can be identified by an isomorphism with vectors in the Euclidean space E^3.

We are trying to find an expression for the rotated vector $\mathbf{p}_1$ in the terms of the original vector $\mathbf{p}_0$, the unit rotation axis $\mathbf{e}$, and the non-negative rotation angle ψ about the rotation axis. The original projector vector $\mathbf{r}_0$ and the rotated projector vector $\mathbf{r}_1$ in the rotation plane are, see Figure5

$$\begin{aligned} &\mathbf{r}_0 = \mathbf{e}\times(\mathbf{p}_0\times\mathbf{e}), \\ &\mathbf{r}_1 = \mathbf{r}_0\cos\psi + \mathbf{e}\times\mathbf{p}_0\sin\psi \qquad \mathbf{p}_0,\mathbf{e},\mathbf{r}_0,\mathbf{r}_1 \in E^3,\ \psi\in R_+, \end{aligned} \tag{2}$$

where $\times$ denotes the cross product on E^3. Now the rotated vector $\mathbf{p}_1$ can be expressed with the aid of (2)

$$\begin{aligned} \mathbf{p}_1 &= \mathbf{p}_0 - \mathbf{r}_0 + \mathbf{r}_1 \\ &= \mathbf{p}_0 + (1-\cos\psi)\mathbf{e}\times(\mathbf{e}\times\mathbf{p}_0) + \mathbf{e}\times\mathbf{p}_0\sin\psi \qquad \mathbf{p}_1,\mathbf{p}_0,\mathbf{e},\mathbf{r}_0,\mathbf{r}_1 \in E^3,\ \psi\in R_+, \end{aligned} \tag{3}$$

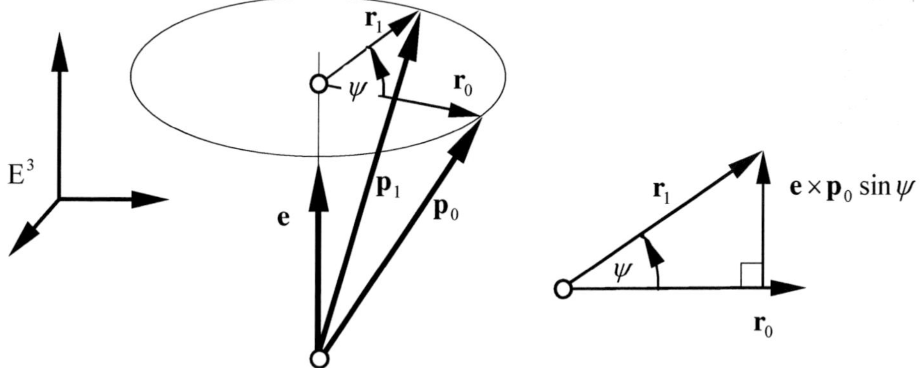

Figure5. A rotational motion about $\mathbf{e}$-axis where $\mathbf{p}_0$ is the original vector and $\mathbf{p}_1$ is the rotated vector. (Note that $\mathbf{e}\times\mathbf{r}_0 = \mathbf{e}\times\mathbf{p}_0$).

13 Definition for Rotation Vector in Euclidean Space E^3

A rotational motion can be represented by a rotation vector defined as

$$\boldsymbol{\Psi} := \psi\mathbf{e} \qquad \|\mathbf{e}\| = 1,\ \mathbf{e}\in E^3,\ \psi\in R_+ \tag{4}$$

where the unit rotation axis vector $\mathbf{e}$ and the non-negative rotation angle ψ are oriented such that they form a right-handed screw, see Figure5. ◊

Note that the length of the rotation vector is equal to the rotation angle, i.e. $\|\boldsymbol{\Psi}\| = \psi$. Here we do not restrict the angle of rotation; it may have any non-negative values. The rotation vector $\boldsymbol{\Psi}$ lives in a three-dimensional vector space that is isomorphic to the Euclidean space E^3. This issue will be realized later.

Now Eqn (3) can be written in terms of the rotation vector, which yields the expression of the rotation operator

$$\begin{aligned} \mathbf{p}_1 &= \mathbf{p}_0 + \frac{\sin\psi}{\psi}\boldsymbol{\Psi}\times\mathbf{p}_0 + \frac{1-\cos\psi}{\psi^2}\boldsymbol{\Psi}\times(\boldsymbol{\Psi}\times\mathbf{p}_0) \\ &= \left(\mathbf{I} + \frac{\sin\psi}{\psi}\tilde{\boldsymbol{\Psi}} + \frac{1-\cos\psi}{\psi^2}\tilde{\boldsymbol{\Psi}}^2\right)\mathbf{p}_0, \end{aligned} \tag{5}$$

where the skew-symmetric tensor $\tilde{\boldsymbol{\Psi}}$ called the rotation tensor, is defined by formula $\tilde{\boldsymbol{\Psi}}\mathbf{a} = \boldsymbol{\Psi}\times\mathbf{a}$, $\forall\mathbf{a}\in E^3$, or more formally $\tilde{\boldsymbol{\Psi}} := \boldsymbol{\Psi}\times$.

14 Definition for Rotation Operator in Euclidean Space E^3

A rotation operator transforms linearly and isometrically a vector into another vector in a rotational motion that is represented by a rotation vector. The rotation operator $\mathbf{R}\in\mathcal{L}iso(E^3,E^3)$ is defined with the aid of the rotation vector $\boldsymbol{\Psi}\in E^3$ by equation:

$$\mathbf{R} := \mathbf{I} + \frac{\sin\psi}{\psi}\tilde{\boldsymbol{\Psi}} + \frac{1-\cos\psi}{\psi^2}\tilde{\boldsymbol{\Psi}}^2, \quad \psi = \|\boldsymbol{\Psi}\|.$$

Then in Figure5, the rotation operator $\mathbf{R}$ transforms the vector $\mathbf{p}_0$ into the vector $\mathbf{p}_1$, i.e. $\mathbf{p}_1 = \mathbf{R}\mathbf{p}_0$. ◊

Now a rotational motion, represented by a rotation vector, is directly involved in a rotation operator. In addition, the definition gives explicitly a canonical[10] parametrization of the rotation operator that is a point of a manifold itself. This rotation manifold, namely *SO*(3), is a three-dimensional smooth manifold embedded in the Euclidean space $E^{3\times3}$ that is isomorphic to the nine-dimensional Cartesian space R^9. We note that a rotation operator is an element of the rotation manifold, i.e. $\mathbf{R}\in SO(3)$.

Expanding the trigonometric terms in *Def. 14* and using the realities that

$$\tilde{\boldsymbol{\Psi}}^{2n-1} = (-1)^{n-1}\psi^{2(n-1)}\tilde{\boldsymbol{\Psi}}, \quad \tilde{\boldsymbol{\Psi}}^{2n} = (-1)^{n-1}\psi^{2(n-1)}\tilde{\boldsymbol{\Psi}}^2 \tag{6}$$

gives the exponential representation of the rotation operator

$$\mathbf{R} = \mathbf{I} + \tilde{\boldsymbol{\Psi}} + \frac{1}{2!}\tilde{\boldsymbol{\Psi}}^2 + \frac{1}{3!}\tilde{\boldsymbol{\Psi}}^3 + \ldots =: \exp(\tilde{\boldsymbol{\Psi}}). \tag{7}$$

---

[10] There exists no effective criterion for the term canonical.

This is a significant property of the rotation operator and offers the shortest relationship between the rotation vector and the rotation operator. We also note that the transpose of the rotation operator is equal to the reverse rotational motion

$$\mathbf{R}^{\mathrm{T}} = \mathbf{I} + \frac{\sin\psi}{\psi}\left(-\widetilde{\Psi}\right) + \frac{1-\cos\psi}{\psi^2}\left(-\widetilde{\Psi}\right)^2 = \mathbf{R}(-\Psi), \tag{8}$$

due to the skew-symmetry of the rotation tensor, $\widetilde{\Psi}^{\mathrm{T}} = -\widetilde{\Psi}$.

This yields the proper orthogonal features of the rotation operator

$$\begin{aligned} &\mathbf{R}^{\mathrm{T}}\mathbf{R} = \mathbf{R}\mathbf{R}^{\mathrm{T}} = \mathbf{I}, \\ &\det(\mathbf{R}) = +1 , \end{aligned} \tag{9}$$

where $\mathbf{I}$ is the identity element. It is evident since the inverse of the rotation operator is the reverse rotation operator. If an operator satisfies Equation (9a) solely, there exist two possible values for its determinant, namely $\det(\mathbf{R}) = \{+1, -1\}$, where the first value (+1) produces the preservation of the orientation.

A rotation operator can be written also with the aid of a rotation axis $\mathbf{e}$, yielding

$$\mathbf{R} = \mathbf{I} + \sin\psi\,\widetilde{\mathbf{e}} + (1-\cos\psi)\widetilde{\mathbf{e}}^2, \quad \psi = \|\Psi\|, \tag{10}$$

This relation makes it comprehensible that the rotation operator does not depend on the multiples of the rotation revolution counts, i.e. $\mathbf{R}(\Psi) = \mathbf{R}(\Psi + 2i\pi\mathbf{e}),\ \forall i \in \mathrm{N}$.

Def. 14 gives a canonical parametrization of the rotation manifold *SO*(3). The parametrization can represent a rotation operator only locally, and there exists no parametrization that is global as well as non-singular. Note that a parametrization is a mapping from an open set of Euclidean space into some open set of the manifold. The rotation vector parametrization is singular at the rotation angle equal to 2π and its multiples. It is clear that the singularity naturally appears in dynamical analysis with large rotations, and cannot be omitted. Singularity should be considered as a non-differentiable hole that must not be omitted by skipping. The singularity is due to fact that the rotation manifold is compact, and there does not exist a single continuous parametrization from an open set of the Euclidean space E^3 onto this compact manifold, see details in [Stuelpnagel 1964].

A rotation operator can be presented by higher dimensional, singularity-free representations where a unit quaternion is a four-parametric example. The coordinates of a quaternion are not independent, in fact, a quaternion produces a three-dimensional manifold into a four-dimensional Euclidean space, that is a unit three-sphere S^3 (surface) embedded in E^4. Hence, we do not speak about parametrization when considering a mapping between different manifolds, for example in the case of a unit quaternion this mapping is $S^3 \to SO(3)$.

The description of a rotation motion has been studied for a long time, so there exists a large number of different representations of a rotation motion. Three-dimensional representations are rotation vectors, Euler angles, Bryant angles, Rodrigues parameters (Gibbs vector), and four-parametric representations are unit quaternions (Euler-Rodrigues parameters), linear parameters, Euler rotation, and Cayley-Klein parameters, see [Spring

1986] and a historical aspect for Euler-Rodrigues parameters in [Cheng & Gupta 1989]. In addition, there exists a higher dimensional representation of a rotation operator, like a rotation matrix, that has a dimension equal to nine.

Four-dimensional descriptions are topologically connected with a unit three-sphere S^3 and to the proper unitary group $SU(2)$ that is a group of complex 2-by-2 matrices, and their joined algebra is an even Clifford subalgebra (quaternion algebra) and an algebra of 2-by-2 skew-Hermitian traceless matrices (Lie-algebra $su(2)$), respectively, see [Choquet-Bruhat *et al.* 1989].

Correspondingly, three-dimensional descriptions are the parametrizations of the rotation manifold and their algebras are the cross product in the Euclidean space E^3 (Lie-algebra in E^3) and an algebra of skew-symmetric tensors (Lie-algebra $so(3)$). We consider three-dimensional descriptions and especially the rotation vector a simple, geometrical significance representation. The major drawback of the rotation vector parametrization, singularity, can be passed by introducing another parametrization chart such that the parametrization mappings cover the rotation manifold globally.

15 Definition for Complement Rotation Vector

Let a rotation vector $\boldsymbol{\Psi}$ with a rotation angle larger than zero and less than perigon (full angle), i.e. $0 < \psi < 2\pi$, then its complement rotation vector $\boldsymbol{\Psi}^{C}$ is defined as

$$\boldsymbol{\Psi}^{C} := \boldsymbol{\Psi} - \frac{2\pi}{\psi}\boldsymbol{\Psi}, \qquad \psi = \|\boldsymbol{\Psi}\|.$$

Then the rotation angle of the complement rotation vector is $\psi^{C} = 2\pi - \psi$ and the rotation axis is $\mathbf{e}^{C} = -\mathbf{e}$. ◊

After substituting the complement rotation vector into *Def.* 14, we notice that the rotation vector and its complement represent the same rotation operator, i.e. $\mathbf{R}(\boldsymbol{\Psi}^{C}) = \mathbf{R}(\boldsymbol{\Psi})$. *Def. 15* is a change of parametrization in the parameter space E^3, see Figure 2 and Figure 6. This change of parametrization is a continuously differentiable mapping on the open domain $0 < \psi < 2\pi$, giving a smooth construction of the rotation manifold $SO(3)$ at this domain. Note that the complement of a complement rotation vector is a rotation vector itself, i.e. $(\boldsymbol{\Psi}^{C})^{C} = \boldsymbol{\Psi}$, hence there is no priority over these parametrization charts.

We could represent the rotation manifold globally with these two parametrization charts. When a rotation angle exceeds straight angle ($\psi > \pi$), we accomplish the change of parametrization according to *Def. 15*, giving a new rotation angle smaller than straight angle. Thus, we never get into trouble with singularity at $\psi = 2\pi$. As it is illustrated in

Figure 6, the change of parametrization maps rotation angle outside of straight angle into inside of straight angle. Note that there exists no other canonical parametrization with rotation less than perigon such as those parametrizations given in *Def. 15*.

The zero rotation vector is an isolated point, the centre of the domain, for the parametrization change. Using a limit process, we find out that the rotation operator approaches to the identity element when the rotation angle is decreased. Hence, we could modify the domain of the parametrization where the rotation angle is less than perigon, i.e. ψ

$< 2\pi$ including the zero rotation angle. This domain is still an open domain in the Euclidean space E^3, indeed, it is an open ball in E^3 with 2π-radius.

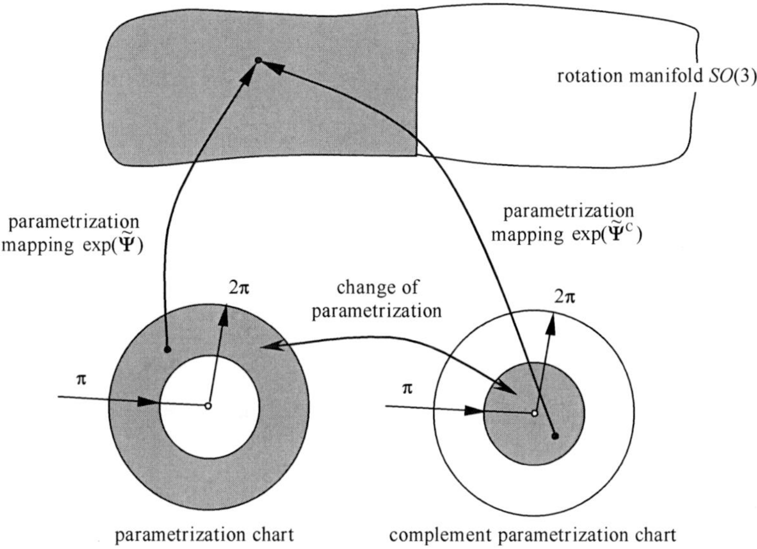

Figure 6. The change of parametrization in the parameter space E^3 for the canonical representation of the rotation manifold.

1.2.1 Lie Group and Lie Algebra

The concept of Lie group and Lie algebra gives an algebraic structure for the rotation manifold. Since general groups and their special cases, Lie groups with corresponding Lie algebra, are quite unfamiliar to engineering literature, we give definition for these objects. We recommend the textbooks like [Choquet-Bruhat *et al.* 1989], [Marsden & Ratiu 1999] and [Selig 1996] for more details. As it was shown that rotation operators form the smooth manifold, called the rotation manifold, we show that the rotation manifold is a Lie group, too. This issue is significant especially in composite rotations.

16 Definition for Group

A group G is a set with an internal operation $G \times G \to G$ by $(\mathbf{A},\mathbf{B}) \mapsto \mathbf{AB}$, such that

[1] the internal operation is associative $\mathbf{A}(\mathbf{BC}) = (\mathbf{AB})\mathbf{C}, \quad \forall \mathbf{A},\mathbf{B},\mathbf{C} \in G$,

[2] there is a unique element $\mathbf{I} \in G$ called identity such that $\mathbf{AI} = \mathbf{IA} = \mathbf{A}, \quad \forall \mathbf{A} \in G$,

[3] for each $\mathbf{A} \in G$ there exists a unique element of G called the inverse of $\mathbf{A}$ such that $\mathbf{A}^{-1}\mathbf{A} = \mathbf{AA}^{-1} = \mathbf{I}$. ◊

The group is called an Abelian group or a commutative group if $\mathbf{AB} = \mathbf{BA}, \ \forall \mathbf{A},\mathbf{B} \in G$. For example, a set of vectors with an internal operation, the vector addition, is an Abelian group where the identity element is the zero vector. In addition, a set of matrices equipped the matrix addition is an Abelian group with a zero matrix as the identity element.

If an internal operation is not commutative, then a group is called a non-Abelian group, or a non-commutative group. For example, a set of invertable *n*-by-*n* matrices with an internal operation the matrix multiplication is a non-Abelian group where the identity element is the identity matrix.

17 Definition for Special Orthogonal Group SO(3)

A special orthogonal (non-commutative) group in the Euclidean space E^3 is defined by

$$SO(3) := \left\{ \mathbf{R}: \mathrm{E}^3 \to \mathrm{E}^3 \text{ linear} \middle| \mathbf{R}^{\mathrm{T}}\mathbf{R} = \mathbf{R}\mathbf{R}^{\mathrm{T}} = \mathbf{I},\ \det(\mathbf{R}) = +1 \right\}.$$

Since $SO(3)$ is a group it has to fulfill all group properties given in *Def. 16.* ◊

A rotation operator defined in *Def.* 14 is also an element of the special orthogonal group as it was shown in Equation (9). The rotation operators form a non-commutative group with the internal operation, called composite mapping, and an identity element as the identity operator $\mathbf{I}$. Hence, we may denote $\mathbf{R} \in SO(3)$.

18 Definition for Lie Group

A Lie group L is a group that is also a differentiable manifold such that an internal operation $L \times L \to L$ by $(\mathbf{A},\mathbf{B}) \mapsto \mathbf{AB}\ \forall \mathbf{A},\mathbf{B} \in L$ is a continuously differentiable mapping. ◊

This is short but not so an easily manageable definition. We have shown that the set of rotation operators form the differentiable manifold, called the rotation manifold; moreover, the set of rotation operators is a non-commutative group, special orthogonal group $SO(3)$. So, only the continuity of internal operation has not been shown. To prove this, it has to be shown that a change of parametrization mappings under composition is continuously differentiable on its domain. A procedure is similar to one for showing a manifold is differentiable. The prove that the group $SO(3)$ is a Lie group is omitted here, but can be found in [Choquet-Bruhat *et el* 1989; pp. 181-182]. It is based on reality that any two composite rotations can be represented by Eulerian angles, giving a differentiable change of parametrization mappings between different sets of Eulerian angles.

19 Definition for Lie Algebra

A Lie algebra l of the Lie group L is a tangent vector space at the identity, $T_{\mathbf{I}}L$, together with a bilinear, skew-symmetric brackets $[\mathbf{a},\mathbf{b}]$ satisfying Jacobi's identity

$$\left[\mathbf{a},\left[\mathbf{b},\mathbf{c}\right]\right] + \left[\mathbf{b},\left[\mathbf{c},\mathbf{a}\right]\right] + \left[\mathbf{c},\left[\mathbf{a},\mathbf{b}\right]\right] = \mathbf{0}, \quad \forall \mathbf{a},\mathbf{b},\mathbf{c} \in l.$$

The skew-symmetry means that $[\mathbf{a},\mathbf{b}] = -[\mathbf{b},\mathbf{a}]$, $\forall \mathbf{a},\mathbf{b} \in l$. ◊

How to obtain Lie brackets is still an open question and we have to define a Lie algebra adjoint transformation.

20 Definition for Lie Algebra Adjoint Transformation

An adjoint transformation $\mathrm{Ad}_{\mathbf{R}}$ of the Lie algebra l is defined by

$$\mathrm{Ad}_{\mathbf{R}}: l \to l, \quad \mathrm{Ad}_{\mathbf{R}}\mathbf{b} := \mathbf{R}\mathbf{b}\mathbf{R}^{-1}, \quad \forall \mathbf{R} \in L.$$

Note that $\mathbf{b} \in l$ is an element of the Lie algebra l and $\mathbf{R}$ is an element of the Lie group L. The Lie algebra adjoint transformation maps an element of the Lie algebra into another Lie algebra element. ◊

21 Determination for Lie Brackets

Lie brackets can be obtained by differentiating the adjoint representation $\mathrm{Ad}_{\mathbf{R}}\mathbf{b}$ with respect to $\mathbf{R}(\eta) \in L$ at the identity in the direction $\mathbf{a} \in l$ such that $\mathbf{R}(\eta = 0) = \mathbf{I}$ and $\mathrm{d}\mathbf{R}(\eta = 0)/\mathrm{d}\eta = \mathbf{a}$ where η is a parameter, giving

$$[\mathbf{a}, \mathbf{b}] = \left. \frac{\mathrm{d}\left(\mathbf{R}(\eta)\mathbf{b}(\mathbf{R}(\eta))^{-1}\right)}{\mathrm{d}\eta} \right|_{\eta = 0}.$$

Lie brackets is a bilinear skew-symmetric form and satisfies Jacobi's identity, given in *Def. 19*. ◊

Especially, let $\mathbf{R}(\eta) \in SO(3)$ be a η-parametrized rotation operator, an element of the special orthogonal group, given by formula $\mathbf{R}(\eta) = \exp(\eta\widetilde{\mathbf{\Psi}})$. Differentiating the expression $\exp(\eta\widetilde{\mathbf{\Psi}})$ with respect to the parameter η at $\eta = 0$ gives the tangent vector space at the identity $\mathbf{I} \in SO(3)$, yielding

$$\left. \frac{\mathrm{d}\exp(\eta\widetilde{\mathbf{\Psi}})}{\mathrm{d}\eta} \right|_{\eta = 0} = \widetilde{\mathbf{\Psi}}. \tag{11}$$

Thus, the skew-symmetric tensor $\widetilde{\mathbf{\Psi}}$ belongs to the tangent space of the rotation manifold $SO(3)$, denoted by $\widetilde{\mathbf{\Psi}} \in T_{\mathbf{I}}SO(3)$, where the identity $\mathbf{I} \in SO(3)$ represents a base point of the rotation manifold. The skew-symmetric tensor $\widetilde{\mathbf{\Psi}}$ is also an element of Lie algebra $so(3)$ for corresponding Lie group $SO(3)$. We could also mark $so(3) = T_{\mathbf{I}}SO(3)$, i.e. Lie algebra is canonical isomorphic to the tangent space of the rotation manifold at the identity. Moreover, we may denote the Lie algebra $so(3)$ as a set of skew-symmetric operators (tensors)

$$so(3) = \left\{ \widetilde{\mathbf{\Psi}}: \mathrm{E}^3 \to \mathrm{E}^3 \text{ linear} \,\middle|\, \widetilde{\mathbf{\Psi}}^{\mathrm{T}} = -\widetilde{\mathbf{\Psi}} \right\}. \tag{12}$$

We obtain the Lie brackets of the Lie algebra $so(3)$ by differentiating the Lie algebra adjoint representation $\mathrm{Ad}_{\mathbf{R}(\eta)}\widetilde{\mathbf{\Theta}}$ with respect to η at $\eta = 0$

$$\frac{\mathrm{d}(\mathrm{Ad}_{\mathbf{R}(\eta=0)}\widetilde{\mathbf{\Theta}})}{\mathrm{d}\eta} = \left. \frac{\mathrm{d}(\exp(\eta\widetilde{\mathbf{\Psi}})\widetilde{\mathbf{\Theta}}\exp(-\eta\widetilde{\mathbf{\Psi}}))}{\mathrm{d}\eta} \right|_{\eta = 0} = \widetilde{\mathbf{\Psi}}\widetilde{\mathbf{\Theta}} - \widetilde{\mathbf{\Theta}}\widetilde{\mathbf{\Psi}}. \tag{13}$$

Hence, the Lie brackets for the Lie algebra $so(3)$ is $[\widetilde{\Psi},\widetilde{\Theta}]=\widetilde{\Psi}\widetilde{\Theta}-\widetilde{\Theta}\widetilde{\Psi},\ \ \widetilde{\Psi},\widetilde{\Theta}\in so(3)$.

The vector cross product $(\cdot\times\cdot):\mathrm{E}^3\times\mathrm{E}^3\to\mathrm{E}^3$ in the Euclidean space E^3 is a Lie algebra with Lie brackets defined by

$$[\mathbf{x},\mathbf{y}]:=\mathbf{x}\times\mathbf{y},\qquad \mathbf{x},\mathbf{y}\in\mathrm{E}^3. \tag{14}$$

The vector cross product $(\cdot\times\cdot)$ is a bilinear, with respect to vector addition and scalar multiplication, and a skew-symmetric operator over E^3 and satisfies Jacobi's identity in *Def. 19*. The Lie algebra $so(3)$ can be identified with the cross product on E^3 by formula

$$\widetilde{\Psi}\mathbf{a}=\Psi\times\mathbf{a},\quad \forall\mathbf{a}\in\mathrm{E}^3, \tag{15}$$

where the vector $\Psi\in\mathrm{E}^3$ is the axial vector for the skew-symmetric tensor $\widetilde{\Psi}\in so(3)$.

22 Definition for Lie Algebra Homomorphism and Isomorphism

Let l and g be two Lie algebras. A mapping $\varphi:g\to l$ is homomorphism, that is

$$\varphi([\Psi,\Theta]_g)=\left[\varphi(\Psi),\varphi(\Theta)\right]_l,\qquad \forall\Psi,\Theta\in g\ .$$

A homomorphism is an isomorphism if the mapping $\varphi:g\to l$ is a vector space isomorphism, i.e. a linear bijection. The Lie algebras as isomorphic, denoted $l\cong g$, if an isomorphism exists between these algebras. ◊

The tilde mapping $\widetilde{\ }:\mathrm{E}^3\to so(3)$ is an homomorphism between the cross product on E^3 and the Lie algebra $so(3)$, i.e.

$$\left(\Psi\times\Theta\right)^{\sim}=\widetilde{\Psi}\widetilde{\Theta}-\widetilde{\Theta}\widetilde{\Psi}, \tag{16}$$

see proof e.g. in [Marsden & Ratiu 1999; p. 290]. The tilde mapping is also a linear bijection giving isomorphic correspondence between the elements of the Lie algebras, denoted by $\mathrm{E}^3\cong so(3)$. For computational purposes, the Lie algebra in the Euclidean space E^3 is simpler than the Lie algebra $so(3)$ and, hence, it will be utilized in following. The Lie algebra in the Euclidean space E^3 equipped with the cross product as the Lie bracket is a rather unusual Lie algebra since, by our knowledge, there does not exist a Lie group whose the Lie algebra it is.

1.2.2 Compound Rotation

A rotation operator is an element of Lie group that is a differentiable manifold as well as a non-commutative group. A compound of rotations is also a rotation itself and induces a Lie group structure. The compound rotation can be defined by two different, nevertheless equivalent ways: the material description, and the spatial description.

23 Definition for Material Description of Compound Rotation

We define the material description of a compound rotation by the left translation mapping $\mathscr{Left}_{\mathbf{R}}: SO(3) \to SO(3)$ as

$$\mathscr{Left}_{\mathbf{R}} \mathbf{R}_{\text{inc}}^{\text{mat}} := \mathbf{R}\mathbf{R}_{\text{inc}}^{\text{mat}} = \mathbf{R}\exp(\widetilde{\boldsymbol{\Theta}}_{\mathbf{R}}), \qquad \mathbf{R}_{\text{inc}}^{\text{mat}}, \mathbf{R} \in SO(3),$$

where $\mathbf{R}_{\text{inc}}^{\text{mat}}$ is a material incremental rotation operator, and $\boldsymbol{\Theta}_{\mathbf{R}}$ is a material incremental rotation vector with respect to the base point $\mathbf{R} \in SO(3)$. This description is called material since the incremental rotation operator acts on a material vector space. ◊

24 Definition for Spatial Description of Compound Rotation

We define the spatial description of a compound rotation by the right translation mapping $\mathscr{Right}_{\mathbf{R}}: SO(3) \to SO(3)$ as

$$\mathscr{Right}_{\mathbf{R}} \mathbf{R}_{\text{inc}}^{\text{spat}} := \mathbf{R}_{\text{inc}}^{\text{spat}} \mathbf{R} = \exp\left(\widetilde{\boldsymbol{\theta}}_{\mathbf{R}}\right)\mathbf{R}, \qquad \mathbf{R}_{\text{inc}}^{\text{spat}}, \mathbf{R} \in SO(3),$$

where $\mathbf{R}_{\text{inc}}^{\text{spat}}$ is a spatial incremental rotation operator, and $\boldsymbol{\theta}_{\mathbf{R}}$ is a spatial incremental rotation vector with respect to the base point $\mathbf{R} \in SO(3)$. This description is called spatial since the incremental rotation operator acts on a spatial vector space. ◊

We use majuscules for material vectors and minuscules for spatial vectors. The material and spatial incremental rotation tensors and their rotation vectors are related by

$$\mathbf{R}_{\text{inc}}^{\text{spat}} = \mathbf{R}\mathbf{R}_{\text{inc}}^{\text{mat}}\mathbf{R}^{\mathrm{T}}, \qquad \widetilde{\boldsymbol{\theta}}_{\mathbf{R}} = \mathbf{R}\widetilde{\boldsymbol{\Theta}}_{\mathbf{R}}\mathbf{R}^{\mathrm{T}}, \text{ and } \quad \boldsymbol{\theta}_{\mathbf{R}} = \mathbf{R}\boldsymbol{\Theta}_{\mathbf{R}}, \tag{17}$$

where the first relation is called an inner automorphism that is an isomorphism onto itself, the second relation is a Lie algebra adjoint transformation $\mathrm{Ad}_{\mathbf{R}}\widetilde{\boldsymbol{\Theta}}_{\mathbf{R}} = \mathbf{R}\widetilde{\boldsymbol{\Theta}}_{\mathbf{R}}\mathbf{R}^{\mathrm{T}}$, see *Def.* 20, and the last relation is another Lie algebra adjoint transformation on the Euclidean space with the vector cross product as the Lie algebra $(\mathrm{E}^3, \cdot\times\cdot)$.

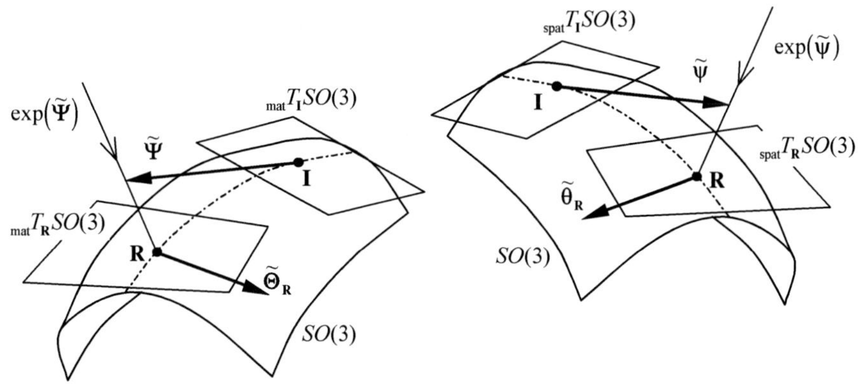

Figure 7. A geometric representation of the material (on the left) and spatial tangent spaces (on the right) on the rotation manifold.

25 Definition for Material Tangent Space[11] of Rotation

Differentiating the material expression of the compound rotation $\mathbf{R}\exp(\eta\widetilde{\Theta})$ with respect to the parameter η and setting $\eta = 0$, yields the material tangent space at the base point $\mathbf{R} \in SO(3)$. This material tangent space on the rotation manifold $SO(3)$ at any base point $\mathbf{R}$ is defined as

$$_{\text{mat}}T_{\mathbf{R}}SO(3) := \left\{ \widetilde{\Theta}_{\mathbf{R}} := (\mathbf{R}, \widetilde{\Theta}) \middle| \text{ with } \mathbf{R}\widetilde{\Theta};\ \mathbf{R} \in SO(3), \widetilde{\Theta} \in so(3) \right\},$$

where an element of the material tangent space $\widetilde{\Theta}_{\mathbf{R}} \in {}_{\text{mat}}T_{\mathbf{R}}SO(3)$ and is a skew-symmetric tensor, i.e. $\widetilde{\Theta}_{\mathbf{R}} \in so(3)$. The notation $(\mathbf{R}, \widetilde{\Theta})$, the pair of the rotation operator $\mathbf{R}$ and the skew-symmetric tensor $\widetilde{\Theta}$, represents the material skew-symmetric tensor at the base point $\mathbf{R} \in SO(3)$, see Figure 7. Hence, we may express that $\widetilde{\Theta}_{\mathbf{R}}$ is a skew-symmetric tensor, or a tangent tensor, at the point $\mathbf{R}$ in the manifold $SO(3)$. For simplicity, we could omit the base point $\mathbf{R}$ by denoting $\widetilde{\Theta} \in {}_{\text{mat}}T_{\mathbf{R}}SO(3)$ if there is no danger of confusion. ◊

26 Definition for Spatial Tangent Space[12] of Rotation

A spatial tangent space on the rotation manifold SO(3) at any base point $\mathbf{R}$ is defined

$$_{\text{spat}}T_{\mathbf{R}}SO(3) := \left\{ \widetilde{\theta}_{\mathbf{R}} := (\mathbf{R}, \widetilde{\theta}) \middle| \text{ with } \widetilde{\theta}\mathbf{R};\ \mathbf{R} \in SO(3), \widetilde{\theta} \in so(3) \right\},$$

where an element of the spatial tangent space $\widetilde{\theta}_{\mathbf{R}} \in {}_{\text{spat}}T_{\mathbf{R}}SO(3)$ and is a skew-symmetric tensor, i.e. $\widetilde{\theta}_{\mathbf{R}} \in so(3)$. The notation $(\mathbf{R}, \widetilde{\theta})$, the pair of the rotation operator $\mathbf{R}$ and the skew-symmetric tensor $\widetilde{\Theta}$, represents a spatial skew-symmetric tensor at the base point $\mathbf{R}$, see Figure 7. Again, we could omit the base point $\mathbf{R}$, i.e. $\widetilde{\theta} \in {}_{\text{spat}}T_{\mathbf{R}}SO(3)$ if there is no danger of confusion. ◊

Rotation operators, the elements of the Lie group $SO(3)$, are defined as linear operators $\mathbf{R} \in \mathscr{L}(\mathrm{E}^3, \mathrm{E}^3)$. Eqns (17b,c) give another interpretation to a rotation operator, it is an adjoint transformation between material and spatial tangent spaces. Additionally, a rotational motion induces the rotation operator, since the rotation operator maps the material place vector $\mathbf{X} \in \mathscr{B}_0$ into the spatial place vector $\mathbf{x} \in \mathscr{B}$ by the equation $\mathbf{x}(t) = \mathbf{R}(t)\mathbf{X}$, i.e. $\mathbf{R} \in \mathscr{L}(\mathscr{B}_0, \mathscr{B})$. More generally, a rotation operator transforms material vectors into spatial vectors, that is $\mathbf{R} \in \mathscr{L}(T_{\mathbf{X}}\mathscr{B}_0, T_{\mathbf{x}}\mathscr{B})$.

1.2.3 Isomorphisms and Tangential Transformations

The Lie algebra $so(3)$, which consists of the skew-symmetric tensors, and the Lie algebra $(\mathrm{E}^3, \cdot\times\cdot)$ in the Euclidean space are isomorphic with the Lie algebra isomorphism

---

[11] Usually this space is called left-invariant vector field.

[12] Usually is called as right-invariant vector field

$\tilde{}: E^3 \to so(3)$ that is the tilde mapping. The spatial and material tangent spaces are isomorphic where the isomorphism is an adjoint transformation $\mathrm{Ad}_{\mathbf{R}}: {}_{\mathrm{mat}}T_{\mathbf{R}}SO(3) \to {}_{\mathrm{spat}}T_{\mathbf{R}}SO(3)$, given in (17b). Additionally, the Lie algebra $so(3)$ is isomorphic in the material tensor space with an isomorphism $so(3) \to {}_{\mathrm{mat}}T_{\mathbf{R}}SO(3)$ by $\tilde{\Theta} \mapsto \mathbf{R}\tilde{\Theta}$. Then we may express

$$E^3 \cong so(3) \cong {}_{\mathrm{mat}}T_{\mathbf{R}}SO(3) \cong {}_{\mathrm{spat}}T_{\mathbf{R}}SO(3) \,. \tag{18}$$

Isomorphism states that for any element from a vector space we can take an element from an isomorphic vector space with a linear one-to-one correspondence. Therefore, the isomorphic spaces have the same structure and we may associate the elements of the isomorphic spaces.

27 Definition for Virtual Rotation Tensor and Virtual Rotation Vector

A virtual rotation tensor $\delta\tilde{\Theta}_{\mathbf{R}}$ is an element of the corresponding tangent space $T_{\mathbf{R}}SO(3)$ for any base point $\mathbf{R} \in SO(3)$ such that it satisfies all linearized constraint equations, which naturally arise from joints and boundary conditions. A virtual rotation vector $\delta\Theta_{\mathbf{R}}$ at the base point $\mathbf{R}$ is the associated axial vector of the virtual rotation tensor $\delta\tilde{\Theta}_{\mathbf{R}}$. ◊

Let us consider the material form of compound rotation, given in *Def.* 23, with the aid of η-parametrized exponential mappings

$$\exp\left(\tilde{\Psi} + \eta\delta\tilde{\Psi}\right) = \exp\left(\tilde{\Psi}\right)\exp\left(\eta\delta\tilde{\Theta}_{\mathbf{R}}\right), \tag{19}$$

where we are finding an incremental rotation tensor, the virtual rotation tensor $\delta\tilde{\Psi}$, such that it belongs to the same tangent space as the rotation tensor $\tilde{\Psi}$, i.e. such that $\delta\tilde{\Psi}, \tilde{\Psi} \in {}_{\mathrm{mat}}T_{\mathbf{I}}SO(3)$ with the identity as a base point omitted for simplicity. Note that $\mathbf{R} = \exp(\tilde{\Psi})$, and $\delta\tilde{\Theta}_{\mathbf{R}} \in {}_{\mathrm{mat}}T_{\mathbf{R}}SO(3)$. We point out that the skew-symmetric tensors $\tilde{\Psi}$ and $\delta\tilde{\Theta}_{\mathbf{R}}$ do not belong to the same tangent space of rotation as it can be verified that $\exp(\tilde{\Psi})\exp(\tilde{\Theta}) \neq \exp(\tilde{\Psi} + \tilde{\Theta})$, generally. The associated rotation vector $\mathbf{\Psi}$ for the skew-symmetric tensor $\tilde{\Psi}$ is called the total material rotation vector whose base point is the identity. Taking the derivatives of (19) with respect to the parameter η at $\eta = 0$ gives after the aid of isomorphism (18), see e.g. [Ibrahimbegović *et. al.* 1995]

$$\begin{aligned} &\delta\Theta_{\mathbf{R}} = \mathbf{T}\cdot\delta\mathbf{\Psi}, \\ &\mathbf{T} = \frac{\sin\psi}{\psi}\mathbf{I} - \frac{1-\cos\psi}{\psi^2}\tilde{\Psi} + \frac{\psi - \sin\psi}{\psi^3}\mathbf{\Psi}\otimes\mathbf{\Psi}^{\wedge}; \\ &\psi = \|\mathbf{\Psi}\|, \quad \mathbf{R} = \exp(\mathbf{\Psi}), \quad \lim_{\mathbf{\Psi}\to 0}\mathbf{T}(\mathbf{\Psi}) = \mathbf{I}, \end{aligned} \tag{20}$$

where the material tangential transformation $\mathbf{T} = \mathbf{T}(\boldsymbol{\Psi})$ is a linear mapping between the virtual material tangent spaces ${}_{\text{mat}}T_{\mathbf{I}}SO(3) \rightarrow {}_{\text{mat}}T_{\mathbf{R}}SO(3)$. Now, we could make another verification that the virtual rotation vector $\delta\boldsymbol{\Theta}_{\mathbf{R}}$ and the virtual total rotation vector $\delta\boldsymbol{\Psi}$ belong to different vector spaces on the manifold. This is because the tangential transformation $\mathbf{T}$ is equal to the identity only at $\boldsymbol{\Psi} = \mathbf{0}$. Note that the transformation $\mathbf{T}$ has an effect on the base points, changing the base point $\mathbf{I}$ into $\mathbf{R}$.

By examining the tangential transformation $\mathbf{T}$ in Eqn (20), we found that the transformation is non-singular when the rotation angle is less than perigon, i.e. $\psi < 2\pi$. It is worth noting that the tangential transformation $\mathbf{T}(\boldsymbol{\Psi})$, the corresponding rotation operator $\mathbf{R}(\boldsymbol{\Psi})$ and the skew-symmetric rotation tensor $\tilde{\boldsymbol{\Psi}}$ have the same eigenvectors. Hence, $\mathbf{T}(\boldsymbol{\Psi})$, $\mathbf{R}(\boldsymbol{\Psi})$, and $\tilde{\boldsymbol{\Psi}}$ are commutative, see [Ibrahimbegović *et. al.* 1995].

28 Definition for Material Vector Space of Rotation

For convenience, we define a material vector space on the rotation manifold at any point $\mathbf{R}$ as

$$ {}_{\text{mat}}T_{\mathbf{R}} := \left\{ \boldsymbol{\Theta}_{\mathbf{R}} := (\boldsymbol{\Psi}, \boldsymbol{\Theta}) \middle| \, \mathbf{R} = \exp\left(\tilde{\boldsymbol{\Psi}}\right) \in SO(3), \boldsymbol{\Theta} \in \mathrm{E}^3 \right\}, $$

where an element of the material vector space is $\boldsymbol{\Theta}_{\mathbf{R}} \in {}_{\text{mat}}T_{\mathbf{R}}$, which is an affine space with the rotation vector $\boldsymbol{\Psi}$ as a base point and the incremental rotation vector $\boldsymbol{\Theta}$ as a tangent vector. ◊

Hence, the tangential transformation $\mathbf{T}$ is a mapping $\mathbf{T}\colon {}_{\text{mat}}T_{\mathbf{I}} \rightarrow {}_{\text{mat}}T_{\mathbf{R}}$. Note that the elements of this material vector space can be added by the parallelogram law only if they occupy the same affine space, i.e. if their associated skew-symmetric tensors belong to the same tangent space of the rotation manifold. *Def. 28* for a material vector space ${}_{\text{mat}}T_{\mathbf{R}}$ should be considered as a useful and simple notation with an equivalence relation with a material tangent space ${}_{\text{mat}}T_{\mathbf{R}}SO(3)$, defined in *Def.* 26.

Respectively, we could determine the spatial tangential transformation, yielding

$$ \delta\boldsymbol{\theta}_{\mathbf{R}} = \mathbf{T}^{\mathrm{T}} \cdot \delta\boldsymbol{\psi}, \quad \mathbf{T} = \mathbf{T}(\boldsymbol{\psi}), \quad \mathbf{R} = \exp(\tilde{\boldsymbol{\psi}}), \quad \psi = \|\boldsymbol{\psi}\| \quad (= \|\boldsymbol{\Psi}\|), \tag{21} $$

where $\mathbf{T}$ is the same linear operator as in the material form (20).

29 Definition for Spatial Vector Space of Rotation

We define a spatial vector space on the rotation manifold at any point $\mathbf{R}$ as

$$ {}_{\text{spat}}T_{\mathbf{R}} := \left\{ \boldsymbol{\theta}_{\mathbf{R}} := (\boldsymbol{\psi}, \boldsymbol{\theta}) \,\middle|\, \mathbf{R} = \exp(\tilde{\boldsymbol{\psi}}) \in SO(3), \boldsymbol{\theta} \in \mathrm{E}^3 \right\}. \tag{22} $$

An element of the spatial vector space is $\boldsymbol{\theta}_{\mathbf{R}} \in {}_{\text{spat}}T_{\mathbf{R}}$. ◊

Hence, the spatial tangential transformation $\mathbf{T}^{\mathrm{T}}$ is a mapping between vector spaces on the rotation manifold ${}_{\text{spat}}T_{\mathbf{I}} \rightarrow {}_{\text{spat}}T_{\mathbf{R}}$. The spatial and the material vector spaces are related by the rotation operator as given in the Eqn (17c),. From Eqn (17c), it follows with the base point $\mathbf{I} \in SO(3)$ (note that $\Psi \in {}_{\mathrm{mat}}T_{\mathbf{I}}$).

$$\psi_{\mathbf{I}} = \mathbf{I}\Psi_{\mathbf{I}} \Rightarrow \psi_{\mathbf{I}} = \Psi_{\mathbf{I}}, \tag{23}$$

where '=' denotes the canonical isomorphism between the spatial and material vector spaces. The identity $\mathbf{I}$ maps between the vector fields ${}_{\mathrm{mat}}T_{\mathbf{I}} \rightarrow {}_{\mathrm{spat}}T_{\mathbf{I}}$. Now, the relation between the spatial and material vectors can be given as $(\psi, \theta) = (\mathbf{I}\Psi, \mathbf{R}\Theta)$ where ψ and Ψ represent the base points in the spatial and material vector spaces, respectively. This relation can be written more compactly as $\theta_{\mathbf{R}} = \mathbf{R}\Theta_{\mathbf{R}}$, called a push-forward, where the rotation operator should be considered as a mapping between the material and the spatial vector spaces of rotation, $\mathbf{R}\colon {}_{\mathrm{mat}}T_{\mathbf{R}} \rightarrow {}_{\mathrm{spat}}T_{\mathbf{R}}$, see Figure 8. A push-forward operator maps a material vector space into a spatial vector space (one-to-one and onto). It makes sense since the rotation operator is a two-point tensor. We note that the push-forward operator $\mathbf{R}$ has no influence on the base point of the rotation. Another push-forward operator for rotation tensors is given in (17b) where $\tilde{\theta}_{\mathbf{R}} = \mathbf{R}\tilde{\Theta}_{\mathbf{R}}\mathbf{R}^{\mathrm{T}}$ is a mapping between the material and spatial tangent spaces of rotation

$$\mathbf{R}(\cdot)\mathbf{R}^{\mathrm{T}}\colon {}_{\mathrm{mat}}T_{\mathbf{R}}SO(3) \rightarrow {}_{\mathrm{spat}}T_{\mathbf{R}}SO(3)\,.$$

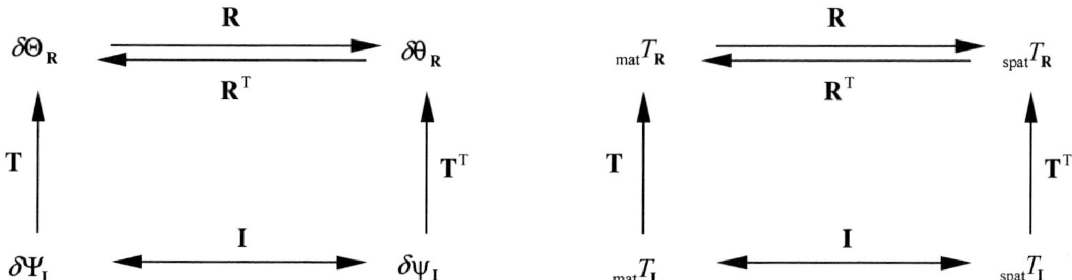

Figure 8. A commutative diagram of virtual material and spatial rotation vectors on the rotation manifold (on the left), and their corresponding vector spaces (on the right).

1.2.4 Angular Velocities, Accelerations and Curvatures

In this Section, we give definitions in the material and spatial representations for angular velocities, angular accelerations, and curvatures.

30 Definition for Material Angular Velocity

A material angular velocity (skew-symmetric) tensor is defined with the aid of rotation operator $\mathbf{R} \in SO(3)$ and its time derivative by

$$\tilde{\Omega}_{\mathbf{R}} := \mathbf{R}^{\mathrm{T}}\dot{\mathbf{R}}$$

where the dot represents to the time derivative. See justification in [Marsden & Ratiu 1999; Ch. 8.6 & 15.2]. ◊

The rotation tensor can be viewed as a mapping, a push-forward of a material vector, $\mathbf{R}: {}_{\text{mat}}T_{\mathbf{R}} \to {}_{\text{spat}}T_{\mathbf{R}}$ between the material and spatial vector spaces. Then the material angular velocity tensor is a mapping $\widetilde{\mathbf{\Omega}}_{\mathbf{R}}: {}_{\text{mat}}T_{\mathbf{R}} \to {}_{\text{mat}}T_{\mathbf{R}}$. Thus, the material angular velocity tensor is indeed a true material tensor. The skew-symmetry can be observed by taking derivative for the equation $\mathbf{R}^{\mathrm{T}}\mathbf{R} = \mathbf{I}$.

If the rotation operator is expressed with the aid of exponential mapping by $\mathbf{R}_{\text{new}} = \mathbf{R}(\mathbf{I} + \widetilde{\Theta}_{\mathbf{R}} + \mathcal{O}(\widetilde{\Theta}_{\mathbf{R}}^2))$, where the fixed rotation $\mathbf{R}$ is superimposed by an infinitesimal rotation $(\mathbf{I} + \widetilde{\Theta}_{\mathbf{R}})$ plus higher order terms and substituting this into *Def.* 30 yields after the limit process $\widetilde{\Theta}_{\mathbf{R}} \to \widetilde{\mathbf{0}}$

$$\widetilde{\mathbf{\Omega}}_{\mathbf{R}} = \dot{\widetilde{\Theta}}_{\mathbf{R}} \Leftrightarrow \mathbf{\Omega}_{\mathbf{R}} = \dot{\Theta}_{\mathbf{R}}. \tag{24}$$

This states that the angular velocity vector is the time derivative of the incremental rotation vector $\Theta_{\mathbf{R}}$; moreover, (if the base point is omitted) $\Theta, \dot{\Theta}, \mathbf{\Omega} \in {}_{\text{mat}}T_{\mathbf{R}}$, which is the material rotation vector space on the rotation manifold. The result in Eqn (24) is often given as definition for the angular velocity vector in the elementary text books.

Similar expression and derivation can be accomplished for the spatial angular velocity tensor and vector, yielding

$$\begin{aligned} &\widetilde{\omega}_{\mathbf{R}} := \dot{\mathbf{R}}\mathbf{R}^{\mathrm{T}}, \\ &\widetilde{\omega}_{\mathbf{R}} = \dot{\widetilde{\theta}}_{\mathbf{R}} \Leftrightarrow \omega_{\mathbf{R}} = \dot{\theta}_{\mathbf{R}}, \end{aligned} \tag{25}$$

where the spatial incremental rotation vector $\theta_{\mathbf{R}}$, its time derivative vector $\dot{\theta}_{\mathbf{R}}$ and the spatial angular vector $\omega_{\mathbf{R}}$ belong to the same spatial vector space on the manifold $\theta, \dot{\theta}, \omega \in {}_{\text{spat}}T_{\mathbf{R}}$, the base point $\mathbf{R}$ omitted.

31 Definition for Angular Accelerations

A material and spatial angular acceleration tensor and vector are defined as the time derivative of corresponding angular velocity term, giving

$$\begin{aligned} &\widetilde{\mathbf{A}}_{\mathbf{R}} := \dot{\widetilde{\mathbf{\Omega}}}_{\mathbf{R}}, && \widetilde{\mathbf{A}}_{\mathbf{R}} \in {}_{\text{mat}}T_{\mathbf{R}}SO(3), \\ &\mathbf{A}_{\mathbf{R}} := \dot{\mathbf{\Omega}}_{\mathbf{R}}, && \mathbf{A}_{\mathbf{R}} \in {}_{\text{mat}}T_{\mathbf{R}}, \\ &\widetilde{\alpha}_{\mathbf{R}} := \dot{\widetilde{\omega}}_{\mathbf{R}}, && \widetilde{\alpha}_{\mathbf{R}} \in {}_{\text{spat}}T_{\mathbf{R}}SO(3), \\ &\alpha_{\mathbf{R}} := \dot{\omega}_{\mathbf{R}}, && \alpha_{\mathbf{R}} \in {}_{\text{spat}}T_{\mathbf{R}}, \end{aligned}$$

where $\mathbf{A}_{\mathbf{R}}$ and $\alpha_{\mathbf{R}}$ are the material and spatial angular acceleration vectors at the base $\mathbf{R}$. ◊

Note that the material incremental rotation vector $\Theta_{\mathbf{R}}$, the material angular velocity vector $\mathbf{\Omega}_{\mathbf{R}}$ and the material angular acceleration vector $\mathbf{A}_{\mathbf{R}}$ (majuscule of alpha-letter) belong to the same material vector space on the rotation manifold, i.e. $\Theta_{\mathbf{R}}, \mathbf{\Omega}_{\mathbf{R}}, \mathbf{A}_{\mathbf{R}} \in {}_{\text{mat}}T_{\mathbf{R}}$ with the base point $\mathbf{R} = \exp(\widetilde{\Psi}_{\mathrm{I}})$. At separate moments, these vectors, however, occupy different vector spaces because the rotation operator depends on time, namely $\mathbf{R} = \mathbf{R}(t)$. The base point is moving in process of time. Vector quantities of this kind may be called *spin vectors*. Spin vectors are rather tricky in numerical sense as they always occupy a distinct vector space on a manifold. Correspondingly, the spatial spin vectors are $\boldsymbol{\theta}_{\mathbf{R}}, \boldsymbol{\omega}_{\mathbf{R}}, \boldsymbol{\alpha}_{\mathbf{R}} \in {}_{\mathrm{spat}}T_{\mathbf{R}}$.

Angular velocity vectors and the time derivative of total rotation vectors are related by, see (20-21)

$$\begin{aligned} &\boldsymbol{\Omega}_{\mathbf{R}} = \mathbf{T}(\boldsymbol{\Psi}_{\mathrm{I}})\cdot\dot{\boldsymbol{\Psi}}_{\mathrm{I}} \quad \text{where } \boldsymbol{\Omega}_{\mathbf{R}} \in {}_{\mathrm{mat}}T_{\mathbf{R}},\ \dot{\boldsymbol{\Psi}}_{\mathrm{I}}, \boldsymbol{\Psi}_{\mathrm{I}} \in {}_{\mathrm{mat}}T_{\mathrm{I}}, \text{ for material description,} \\ &\boldsymbol{\omega}_{\mathbf{R}} = \mathbf{T}^{\mathrm{T}}(\boldsymbol{\psi}_{\mathrm{I}})\cdot\dot{\boldsymbol{\psi}}_{\mathrm{I}} \quad \text{where } \boldsymbol{\omega}_{\mathbf{R}} \in {}_{\mathrm{spat}}T_{\mathbf{R}},\ \dot{\boldsymbol{\psi}}_{\mathrm{I}}, \boldsymbol{\psi}_{\mathrm{I}} \in {}_{\mathrm{spat}}T_{\mathrm{I}}, \text{ for spatial description,} \end{aligned} \tag{26}$$

where the tangential transformation depends on the total rotation vector, and the rotation operator is $\mathbf{R} = \exp(\widetilde{\Psi}_{\mathrm{I}}) = \exp(\widetilde{\psi}_{\mathrm{I}})$. Similar expression for the angular acceleration vector can be obtained by differentiating the above formulas, giving

$$\begin{aligned} &\mathbf{A}_{\mathbf{R}} = \mathbf{T}\cdot\ddot{\boldsymbol{\Psi}}_{\mathrm{I}} + \dot{\mathbf{T}}\cdot\dot{\boldsymbol{\Psi}}_{\mathrm{I}} \quad \text{where } \mathbf{A}_{\mathbf{R}} \in {}_{\mathrm{mat}}T_{\mathbf{R}},\ \boldsymbol{\Psi}_{\mathrm{I}}, \dot{\boldsymbol{\Psi}}_{\mathrm{I}}, \ddot{\boldsymbol{\Psi}}_{\mathrm{I}} \in {}_{\mathrm{mat}}T_{\mathrm{I}} \quad \text{for material description,} \\ &\boldsymbol{\alpha}_{\mathbf{R}} = \mathbf{T}^{\mathrm{T}}\cdot\ddot{\boldsymbol{\psi}}_{\mathrm{I}} + \dot{\mathbf{T}}^{\mathrm{T}}\cdot\dot{\boldsymbol{\psi}}_{\mathrm{I}} \quad \text{where } \boldsymbol{\alpha}_{\mathbf{R}} \in {}_{\mathrm{spat}}T_{\mathbf{R}},\ \boldsymbol{\psi}_{\mathrm{I}}, \dot{\boldsymbol{\psi}}_{\mathrm{I}}, \ddot{\boldsymbol{\psi}}_{\mathrm{I}} \in {}_{\mathrm{spat}}T_{\mathrm{I}} \quad \text{for spatial description.} \end{aligned} \tag{27}$$

Note that the tangential transformations $\mathbf{T}, \dot{\mathbf{T}} \in \mathscr{L}({}_{\mathrm{mat}}T_{\mathrm{I}}, {}_{\mathrm{mat}}T_{\mathbf{R}})$ and $\mathbf{T}^{\mathrm{T}}, \dot{\mathbf{T}}^{\mathrm{T}} \in \mathscr{L}({}_{\mathrm{spat}}T_{\mathrm{I}}, {}_{\mathrm{spat}}T_{\mathbf{R}})$ operate with the different base points.

32 Definition for Curvatures

A material curvature tensor $\widetilde{\mathbf{K}}_{\mathbf{R}}$ of s-parametrized curve is defined as

$$\widetilde{\mathbf{K}}_{\mathbf{R}} := \mathbf{R}^{\mathrm{T}}\frac{\mathrm{d}\mathbf{R}}{\mathrm{d}s} = \mathbf{R}^{\mathrm{T}}\mathbf{R}',$$

whose axial vector is called the material curvature vector $\mathbf{K}_{\mathbf{R}}$ (majuscule of kappa-letter). A spatial curvature tensor $\widetilde{\boldsymbol{\kappa}}$ is defined, respectively

$$\widetilde{\boldsymbol{\kappa}}_{\mathbf{R}} := \mathbf{R}'\mathbf{R}^{\mathrm{T}},$$

where prime denotes the derivative with respect to the length parameter s. ◊

Material and spatial curvature tensors and corresponding curvature vectors are related by

$$\widetilde{\boldsymbol{\kappa}}_{\mathbf{R}} = \mathrm{Ad}_{\mathbf{R}}\widetilde{\mathbf{K}}_{\mathbf{R}} = \mathbf{R}\mathbf{K}_{\mathbf{R}}\mathbf{R}^{\mathrm{T}}, \quad \boldsymbol{\kappa}_{\mathbf{R}} = \mathrm{ad}_{\mathbf{R}}\mathbf{K}_{\mathbf{R}} = \mathbf{R}\mathbf{K}_{\mathbf{R}}, \tag{28}$$

where the rotation operator $\mathbf{R}$ is a push-forward operator $\mathbf{R} \in \mathscr{L}({}_{\mathrm{mat}}T_{\mathbf{R}}, {}_{\mathrm{spat}}T_{\mathbf{R}})$ between material and spatial vector spaces and keeps the base point unaltered. A relation between curvature vectors and total rotation vectors becomes from an analogy of Eqns (26)

$$\mathbf{K}_{\mathbf{R}} = \mathbf{T}(\boldsymbol{\Psi}_{\mathbf{I}})\cdot\boldsymbol{\Psi}'_{\mathbf{I}} \quad \text{where} \quad \mathbf{K}_{\mathbf{R}} \in {}_{\text{mat}}T_{\mathbf{R}},\ \boldsymbol{\Psi}'_{\mathbf{I}} \in {}_{\text{mat}}T_{\mathbf{I}}, \quad \text{for material description,}$$
$$\boldsymbol{\kappa}_{\mathbf{R}} = \mathbf{T}^{\mathrm{T}}(\boldsymbol{\psi}_{\mathbf{I}})\cdot\boldsymbol{\psi}'_{\mathbf{I}} \quad \text{where} \quad \boldsymbol{\kappa}_{\mathbf{R}} \in {}_{\text{spat}}T_{\mathbf{R}},\ \boldsymbol{\psi}'_{\mathbf{I}} \in {}_{\text{spat}}T_{\mathbf{I}}, \quad \text{for spatial description,} \tag{29}$$

where the prime denotes the derivative with respect to the length parameter s, and $\mathbf{R}$ is the base point.

1.3 Constraint Point-Manifolds

So far we have studied only finite-dimensional manifolds. The placement field of continuum medium takes values in a Hilbert space[13], where chart parametrization maps vector-valued functions into vector-valued functions. The placement field needs an infinite number of basis functions in order to present an arbitrary placement field on continuum, yielding infinite-dimensional manifolds.

In multibody mechanics, constraint equations naturally arise from kinematic relations between bodies, boundary conditions and kinematic assumptions. Different joints like revolute and prismatic joints are examples of point-wise kinematic relations that can be described geometrically, i.e. these joints can be presented by holonomic finite-dimensional constraints. A kinematic assumption like Timoshenko-Reissner hypothesis is correspondingly a continuous kinematic relation (infinite dimensional) and it reduces an internal dimensionality by mapping a three-dimensional solid into a one-dimensional solid, called a beam. Beams are internally one-dimensional but infinite manifolds, whose generalized placement fields are presented by functions. Hence, beam models have only one spatial parameter, called a length parameter.

A beam can be considered as a vector bundle on a manifold, where for each position of the length parameter there is a two-dimensional vector plane where a cross-section belongs. This vector bundle occupies a set, the volume of the beam, in the Euclidean space.

33 Definition for (Holonomic) Point-Constraints

A holonomic point-wise constraint equation in a vectorial form is defined as a smooth mapping $\mathbf{h}\colon \mathrm{R}\times\mathrm{E}^n \to \mathrm{E}^{n-d}$ $(d < n)$ by

$$\mathbf{h}(t,\mathbf{x}) = \mathbf{0},$$

where the arguments are time t and the generalized place vector $\mathbf{x}(t) \in \mathrm{E}^n$. Constraint equations are assumed to be an independent set of equations. ◊

A constraint equation, which is impossible to present in a holonomic form, *Def.* 33, is nonholonomic, i.e. it is not and cannot be integrated into a holonomic form. It is clear that we cannot describe nonholonomic constraint equations, they are just kinematic relations, which are not holonomic. Different kinematic relations are shown in

Figure 9, where geometric constraints include all holonomic and the so called unilateral constraints, which are given by inequality equations with the function of time and a

---

[13] Hilbert space is a complete inner-product space, and here especially a complete infinite-dimensional inner-product vector-valued function space, see Hilbert spaces e.g. in [Debnath & Mikusinski 1990]

generalized place vector only. Unilateral constraints arise when modeling a kinematic relation between bodies in a contact. Especially in multibody systems, a joint clearance (play) and collision problems may be modeled by a contact formulation.

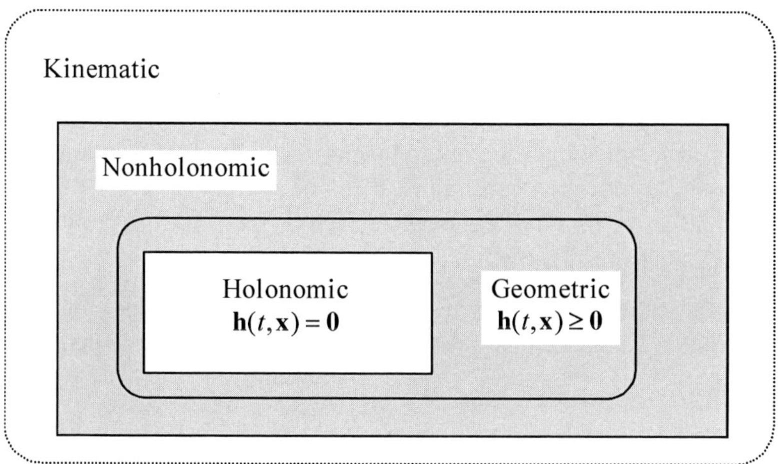

Figure 9. Different types of kinematic constraints and their occupying areas.

34 Definition for (Holonomic) Constraint Point-Manifold

A holonomic point-wise constraint equation (*Def.* 33) induces a d-manifold that is defined by

$$\mathcal{M}:=\left\{t\times\mathbf{x}\in\mathrm{R}\times\mathrm{E}^{n}\middle|\ \mathbf{h}(t,\mathbf{x})=\mathbf{0}\in\mathrm{E}^{n-d}\right\}.$$

The constraint point-manifold $\mathcal{M}$ is a d-dimensional smooth manifold with time as 1-parameter family. The constraint manifold at fixed time $t=t_0$ is denoted by $\mathcal{M}_{t0}$. ◊

Although a constraint point-manifold is smooth, its master, an infinite-dimensional constraint field-manifold in a multibody system, is usually smooth inside a body but has nonsmooth points at various joints, elbows, the sudden change of cross-section, etc, depending on different types of models. It is also informative to find out that a holonomic constraint manifold does not have a boundary at all, on the contrary geometric constraints with at least one unilateral constraint do, see Fig. 10.

A particular interesting nonholonomic constraint is presented by an equality equation (bilateral) with the function of time, a placement vector, and linearly on a velocity, see e.g. nonholonomic discrete systems in [Rabier & Rheinboldt 2000] and [Rosenberg 1980]. A rolling coin on the surface without sliding is a classic example, see Fig. 10 where the nonholonomic constraint describes the rolling coin problem. A virtual displacement is closely associated with constraint manifold, see Fig.10, and needs a proper definition. We note that virtual quantities are finite, not necessary infinitesimal.

35 Definition for Tangent Point-Bundle

A virtual displacement $\delta\mathbf{x}$ at any generalized place vector $\mathbf{x}\in\mathrm{E}^{n}$ and a fixed time $t=t_0$

$$T\mathcal{M}_{t_0} := \left\{ (\mathbf{x}, \delta\mathbf{x}) \in \mathrm{E}^n \times \mathrm{E}^n \,\middle|\, \mathbf{x} \in \mathcal{M}_{t_0},\ \mathrm{D}_{\mathbf{x}}\mathbf{h}(t_0, \mathbf{x}) \cdot \delta\mathbf{x} = \mathbf{0},\ \ \mathrm{D}_{\mathbf{x}}\mathbf{h} \text{ is surjection} \right\},$$

where $\mathrm{D}_{\mathbf{x}}\mathbf{h}$ is a Fréchet partial derivative of holonomic constraints with respect to $\mathbf{x}$ at $t = t_0$. The definition limits the singular (nonregular) points of the constraint manifold out by demanding the derivative of the constraints is a surjective (onto) mapping. ◊

This surjectivity request yields that dimensionality do not vary in the manifold that has importance when accomplishing the finite element method, the constraint manifold has a fixed dimensional independent on constraints. It can be proven, see e.g. [Rheinboldt 1986; p. 44-45], that the null-space of $\mathrm{D}_{\mathbf{x}}\mathbf{h}$, denoted $\ker \mathrm{D}_{\mathbf{x}}\mathbf{h}$ (kernel), is equal to the tangent space. Thus, $T\mathcal{M}_{t_0}$ is indeed a tangent bundle.

A vector bundle establishes a tangent vector space for each regular point of the manifold $\mathcal{M}$ at the fixed time t_0. For practical reasons, we need a tangent space that is an element of the tangent bundle.

36 Definition for Tangent Point-Space

For the fixed time $t = t_0$, we could set $\mathbf{x}_0 = \mathbf{x}(t_0)$, giving a tangent point-space at the point $\mathbf{x}_0 \in \mathcal{M}_{t_0}$

$$T_{\mathbf{x}_0}\mathcal{M} := \left\{ \delta\mathbf{x} \in \mathrm{E}^n \,\middle|\, (\mathbf{x}_0, \delta\mathbf{x}) \in T\mathcal{M}_{t_0} \right\}.$$

Then we may denote for any virtual displacement vector $\delta\mathbf{x} \in T_{\mathbf{x}_0}\mathcal{M}$, where the base point $\mathbf{x}_0$ is included in the notation as a subscript. ◊

A geometric interpretation of tangent space and its element virtual displacement have been illustrated with the holonomic and nonholonomic cases in Fig.10..

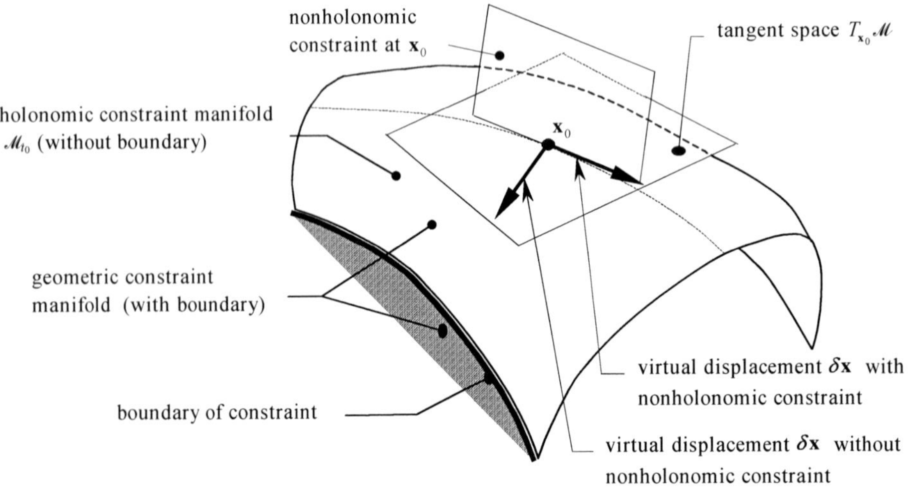

Figure 10. A geometric interpretation of kinematic constraints.

1.4 Constraint Field-Manifolds

In this Section, we give definitions for Gâteaux and Fréchet differentials, and for a constraint manifold modeled in infinite-dimensional Hilbert spaces, called field-manifolds[14] here. We assume that any operator we consider is Fréchet differentiable, which is stronger than Gâteaux differentiable. Hence, we use Gâteaux differential for more useful and simple way to calculate a Fréchet differential or derivative. We note that if an operator is Fréchet differentiable then its Fréchet and Gâteaux differential are equal, see more details in [Oden & Reddy 1976; Ch 2].

In following definition, we consider an operator $\mathbf{f}:\mathscr{X} \subset \mathscr{H}_1 \to \mathscr{H}_2$, later called a vector, from a set $\mathscr{X}$ of the Hilbert space $\mathscr{H}_1$ into the Hilbert space $\mathscr{H}_2$. The vector $\mathbf{f}$ is a general vector-valued nonlinear mapping between function spaces. We also assume that the vector $\mathbf{f}(\mathbf{x})$ is Fréchet differentiable, i.e. it has a unique Fréchet derivative.

37 Definition for Fréchet Derivative and Differential

The Fréchet derivative of the vector $\mathbf{f}:\mathscr{X} \subset \mathscr{H}_1 \to \mathscr{H}_2$ at fixed $\mathbf{x} \in \mathscr{X}$ is defined as a continuous linear operator $\mathrm{D}\mathbf{f}(\mathbf{x}):\mathscr{H}_1 \to \mathscr{H}_2$ such that

$$\mathbf{f}(\mathbf{x}+\mathbf{u}) - \mathbf{f}(\mathbf{x}) = \mathrm{D}\mathbf{f}(\mathbf{x})\cdot\mathbf{u} + \mathbf{r}(\mathbf{x},\mathbf{u}),$$

where the remainder obeys the condition $\lim_{\mathbf{u}\to\mathbf{0}} \frac{\|\mathbf{r}(\mathbf{x},\mathbf{u})\|_{\mathscr{H}_2}}{\|\mathbf{u}\|_{\mathscr{H}_1}} = 0$. $\mathrm{D}\mathbf{f}(\mathbf{x})\cdot\mathbf{u}$ is called Fréchet differential. ◊

A vector is called Fréchet differentiable if its Fréchet derivative exists. This derivative is also a linearized form, or better its affine form with together **f**(**x**), for a nonlinear vector $\mathbf{f}(\mathbf{x}+\mathbf{u})$ at **x**. *Def.* 37 is rather simple but not so practical way to calculate Fréchet derivative, hence we define Gâteaux differential for a more practical formula to calculate Fréchet differential and derivative for Fréchet differentiable vector.

38 Definition for Gâteaux Differentia

The Gâteaux differential of the vector $\mathbf{f}:\mathscr{X} \subset \mathscr{H}_1 \to \mathscr{H}_2$ at fixed $\mathbf{x} \in \mathscr{X}$ is defined as a limit

$$\mathrm{D}\mathbf{f}(\mathbf{x})\cdot\mathbf{u} := \lim_{\eta\to 0} \frac{\mathbf{f}(\mathbf{x}+\eta\mathbf{u}) - \mathbf{f}(\mathbf{x})}{\eta} = \left.\frac{\mathrm{d}\,\mathbf{f}(\mathbf{x}+\eta\mathbf{u})}{\mathrm{d}\,\eta}\right|_{\eta=0},$$

where the limit is to be interpreted in the norm of $\mathscr{H}_2$. The later formula is a practical and simple way to compute the directional derivative that is the term $\mathrm{D}\mathbf{f}(\mathbf{x})\cdot\mathbf{u}$ where $\mathbf{u} \in \mathscr{H}_1$ indicates direction. ◊

---

[14] We use the name (infinite-dimensional) field-manifold contrast to a (finite-dimensional) point-manifold.

We have assumed that $\mathscr{X}$ is a set, but next we give a more structure. We denote this structured set as $\mathscr{C}$.

39 Definition for Constraint Field-Manifold

A set $\mathscr{C}$ of a Hilbert space $\mathscr{H}_1$ is defined as infinite-dimensional constraint manifold embedded in Hilbert space such that it satisfies all the kinematic constraints of the holonomic type by

$$\mathscr{C} := \left\{ t \times \mathbf{x} \in \mathrm{R} \times \mathscr{H}_1 \;\middle|\; \mathbf{h}(t, \mathbf{x}) = \mathbf{0} \in \mathscr{H}_2 \right\},$$

where $\mathbf{h}(t, \mathbf{x})$ indicates an independent set of the holonomic constraint equations. A constraint manifold at a fixed time $t = t_0$ is denoted by $\mathscr{C}_{t0}$. ◊

Compare the constraint field-manifold with the constraint point-manifold defined in *Def.* 34. A similar way as in the point-wise case, we could define a tangent field-space which is a space of vector-valued functions

40 Definition for Tangent Field-Bundle

A virtual displacement field[15] $\delta\mathbf{x}$ at any place field $\mathbf{x} \in \mathscr{X}_0$ and the fixed time $t = t_0$ is defined as

$$T\mathscr{C}_{t_0} := \left\{ (\mathbf{x}, \delta\mathbf{x}) \in \mathscr{H}_1 \times \mathscr{H}_1 \;\middle|\; \mathbf{x} \in \mathscr{C}_{t_0},\ \mathrm{D}_{\mathbf{x}} \mathbf{h}(t_0, \mathbf{x}) \cdot \delta\mathbf{x} = \mathbf{0},\ \ \mathrm{D}_{\mathbf{x}} \mathbf{h} \text{ is surjection} \right\},$$

where $\mathrm{D}_{\mathbf{x}}\mathbf{h}$ is the Fréchet partial derivative of the holonomic constraints with respect to $\mathbf{x}$ at $t = t_0$. The definition limits the nonsmooth isolated points of the constraint manifold out by demanding the existence of the Fréchet derivative. ◊

41 Definition for Tangent Field-Space

For the fixed time $t = t_0$, a tangent field-space at the base point $\mathbf{x}_0 \in \mathscr{C}_{t0}$ is defined

$$T_{\mathbf{x}_0}\mathscr{C} := \left\{ \delta\mathbf{x} \in \mathscr{H}_1 \;\middle|\; (\mathbf{x}_0, \delta\mathbf{x}) \in T\mathscr{C}_{t_0} \right\},$$

where the tangent field-bundle is defined in *Def.* 40. We may denote any virtual displacement field $\delta\mathbf{x} \in T_{\mathbf{x}_0}\mathscr{C}$, where the base $\mathbf{x}_0$ is included in the notation as a subscript. ◊

Note that the place field $\mathbf{x}$ is a vector-valued function satisfying all the constraints equations (holonomic), called the constraint field-manifold $\mathscr{C}$.

42 Definition for Velocity Field-Space

A velocity field-space is closely related with the tangent field-space and is defined by formula

---

[15] a vector-valued function, more precisely

$$T_{\mathbf{x}}\mathscr{C} := \left\{ \dot{\mathbf{x}} \in \mathscr{H}_1 \,\middle|\, (\mathbf{x}, \dot{\mathbf{x}}) \in T\mathscr{C} \right\},$$

where now time is free, not fixed, like in the virtual displacement. Compare with *Def.* 9, a tangent vector in a finite-dimensional case. The velocity field that is an element of the velocity field-space is also denoted by $\mathbf{v} := \dot{\mathbf{x}} \in T_{\mathbf{x}}\mathscr{C}$ ◊

1.5 Variation, Lie Derivative and Lie Variation

In this Section, we give definition for the variation, Lie derivative and Lie variation. The concept of push-forward and pull-back operators is essential for understanding Lie derivatives and variations.

43 Definition for Variation Operator

The variation operator $\mathrm{\delta}$ is defined as the special case of Fréchet differential at the fixed time $t = t_0$ by

$$\mathrm{\delta}\mathbf{h}(t_0, \mathbf{x}, \mathbf{v}) := \mathrm{D}_{\mathbf{x}}\mathbf{h}(t_0, \mathbf{x}, \mathbf{v}) \cdot \delta\mathbf{x} + \mathrm{D}_{\mathbf{v}}\mathbf{h}(t_0, \mathbf{x}, \mathbf{v}) \cdot \delta\mathbf{v},$$

where $\mathbf{x} \in \mathscr{C}_{t0}$ is a place field, $\delta\mathbf{x} \in T_{\mathbf{x}0}\mathscr{C}$ is a virtual displacement field, $\mathbf{v} \in T_{\mathbf{x}}\mathscr{C}$ is a velocity field, and $\delta\mathbf{v} := \delta\dot{\mathbf{x}} \in T_{\mathbf{x}0}\mathscr{C}$ is a virtual velocity field. Moreover, $\mathrm{D}_{\mathbf{x}}, \mathrm{D}_{\mathbf{v}}$ are Fréchet partial derivatives with respect to place and velocity, correspondingly. ◊

The variation operator $\mathrm{\delta}$ depends linearly on the virtual displacement and the virtual velocity. Note a minor notational difference between the virtual and variation operators, δ and $\mathrm{\delta}$. Calculating the place and velocity variation, after *Def. 43*, yields

$$\mathrm{\delta}\mathbf{x} = \delta\mathbf{x} \text{ and } \mathrm{\delta}\mathbf{v} = \delta\mathbf{v}. \tag{30}$$

This should be interpreted: the variation of place vector $\mathrm{\delta}\mathbf{x}$ is equal, not the same thing, as the virtual displacement $\delta\mathbf{x}$. The variation has an operational meaning whereas the virtual displacement is a geometrical quantity. In generally, the variation of 'something' and the virtual 'something' are not equal, e.g. a virtual work may exists although there does not exist a work function at all and neither the work variation.

44 Theorem

Generally, the variation operator and the time derivative operator do not commutate.

Proof: We will prove this theorem by a counter example. Let consider the constraint equation $\dot{\mathbf{x}} + t\dot{\mathbf{y}} = \mathbf{0}$, where t represents time. Its variation is $\mathrm{\delta}\dot{\mathbf{x}} + t\mathrm{\delta}\dot{\mathbf{y}} = \mathbf{0}$. On the other hand, the virtual displacement of the constraint equation is $\delta\mathbf{x} + t\delta\mathbf{y} = \mathbf{0} \Leftrightarrow \mathrm{\delta}\mathbf{x} + t\mathrm{\delta}\mathbf{y} = \mathbf{0}$, whose time derivative is respectively

$$\frac{d\delta\mathbf{x}}{dt}+t\frac{d\delta\mathbf{y}}{dt}+\delta\mathbf{y}=\mathbf{0}$$

This is clearly different from $\delta\dot{\mathbf{x}}+t\delta\dot{\mathbf{y}}=\mathbf{0}$, which is the variation of the original constraint equation. ◊

Usually in the textbooks, it is proven that reverse relation is true, but they implicitly or explicitly assume that constraint equations are holonomic when the time derivative and variation operators are commutative. In the proof, the example is a nonholonomic one. Property that the time derivative and the variation do not generally commutate has been noticed at least in [Burke 1996; p. 314]. If it is assumed that the time derivative and variation operators commutate, then nonholonomic constraints must not be substituted into a kinetic energy function, see e.g. [Rosenberg 1980; p. 172-173].

When deriving Hamilton's principle from the principle of virtual work[16] (d'Alembert's principle), there is exploited the commutative assumption between the time derivative and variation. Hence, in the case of nonholonomic constraints and monogenic forces[17] Hamilton's principle is not a true variational principle, i.e. an extremum principle. This problem may be solved by considering only holonomic constraints and monogenic forces, then Hamilton's principle is a true variational principle, see the textbook [Lanczos 1966] for more details.

Here we follow the paper [Stumpf & Hoppe 1997] for deriving push-forward and pull-back operators and Lie derivatives. The push-forward and pull-back notation has been comprehensively utilized in the textbook [Marsden & Hughes 1983] which gives a differential geometric foundation of elasticity. In this textbook, it is assumed that there exists a smooth diffeomorphism between manifolds. However, the existence of such mapping is rather rarely possible. Hence, we define push-forward and pull-back operators according to the paper [Stumpf & Hoppe 1997], where no diffeomorphism between manifolds are assumed to be present.

Let the operator $\mathbf{R}: T_{\mathbf{X}}\mathscr{B}_0 \to T_{\mathbf{x}}\mathscr{B}$ be an invertible linear mapping between the tangent spaces of the material and spatial manifolds. The material manifold is denoted by $\mathscr{B}_0$ and the spatial manifold by $\mathscr{B}$. Moreover, let $\{\mathbf{g}_i\}$ and $\{\mathbf{G}_i\}$ be the bases for the spatial and material tangent spaces $T_{\mathbf{X}}\mathscr{B}_0$ and $T_{\mathbf{x}}\mathscr{B}$, respectively, and let $\{\mathbf{g}_i^*\}$ and $\{\mathbf{G}_i^*\}$ be the corresponding dual bases for the spatial and material cotangent spaces $T_{\mathbf{X}}^*\mathscr{B}_0$ and $T_{\mathbf{x}}^*\mathscr{B}$, see Section 0.

45 Definition for Pull-Back Operator

The pull-back operator by the isomorphism $\mathbf{R} \in \mathscr{Iso}(T_{\mathbf{X}}\mathscr{B}_0, T_{\mathbf{x}}\mathscr{B})$ for the spatial vector $\mathbf{a} = a_i\mathbf{g}_i \in T_{\mathbf{x}}\mathscr{B}$ is defined by

$$\mathbf{R}^{\triangleleft}\mathbf{a} := a_i\left(\mathbf{R}^{-1}\mathbf{g}_i\right) \in T_{\mathbf{X}}\mathscr{B}_0,$$

---

16 We also include inertial forces in the form of virtual work

17 a force is monogetic if there exists single scalar work potential function that depends on place, velocity and time

where $\mathbf{R}^{-1} \in \mathscr{L}iso(T_{\mathbf{x}}\mathscr{B}, T_{\mathbf{X}}\mathscr{B}_0)$ is the inverse operator of $\mathbf{R}$. The pull-back operator by $\mathbf{R} \in \mathscr{L}(T_{\mathbf{X}}\mathscr{B}_0, T_{\mathbf{x}}\mathscr{B})$ for the spatial covector $\mathbf{f} = f_i \mathbf{g}_i^* \in T_{\mathbf{x}}^*\mathscr{B}$ is defined by

$$\mathbf{R}^{\triangleleft}\mathbf{f} := f_i\left(\mathbf{R}^*\mathbf{g}_i^*\right) \in T_{\mathbf{X}}^*\mathscr{B}_0,$$

where $\mathbf{R}^* \in \mathscr{L}(T_{\mathbf{x}}^*\mathscr{B}, T_{\mathbf{X}}^*\mathscr{B}_0)$ is the adjoint operator of $\mathbf{R}$. ◊

The definition of the pull-back operator makes clear why we distinguish between vectors and covectors, their pull-back operators are quite different. A pull-back operator maps spatial vectors into material vectors, and spatial covectors into material covectors, hence we could also name the pull-back operator as the materializer operator. Especially when considering objectivity and/or two-point tensors the later name, the materializer operator, makes more sense. Note that the associated operator $\mathbf{R}$ does not need to be an isomorphism in the pull-back operation for covectors,

46 Definition for Push-Forward Operator

The push-forward operator by $\mathbf{R} \in \mathscr{L}(T_{\mathbf{X}}\mathscr{B}_0, T_{\mathbf{x}}\mathscr{B})$ for the material vector $\mathbf{A} = A_i \mathbf{G}_i \in T_{\mathbf{X}}\mathscr{B}_0$ is defined by

$$\mathbf{R}_{\triangleright}\mathbf{A} := A_i\left(\mathbf{R}\mathbf{G}_i\right) \in T_{\mathbf{x}}\mathscr{B}.$$

The push-forward operator by the isomorphism $\mathbf{R} \in \mathscr{L}iso(T_{\mathbf{X}}\mathscr{B}_0, T_{\mathbf{x}}\mathscr{B})$ for the material covector $\mathbf{F} = F_i \mathbf{G}_i^* \in T_{\mathbf{X}}^*\mathscr{B}_0$ is defined by

$$\mathbf{R}_{\triangleright}\mathbf{F} := F_i\left(\mathbf{R}^{-*}\mathbf{G}_i^*\right) \in T_{\mathbf{x}}^*\mathscr{B},$$

where $\mathbf{R}^{-*} \in \mathscr{L}iso(T_{\mathbf{X}}^*\mathscr{B}_0, T_{\mathbf{x}}^*\mathscr{B})$ is the inverse of $\mathbf{R}^*$. ◊

If the operator $\mathbf{R}$ is invertible between the material and spatial tangent spaces $\mathbf{R} \in \mathscr{L}iso(T_{\mathbf{X}}\mathscr{B}_0, T_{\mathbf{x}}\mathscr{B})$, then its adjoint, its inverse and the inverse of adjoint operators are $\mathbf{R}^* \in \mathscr{L}iso(T_{\mathbf{x}}^*\mathscr{B}, T_{\mathbf{X}}^*\mathscr{B}_0)$, $\mathbf{R}^{-1} \in \mathscr{L}iso(T_{\mathbf{x}}\mathscr{B}, T_{\mathbf{X}}\mathscr{B}_0)$ and $\mathbf{R}^{-*} \in \mathscr{L}iso(T_{\mathbf{X}}^*\mathscr{B}_0, T_{\mathbf{x}}^*\mathscr{B})$, respectively.

As for the pull-back operator, the definition of the push-forward operator makes clear why we distinguish between vectors and covectors. The push-forward operator maps the material (co)vectors into the spatial (co)vectors. The push-forward (or pull-back) operator is defined for a higher order tensor such as the push-forward (or pull-back) operator for each basis vector separately.

For example, the push-forward of the second order tensor $\mathbf{G} \in \mathscr{L}iso(T_{\mathbf{X}}\mathscr{B}_0, T_{\mathbf{X}}^*\mathscr{B}_0)$, a material metric tensor, in the tensor space $\mathscr{T}(0,2;0,0)$ by an isomorphism $\mathbf{F} \in \mathscr{L}iso(T_{\mathbf{X}}\mathscr{B}_0, T_{\mathbf{x}}\mathscr{B})$, a deformation gradient, is

$$\mathbf{F}_{\triangleright}\mathbf{G} = \mathbf{F}_{\triangleright}\left(G_{ij}\mathbf{G}_i^* \otimes \mathbf{G}_j^*\right) = G_{ij}(\mathbf{F}^{-*}\mathbf{G}_i^*) \otimes (\mathbf{F}^{-*}\mathbf{G}_j^*) = \mathbf{F}^{-*}\mathbf{G}\mathbf{F}^{-1} \in \mathscr{L}iso(T_{\mathbf{x}}\mathscr{B}, T_{\mathbf{x}}^*\mathscr{B}), \quad (31)$$

where we have used the relation $\mathbf{a}\otimes\mathbf{F}\mathbf{b}=(\mathbf{a}\otimes\mathbf{b})\mathbf{F}^{\star},\ \forall\mathbf{b}\in T_{\mathbf{X}}\mathscr{B}_0$. The resulting tensor $\mathbf{F}^{-\star}\mathbf{G}\mathbf{F}^{-1}$ is a spatial deformation tensor (Cauchy deformation tensor), often denoted by $\mathbf{c}$. This corresponds to classical tensor analysis when identifying the material metric tensor by the identity $\mathbf{G}\to\mathbf{I}$ and the adjoint operator by the transpose operator $\mathbf{F}^{-\star}\to\mathbf{F}^{-\mathrm{T}}$. Hence, we get the formula for the classical spatial deformation tensor $\mathbf{c}=\mathbf{F}^{-\mathrm{T}}\mathbf{F}^{-1}$

The deformation gradient is itself a two-point tensor, i.e. $\mathbf{F}=F_{ij}\mathbf{g}_i\otimes\mathbf{G}_j^{\star}\in T_{\mathbf{x}}\mathscr{B}\otimes T_{\mathbf{X}}^{\star}\mathscr{B}_0$ is a type of $\mathscr{T}(0,1;1,0)$. The pull-back operator of a two-point tensor is defined as the pull-back of spatial basis vectors and covectors. In the case of the deformation gradient, this pull-back operator by $\mathbf{R}\in\mathscr{L}iso(T_{\mathbf{X}}\mathscr{B}_0,T_{\mathbf{x}}\mathscr{B})$ reads

$$\mathbf{R}^{\triangleleft}\mathbf{F}=\mathbf{R}^{\triangleleft}\left(F_{ij}\mathbf{g}_i\otimes\mathbf{G}_j^{\star}\right)=F_{ij}(\mathbf{R}^{\triangleleft}\mathbf{g}_i)\otimes\mathbf{G}_j^{\star}=F_{ij}(\mathbf{R}^{-1}\mathbf{g}_i)\otimes\mathbf{G}_j^{\star}\quad\in T_{\mathbf{X}}\mathscr{B}_0\otimes T_{\mathbf{X}}^{\star}\mathscr{B}_0\,. \tag{32}$$

Thus the resulting tensor $\mathbf{R}^{\triangleleft}\mathbf{F}$ is a type of $\mathscr{T}(1,1;0,0)$ tensor, which is purely a material tensor. We note that push-forward and pull-back operators do not effect on tensor components, it just changes the bases.

47 Definition for Lie Derivative

The Lie derivative $\mathrm{L}_{\mathbf{R}}\mathbf{c}$ of the general tensor $\mathbf{c}(\eta)\in\mathscr{T}$ with respect to the isomorphic mapping $\mathbf{R}(\eta)\in\mathscr{L}iso(T_{\mathbf{X}}\mathscr{B}_0,T_{\mathbf{x}}\mathscr{B})$ and the parameter η is defined by

$$\mathrm{L}_{\mathbf{R}}\mathbf{c}:=\mathbf{R}_{\triangleright}\left(\frac{\mathrm{d}\,\mathbf{R}^{\triangleleft}\mathbf{c}}{\mathrm{d}\,\eta}\right)\quad\in\mathscr{T}\,.$$

Note that $\mathbf{c}(\eta)$ and $\mathbf{R}(\eta)$ depend on the parameter η. ◊

In Def. 47, the pull-back operator $\mathbf{R}^{\triangleleft}$ materializes a spatial or two-point tensor. It is known that the derivative of an objective material tensor is an objective tensor, see e.g. [Ogden 1984; Sec. 2.4]. The push-forward operator $\mathbf{R}_{\triangleright}$ is considered as the inverse of the pull-back operation where the resulting Lie derivative tensor $\mathrm{L}_{\mathbf{R}}\mathbf{c}$ belongs to the same tensor space $\mathscr{T}$ as the original tensor $\mathbf{c}$.

48 Definition for Lie Variation

The Lie variation $\delta_{\mathbf{R}}\mathbf{c}$ of the general tensor $\mathbf{c}\in\mathscr{T}$ with respect to the isomorphic mapping $\mathbf{R}\in\mathscr{L}iso(T_{\mathbf{X}}\mathscr{B}_0,T_{\mathbf{x}}\mathscr{B})$ is defined by

$$\delta_{\mathbf{R}}\mathbf{c}:=\mathbf{R}_{\triangleright}\delta(\mathbf{R}^{\triangleleft}\mathbf{c})\quad\in\mathscr{T}\,,$$

where the variation operator δ is given in *Def.* 43, which is accomplished at the fixed time $t=t_0$. ◊

As the Lie derivative, the Lie variation is an objective quantity if the original tensor is objective. The definition of Lie variation is connected with a virtual displacement that can be seen by writing the Lie variation with the aid of Gâteaux differential at the point $(\mathbf{x},\mathbf{v})$ and the fixed time $t = t_0$

$$\delta_{\mathbf{R}}\mathbf{c} = \mathbf{R}_{\triangleright}\left(\frac{\mathrm{d}\,\mathbf{R}^{\triangleleft}\mathbf{c}}{\mathrm{d}\eta}\right)\Bigg|_{\eta=0}, \tag{33}$$

where the tensor $\mathbf{c}(t_0,\mathbf{x}+\eta\delta\mathbf{x},\mathbf{v}+\eta\delta\mathbf{v})$ and the operator $\mathbf{R}(t_0,\mathbf{x}+\eta\delta\mathbf{x},\mathbf{v}+\eta\delta\mathbf{v})$ depend on the virtual displacement $\delta\mathbf{x}$ and the virtual velocity $\delta\mathbf{v}$. Note that e.g. the virtual displacement belongs to the tangent point-space $T_{\mathbf{x}_0}\mathcal{M}$ in the finite-dimensional case and to the tangent field-space $T_{\mathbf{x}_0}\mathcal{C}$ in the infinite-dimensional case.

For example, the Lie variation of the deformation tensor $\mathbf{F} = F_{ij}\mathbf{g}_i \otimes \mathbf{G}_j^* \in T_{\mathbf{x}}\mathcal{B} \otimes T_{\mathbf{X}}^*\mathcal{B}_0$, a type of $\mathcal{T}(0,1;1,0)$ tensor, by the rotation operator $\mathbf{R} \in \mathcal{L}iso(T_{\mathbf{X}}\mathcal{B}_0, T_{\mathbf{x}}\mathcal{B}) = SO(3)$ reads

$$\begin{aligned}\delta_{\mathbf{R}}\mathbf{F} &= \mathbf{R}_{\triangleright}\delta(\mathbf{R}^{\triangleleft}\mathbf{F}) = \mathbf{R}\,\delta\left(\mathbf{R}^{\mathrm{T}}\mathbf{F}\right) = \mathbf{R}\left(\delta\mathbf{R}^{\mathrm{T}}\mathbf{F} + \mathbf{R}^{\mathrm{T}}\delta\mathbf{F}\right)\\ &= \delta\mathbf{F} + \mathbf{R}\delta\mathbf{R}^{\mathrm{T}}\mathbf{F},\end{aligned} \tag{34}$$

where we have used the result of (32). The variation of rotation operator in the material and spatial description is

$$\begin{aligned}\delta\mathbf{R} &= \frac{\mathrm{d}\mathbf{R}\exp(\eta\delta\widetilde{\Theta}_{\mathbf{R}})}{\mathrm{d}\eta}\Bigg|_{\eta=0} = \mathbf{R}\delta\widetilde{\Theta}_{\mathbf{R}} \quad \text{for material description,}\\ \delta\mathbf{R} &= \frac{\mathrm{d}\exp(\eta\delta\widetilde{\theta}_{\mathbf{R}})\mathbf{R}}{\mathrm{d}\eta}\Bigg|_{\eta=0} = \delta\widetilde{\theta}_{\mathbf{R}}\mathbf{R} \quad \text{for spatial description,}\end{aligned} \tag{35}$$

hence the term $\mathbf{R}\delta\mathbf{R}^{\mathrm{T}}$ in (34) is equal to $-\delta\widetilde{\theta}_{\mathbf{R}}$ in both descriptions because $\delta\widetilde{\theta}_{\mathbf{R}} = \mathbf{R}\delta\widetilde{\Theta}_{\mathbf{R}}\mathbf{R}^{\mathrm{T}}$ according to (17). Finally, we have the Lie variation of the deformation tensor $\mathbf{F}$ with respect to the rotation operator $\mathbf{R}$ as

$$\delta_{\mathbf{R}}\mathbf{F} = \delta\mathbf{F} - \delta\widetilde{\theta}_{\mathbf{R}}\mathbf{F} \quad \in T_{\mathbf{x}}\mathcal{B} \otimes T_{\mathbf{X}}^*\mathcal{B}_0, \tag{36}$$

that is also called a corotational variation operator. Although the spatial virtual rotation tensor $\delta\widetilde{\theta}_{\mathbf{R}} \in {}_{\mathrm{spat}}T_{\mathbf{R}}SO(3)$, i.e. it occupies a spatial tangent space, see *Def.* 26, it is also an element of the tensor space $T_{\mathbf{x}}\mathcal{B} \otimes T_{\mathbf{x}}^*\mathcal{B}$.

2. Virtual Work Forms

The same equations of motion may be derived by very different principles and approaches. One of the oldest approaches is Newton's second law of motion that is a fundamental principle. In Newtonian mechanics, forces are of two kinds, internal forces and external forces with respect to a corresponding mechanical system. Moreover, Newtonian mechanics is rather involved with vectorial representation and is sometimes named as vectorial mechanics. As we have shown in the previous section, vectors, tensor, and corresponding spaces are fundamental elements described differential geometry. However, virtual displacements and rotations are not included in Newtonian mechanics, but rather in Lagrangian mechanics.

In Lagrangian mechanics, forces are divided differently into constraint forces and applied forces. In addition, rich mathematical methods like variational calculus and other mechanical principles are included. In terms of differential geometry, Lagrangian mechanics describes a motion on an event manifold[18] with a Lagrangian functional on the tangent bundle of the event manifold[19]. Lagrangian mechanics, and more generally analytical mechanics, is well suited for closed mechanical systems, but eligibility for open systems is not so clear. Moreover, the Lagrangian equations of motion as well as Hamilton's principle have a strong algebraic and analytic nature, but their geometric meaning is not so obvious. Also, a Lagrangian functional blacks out the interdependency of variables and their computational structure, which is needed in the finite element method.

We think that there is a need for another type of mechanics between Newtonian and Lagrangian mechanics, as it is sometimes done. This mechanics could be named d'Alembertian mechanics[20] where the principle of virtual work is its cornerstone. Here we include also inertial forces in the virtual work form. The virtual work may be viewed as a linear form on the tangent field-bundle $T\mathscr{C}_{t0}$, see *Def. 40.* This field-bundle is also a tangent bundle of the placement manifold at fixed time. We give definitions for the virtual work in the finite-dimensional and infinite-dimensional cases..

49 Definition for Virtual Work on Constraint Point-Manifold

Virtual work on the tangent point-bundle $T\mathscr{M}_{t0}$ at the fixed time $t = t_0$ and the place vector $\mathbf{x}(t_0) =: \mathbf{x}_0 \in \mathscr{M}_{t0}$ is defined as a linear form by

$$\delta W := \mathbf{f} \cdot \delta \mathbf{x},$$

where the virtual displacement $\delta\mathbf{x} \in T_{\mathbf{x}0}\mathscr{M}$, and the force vector $\mathbf{f} := \mathbf{f}(t_0, \mathbf{x}_0) \in T^*_{\mathbf{x}0}\mathscr{M}$ which belongs to the cotangent point-space. The tangent point-space $T_{\mathbf{x}0}\mathscr{M}$ is defined in *Def.* 36. ◊

Note that we have restricted to holonomic constraints. Moreover, the virtual displacement $\delta\mathbf{x} \in T_{\mathbf{x}0}\mathscr{M}$ truly occupies the subspace of E^n which is a tangential space at the base $\mathbf{x}_0$ with dimension d. The dimension d is also the dimension of the constraint point-manifold, see *Def. 34.*

---

[18] An event manifold is a time-placement manifold which is also a constraint manifold, i.e. the time and placement variables satisfy all constraints.

[19] This tangent bundle is called time-state manifold.

[20] Due to Jean Le Rond d'Alembert often a misunderstood scientist who regarded mechanics as much a part of mathematics as geometry or algebra.

Forces may be classified into external, internal (like in Newtonian mechanics) and additionally into inertial forces. We name the terms like $-m\ddot{\mathbf{x}}$ as inertial forces while the terms like $m\ddot{\mathbf{x}}$ are called acceleration forces. This may clarify the notational mess in the literature of applied mechanics that only is a controversy on words. We note that an inertial force may be regarded as an effective force. Indeed, if an external force is acting on a particle, which is otherwise free, then the inertial force may be regarded as the reaction force, hence the force equilibrium in the dynamical sense is achieved in this mechanical system.

Another way to classify forces is used in Lagrangian mechanics where forces are separated into constraint and applied forces. Constraint forces can be verified with the aid of the virtual work since they are workless. Then we may denote that constraint forces occupy $\mathbf{f}^{\text{con}} \in T^*_{\mathbf{x}0}\mathcal{M}^{\perp}$ that is orthogonal with $T_{\mathbf{x}0}\mathcal{M}$ via duality pairing. Hence, we may neglect the constraint forces in the virtual work forms.

50 Definition for Virtual Work on Field-Manifold

The virtual work on the tangent field-bundle $T\mathcal{C}_{t0}$ at the fixed time $t = t_0$ and the place field $\mathbf{x}_0 := \mathbf{x}(t_0) \in \mathcal{C}_{t0}$ is defined as an integral over the domain of the body B

$$\delta W := \int_B \mathbf{f} \cdot \delta\mathbf{x}\, \mathrm{d}V ,$$

where the virtual displacement field $\delta\mathbf{x} \in T_{\mathbf{x}0}\mathcal{C}$, and the force field $\mathbf{f} = \mathbf{f}(t_0, \mathbf{x}_0) \in T^*_{\mathbf{x}0}\mathcal{C}$ which belongs to the cotangent field-space. The tangent field-space $T_{\mathbf{x}0}\mathcal{C}$ is defined in *Def. 41*. ◊

Similarly as in the finite-dimensional case, the same classification may be realized. Especially, constraint forces occupy $\mathbf{f}^{\text{con}} \in T^*_{\mathbf{x}0}\mathcal{C}^{\perp}$ that is orthogonal with the tangent field-space $T_{\mathbf{x}0}\mathcal{C}$ via duality pairing.

51 Principle of Virtual Work

The principle of virtual work states that at a dynamical equilibrium, the virtual work with respect to any virtual displacement, at the fixed time $t = t_0$ and the place vector $\mathbf{x}_0$, vanishes, i.e.

$$\delta W = 0, \quad \forall \delta\mathbf{x} \in T\mathcal{M}_{t_0} \quad \text{or} \quad \forall \delta\mathbf{x} \in T\mathcal{C}_{t_0} ,$$

where the alternatives correspond the virtual displacements on the tangent space of the constraint point-manifold, $T_{\mathbf{x}0}\mathcal{M}$, and on the tangent space of the constraint field-manifold, $T_{\mathbf{x}0}\mathcal{C}$, see Figure11. ◊

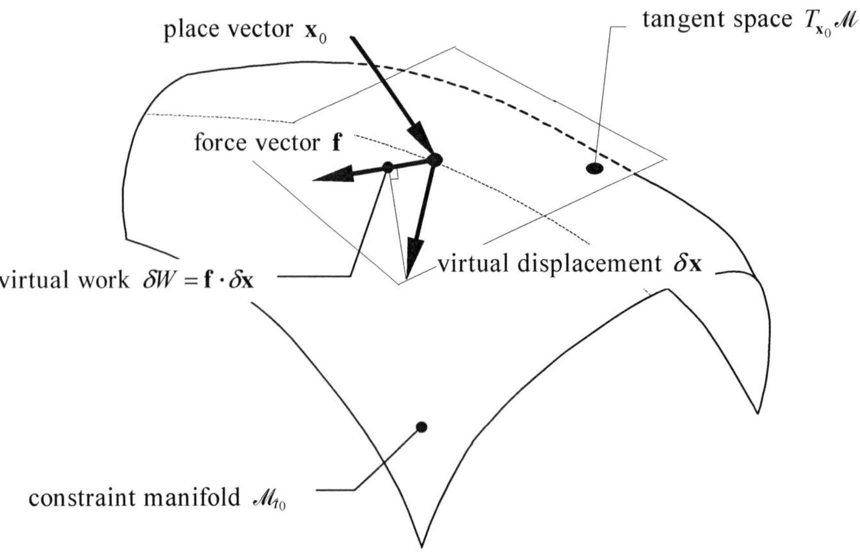

Figure11. A geometric representation of the virtual work.

2.1 Weak Balance Equations for Continuum

We could choose the virtual work of internal forces in many ways by taking different stress tensors and their corresponding virtual work conjugates. Although these work pairs are equivalent, the resulting computational models may give slightly different results. The reason for this arises from material constitutive equations. If we choose the same constant material stiffness tensors for the virtual works (which means the linear constitutive relations between the strain and stress quantities), we get the virtual works that are not equal. This is because the connections between different stress tensors include deformation gradients and their Jacobians which will complicate the equivalent constitutive relations.

Strain measures usually depend on logarithmically, linearly, or quadratically on the principal stretches. If strains are infinitesimal, all these strain measurements may be adjusted as equal. For the virtual work pair as a force quantity, we select the first Piola-Kirchhoff stress tensor. The work pair for the first Piola-Kirchhoff stress tensor is the virtual deformation gradient yielding rather a simple formulation. This strain measure depends linearly on the principal stretches and is therefore more closely connected with logarithmic (natural) strains than Lagrangian strains where the dependency is quadratic.

52 Definition for the First Piola-Kirchhoff Stress Tensor

The first Piola-Kirchhoff stress tensor $\mathbf{P} \in T_{\mathbf{x}}\mathcal{B} \otimes T_{\mathbf{X}}\mathcal{B}_0$ is usually defined by relation

$$\mathbf{p}\,\mathrm{d}a = \mathbf{P} \cdot \mathbf{N}_0\,\mathrm{d}A\,,$$

where $\mathbf{N}_0 \in T^*_{\mathbf{X}}\mathcal{B}_0$ is the unit normal covector in the cotangent of the material placement, $\mathrm{d}A$ and $\mathrm{d}a$ are differential areas in the material and spatial placements, and $\mathbf{p} \in T_{\mathbf{x}}\mathcal{B}$ is a stress vector, see Figure 12. ◊

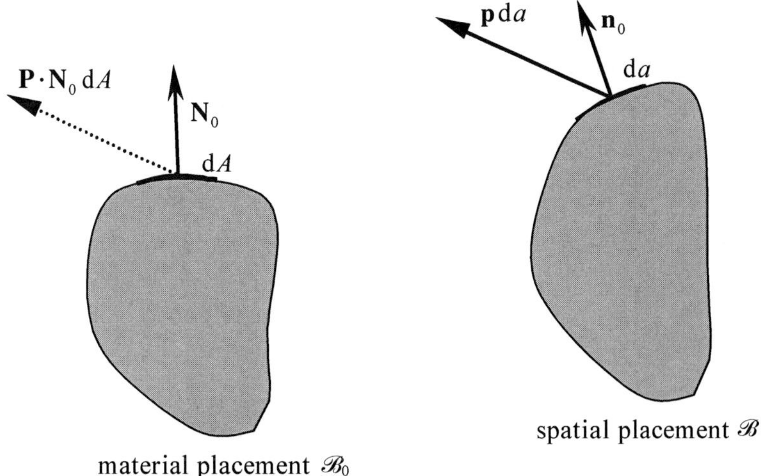

Figure 12. An interpretation of the first Piola-Kirchhoff stress tensor, note that $\mathbf{P}\cdot\mathbf{N}_0\,\mathrm{d}A \in T_{\mathbf{x}}\mathscr{B}$ although it is drawn on the material placement.

The stress vector $\mathbf{p} \in T_{\mathbf{x}}\mathscr{B}$ lies in the tangent space of the spatial placement $\mathscr{B}$. We note that the material basis vectors $\{\mathbf{t}_1, \mathbf{t}_2, \mathbf{t}_3\}$ span the tangent space $T_{\mathbf{x}}\mathscr{B}$. The first Piola-Kirchhoff stress tensor $\mathbf{P}$ is an example of the two-point tensor such that its stress vector belongs to the spatial vector space, its normal vector to the material vector space, and its differential area to the material placement.

We could write the first Piola-Kirchhoff (PK-I) stress tensor by a linear combination of stress vectors and basis vectors, as

$$\mathbf{P} = \mathbf{T}_1 \otimes \mathbf{E}_1 + \mathbf{T}_2 \otimes \mathbf{E}_2 + \mathbf{T}_3 \otimes \mathbf{E}_3 \quad \in T_{\mathbf{x}}\mathscr{B} \otimes T_{\mathbf{X}}\mathscr{B}_0, \tag{37}$$

where $\mathbf{T}_i = \mathbf{T}_i(\mathbf{x})$ is the spatial stress vector acting on the tangent space of spatial placement $T_{\mathbf{x}}\mathscr{B}$ and $\mathbf{E}_i$ is the material basis vector on the material placement, compare Figure 12. The material and spatial place vectors are related by a regular deformation mapping

$$\mathbf{x} = \chi_t\left(\chi_0^{-1}(\mathbf{X})\right) =: \chi_t(\mathbf{X}), \tag{38}$$

where we have used the same symbol for convenience, see *Def.* 11.

Before giving the balance equations of continuum, we define gradient and divergence operators. This is necessary since in the literature of applied mechanics, the gradient and divergence operators are defined differently.

53 Definitions for Gradient and Divergence

We define the gradient of the tensor $\mathbf{T} \in \mathscr{T}$ at the place point $\mathbf{x} \in \mathscr{V}$ as a directional derivative in the direction $\mathbf{u} \in \mathscr{V}$ by formula

$$\nabla \mathbf{T}(\mathbf{x}) \cdot \mathbf{u} = \left. \frac{\mathrm{d}\, \mathbf{T}(\mathbf{x} + \eta \mathbf{u})}{\mathrm{d}\eta} \right|_{\eta = 0} = \mathrm{D}\mathbf{T}(\mathbf{x}) \cdot \mathbf{u} ,$$

where the gradient $\nabla\mathbf{T}$ is an element of the tensor space $\mathscr{T} \otimes \mathscr{V}^*$. Moreover, we may define $\nabla\mathbf{T} := \mathrm{D}\mathbf{T}$. A gradient may be taken with respect to the spatial or material place vector where the material gradient is denoted by ${}^0\nabla$. The divergence operator of the tensor $\mathbf{T} \in \mathscr{T} \otimes \mathscr{W}$ at the point $\mathbf{x} \in \mathscr{V}$ is defined as a contraction by the double-dot product of the gradient:

$$\nabla \cdot \mathbf{T} := \nabla\mathbf{T} : \mathbf{I} \in \mathscr{T} ,$$

where the identity $\mathbf{I} \in \mathscr{W}^* \otimes \mathscr{V}$. ◊

For example, the deformation gradient $\mathbf{F}$ can be defined as the material gradient of the deformation $\mathbf{x} := \chi_t(\mathbf{X})$ by the formula $\mathbf{F} := {}^0\nabla\mathbf{x}(\mathbf{X})$. However, the deformation $\chi_t : \mathscr{B}_0 \rightarrow \mathscr{B}$ is more like a point mapping than a vector. Hence, the deformation gradient is usually defined by the tangent of the deformation:

$$\mathbf{F} := T_{\mathbf{X}}\chi(\mathbf{X}) \in T_{\mathbf{x}}\mathscr{B} \otimes T^*_{\mathbf{X}}\mathscr{B}_0 . \tag{39}$$

This is not a contradiction because in *Def.* 53 we have assumed that $\mathbf{T}$ is a tensor (or a vector in this case).

The equations of motion of continuum with boundary conditions, in the terms of the first Piola-Kirchhoff stress tensor, can be written as

$$\begin{aligned} &\left. \begin{aligned} {}^0\nabla \cdot \mathbf{P} + \mathbf{b} &= \rho_0 \ddot{\mathbf{x}} \\ \mathbf{P}\mathbf{F}^* &= \mathbf{F}\mathbf{P}^* \end{aligned} \right\} \text{ in } \mathscr{B}_0, \\ &\mathbf{P} \cdot \mathbf{N}_0 = \overline{\mathbf{T}}_\sigma \quad \text{on } \partial\mathscr{B}_{0\sigma}, \\ &\mathbf{x} = \overline{\mathbf{x}} \quad \text{on } \partial\mathscr{B}_{0u}, \end{aligned} \tag{40}$$

where $\mathbf{b}, \rho_0, \mathbf{N}_0, \overline{\mathbf{T}}_\sigma, \overline{\mathbf{x}}$ are the body force vector, the density of the material body, the normal vector of the traction boundary, the given traction vector and the given placement vector, respectively. The base points are given in the material placement $\mathscr{B}_0$, but they occupy in the tangent spaces of the spatial placement $T_{\mathbf{x}}\mathscr{B}$. E.g. $\mathbf{b} := \mathbf{b}(\chi_t(\mathbf{X})) \in T_{\mathbf{x}}\mathscr{B}$ and $\mathbf{P}\mathbf{F}^*(\chi_t(\mathbf{X})) \in T_{\mathbf{x}}\mathscr{B} \otimes T_{\mathbf{x}}\mathscr{B}$.

The virtual work can be decomposed into three terms: external, internal and inertial virtual works with the equation

$$\delta W = \delta W_{\text{ext}} - \delta W_{\text{int}} + \delta W_{\text{inert}} , \tag{41}$$

where the subscripts correspond to ‘external’, ‘internal’ and ‘inertial’. In the virtual work of internal forces, the minus sign indicates that internal forces work against the virtual displacements. Additionally, the inertial virtual work δW_{inert} includes the minus sign inside its form. Sometimes it is convenient to avoid additional minus signs by introducing the virtual work of acceleration forces by the formula

$$\delta W_{\text{acc}} := -\delta W_{\text{inert}} \,. \tag{42}$$

When we apply the principle of virtual work, the governing equations are derived from the equation

$$\delta W_{\text{accA}} = \delta W_{\text{ext}} - \left(\delta W_{\text{int}} + \delta W_{\text{accB}}\right) \tag{43}$$

where δW_{accA} is the virtual work of acceleration forces, which depends on the acceleration vector, and δW_{accB} corresponds to the terms of the acceleration virtual work like the virtual work of centrifugal forces.

Next, we give the form of virtual work and we show that the principle of virtual work satisfies the equations of motion (40). Let the virtual work be stated as

$$\delta W = \int_{\mathscr{B}_0} \left(\langle \delta \mathbf{x}, \mathbf{b} \rangle_{\mathbf{g}}\right) \mathrm{d}V + \int_{\partial\mathscr{B}_0} \langle \delta \mathbf{x}, \overline{\mathbf{T}}_\sigma \rangle_{\mathbf{g}} \, \mathrm{d}A_\sigma - \int_{\mathscr{B}_0} (\delta_{\mathbf{R}} \mathbf{F} : \mathbf{g}\mathbf{P}) \, \mathrm{d}V - \int_{\mathscr{B}_0} \langle \delta \mathbf{x}, \rho_0 \ddot{\mathbf{x}} \rangle_{\mathbf{g}} \, \mathrm{d}V = 0 \,, \tag{44}$$

where the Lie derivative of the deformation gradient has been calculated in (36). In Eqn (44), the first two terms correspond to the external virtual work δW_{ext}, the third term to the internal virtual work δW_{int}, and the last term to the acceleration virtual work δW_{acc}.

2.2 Constitutive Relations

Let W_{str} be a strain energy function per unit volume of a hyperelastic material. The first Piola-Kirchhoff stress tensor $\mathbf{P}$ can also be defined from the energy function $W_{\text{str}}(\mathbf{F})$ [Stumpf & Hoppe 1997] by

$$\mathbf{P} := \mathbf{g}^{-1} \frac{\partial W_{\text{str}}(\mathbf{F})}{\partial \mathbf{F}} \quad \in T_{\mathbf{x}}\mathscr{B} \otimes T_{\mathbf{X}}\mathscr{B}_0 \,. \tag{45}$$

We also assume that the strain energy function is frame-indifferent under the orthogonal transformation $\mathbf{F}^{+} = \mathbf{QF}$ by obeying the identity

$$W_{\text{str}}(\mathbf{F}^{+}) = W_{\text{str}}(\mathbf{QF}) = W_{\text{str}}(\mathbf{F}) \,. \qquad 46)$$

This means that the strain energy function W_{str} is invariant under rigid body rotation. If we set $\mathbf{Q} = \mathbf{R}^{\mathrm{T}}$, then we have via a pull-back operator the strain energy function purely in the material domain, which provides a different stress quantity.

The Lie variation of the energy function $W_{\text{str}}(\mathbf{F})$ with respect to the rotation operator $\mathbf{R}$ can be written with using the above relation as

$$\delta_{\mathbf{R}} W_{\text{str}} = \frac{\partial W_{\text{str}}(\mathbf{F})}{\partial \mathbf{F}} : \delta_{\mathbf{R}} \mathbf{F} = \mathbf{gP} : \delta_{\mathbf{R}} \mathbf{F} \ \in \mathrm{R} \tag{47}$$

that is equal to the virtual work of internal forces δW_{int}, see Eqn (44). We get the same result for $\delta W_{\text{str}}(\mathbf{R}^{\mathrm{T}}\mathbf{F})$. The first Piola-Kirchhoff stress tensor as well as tensor $\mathbf{gP}$ is a two-point tensor defined on material and spatial placement manifolds. We use Lie variation, push-forward and pull-back relations to derive one-point stress tensor that corresponds Eqn (47), yielding

$$\begin{aligned} \delta_{\mathbf{R}} W_{\text{str}} &= \mathbf{gP} : \left(\mathbf{R}\delta(\mathbf{R}^{\mathrm{T}}\mathbf{F})\right) = (\mathbf{R}^{*}\mathbf{gP}) : \delta(\mathbf{R}^{\mathrm{T}}\mathbf{F}) \\ &= (\mathbf{GR}^{\mathrm{T}}\mathbf{P}) : \delta(\mathbf{R}^{\mathrm{T}}\mathbf{F}) = (\mathbf{GR}^{\mathrm{T}}\mathbf{P}) : \delta(\mathbf{R}^{\mathrm{T}}\mathbf{F} - \mathbf{I}) \end{aligned} \tag{48}$$

where we have used $\mathbf{R}^{*}\mathbf{g} = \mathbf{GR}^{\mathrm{T}}$ according to the definition of the transpose operator, see *Def. 12*.

The Lie variation of strain energy function (48) introduces new material stress and strain tensor, defined by

$$\begin{aligned} \Sigma &:= \mathbf{GR}^{\mathrm{T}}\mathbf{P} \quad \in T_{\mathbf{X}}^{*}\mathscr{B}_0 \otimes T_{\mathbf{X}}\mathscr{B}_0 , \\ \mathbf{H} &:= \mathbf{R}^{\mathrm{T}}\mathbf{F} - \mathbf{I} \quad \in T_{\mathbf{X}}\mathscr{B}_0 \otimes T_{\mathbf{X}}^{*}\mathscr{B}_0 . \end{aligned} \tag{49}$$

The material stress tensor $\Sigma = \Sigma_{ij}\mathbf{E}_i^{*} \otimes \mathbf{E}_j$, as well as, its work conjugate strain tensor $\mathbf{H} = H_{ij}\mathbf{E}_i \otimes \mathbf{E}_j^{*}$ are both unsymmetrical tensors and are not named in continuum mechanics.

Let us consider a following linear constitutive relation between the stress components of the tensor Σ and the strain components of the tensor $\mathbf{H}$ given by

$$\Sigma = \mathbf{C} : \mathbf{H} \tag{50}$$

where the elasticity tensor $\mathbf{C} \in T_{\mathbf{X}}^{*}\mathscr{B}_0 \otimes T_{\mathbf{X}}\mathscr{B}_0 \otimes T_{\mathbf{X}}^{*}\mathscr{B}_0 \otimes T_{\mathbf{X}}\mathscr{B}_0$ is a fourth order material tensor.

2.3 On Symmetry of Second Variation

Next, we follow the presentation given in the papers [Makowski & Stumpf 1995] and [Simo 1992] to show why a tangent stiffness tensor can be nonsymmetrical. This is somehow rather involved issue and we give a simple explanation for the nonsymmetry of the second variation. Detailed derivation could be found in the above papers. We consider a finite dimensional point-manifold $\mathscr{M}$ without any restriction. Let us introduce a r-parametrized curve on the manifold $\mathscr{M}$ such that

$$\alpha : \mathrm{E}^1 \to \mathcal{M}, \quad r \mapsto \alpha(r), \quad \mathbf{q} = \alpha(0), \quad \delta\mathbf{q} = \alpha'(0), \tag{51}$$

where $\delta\mathbf{q}$ is the tangent to the curve α at the point $\mathbf{q}$ of the manifold $\mathcal{M}$, see Figure 13. Let $W : \mathcal{M} \to \mathrm{R}$ be a work function on the manifold $\mathcal{M}$. The work function depends on the displacement vector $\mathbf{q} \in \mathcal{M}$. The first variation of the work function W at the point $\mathbf{q} \in \mathcal{M}$ in the direction $\delta\mathbf{q} \in T_{\mathbf{q}}\mathcal{M}$ reads

$$\delta W(\mathbf{q}; \delta\mathbf{q}) := \left.\frac{\mathrm{d}W(\alpha(r))}{\mathrm{d}r}\right|_{r=0} = \mathrm{D}_\alpha W(\alpha(0)) \cdot \delta\mathbf{q}, \tag{52}$$

that is very similar to the variation defined on a vector space. The first variation δW depends linearly on the direction $\delta\mathbf{q}$, hence we may denote $\delta W = \mathbf{f} \cdot \delta\mathbf{q}$, where the force vector $\mathbf{f}(\mathbf{q})$ belongs to the cotangential space $T_{\mathbf{q}}^{*}\mathcal{M}$. As usual, we call the point $\mathbf{q}_0 \in \mathcal{M}$ as a critical point of function, an equilibrium, if the variation $\delta W(\mathbf{q}_0; \delta\mathbf{q})$ vanishes for arbitrary $\delta\mathbf{q} \in T_{\mathbf{q}}\mathcal{M}$. At the critical point, the corresponding force vector $\mathbf{f}(\mathbf{q}_0)$ vanishes, i.e. $\mathbf{f}(\mathbf{q}_0) = \mathbf{0}$.

In order to give the second variation of function on the manifold, we introduce another curve with the following properties

$$\beta : \mathrm{E}^1 \to \mathcal{M}, \quad s \mapsto \beta(s), \quad \mathbf{q} = \beta(0), \quad \Delta\mathbf{q} = \beta'(0), \tag{53}$$

where $\Delta\mathbf{q} \in T_{\mathbf{q}}\mathcal{M}$ is the tangent to the curve β. We denoted the tangency by Δ in order to indicate the difference from $\delta\mathbf{q}$. Note that the virtual displacement $\delta\mathbf{q}(s) \in T_{\beta(s)}\mathcal{M}$ depends on the curve β if the base point varies according to the curve β, see Figure 13.

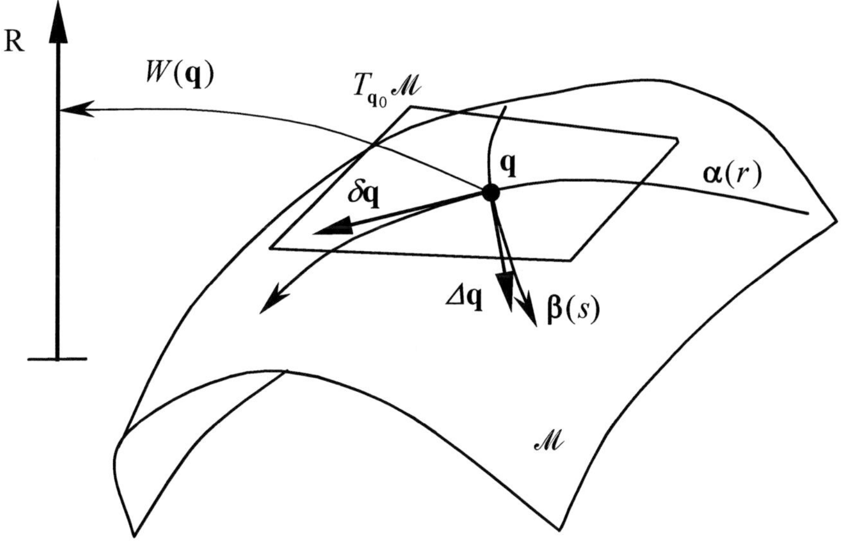

Figure 13. A geometric presentation of the parametrized curve $\alpha : \mathrm{E}^1 \to \mathcal{M}$ and the work function $W : \mathcal{M} \to \mathrm{R}$ on the manifold $\mathcal{M}$.

Hence, we could write the second variation of the work function W on the manifold $\mathcal{M}$

$$\delta^2 W(\mathbf{q};\delta\mathbf{q},\Delta\mathbf{q}) := \left.\frac{\mathrm{d}\,\delta W(\beta(s))}{\mathrm{d}s}\right|_{s=0} = \left.\frac{\mathrm{d}\big(\mathbf{f}(\beta(s))\cdot\delta\mathbf{q}(\beta(s))\big)}{\mathrm{d}s}\right|_{s=0}$$
$$= \big(\mathrm{D}_\beta\,\mathbf{f}(\beta(0))\cdot\Delta\mathbf{q}\big)\cdot\delta\mathbf{q}(\beta(0)) + \mathbf{f}(\beta(0))\cdot\big(\mathrm{D}_\beta\,\delta\mathbf{q}(\beta(0))\cdot\Delta\mathbf{q}\big), \tag{54}$$

where the first term is often denoted by $H(\mathbf{q};\delta\mathbf{q},\Delta\mathbf{q})$ that is the Hessian of the function W.

The Hessian of the function W could be denoted by

$$H(\mathbf{q};\delta\mathbf{q},\Delta\mathbf{q}) := \mathrm{D}_{\alpha\beta}\,W\big(\alpha(0),\beta(0)\big) : \big(\delta\mathbf{q}\otimes\Delta\mathbf{q}\big), \tag{55}$$

that is always a symmetric form on the Riemannian manifold $\mathcal{M}$, i.e. $H(\mathbf{q};\delta\mathbf{q},\Delta\mathbf{q}) = H(\mathbf{q};\Delta\mathbf{q},\delta\mathbf{q})$ since the Fréchet partial derivatives commute $\mathrm{D}_{\alpha\beta}\,W = \mathrm{D}_{\beta\alpha}\,W$ for the smooth function W. However, the second term in the formula (54) that reads

$$\mathbf{f}(\beta(0))\cdot\big(\mathrm{D}_\beta\,\delta\mathbf{q}(\beta(0))\cdot\Delta\mathbf{q}\big) \tag{56}$$

is generally nonsymmetric, unless $\mathbf{q}_0 = \mathbf{q}(\beta(0))$ is a critical point (an equilibrium). At the critical point, we have the force vector $\mathbf{f}(\mathbf{q}_0) = \mathbf{0}$, giving the symmetric second variation of the function W. The nonsymmetrical term (56) vanishes also if Riemannian manifold $\mathcal{M}$ is a flat manifold. Then we get for the derivative $\mathrm{D}_\beta\,\delta\mathbf{q}(\beta(0)) \equiv \mathbf{O}$.

We conclude that the second variation $\delta^2 W(\mathbf{q};\delta\mathbf{q},\Delta\mathbf{q})$ is a symmetric form if $\mathbf{q}_0 \in \mathcal{M}$ is a critical point of the function W, or if the manifold $\mathcal{M}$ is a flat (Euclidean) manifold. The beam and shell placements can be identified by the manifold $(\mathbf{d},\mathbf{R}) \in \mathrm{E}^3 \times SO(3)$, where $\mathbf{d}$ is the translational displacement and $\mathbf{R}$ is the rotation operator. If we present rotation by the (material) incremental rotation vector $\mathbf{\Theta}_\mathbf{R} \in {}_{\mathrm{mat}}T_\mathbf{R}$, where the base point $\mathbf{R}$ depends on solution, we have a nonsymmetric stiffness tensor away from a critical point. However, at a critical point , i.e. an equilibrium, we have a symmetric stiffness tensor.

The symmetry of stiffness tensor can be achieved at arbitrary point by the parametrization of the manifold. The parametrization mapping φ maps from an open set in an Euclidean space into an open set of the manifold $\mathcal{M}$, see Figure **14**. The parametrized work function $W\circ\varphi : \mathcal{U} \subset \mathrm{E} \to \mathrm{R}$ is a mapping from an Euclidean set into the set of real numbers R. Since the set of an Euclidean space is a flat manifold, the nonsymmetric term of the second variation (56) will always vanish. Especially, the rotation manifold $SO(3)$ could be parametrized by the rotation vector $\mathbf{\Psi} \in {}_{\mathrm{mat}}T_\mathbf{I} = \mathrm{E}^3$, where the parametrization mapping is the exponential mapping, $\mathbf{\Psi} \mapsto \exp(\tilde{\mathbf{\Psi}}) \in SO(3)$.

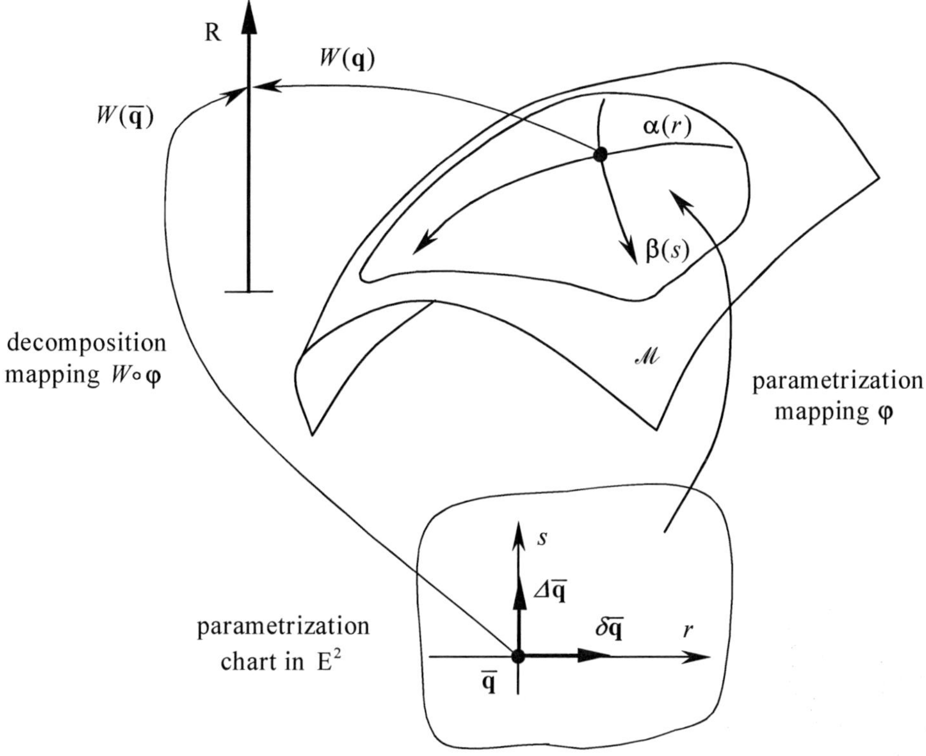

Figure 14. A geometric presentation for the parametrization of the two-manifold $\mathcal{M}$.

3. Parametrization of Constraint Manifold

In this Chapter, we consider constraint point-manifolds defined in *Def. 34* that arise from point-wise holonomic constraint equations. The usual geometric joints of a multibody system like spherical, revolute, cylindrical, universal, helical, prismatic, and sliding joints can be presented via holonomic constraint equations that only depend on displacement at corresponding geometric points. All these constraints generate a smooth point-manifold that can be parametrized. Also, the principle of virtual work and its geometric structure are naturally related with the parametrization of the constraint manifold.

Let us consider a Newtonian problem in a n-dimensional Euclidean space E^n

$$\mathbf{f}(\mathbf{q},\dot{\mathbf{q}}) - \mathbf{M} \cdot \ddot{\mathbf{q}} = \mathbf{0}\,, \tag{57}$$

where $\mathbf{f}(\mathbf{q},\dot{\mathbf{q}})$ is a generalized force vector, $\mathbf{M}$ is a mass tensor, and $\ddot{\mathbf{q}}$ is an acceleration vector. The problem is also subjected to holonomic constraints $\mathbf{h}: E^n \to E^{n-d}$ that generate a d -manifold embedded in E^n given by

$$\mathcal{M} := \left\{ t \in R, \mathbf{q} \in E^n \middle| \mathbf{h}(t,\mathbf{q}) = \mathbf{0} \in E^{n-d} \right\}. \tag{58}$$

We assume that the manifold $\mathscr{M}$ is regular at every point, i.e. the derivative $D_{\mathbf{q}}\mathbf{h} \in \mathscr{L}(E^n, E^{n-d})$ is surjective at every point of the manifold. This assumption is equal to the full-rank condition and is not restricting assumption. In differential geometry, the mapping $\mathbf{h}$ which satisfies the above assumption is called a submersion.

Now, applying the principle of virtual work into the problem (57) on the manifold (58) yields

$$\delta W = (\mathbf{f}(\mathbf{q},\dot{\mathbf{q}}) - \mathbf{M}\cdot\ddot{\mathbf{q}})\cdot\delta\mathbf{q} = 0 . \tag{59}$$

The generalized force vector $\mathbf{f}$ can be split into the applied force vector $\mathbf{f}^{\text{appl}} \in T^*_{\mathbf{q}_0}\mathscr{M}$ and into the constraint force vector $\mathbf{f}^{\text{con}} \in T^*_{\mathbf{q}_0}\mathscr{M}^{\perp}$, whose virtual work vanishes. The virtual displacement $\delta\mathbf{q}$ belongs to the tangent space $T_{\mathbf{q}_0}\mathscr{M}$, given by

$$T_{\mathbf{q}_0}\mathscr{M} = \left\{\delta\mathbf{q} \in E^n \middle| \, D_{\mathbf{q}}\mathbf{h}(t_0,\mathbf{q}_0)\cdot\delta\mathbf{q} = \mathbf{0}\right\}, \tag{60}$$

see Section 0 for more details.

Suppose that we can divide the displacement vector $\mathbf{q} \in E^n$ into the master-released displacement vector $\mathbf{q}_{\text{mr}} \in E^d$ and into the slave displacement vector $\mathbf{q}_{\text{s}} \in E^{n-d}$ such that the derivative $D_{\mathbf{q}_s}\mathbf{h}(t,\mathbf{q}_{\text{mr}},\mathbf{q}_{\text{s}})$ is an isomorphism, i.e. a linear bijection. Here, we utilize the terminology: 'master', 'released', and 'slave' that is conventionally used in finite element literature. After the implicit function theorem, there exists a unique mapping $\boldsymbol{\phi}(t,\mathbf{q}_{\text{mr}}): \mathrm{R} \times E^d \to E^{n-d}$ on some neighborhood such that

$$\hat{\mathbf{h}}(t,\mathbf{q}_{\text{mr}}) := \mathbf{h}(t,\mathbf{q}_{\text{mr}},\boldsymbol{\phi}(t,\mathbf{q}_{\text{mr}})) = \mathbf{0} . \tag{61}$$

Hence, the slave displacement can be given in terms of the master-released displacement vector, i.e. $\mathbf{q}_{\text{s}} = \boldsymbol{\phi}(t,\mathbf{q}_{\text{mr}})$. Moreover, the parametrization of the manifold $\mathscr{M}$, Eqn (58), can be now written as

$$\boldsymbol{\varphi} : \mathrm{R} \times E^d \to \mathscr{M} \qquad (t,\mathbf{q}_{\text{mr}}) \mapsto \boldsymbol{\varphi}(t,\mathbf{q}_{\text{mr}}) := (\mathbf{q}_{\text{mr}},\boldsymbol{\phi}(t,\mathbf{q}_{\text{mr}})) . \tag{62}$$

The parametrization mapping $\boldsymbol{\varphi}$ realizes the constraints $\mathbf{h}(t,\boldsymbol{\varphi}(t,\mathbf{q}_{\text{mr}})) = \mathbf{0}$ according to Eqn (61), see Figure 15. Sometimes, constraints are given by a time-independent equation as $\mathbf{h}(\mathbf{q}_{\text{mr}},\mathbf{q}_{\text{s}}) := \mathbf{q}_{\text{s}} - \boldsymbol{\phi}(\mathbf{q}_{\text{mr}}) = \mathbf{0}$ that gives a natural global parametrization of the manifold $\mathscr{M}$, since $D_{\mathbf{q}_s}\mathbf{h}(t,\mathbf{q}_{\text{mr}},\mathbf{q}_{\text{s}}) = \mathbf{I}$. We could view the time variable as an independent parameter that usually arises from displacement boundary conditions. We note that the derivative of the parametrization (62) with respect to the master-released displacement vector $\mathbf{q}_{\text{mr}}$, that is $D_{\mathbf{q}_{\text{mr}}}\boldsymbol{\varphi}(\mathbf{q}_{\text{mr}})$, is an injective (one-to-one) operator everywhere. In differential geometry, this kind of mapping is called an immersion.

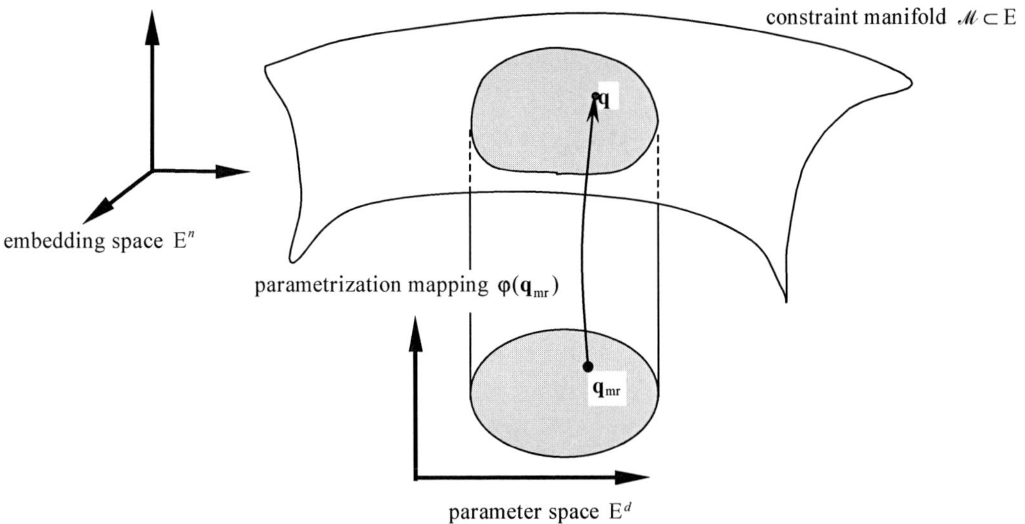

Figure 15. A geometric representation of the parametrization of the constraint manifold.

Next, we consider a time-independent parametrization mapping $\varphi : E^d \rightarrow \mathcal{M} \subset E^n$

$$\mathbf{q} = \varphi(\mathbf{q}_{mr}), \tag{63}$$

that we use for pull-backing the virtual work form (59) on the manifold $\mathcal{M}$ into the Euclidean space E^d. The variation of the displacement vector $\mathbf{q} \in \mathcal{M}$ in terms of the master-released displacement vector $\mathbf{q}_{mr} \in E^d$ reads

$$\begin{aligned} \delta\mathbf{q} &= \mathbf{B} \cdot \delta\mathbf{q}_{mr}, \\ \mathbf{B} &:= D_{\mathbf{q}_{mr}} \varphi(\mathbf{q}_{mr}), \end{aligned} \tag{64}$$

where we have defined an injective kinematic operator $\mathbf{B} : T_{\mathbf{q}_{mr}} E^d = E^d \rightarrow T_{\mathbf{q}}\mathcal{M} \subset E^n$ between the tangent spaces. We also need the time derivatives for $\mathbf{q}$ that are given via the kinematic operator

$$\begin{aligned} \dot{\mathbf{q}} &= \mathbf{B} \cdot \dot{\mathbf{q}}_{mr} \quad \in T_{\mathbf{q}}\mathcal{M}, \\ \ddot{\mathbf{q}} &= \mathbf{B} \cdot \ddot{\mathbf{q}}_{mr} + \dot{\mathbf{B}} \cdot \dot{\mathbf{q}}_{mr} \quad \in T_{\mathbf{q}}\mathcal{M}, \end{aligned} \tag{65}$$

where the time derivative of kinematic operator $\dot{\mathbf{B}}$ depends on the displacement $\mathbf{q}_{mr}$ and linearly on the velocity vector $\dot{\mathbf{q}}_{mr}$. This can be noticed by observing

$$\begin{aligned} \dot{\mathbf{B}} &= D_{\mathbf{q}_{mr}} \mathbf{B} \cdot \dot{\mathbf{q}}_{mr}, \\ \dot{\mathbf{B}} \cdot \dot{\mathbf{q}}_{mr} &= D^2_{\mathbf{q}_{mr}} \varphi(\mathbf{q}_{mr}) : (\dot{\mathbf{q}}_{mr} \otimes \dot{\mathbf{q}}_{mr}), \end{aligned} \tag{66}$$

where the latter equation expresses a quadratic dependency of the velocity vector $\dot{\mathbf{q}}_{mr}$.

Now, the principle of virtual work (59) on the manifold $\mathcal{M}$ can be written in the parameter space E^d according to the relations (63) and (64), yielding

$$\delta\mathbf{q}_{\mathrm{mr}}\cdot\mathbf{B}^*\left(\mathbf{f}(\mathbf{q}_{\mathrm{mr}},\dot{\mathbf{q}}_{\mathrm{mr}})-\mathbf{M}\mathbf{B}\ddot{\mathbf{q}}_{\mathrm{mr}}-\mathbf{M}\dot{\mathbf{B}}\dot{\mathbf{q}}_{\mathrm{mr}}\right)=0, \qquad \forall\,\delta\mathbf{q}_{\mathrm{mr}}\in E^d, \tag{67}$$

where the constraint equations (58) are satisfied automatically because of the parametrization (63). The above equation can be viewed as a pull-back operator for the covector $\mathbf{f}\in E^{*n}$, see *Def. 45*.

Linearizing the virtual work (67) at the fixed time $t=t_0$ around the state point $(\mathbf{q}_{\mathrm{mr}0},\dot{\mathbf{q}}_{\mathrm{mr}0})\in E^d\times E^d$ in the direction $(\Delta\mathbf{q}_{\mathrm{mr}},\Delta\dot{\mathbf{q}}_{\mathrm{mr}})$ gives

$$\delta\mathbf{q}_{\mathrm{mr}}\cdot(\mathbf{f}_{\mathrm{mr}0}-\mathbf{M}_{\mathrm{mr}}\ddot{\mathbf{q}}_{\mathrm{mr}}-\mathbf{C}_{\mathrm{mr}}\cdot\Delta\dot{\mathbf{q}}_{\mathrm{mr}}-\mathbf{K}_{\mathrm{mr}}\cdot\Delta\mathbf{q}_{\mathrm{mr}})=0, \qquad \forall\,\delta\mathbf{q}_{\mathrm{mr}}\in E^d, \tag{68}$$

where the generalized force vector $\mathbf{f}_{\mathrm{mr}0}$, the generalized stiffness, damping, and mass tensors are, respectively

$$\begin{aligned}
&\mathbf{f}_{\mathrm{mr}0}:=\mathbf{B}^*\left(\mathbf{f}(\mathbf{q}_{\mathrm{mr}0},\dot{\mathbf{q}}_{\mathrm{mr}0})-\mathbf{M}\dot{\mathbf{B}}\dot{\mathbf{q}}_{\mathrm{mr}0}\right) \quad \in E^{*d},\\
&\mathbf{K}_{\mathrm{mr}}:=\mathrm{D}_{\mathbf{q}_{\mathrm{mr}}}\left(\mathbf{B}^*(-\mathbf{f}(\mathbf{q}_{\mathrm{mr}},\dot{\mathbf{q}}_{\mathrm{mr}})+\mathbf{M}\mathbf{B}\ddot{\mathbf{q}}_{\mathrm{mr}}+\mathbf{M}\dot{\mathbf{B}}\dot{\mathbf{q}}_{\mathrm{mr}})\right) \quad \in E^{*d\times *d},\\
&\mathbf{C}_{\mathrm{mr}}:=\mathrm{D}_{\dot{\mathbf{q}}_{\mathrm{mr}}}\left(\mathbf{B}^*(-\mathbf{f}(\mathbf{q}_{\mathrm{mr}},\dot{\mathbf{q}}_{\mathrm{mr}})+\mathbf{M}\dot{\mathbf{B}}\dot{\mathbf{q}}_{\mathrm{mr}})\right) \quad \in E^{*d\times *d},\\
&\mathbf{M}_{\mathrm{mr}}:=\mathbf{B}^*\mathbf{M}\mathbf{B} \quad \in E^{*d\times *d}.
\end{aligned} \tag{69}$$

The above equations are the most fundamental relations to derive force vectors and their tangent tensors when the parametrization mapping exists: $\varphi: E^d\to\mathcal{M}$. In Figure 16, there is shown a geometric structure of the virtual work and the corresponding tangential space $T^*_{\mathbf{q}_0}\mathcal{M}$ where the applied force $\mathbf{f}^{\mathrm{appl}}$ belongs. Usually, we could give the slave displacement vector $\mathbf{q}_{\mathrm{s}}$ in terms of the master-released displacement vector $\mathbf{q}_{\mathrm{ms}}$ via mapping $\mathbf{q}_{\mathrm{s}}=\boldsymbol{\phi}(\mathbf{q}_{\mathrm{mr}})$, hence the parametrization mapping φ yields from Eqn (62).

We note that the generalized force vector $\mathbf{f}$ includes as well as external, internal, and velocity dependent acceleration forces via the relation

$$\mathbf{f}(\mathbf{q},\dot{\mathbf{q}})=\mathbf{f}^{\mathrm{ext}}(\mathbf{q})-\mathbf{f}^{\mathrm{int}}(\mathbf{q},\dot{\mathbf{q}})-\mathbf{f}^{\mathrm{accB}}(\mathbf{q},\dot{\mathbf{q}}), \quad \in E^{*n}, \tag{70}$$

where $\mathbf{f}^{\mathrm{ext}}, \mathbf{f}^{\mathrm{int}}$ and $\mathbf{f}^{\mathrm{accB}}$ denote external, internal, and velocity dependent acceleration force vectors. Note that $\mathbf{q}=\varphi(\mathbf{q}_{\mathrm{mr}})$ by the parametrization (63). In Newtonian mechanics, it is convenient to separate forces into external and internal forces, in d'Alembertian mechanics into applied and constraint forces.

We could define the following relations

$$\begin{aligned}
&\mathbf{K} := \mathrm{D}_{\mathbf{q}}\mathbf{f}^{\mathrm{int}}(\mathbf{q},\dot{\mathbf{q}}), \quad \mathbf{C} := \mathrm{D}_{\dot{\mathbf{q}}}\mathbf{f}^{\mathrm{int}}(\mathbf{q},\dot{\mathbf{q}})\\
&\mathbf{K}_{\mathrm{load}} := \mathrm{D}_{\mathbf{q}}\mathbf{f}^{\mathrm{ext}}(\mathbf{q})\\
&\mathbf{K}_{\mathrm{cent}} := \mathrm{D}_{\mathbf{q}}\mathbf{f}^{\mathrm{acc}}(\mathbf{q},\dot{\mathbf{q}},\ddot{\mathbf{q}}), \quad \mathbf{C}_{\mathrm{gyro}} := \mathrm{D}_{\dot{\mathbf{q}}}\mathbf{f}^{\mathrm{acc}}(\mathbf{q},\dot{\mathbf{q}},\ddot{\mathbf{q}})
\end{aligned} \tag{71}$$

that belong to the Euclidean space $\mathrm{E}^{*n\times *n}$, the embedding space.

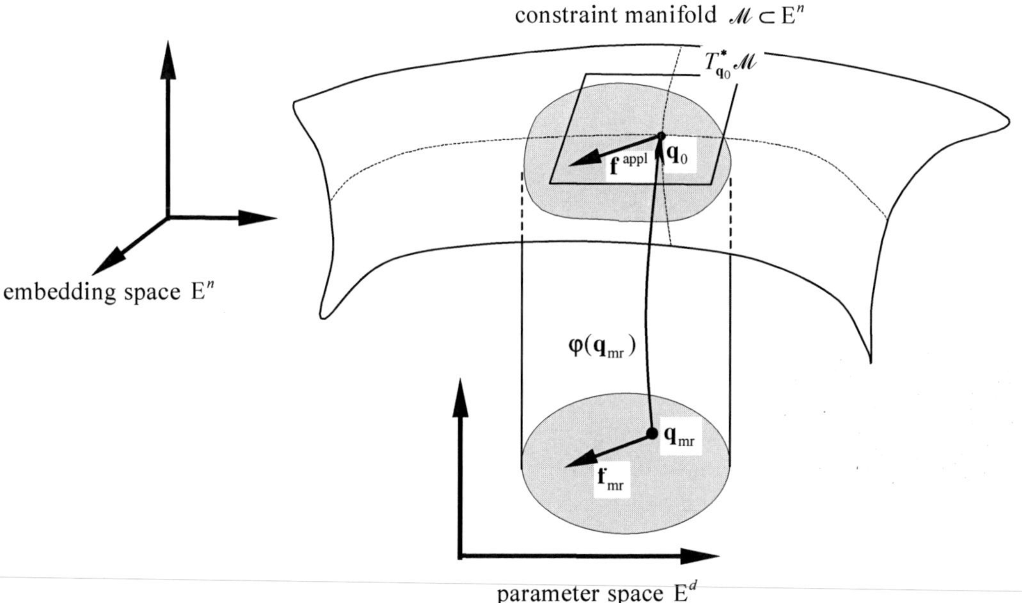

Figure 16. A Geometric structure of the applied force vector on the manifold and in the parameter space.

Substituting relations (70-71) and into (69) yields

$$\begin{aligned}
&\mathbf{f}_{\mathrm{mr0}} := \mathbf{B}^*\big(\mathbf{f}^{\mathrm{ext}}(\mathbf{q}_{\mathrm{mr}}) - \mathbf{f}^{\mathrm{int}}(\mathbf{q}_{\mathrm{mr}},\dot{\mathbf{q}}_{\mathrm{mr}}) - \mathbf{f}^{\mathrm{accB}}(\mathbf{q}_{\mathrm{mr}},\dot{\mathbf{q}}_{\mathrm{mr}}) - \mathbf{M}\dot{\mathbf{B}}\dot{\mathbf{q}}_{\mathrm{mr0}}\big) \quad \in \mathrm{E}^{*d},\\
&\mathbf{K}_{\mathrm{mr}} := \mathbf{B}^*\big((-\mathbf{K}_{\mathrm{load}} + \mathbf{K} + \mathbf{K}_{\mathrm{cent}})\mathbf{B} + (\mathbf{C}+\mathbf{C}_{\mathrm{gyro}})\dot{\mathbf{B}} + \mathbf{M}\ddot{\mathbf{B}}\big) + \mathbf{K}_{\sigma\mathrm{mr}}(-\mathbf{f}^{\mathrm{ext}} + \mathbf{f}^{\mathrm{int}} + \mathbf{f}^{\mathrm{acc}}) \quad \in \mathrm{E}^{*d\times *d},\\
&\mathbf{C}_{\mathrm{mr}} := \mathbf{B}^*\big((\mathbf{C}+\mathbf{C}_{\mathrm{gyro}})\mathbf{B} + 2\mathbf{M}\dot{\mathbf{B}}\big) \quad \in \mathrm{E}^{*d\times *d},\\
&\mathbf{M}_{\mathrm{mr}} := \mathbf{B}^*\mathbf{M}\mathbf{B} \quad \in \mathrm{E}^{*d\times *d},
\end{aligned} \tag{72}$$

where the acceleration force $\mathbf{f}^{\mathrm{acc}} = \mathbf{f}^{\mathrm{accA}} + \mathbf{f}^{\mathrm{accB}} \in \mathrm{E}^{*n}$, and the geometric stiffness tensor $\mathbf{K}_{\sigma\mathrm{mr}}$ is defined by

$$\mathbf{K}_{\sigma\mathrm{mr}}(\mathbf{f}) := \mathrm{D}_{\mathbf{q}_{\mathrm{mr}}}(\mathbf{B}^*\bar{\mathbf{f}}), \tag{73}$$

such that the vector $\mathbf{f}$ is kept constant under the differentiation, denoted by $\bar{\mathbf{f}}$. The tensor $\mathbf{K}_{\sigma\mathrm{mr}}$ is a symmetric tensor since according to the definition (64b) we have $\mathbf{K}_{\sigma\mathrm{mr}}(\mathbf{f}) = \mathrm{D}^2_{\mathbf{q}_{\mathrm{mr}}}(\varphi\cdot\bar{\mathbf{f}})$.

Equations (72) are important relations when deriving the generalized force vector and the corresponding tangent tensors. This is because the tangent tensors in the parameter space E^d are given in terms of the tangent tensors in the embedding space E^n where the tangent tensors are known. Also, we need the kinematic operator $\mathbf{B}$ and its time derivatives and its Frechét derivative. We note that the force vector $-\mathbf{f}^{\text{ext}} + \mathbf{f}^{\text{int}} + \mathbf{f}^{\text{acc}}$ in the tensor $\mathbf{K}_{\sigma\text{mr}}$ has an opposite direction compared with the constraint force vector $\mathbf{f}^{\text{con}} \in T^*_{\mathbf{q}_0}\mathcal{M}^{\perp}$. This naturally follows from the principle of virtual work (59).

4. Conclusions

We have studied differential geometry very elementarily way. We have divided vector spaces into material and spatial spaces are necessary since these spaces behave differently in observer transformation and in objective derivatives (Lie-derivatives). All the vector spaces, which we have considered, have a metric tensor thus they are metric vector spaces, and all the finite dimensional manifolds are Riemannian manifolds that are embedded in an Euclidean space. Additionally, we may identify a dual vector space by its primary vector space. In classical tensor analysis, this identification is applied, but here we make distinction between primary and dual spaces in the formulation, and the identification is accomplished later in the finite element implementation. If the identification of dual and primary vector spaces is done a priori, then push-forward and pull-back operations are not uniquely defined. Definitions for push-forward and pull-backs operation likewise Lie-derivatives have been given. As an example, we have studied rotation manifold and its underlying geometric structure in the terms of differential geometry.

In continuum mechanics, we have different manifolds. The placement field of continuum medium takes values in a Hilbert space, where a chart parametrization maps vector-valued functions into vector-valued functions. The placement field needs an infinite number of basis functions in order to present an arbitrary placement field on continuum, yielding infinite-dimensional manifolds.

In Lagrangian mechanics, forces are divided differently into constraint forces and applied forces. In addition, rich mathematical methods like variational calculus and other mechanical principles are included. In terms of differential geometry, Lagrangian mechanics describes a motion on an event manifold with a Lagrangian functional on the tangent bundle of the event manifold. An event manifold is a time-placement manifold which is also a constraint manifold, i.e. the time and placement variables satisfy all constraints.

The virtual work has been viewed as a linear form on the tangent field-bundle. This field-bundle is also a tangent bundle of the placement manifold at fixed time. We have given definitions for the virtual work in the finite-dimensional and infinite-dimensional cases. In addition, we have given definitions for the variation, Lie derivative and Lie variation. The concept of push-forward and pull-back operators is essential for understanding Lie derivatives and variations.

We have considered constraint point-manifolds that arise from point-wise holonomic constraint equations. The usual geometric joints of a multibody system like spherical, revolute, cylindrical, universal, helical, prismatic, and sliding joints can be presented via holonomic constraint equations that only depend on displacement at corresponding geometric points. All these constraints generate a smooth point-manifold that can be parametrized. Also, the principle of virtual work and its geometric structure are naturally related with the parametrization of the constraint manifold. We have also given a general procedure how a point-manifold can be parametrized.

REFERENCES

[1] Abraham, R., Marsden, J. E., Ratiu, T., (1983), *Manifolds, Tensor Analysis, and Applications*, Addison-Wesley, Massachusetts.

[2] Argyris, J., (1982), "An Excursion into Large Rotations", *Computer Methods in Applied Mechanics and Engineering*, 32, pp. 85-155.

[3] Arnold, V.I., (1978), *Mathematical Methods of Classical Mechanics*, Springer-Verlag, New York.

[4] Bonet J., Wood, R.D., (1997), *Nonlinear Continuum Mechanics for Finite Element Analysis*, Cambridge University Press, Cambridge.

[5] Burke, W.L., (1996), *Applied Differential Geometry*, Cambridge University Press, Cambridge.

[6] Cheng, H., Gupta, K.C., (1989), "An Historical Note on Finite Rotations", *ASME Journal of Applied Mechanics*, 56, pp. 139-145.

[7] Choquet-Bruhat, Y., DeWitt-Morette, C., Dillard-Bleick, M., (1989), *Analysis, Manifolds and Physics, Part I: Basics*, North-Holland, Amsterdam.

[8] Debnath, L., Mikusinski, P., (1990), *Introduction to Hilbert Spaces with Applications*, Academic Press, Boston.

[9] Ibrahimbegović, A., Frey, F., Kozar, I., (1995), "Computational Aspects of Vector-Like Parametrization of Three-Dimensional Finite Rotations", *International Journal for Numerical Methods in Engineering*, 38, pp. 3653-3673.

[10] Lanczos, C., (1966), *The Variational Principles of Mechanics*, University of Toronto Press, Toronto.

[11] Makowski, J., Stumpf, H., (1995), "On the 'Symmetry' of Tangent Operators in Nonlinear Mechanics", *ZAMM (Z. Angew. Math. Mech.) Applied Mathematics and Mechanics*, 75(3), pp. 189-198.

[12] Marsden J.E., Hughes, T.J.R., (1983), *Mathematical Foundations of Elasticity*, Prentice-Hall, Englewood Cliffs.

[13] Marsden, J.E., Ratiu, T.S., (1999), *Introduction to Mechanics and Symmetry : A Basic Exposition of Classical Mechanical Systems*, Springer-Verlag, New York.

[14] Oden , J.T., Reddy, J.N., (1976), *Variational Methods in Theoretical Mechanics*, Springer-Verlag, Berlin.

[15] Ogden, R.W., (1984), *Non-Linear Elastic Deformations*, Ellis Horwood, Chichester.

[16] Rabier, P.J., Rheinboldt, W.C., (2000), *Nonholonomic Motion of Rigid Mechanical Systems from a DAE Viewpoint*, SIAM Society for Industrial and Applied Mathamatics, Philadelphia.

[17] Rheinboldt, W.C., (1986), *Numerical Analysis of Parametrized Nonlinear Equations*, John Wiley & Sons, New York.

[18] Rosenberg, R.M., (1980), *Analytical Dynamics of Discrete Systems*, Plenum Press, New York.

[19] Selig, J.M., (1996), *Geometrical Methods in Robotics*, Springer-Verlag, New York.

[20] Simo, J.C., (1992), "The (Symmetric) Hessian for Geometrically Nonlinear Models in Solid Mechanics: Intrinsic Definition and Geometric Interpretation", *Computer Methods Applied Mechanics Engineering*, 96, pp. 189-200.

[21] Spring, K., (1986), "Euler Parameters and the Use of Quaternion Algebra in the Manipulation of Finite Rotations: A Review", *Mechanism and Machine Theory*, 21(5), pp. 365-373.

[22] Stuelpnagel, J., (1964), "On the Parametrization of the Three-Dimensional Rotation Group", *SIAM Review*, 6(4), pp. 422-430.

[23] Stumpf, H., Hoppe, U., (1997), "The Application of Tensor Algebra on Manifolds to Nonlinear Continuum Mechanics – Invited Survey Article", *ZAMM (Z. Angew. Math. Mech.) Applied Mathematics and Mechanics*, 77(5), 327-339.

[24] Truesdell, C., (1977), *A First Course in Rational Continuum Mechanics*, Academic Press, New York.

[25] Wang, C.-C., Truesdell, C., (1973), *Introduction to Rational Elasticity*, Noordhoff, Leyden.

In: Continuum Mechanics
Editors: Andrus Koppel and Jaak Oja, pp.53-83
ISBN: 978-1-60741-585-5

Chapter 2

Analysis of Shell Structures Applying Triangular Finite Elements

C. W. S. To[*]
University of Nebraska, Lincoln, Nebraska, USA

Abstract

Over the years many shell finite elements has been developed for applications in aerospace, automotive and shipbuilding industries. Misuse or abuse of some of these shell finite elements in the relatively mature shell finite element technology is not uncommon. Some of these shell finite elements were based on the principles of classical shell theory in which the simplest one is the theory of Love. Others were based on intuitive or heuristic arguments. For reasons of economy, mathematical simplicity and accuracy, lower order flat triangular shell finite elements are popular among designers. This article is concerned with the review, development and application of triangular shell finite elements. Emphasis is on mixed formulation based lower order flat triangular shell finite elements. Finite element representation of shell structures is introduced. Linear analysis of static and dynamic, and nonlinear analysis of static and dynamic problems, are included in this article.

1. Introduction

Over the last four decades many shell finite elements have been developed [1-6] for use in the aerospace, automotive, mechanical and structural, and shipbuilding industries. More recently, some of these shell finite elements have been adopted for analysis in the field of biomechanics and bioengineering. Some of these elements were based on classical and refined shell theories, while others were hinged on heuristic or intuitive arguments. Still others were derived with a combination of mathematical or numerical bases without due consideration of the mechanical behavior and properties of the particular class of shell structures. While finite element analysis of shell structures is generally regarded as a mature technology there is, however, a considerable amount of effort being exerted on the development of new shell finite elements which have to be examined with a view to identify the simple and efficient ones for general linear and nonlinear shell analysis. This article is exclusively concerned with the development and application of flat and curved triangular shell finite elements presented in the literature over the last four decades. They are considered for the following three important reasons. First, in the derivation of consistent element matrices the amount of algebraic manipulation for triangular elements is considerably less than the corresponding rectangular elements or elements with shapes other than triangular. Second, in term of complex geometries triangular shell elements are relatively simpler to use for mesh constructions and therefore the overall computational time for a particular problem can be considerably shorter. Third, linear triangular shell elements are relatively easier to be extended to highly nonlinear ones. For the latter class of problems when severe mesh distortions occur one can apply the meshless method which is beyond the present scope and therefore will not be included in this article. Furthermore, to limit the scope of the present article, solid shell finite elements and triangular laminated composite as well as sandwich shell elements are not considered.

In the next section displacement formulation-based flat triangular shell finite elements are reviewed. Section 3 deals with displacement formulation-based curved triangular shell elements. Section 4 is concerned with the discrete Kirchhoff theory (DKT) elements. Mixed and hybrid formulation-based triangular elements are considered in Section 5. Spline function-based triangular elements are included in Section 6. Finally, concluding remarks are presented in Section 7.

2. Displacement Formulation-Based Flat Triangular Shell Elements

The development of lower order (C^0) flat triangular elements for shell analysis dates back to the early 1960s. Such flat elements model shells by superposing stretching behavior (representing by membrane element) and bending feature (representing by plate bending element). The main advantages of modeling shells this way are: (1) simplicity of formulation, (2) ease of data input for shell geometry, (3) convenience of mixing with other elements, (4) ability to include rigid body motion, (5) treatment of general shell geometry, and (6) convenience in incorporating complex loading and boundary conditions. Early main reservations and concerns which have been regarded as disadvantages are: (1) there were reservations on their ability to represent the true behavior of curved shells [7], and (2) there were also concern that convergence to deep shell solutions could not be achieved with flat element size refinement. Of course, as pointed out in [8], subsequent theoretical developments of Idelsohn [9] and Morley [10] have largely put the concern of the aforementioned disadvantages to rest.

The displacement formulation-based triangular flat finite element for shell analysis appeared to have been pioneered almost simultaneously in 1968 by Clough and Johnson [11],

* Department of Mechanical Engineering, University of Nebraska, N104 Scott Engineering Center, Lincoln, Nebraska 68588-0656, U.S.A., E-mail: cto2@unl.edu

and by Zienkiewicz *et al.* [12]. Each of these two elements has three nodes. Every node has five degrees-of-freedom (dof). The latter include two in-plane displacements, one transversal displacement, and two first derivatives of the transversal displacement with respect to the two axes perpendicular to the transversal displacement. In 1972 Dawe [13] presented a three-node element which has twelve dof. The latter include three translational dof at every corner node and one dof at midside of every side.

Some five years later Argyris *et al.* [14] presented a three-node element for linear and nonlinear analysis. This element has eighteen dof and was derived starting with three layers in which the middle layer is the neutral plane. At every node there are three translational dof u, v and w, and three additional dof which are w_x, w_y and w_{xy}. While it was based on the displacement formulation the so-called natural approach was adopted in the derivation.

A three node eighteen dof displacement formulation-based shell element was presented by Olson and Bearden [15]. This element combined the bending triangle with a plane stress triangle including in-plane rotations at every vertex. In the plane stress element the displacement interpolation is *incomplete*.

A faceted shell element with Loof nodes was presented by Meek and Tan [16]. This element has twenty four dof.

During the beginning of the 1990's the development of flat triangular shell elements took a new turn in the sense that the triangular flat facet approximation obtained by combining the bending and membrane elements was formulated entirely from simple cubic polynomial displacement fields [17]. This three node element has eighteen dof. Thus, every vertex node has six dof which including three translational dof and three rotational dof. However, the so-called drilling dof (ddof) are not true drilling rotations in the sense of plane elasticity. An improvement over [17] by Allman was presented in 1994 [18]. The improvement was that the theoretical development was based on a variational principle. The latter allows a direct formulation of the element stiffness matrix without performing the matrix inversion during the derivation. It may be appropriate to mention that these two elements [17,18] employed Kirchhoff's hypothesis that the transverse shear is zero.

Around the same time Cook [19] attempted to improve the eighteen dof triangular shell element by, essentially, modifiying the membrane stiffness, and including the simple membrane-bending coupling device. He seemed to suggest that further improvement of performance of the lower-order triangular shell elements is possible by the introduction of membrane-bending coupling device.

In 1994 two shell finite elements adopted the Reissner and Mindlin theory for the bending component [20,21]. Specifically, a shear-locking free isoparametric three-node triangular finite element for thick and thin shell analysis was proposed by Kabir [20]. The transverse shear deformation in the shell formulation was considered. This element has fifteen dof. That is, every corner node has three translational dof and two rotation dof. The shear locking problem of this element was eliminated by imposing a constant transverse shear strain condition and introducing a shear correction expression in the formulation. In [21], however, the shell element has a standard linear deflection field and an incompatible linear rotation field expressed in terms of the mid-side rotations. In this element locking is circumvent by introducing an assumed linear shear strain field based on the tangential shear strains at the mid-sides. It is interesting to note that the element is free of spurious modes, satisfies the patch test and behaves correctly for thick and thin plates and shells. It was stated that the element degenerated in an explicit manner to a simple dicrete Kirchhoff form [21].

In another direction, refined non-conforming triangular elements were proposed by Chen and Cheung [22] for the analysis of shell structures. Essentially, these elements were based on the Bazley, Cheung, Irons and Zienkiewicz (BCIZ) bending conforming and non-conforming components and the constant strain triangle (CST). It may be appropriate to note that the BCIZ failed to pass the patch test in its original formulation. Another feature of these element is that Allman's triangular plane element [23] with ddof was incorporated. Two main versions were derived. One has fifteen dof and the other has eighteen dof. A simple reduced higher-order membrane strain matrix was proposed to circumvent membrane locking of the eighteen dof shell element.

Bletzinger *et al.* [24] presented a unified approach for shear-locking-free triangular and rectangular shell finite elements. The concept of discrete shear gap (DSG) was introduced. These elements are purely displacement formulation based. They used only the usual displacement and rotation dof at the nodes, without additional internal parameters. What is not clear here is the meaning of "usual". Does it mean three displacement dof and two rotational dof or it should be interpreted as three displacement and three rotational dof ? It was claimed that the elements passed the patch test. Besides, only one, the Scordelis-Lo roof, of the suite of benchmark problems was mentioned but results could not be found. It is also relevant to point out that in the bottom of second column on page 332, it was recognized that the "overall results of all elements tested in this example are relatively poor" [24].

Finally, Kuznetsov and Levyakov [25] proposed a refined geometrically nonlinear formulation of a thin-shell triangular finite element. It is based on the Kirchhoff-Love hypotheses. The strain relations were obtained by integrating the differential equations of a planar curve. This element has fifteen dof. Static geometrically nonlinear problems were studied.

3. Displacement Formulation-Based Curved Triangular Shell Elements

Aside from the foregoing displacement formulation-based flat triangular shell elements, curved triangular shell elements were developed and presented in the literature. During 1968 several doubly curved shell finite elements were presented. One doubly curved shallow shell element was presented by Strickland and Loden [26]. It has three nodes and fifteen dof. Another was introduced by Bonnes *et al.* [27]. It has six nodes and thirty six dof. Argyris *et al.* [28] presented the so-called SHEBA family of shell elements. Doubly curved shallow shell element with six nodes and sixty three dof was included.

Another doubly curved shallow shell element was presented by Ford in 1969 [29]. This element has three nodes and twenty four dof. It is interesting to note that this element is identical, in the sense that it has the same number of nodes and dof together with identical interpolation polynomials, to that by Megard [30].

In 1970 Cowper *et al.* [31] derived another doubly curved triangular shell element. This one has three nodes and thirty six dof. This is identical, in terms of node number, dof and interpolation polynomials, to that presented by Morin [32]. In the same year Dupuis *et al.* [33] developed another doubly curved triangular shell element for thin elastic shell analysis. This element has three nodes and fifty four dof.

The following year saw the introduction of another two doubly curved three-node shallow shell elements by Brebbia *et al.* [34]. One of these two elements has fifteen dof while the other has twenty seven dof.

Dawe [35] in 1975 presented a higher-order doubly curved triangular shell element. It has three nodes and fifty four dof. This element seems to be identical to that presented in [33] in that it has identical number of nodes and dof.

Subsequently, Thomas and Gallagher [36] derived a doubly curved shallow shell element. The latter has three nodes and thirty dof. The nodal dof includes u, u_x, u_y, v, ...,w_x, w_y and at the central node the dof are u, v and w.

4. Shell Elements Based on Discrete Kirchhoff Theory

While the displacement formulation based triangular shell elements reviewed in the last two sections were popular because they involved with consideration of displacement field two parallel developments took place in 1970. One made use of the discrete Kirchhoff theory (DKT) and the other was the degenerated three dimensional isoparametric formulations incorporating the Kirchhoff hypothesis. In the DKT it involves with the application of the hypothesis at a discrete number of points in an element. This approach was first proposed by Wempner *et al.* [37]. Dhatt [38] appeared to be the first in adopting this approach in his development of a triangular shell element that has three nodes and twenty seven dof. Two years later Batoz and Dhatt [39] presented two simple three node shell elements. Of these one has fifteen dof while the other has twenty dof. For large deflection analysis Batoz *et al.* [40] proposed another three node DKT based shell element which has twenty seven dof. In 1981 Bathe and Ho [41] presented a three node eighteen dof curved shell element. The nodal dof are u, v, w, w_x, w_y and w_{xy}. Every of the foregoing three node elements essentially consists of the DKT bending element and the CST.

In 1983 Murphy and Gallagher [42] presented an anisotropic cylindrical shell element based on the DKT. It has three nodes and twenty seven dof.

Dhatt *et al.* [43] developed a six node and twenty seven dof triangular shell element.

Around the same time Carpenter *et al.* [44] reported a flat triangular shell element with improved membrane interpolation. This was apparently further developed and the resulting improvements were presented in [45]. The curved triangular element in [45] has three nodes every one of which has five dof. Two versions of the membrane component of the shell element were derived and results examined. The first version was based on the Marguerre shallow shell theory and strain projection method that eliminates spurious membrane strain energy. The second version was based on a linear membrane field governed by normal rotations and reduced quadrature.

Levy and Gal [46] applied the DKT flat triangular and the CST membrane element in their geometrically nonlinear shell analysis. The geometric stiffness matrix, however, is derived by load perturbing the linear equilibrium equations. It was claimed that finite rotations were considered and rigid body motion was eliminated.

5. Mixed/Hybrid and Assumed Natural Strain Formulation-Based Elements

The lower order displacement formulation based shell elements and elements based on the degenerated three dimensional isoparametric formulations incorporating the Kirchhoff hypothesis are frequently plagued with the shear locking problem [47]. A scheme for dealing with this problem is the hybrid/mixed formulation. In terms of handling shear locking it has been shown that [2,3] the hybrid/mixed approach is equivalent to the displacement based formulation with reduced integration. Further, in term of providing continuity for both the displacement and strain or stress fields, hybrid/mixed formulation is unique.

For convenience, the development of mixed/hybrid and assumed natural strain (ANS) formulation based triangular shell finite elements are divided into two phases which will be considered in the following sub-sections. The first phase covers the period from the 1960's through late 1980's whereas the second phase spans the duration between the early 1990's to the present.

5.1. First Phase Developments

Dungar *et al.* [48,49] applied the hybrid formulation based right-angle triangle for the vibration and stress analyses of plate and shell structures. This hybrid formulation based element was derived with the stress components assumed within the element and the displacements are assumed on the boundaries of the element. This formulation is, of course, originated by Pian [50].

Subsequently, Edwards and Webster [51] provided a hybrid formulation based triangular cylindrical element. Rigby *et al.* [52] presented a Hellinger-Reissner principle based six node doubly curved triangular thin shell element.

Meanwhile, Lee *et al.* [53], using the Hellinger-Reissner principle in which assumed displacement and strain fields were included, and degenerated solid and isoparametric formulation were adopted, provided a curved triangular shell element for thin shell analysis. In the derivation of element stiffness matrix, seven point numerical integration was applied. This element has nine nodes and forty five dof. Thus, every node has five dof.

Based on the mixed formulation Saleeb *et al.* [54] presented a C^0-linear triangular plate/shell element. The particular issue addressed was the role of edge shear constraints.

5.2. Second Phase Developments

Developments in the second phase can further be divided into two categories. The first category includes those having non-optimal properties and those possess optimal features.

Those elements with one or more defects/disadvantages such as the spurious zero energy mode that requires some form of stabilization scheme are included in the first category, whereas those with optimal properties are dealt with in the second category. The optimal properties should at least include: (a) rank sufficiency, (b) no spurious mode and no shear locking, and (c) the element having only three nodes every one of which should possess six dof. The latter are three translational and three rotational dof.

5.2.1. Elements with Non-Optimal Properties

Fish Belytschko [55] derived a rank-sufficient flat triangular shell element with ddof based on a three field variational principle with relaxed interelement compatibility and traction continuity conditions. This variational principle is known as the Pian-Chen variational principle for incompatible elements. The weak form is similar to the Hu-Washizu variational principle except that the interelement continuity conditions are relaxed and satisfied only in the weak sense. A generalized spurious mode control procedure based on the assumed strain method [56] was developed to stabilize zero energy kinematical or spurious mode.

Boisse *et al.* [57] presented a C^0 three-node shell element for nonlinear structural analysis. It adopted Mindlin kinematics, degenerated solid approach and linear Lagrange functions for geometry and displacement interpolations. The variational principle is eqivalent to the Hu-Washizu principle. Nonlinear deformation with small rotation in loading step was applied. Every node of the element has five dof.

Sze and Zhu [58] reported the derivation of a quadratic ANS curved triangular shell element. The latter consists of six nodes. All the sampled natural strains are optimal with respect to the derivative of a prescribed cubic field in the subparametric element. It was found that the element indicated no sign of locking, passing all the patch tests and provided satisfactory accuracy.

In another report [59] Kim and Kim proposed a three-node macro triangular shear deformable shell element based on the ANS. It was pointed out that the element has a spurious zero energy mode due to the fact that the strain field is set to be constant in each sub-element for macro-ANS element and in simple ANS element.

Lee and Bathe [60] reported the development of the mixed interpolation of tensorial components (MITC) isotropic triangular shell finite elements. The latter include one three-node and two versions of a six-node element. A deficiency, in the sense that some locking is present, of the three-node triangular shell element was candidly mentioned by Lee *et al.* [61] and provided the motivation to further study the element behavior. Three different isotropic three-node elements, identified as QUAD3, MITC3 and SRI3, and one non-isotropic three-node element, called NIT3 were developed [61]. It was pointed out that the NIT3 element can be derived using the DSG concept of [24]. A clamped plate problem and a hyperboloid shell problem with various mesh topologies were studied for convergence.

5.2.2. Elements with Optimal Properties

During the early part of 1990's several lower order three node eighteen dof flat triangular shell elements were developed and reported by To and Liu [62,63]. In these elements the ddof were included. Thus, at every node there are three translational dof and three rotational dof. They were based on the hybrid strain and displacement formulation. That is, it is a mixed formulation in the sense that the bending, membrane and transverse shear components denoted respectively by k_b , k_m and k_s were derived by the Hellinger-Reissner variational principle with assumed strain and assumed displacement fields while the torsional component associated with the ddof and designated by k_t was obtained by the displacement formulation. Therefore, the element stiffness matrix is given by $k = k_b + k_m + k_s + k_t$ .Of the sixteen versions, two identified as NFORMU 8 and NFORUM 16 [63] are rank sufficient and capable of producing the six rigid body modes correctly. NFORMU 8 is identical to NFORMU 16 except that in the latter version the transverse displacement w is quadratic instead of linear. The two optimal elements also passed the patch test. Shear locking is not present in these elements. When the inplane and the ddof are constrained the single element test provides correctly the three rigid body modes.

Furthermore, the flat triangular plates of the corresponding sixteen flat triangular shell finite elements were shown [64] to pass the so-called *inf-sup* condition [65]. Specifically, the *inf-sup* condition defined by Eq. (f) in page 325 of [65] for mixed formulation finite element is satisfied.

Two nonlinear flat triangular shell elements based on the foregoing two linear versions were derived. Nonlinear static and dynamic analyses were performed. The theoretical development and computed numerical results were reported in [66,67]. In parallel, the linear and nonlinear shell elements were extended to application of laminated composite shell structures by To and Wang [68-70]. In [69,70] the importance of incorporating the concept of directors in the formulation was demonstrated as the directors are important parameters that constitute the so-called "exact geometry" for large rotation problems.

One common feature of these elements presented by To and associates was the derivation of explicit expressions for consistent mass and stiffness matrices by using the symbolic manipulation packages, MAPLE and MACSYMA. Thus, numerical matrix inversion and numerical integration in the derivation of element matrices were not necessary. This can very likely improve the computational efficiency drastically as previous investigation by To and Liu [71] demonstrates that a large reduction in computational time can be achieved using explicit expressions instead of numerical integration. In the latter reference, a nonlinear analysis of the nine-layered simply supported square plate under bisinusoidal pressure was performed to provide such a demonstration.

Three different routes of development and application of these elements were performed. The first route is concerned with linear and nonlinear static and dynamic analyses of isotropic and laminated composite shell structures to deterministic loads [69-73]. The second route dealt with (a) shell structures disturbed by nonstationary Gaussian random excitations [74,75] and nonstationary non-Gaussian random excitations [76], and (b) shell structures having spatially stochastic properties and under nonstationary random excitations [77,78]. The third route is concerned with dynamic and optimal random vibration control of laminated composite shell structures with piezoelectric components [79,80].

Owing to their optimal properties and potential adoption by others for shell structure analysis and for completeness, the mixed formulation and derivation of element stiffness matrices k for the two optimal linear shell finite elements are included in the following sub-section. Representative linear results are presented in Sub-section 5.2.4.

5.2.3. Mixed Formulation and Derivation of Optimal Shell Elements

The mixed formulation adopted in the development of the optimal shell elements consists of two parts. The first part has to do with the application of the hybrid strain formulation that is based on the assumed strain and assumed displacement fields of Hellinger-Reissner variational principle. The second part deals with displacement formulation for the element stiffness component associated with the ddof.

5.2.3.1. Hellinger-Reissner Variational Principle

The functional for the linear version of the Hellinger-Reissner principle can be written as:

$$\pi_{HR}(u,\varepsilon) = \int_{V_b}\left[-\frac{1}{2}\varepsilon^T D\varepsilon + \varepsilon^T D(Lu) - u^T f \right] dv - \int_{S_t} u^T \bar{t}\, ds - \int_{S_u} (u - \bar{u})^T \left[\Gamma(D\varepsilon)\right] ds \tag{1}$$

where

- u is the displacement vector;
- σ is the stress vector;
- ε is the strain vector;
- f is the body force vector;
- D is the elastic matrix of the material, such that $\sigma = D\varepsilon$;
- C is the compliance matrix of the material, such that $\varepsilon = C\sigma$;
- L is the linear operator to calculate strain from displacement;
- Γ is the linear operator to evaluate surface traction from stress;
- $\bar{u}$ is the vector of prescribed displacement on boundary;
- $\bar{t}$ is the vector of prescribed surface traction;
- V_b is the volume of the body;
- S_t is the portion of the surface of the body where $\bar{t}$ is applied;
- S_u is the portion of the surface of the body where $\bar{u}$ is applied;

and the superscript T denotes transpose.

Since in hybrid strain formulation the final unknowns are nodal displacements, the satisfaction of displacement boundary condition, $u = \bar{u}$, is easily met and the term with $(u - \bar{u})$ can thus be disregarded. Equation (1) then becomes

$$\pi_{HR}(u,\varepsilon) = \int_{V_b}\left[-\frac{1}{2}\varepsilon^T D\varepsilon + \varepsilon^T D(Lu) - u^T f \right] dv - \int_{S_t} u^T \bar{t}\, ds. \tag{2}$$

Equation (2) is the foundation of the present hybrid strain shell element formulation.

5.2.3.2. Discretized Functional and Equilibrium Equation

Assuming that at the element level

$$u = Nq, \qquad \varepsilon = P\beta, \tag{3a, b}$$

where q and β are vectors of nodal displacement, and strain parameter, respectively, while N and P are the corresponding interpolation or shape function matrices.

Substituting Eq. (3) into (2) gives

$$\pi_{HR}(u,\varepsilon)=\sum \pi_{HR}(q,\beta)$$
$$=\sum\left\{\int_V\left[-\tfrac{1}{2}\beta^T P^T DP\beta+\beta^T P^T D(LNq)\right]dV\right\}$$
$$-\sum\left[\int_V (q^T N^T f)\,dV-\int_A (q^T N^T \bar{t})\,dA\right], \tag{4}$$

where the subscripts V and A of the integration symbol denote the volume and area of an element, respectively, and the summation is performed over all the elements in the usual manner.

Applying the stationarity condition to the functional in Eq. (4) with respect to β and defining one can obtain

$$H=\int_V P^T DP\,dV\,,\quad G_e=\int_V P^T DB\,dV \tag{5a, b}$$

$$-H\beta+G_e q=0\,,\quad \beta=H^{-1}G_e q\,. \tag{6a, b}$$

Substituting the last equation into (4) results in

$$\pi_{HR}(u,\varepsilon)=\sum \pi_{HR}(q)$$
$$=\sum\left[\frac{1}{2}q^T G_e^T H^{-1} G_e q\right]-\sum q^T\left[\int_V (N^T f)\,dV\right] \tag{7}$$
$$-\sum q^T\left[\int_A (N^T \bar{t})\,dA\right]$$

Applying the stationarity condition to Eq. (7) with respect to q gives

$$G_e^T H^{-1} G_e q-\int_V N^T f\,dV-\int_A N^T \bar{t}\,dA=0\,. \tag{8}$$

Defining

$$k_h=G_e^T H^{-1} G_e\,,\quad f_c=\int_V N^T f\,dV+\int_A N^T \bar{t}\,dA \tag{9a.b}$$

where k_h is the element stiffness matrix based on the hybrid strain formulation, and f_c the consistent load vector due to body force and surface traction Eq. (8) can then be written as

$$k_h q=f_c\,.$$

Assembling all elements in the usual manner yields the equilibrium equation for the system

$$K_h Q=F_c \tag{10}$$

with K_h, Q and F_c being the hybrid strain based assembled stiffness matrix, assembled nodal displacement and consistent load vectors, respectively. The strains or stresses are recovered through the following relations

$$\varepsilon = P\beta = PH^{-1}G_e q, \quad \sigma = D\varepsilon = DPH^{-1}G_e q. \tag{11}$$

5.2.3.3 Element Geometry

The three nodes of the flat triangular shell element are allocated at the three corners of the mid-surface of the. A local rectangular co-ordinate system is attached to node 1, with its r-axis coinciding with the side 1-2, its t-axis being parallel to the normal of the element and its s-axis perpendicular to the r-t plane as shown in Figure 1. With such a co-ordinate system, the r and s co-ordinates of nodes 1, 2, and 3 are: (0,0), $(r_2,0)$ and (r_3, s_3), respectively. There are 6 dof at each node, which are: u, v, and w being the displacements in the r-, s- and t-directions, respectively; and θ_r, θ_s and θ_t being the rotations about r-, s- and t-axes, respectively. They are positive if along the positive directions of r, s and t-axes.

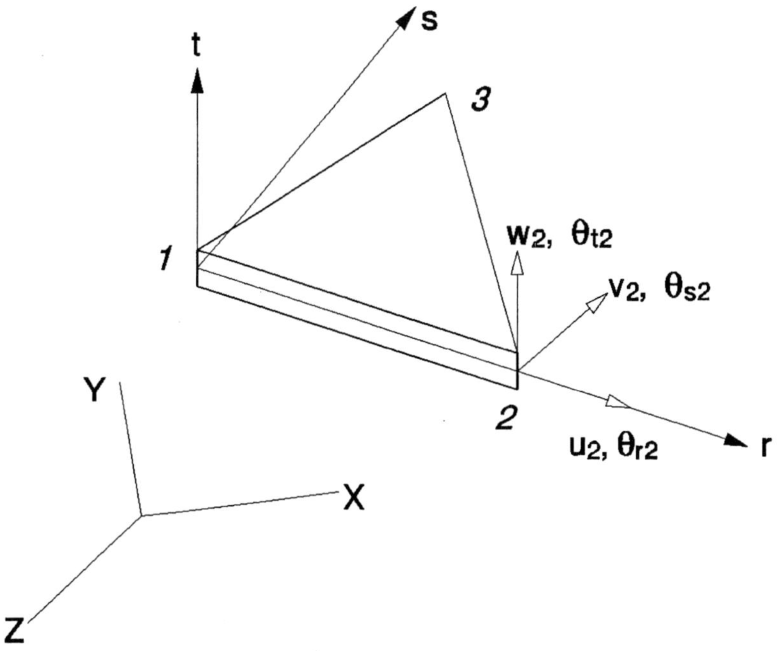

Figure 1. Flat triangular shell finite element in local and global co-ordinate systems.

The relation between the natural and the r-s co-ordinates is given by

$$\begin{Bmatrix} \xi_1 \\ \xi_2 \\ \xi_3 \end{Bmatrix} = \frac{1}{r_2 s_3} \begin{bmatrix} r_2 s_3 & -s_3 & r_3 - r_2 \\ 0 & s_3 & -r_3 \\ 0 & 0 & r_2 \end{bmatrix} \begin{Bmatrix} 1 \\ r \\ s \end{Bmatrix}. \tag{12}$$

In the computation the three quantities r_2, r_3 and s_3 need be determined from known global position vectors of the three nodes: $P_i = (X_i, Y_i, Z_i)$, where i runs from 1 to 3.

5.2.3.4 Displacement and Strain Interpolations

The linear interpolation relation can be written in terms of natural co-ordinates:

$$\begin{aligned} &\left[u \quad v \quad w \quad \theta_r \quad \theta_s \quad \theta_t\right]^T = \\ &N_5\left[u_1 \quad v_1 \quad w_1 \quad \theta_{r1} \quad \theta_{s1} \quad \theta_{t1} \quad u_2 \quad \ldots \quad \theta_{s3} \quad \theta_{t3}\right]^T, \end{aligned} \tag{13}$$

where the first subscript in the displacements and the second subscript in the rotations denote the nodal number, and

$$N_5 = \left[\,[N_1]_{6\times 5} \quad [0]_{6\times 1} \quad [N_2]_{6\times 5} \quad [0\,]_{6\times 1} \quad [N_3]_{6\times 5} \quad [0]_{6\times 1}\right]_{6\times 18}, \tag{14}$$

where

$$[N_i] = \begin{bmatrix} \xi_i & 0 & 0 & 0 & 0 \\ 0 & \xi_i & 0 & 0 & 0 \\ 0 & 0 & \xi_i & 0 & 0 \\ 0 & 0 & 0 & \xi_i & 0 \\ 0 & 0 & 0 & 0 & \xi_i \\ 0 & 0 & 0 & 0 & 0 \end{bmatrix}_{6\times 5}, \quad i = 1,\ 2,\ 3.$$

The displacement and rotation interpolations are included in Appendix A for completeness.

The assumed strain field is

$$\left[\varepsilon_r \quad \varepsilon_s \quad \varepsilon_{rs} \quad \varepsilon_{st} \quad \varepsilon_{tr}\right]^T = P_5\left[\beta_1 \quad \beta_2 \quad \beta_3 \quad \ldots \quad \beta_8 \quad \beta_9\right]^T \tag{15}$$

in which P_5 has been defined in [62,66] and is included in the following for completeness

$$P_5 = \begin{bmatrix} 1 & 0 & 0 & t & 0 & 0 & 0 & 0 & 0 \\ 0 & 1 & 0 & 0 & t & 0 & 0 & 0 & 0 \\ 0 & 0 & 1 & 0 & 0 & t & 0 & 0 & 0 \\ 0 & 0 & 0 & 0 & 0 & 0 & p_5^{47} & p_5^{48} & 0 \\ 0 & 0 & 0 & 0 & 0 & 0 & p_5^{57} & p_5^{58} & p_5^{59} \end{bmatrix}_{5\times 9},$$

$$p_5^{47} = -\, s_3(1 - 2\xi_2), \quad p_5^{48} = s_3(2\xi_2 + 2\xi_3 - 1),$$

$$p_5^{57} = -r_3(1 - 2\xi_2), \quad p_5^{58} = (r_3 - r_2)(2\xi_2 + 2\xi_3 - 1),$$

$$p_5^{59} = r_2(1 - 2\xi_3),$$

and t is the thickness co-ordinate ranging from - $h/2$ to $h/2$ with h being the thickness of the shell element.

With Eq. (15) the stresses calculated from the assumed strain functions satisfy the point-wise equilibrium condition; all kinematic deformation modes are suppressed; and the element properties are invariant. In fact P_5 was employed in [54] for a hybrid stress formulation. It argued that, since transversal shear strains ε_{st} and ε_{tr} are written in terms of natural co-ordinates ξ_2 and ξ_3 only, the resulting shear constraints, when shear strains diminishes to zero for large length/thickness ratio, will all be of discrete-Kirchhoff type. This effectively prevents shear-locking from occurring.

5.2.3.5. Hybrid Strain Based Element Stiffness Matrices

The strain-displacement relations for membrane strains

$$\varepsilon_r = u_{,r}, \quad \varepsilon_s = v_{,s}, \quad \varepsilon_{rs} = u_{,s} + v_{,r}. \tag{16}$$

Similarly, for the bending strains the strain-displacement relations are

$$\varepsilon_r = t\,\theta_{s,r}, \quad \varepsilon_s = -\,t\,\theta_{r,s}, \quad \varepsilon_{rs} = t\,(\theta_{s,s} - \theta_{r,r}). \tag{17}$$

For the transverse shear strains the strain-displacement relations are

$$\varepsilon_{st} = w_{,s} - \theta_r, \quad \varepsilon_{tr} = w_{,r} + \theta_s. \tag{18}$$

In the foregoing equations, Eqs. (16) through (18), the subscripts ,s and ,r denote $\partial/\partial s$ and $\partial/\partial r$, respectively.

Then the strain-displacement matrix B can be obtained as [62,66]

$$B_m = [\,[B_{m1}]_{5\times 6}\,[B_{m2}]_{5\times 6}\,[B_{m3}]_{5\times 6}\,]_{5\times 18},$$

$$B_b = [\,[B_{b1}]_{5\times 6}\,[B_{b2}]_{5\times 6}\,[B_{b3}]_{5\times 6}\,]_{5\times 18}, \tag{19}$$

$$B_s = [\,[B_{s1}]_{5\times 6}\,[B_{s2}]_{5\times 6}\,[B_{s3}]_{5\times 6}\,]_{5\times 18},$$

with submatrices

$$[B_{mi}] = \begin{bmatrix} \xi_{i,r} & 0 & 0 & 0 & 0 & 0 \\ 0 & \xi_{i,s} & 0 & 0 & 0 & 0 \\ \xi_{i,s} & \xi_{i,r} & 0 & 0 & 0 & 0 \\ 0 & 0 & 0 & 0 & 0 & 0 \\ 0 & 0 & 0 & 0 & 0 & 0 \end{bmatrix}_{5\times 6},$$

$$[B_{bi}] = \begin{bmatrix} 0 & 0 & 0 & 0 & t\,\xi_{i,r} & 0 \\ 0 & 0 & 0 & -t\,\xi_{i,s} & 0 & 0 \\ 0 & 0 & 0 & -t\,\xi_{i,r} & t\,\xi_{i,s} & 0 \\ 0 & 0 & 0 & 0 & 0 & 0 \\ 0 & 0 & 0 & 0 & 0 & 0 \end{bmatrix}_{5\times 6},$$

$$[B_{si}] = \begin{bmatrix} 0 & 0 & 0 & 0 & 0 & 0 \\ 0 & 0 & 0 & 0 & 0 & 0 \\ 0 & 0 & 0 & 0 & 0 & 0 \\ 0 & 0 & \xi_{i,s} & -\xi_i & 0 & 0 \\ 0 & 0 & \xi_{i,r} & 0 & \xi_i & 0 \end{bmatrix}_{5\times 6}, \quad i = 1, 2, 3$$

and the first subscripts m, b, and s designate membrane, bending and transverse shear, respectively.

Upon application of Eq. (5) and assuming linear isotropic material, one can show that

$$H_5 = \int_V P_5^T D P_5 \, dV, \qquad (G_e)_m = \int_V P_5^T D B_m \, dV,$$
$$(G_e)_b = \int_V P_5^T D B_b \, dV, \qquad (G_e)_s = \int_V P_5^T D B_s \, dV \tag{20}$$

with elastic material matrix

$$D=\begin{bmatrix} \frac{E}{1-\nu^2} & \frac{\nu E}{1-\nu^2} & 0 & 0 & 0 \\ \frac{\nu E}{1-\nu^2} & \frac{E}{1-\nu^2} & 0 & 0 & 0 \\ 0 & 0 & G & 0 & 0 \\ 0 & 0 & 0 & \kappa_s G & 0 \\ 0 & 0 & 0 & 0 & \kappa_s G \end{bmatrix}_{5\times 5},$$

where E is the Young's modulus, G shear modulus, ν Poisson's ratio and κ_s the form factor of shear which is equal to 5/6.

By making use of Eqs. (20) and (9a), one has

$$k_m=(G_e)_m^T H_5^{-1}(G_e)_m, \qquad k_b=(G_e)_b^T H_5^{-1}(G_e)_b,$$
$$k_s=(G_e)_s^T H_5^{-1}(G_e)_s, \qquad k_h=k_m+k_b+k_s. \tag{21}$$

Note that the arrangement of the non-zero entries (elements) in matrices B_m, B_b and B_s leads to decoupling between k_m and k_b. This is a consequence of the fact that the element is flat. However, k_b and k_s are still coupled.

Equations (20) and (21) indicate that, to obtain the hybrid strain based element stiffness matrix k_h, four integrals have to be evaluated. These integrations can be performed numerically. However, as the triangular natural co-ordinates are employed it is possible to perform the integrations analytically. The integrations were performed with a symbolic manipulation package, MAPLE. The strategy of integration employing MAPLE was outlined in page 180 of [62]. Explicit expressions for the element stiffness matrices have been obtained and are not included here for brevity.

5.2.3.6. Element Stiffness Matrices Incoporating Ddof

Various schemes for dealing with the problem due to ddof have been reported in the literature. As the flat triangular shell element is a combination of a triangular plane stress element and a triangular bending element, Allman's proposal of incorporating the ddof into a triangular plane stress/strain element [23,17,18] provides a starting point.

Symbolically, the consistent element stiffness matrix that can be written as

$$[k]_{18\times 18}=\begin{bmatrix} [G_e^T H^{-1} G_e]_{15\times 15} & [0]_{3\times 15} \\ [0]_{15\times 3} & [k_{\theta_t}]_{3\times 3} \end{bmatrix}_{18\times 18}$$

where $[0]_{15\times 3}$ and $[0]_{3\times 15}$ are null matrices with dimensions 15×3 and 3×15, and the 3×3 matrix $[k_{\theta t}]$ is diagonal and contains three equally valued entries $k_{\theta t}$. In fact, the latter is related to the skew symmetric component of the in-plane torsional strain,

$$\varepsilon_{rs}^{s} = \theta_t - \frac{1}{2}\left(v_{,r} - u_{,s}\right), \tag{22}$$

where the superscript s denotes skew symmetric.

Introducing the following complete interpolation function matrix N

$$N = N_5 + N_t$$

where N_5 has been defined in Eq. (14), and

$$N_t = \begin{bmatrix} [0]_{5\times5} & [0]_{5\times1} & [0]_{5\times5} & [0]_{5\times1} & [0]_{5\times5} & [0]_{5\times1} \\ [0]_{1\times5} & \xi_1 & [0]_{1\times5} & \xi_2 & [0]_{1\times5} & \xi_3 \end{bmatrix}_{6\times18} \tag{23}$$

such that the strain energy functional due to torsional deformation, π_t may be expressed as a function of nodal displacement q [62,66],

$$\pi_t(q) = \frac{1}{2} G h \Sigma\left(\int_A q^T \bar{r}^T \bar{r} q \, dA \right) \tag{24}$$

where h and G are the thickness and shear modulus, respectively; whereas the vector $\bar{r}$ will be defined later in Eq. (29).

The last row of Eq. (23) represents a linear variation of θ_t over an element,

$$\theta_t = \xi_1 \theta_{t1} + \xi_2 \theta_{t2} + \xi_3 \theta_{t3}. \tag{25}$$

Applying Eqs. (4) and (24) one has the following functional

$$\pi(q,\beta) = \pi_{HR}(q,\beta) + \pi_t(q). \tag{26}$$

Following the steps in Sub-sub-section 5.2.3.2 for applying the stationarity condition of Eq. (26) with respect to β, and then with respect to q, one has the element stiffness matrix

$$k = k_m + k_b + k_s + k_t \tag{27}$$

where k_m, k_b and k_s have been defined by Eq. (21) and the element stiffness component matrix associated with the torsional energy by displacement formulation is

$$k_t = G h \int_A \bar{r}^T \bar{r} \, dA. \tag{28}$$

Since the integrands in the last equation are polynomials up to degree 2, the exact integration of k_t requires 3-point quadrature. Such integration results are denoted by k_t^3 the non-zero elements or entries of which have been derived but not included here for brevity.

Denoting the element membrane component stiffness matrix as $(k_m)'$ which is defined as $(k_m)' = \left((G_e)'_m\right)^T H_5^{-1} (G_e)'_m$, in which the elements of $(G_e)'_m$ are obtained and not included here for brevity. Associated with this stiffness matrix, $(k_m)'$, the row matrix $\bar{r}$ in the torsional component becomes

$$\bar{r} = \frac{1}{2}\left[\bar{r}_{1,1}, \bar{r}_{1,2}, 0, 0, 0, \bar{r}_{1,6}, \bar{r}_{1,7}, \bar{r}_{1,8}, 0, 0, 0, \bar{r}_{1,12}, \bar{r}_{1,13}, 0, 0, 0, 0, \bar{r}_{1,18}\right]_{1\times 18} \tag{29}$$

where

$$\bar{r}_{1,1} = \frac{r_3 - r_2}{r_2 s_3}, \quad \bar{r}_{1,2} = \frac{1}{r_2}, \quad \bar{r}_{1,6} = 2\xi_1 + \bar{p}_{1,s} - \bar{q}_{1,r},$$

$$\bar{r}_{1,7} = -\frac{r_3}{r_2 s_3}, \quad \bar{r}_{1,8} = -\frac{1}{r_2}, \quad \bar{r}_{1,12} = 2\xi_2 + \bar{p}_{2,s} - \bar{q}_{2,r},$$

$$\bar{r}_{1,13} = \frac{1}{s_3}, \quad \bar{r}_{1,18} = 2\xi_3 + \bar{p}_{3,s} - \bar{q}_{3,r},$$

while $\bar{p}_i$ and $\bar{q}_i$ are defined in Appendix A.

The in-plane displacements u and v are now coupled with the ddof through quantities such as $\bar{p}_i$ and $\bar{q}_i$. The consequence of this coupling is that one has to replace B_{mi} (i = 1,2 and 3) in Eq. (19) with a new bending part, $(B_{mi})'$ of the strain-displacement matrix. Accordingly, $(G_e)_m$ in Eq. (20*b*) and k_m in Eq. (21*a*) have to be changed to $(G_e)_m'$ and $(k_m)'$. Note that the coupling between the in-plane displacements and the ddof does not alter the two remaining parts, namely, the bending and transverse shear components k_b and k_s of the hybrid strain based element stifffness matrix k_h. Furthermore, the interpolation function for θ_t itself does not enter into $(B_{mi})'$. Thus, it has no contribution to k_h.

Substituting Eq. (29) into (28) and performing the integration leads to the desired stiffness matrices. Note that the exact integration needs 3-point quadrature for applying Eq. (29). Only two optimal versions are considered here. The result for the optimal version by exact integration upon applying Eqs. (29), (A1) and (A3) designated as k_t^{o1}, and for the other optimal version by exact integration upon using Eqs. (29), (A2) and (A3) as k_t^{o2}. The explicit expressions for k_t^{o1} and k_t^{o2} have been obtained but not included here for brevity.

In short, the two optimal versions of the element stiffness matrix k, depending on how the ddof are treated, are identified as:

$$k = k_h + k_t^{o1}, \text{ and } k = k_h + k_t^{o2}$$

in which the membrane component of the first version is based on Allman triangle (AT) formulation and the *w* displacement function is linear. In the second version, the membrane component is based on Allman triangle (AT) formulation but the *w* displacement function is quadratic.

In summary, the main difference between Allman's rotations and the true rotations is the part associated with the torsional energy due to the "skew symmetric" component. Thus, for a three-dimensional problem one has

$$\varepsilon^{s} = \theta - L^{*} u^{*}, \tag{30}$$

where the superscript *s* designates skew symmetric,

$$u^{*} = \begin{bmatrix} u & v & w \end{bmatrix}^{T}, \quad \theta = \begin{bmatrix} \theta_r & \theta_s & \theta_t \end{bmatrix}^{T},$$

$$L^{*} = \frac{1}{2}\begin{bmatrix} 0 & -\partial/\partial t & \partial \\ \partial & 0 & -\partial/\partial r \\ -\partial/\partial s & \partial & 0 \end{bmatrix}.$$

The corresponding relation for two dimensional problems, of course, is that defined by Eq. (22). To provide a simple and better understanding of the physical meaning of Eq. (30), one can apply it to the Timoshenko beam theory with due modification to symbols. Then, the term on the left-hand side (lhs) of Eq. (30) is the transverse shear whereas the first term on the right-hand side (rhs) of Eq. (30) is the rotation of the cross-section of the beam, and the second term on the rhs of Eq. (30) is the slope of the transverse deflection with respect to the axial co-ordinate.

5.2.4. Representative Computed Results

To highlight the accuracy, reliability and range of application as well as efficiency results were presented in [8,62-64,81]. In this sub-section selected results from these references are included. They are: (a) the ring shell under its own weight, (b) the pinched cylinder, and (c) hemispherical shell subjected to alternating loads applied at the equator.

5.2.4.1 Spherical Ring Loaded with its Own Weight

The spherical ring, the meridian of which is shown in Figure 2(b), is clamped at its lower edge and free at the upper one. The material properties of the ring are: $E = 30.0\times10^{9}$ N/m^2, $\nu = 1/6$ and the weight is along the vertical direction and measured at 3×10^{3} N/m^2. With a thickness of 60.0 mm, the radial displacement of the free edge, quoted from [82], is - 238.0 mm, and the vertical displacement of that edge is - 220.2 mm. By making use of symmetry, only a quadrant of the ring was modeled by the optimal finite elements since results by each cannot be distinguished on the plot. The boundary conditions are, $U = V = W = \Theta_x = \Theta_y = \Theta_z = 0.0$ along the clamped edge, $U = \Theta_y = \Theta_z = 0.0$ for nodes with $X = 0.0$, and $V = \Theta_x = \Theta_z = 0.0$ for nodes with $Y = 0.0$. Four meshes employed were 4×4A, 4×4C, 6×6A and 6×6C. It was observed that the finite element results obtained indicated that using the proposed shell elements the problem can be adequately represented by relatively coarse meshes such as 4×4 and 6×6 [81]. One possible explanation for the excellent results with these coarse meshes is that the stress distribution is relatively even in this problem.

Some of the computed results in [81] are given in Figure 2(d) in which the shallow shell theory results are from [83].

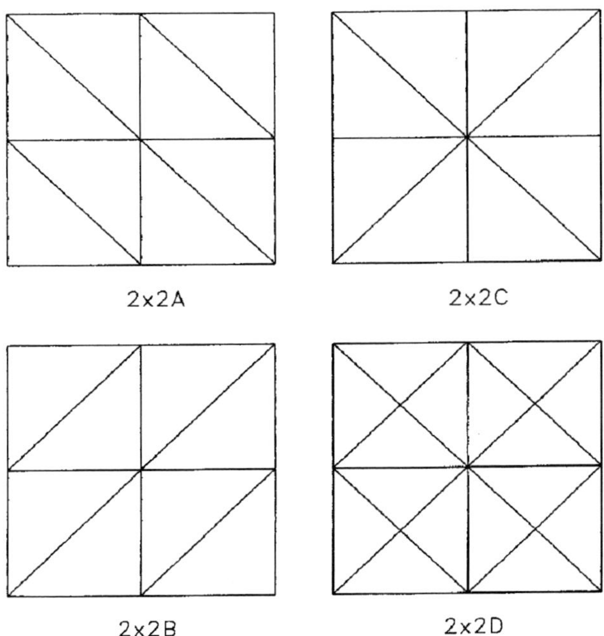

Figure 2(a). Various mesh representations.

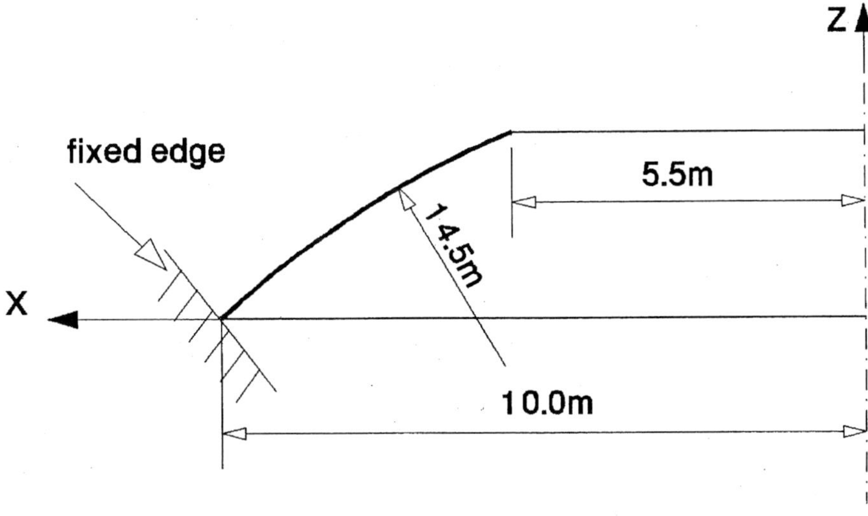

Figure 2(b). Spherical ring shell showing its meridian.

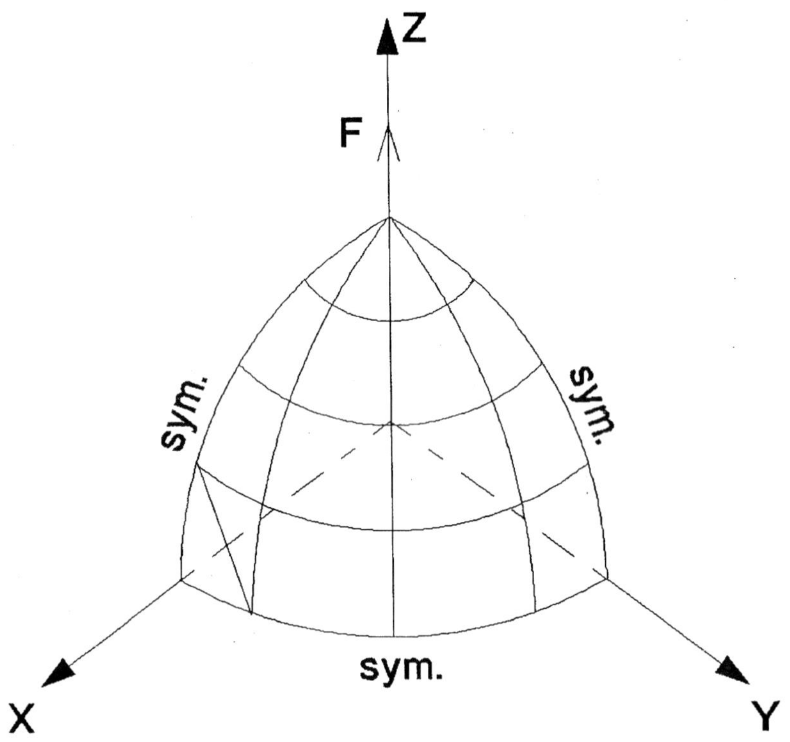

Figure 2(c). A quadrant of ring shell showing a A-mesh at bottom left-hand corner.

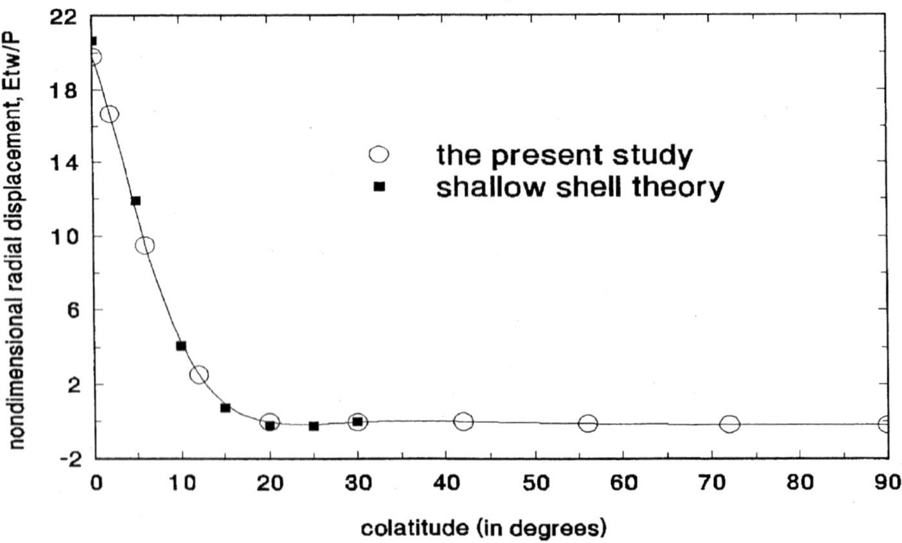

Figure 2(d). Non-dimensional radial displacement at free edge.

5.2.4.2 Pinched Cylinder

The pinched cylinder has rigid end diaphragms and is subjected to point loads as shown in Fig. 3(a). The exact deflection at the load points is 0.463×10^{-3} mm (0.18248×10^{-4} in) as quoted in [44,45,54], for example. Only one-eighth of the cylinder is considered due to symmetry. The boundary conditions are therefore, $U = V = \Theta_z = 0$ for arc AD, $V = \Theta_x = \Theta_z = 0$ for side DC, $U = \Theta_y = \Theta_z = 0$ for side AB, and $W = \Theta_x = \Theta_y = 0$ for arc BC. The geometrical and material properties are: R = 0.762 m (300 in), L = 1.524 m (600 in), h = 76.2 mm (3 in), $E = 20.7\times10^{10}$ Pa (3.0×10^{6} psi) and v = 0.3. The point load is F = 4.455 N (1.0 lb) for the problem, or 1.115 N (0.25 lb) for the finite element models.

This problem is not part of the obstacle course [84] but has been studied by a number of authors [44,45,54]. It is one of the most severe tests available to determine the ability of shell elements to represent both the inextensional bending actions and very complex membrane stress states. In this problem there is also significant coupling between membrane and bending actions. Several meshes were generated. These meshes have approximately equal nodal spacing along side AB and arc AD. Thus, all elements are right-angle triangles with their two right-angled sides having approximately equal side lengths. Mesh types A, B and D were attempted. In mesh type D more nodes have to be added accordingly. Computed results are included in Figure 3(b). Note that the data labelled Ref.[62] in the legend of Figure 3(b) designated computed results with mesh type A defined in [62] which donot possess approximately equal nodal spacing along side AB and arc AD. As can be observed the difference of results between those applying the present mesh type A and using mesh type A in [62] is insignificantly small. More results and discussion can be found in Liu and To [8].

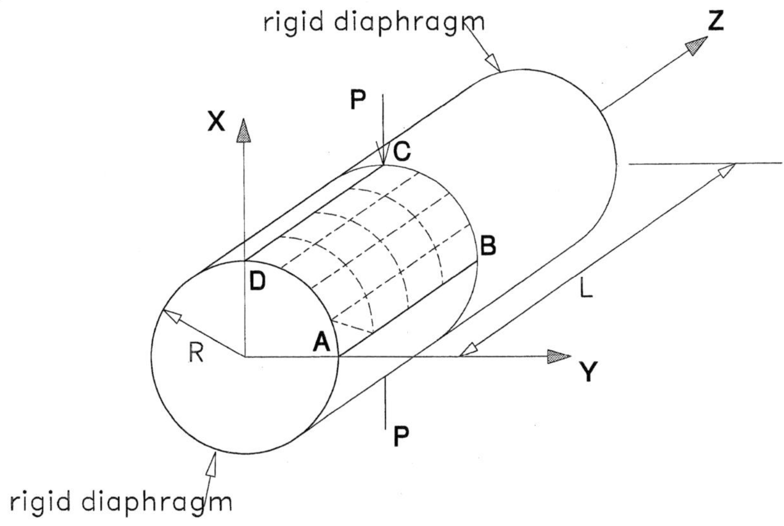

Figure 3(a). Pinched cylinder.

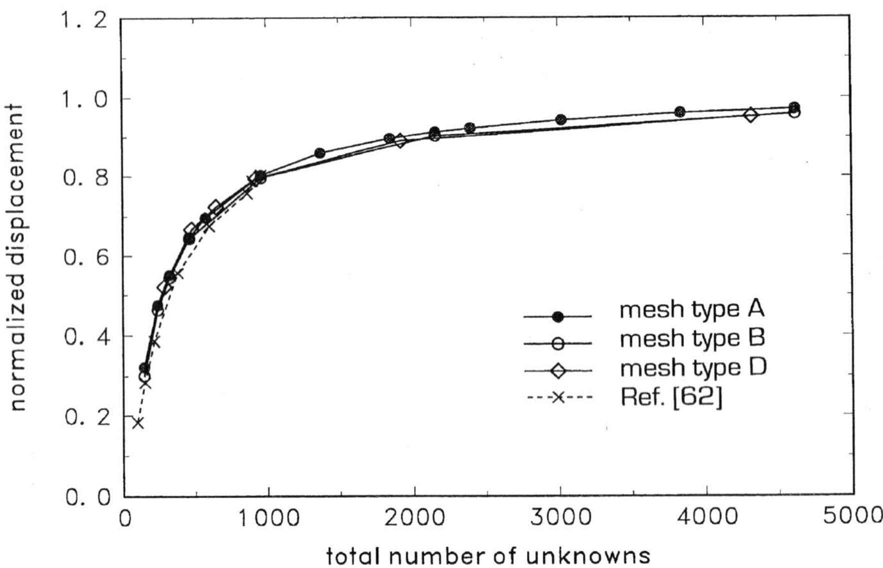

Figure 3(b). Convergence study of pinched cylinder.

5.2.4.3 Hemispherical Shell

The hemispherical shell subjected to alternating loads at the equator is a very challenging problem for several reasons. First, it is a test for an element's ability to represent inextensional modes since the shell exhibits almost no membrane strains. In addition, this problem is useful in checking the ability of an element to handle rigid body rotations about normal to the shell surface, because large sections of the shell rotate almost as a rigid body under the given loads. Secondly, there has been inconsistency in the finite element literature in regard to the value of the alternating load F when applied to one quadrant of the shell. A number of publications used 8.91 N or 2.0 lb, see for example [62, 44,45,84]. On the other hand, the value 4.455 N or 1.0 lb for the alternating load has also been employed in [19,54]. In the present investigation F has been changed from 8.91 N as in [62] to 4.455 N. Finally, the shell may be analyzed with or without a central hole of 18E, see Figure 4(a) and (b).

The problem has been included in the obstacle course [84]. Its pertinent data are: mean radius of the shell R = 254 mm (10.0 in), thickness h = 1.016 mm (0.04 in), Young's modulus $E = 4.706\times10^{11}$ Pa (6.825×10^{7} psi) and Poisson's ratio v = 0.3. Reference [84] found that the solution for the displacements of the points under the loads is 2.347 mm (0.0924 in) for the case of no central hole, and 2.388 mm (0.094 in) for the case of a 18E central hole. Note that these values are a theoretical lower bound [84].

By making use of symmetry of geometry, only one quadrant of the hemisphere is analyzed and each of the two alternating point loads on the quadrant is F = 4.455 N (1.0 lb). The boundary conditions for this example are: free for nodes on the X-Y plane, $V = \Theta_x = \Theta_z = 0$ for nodes on the X-Z plane and $U = \Theta_y = \Theta_z = 0$ for nodes on the Y-Z plane. Finally for the case of no central hole, the pole is fixed. That is, all of its six dof are set to zero. Mesh types A, B, C and D were used. The numbers of nodes along the longitudinal and latitudinal directions are equal. Computed results normalized by their corresponding theoretical values quoted above were presented in Figure 4(c) for mesh type C since it was found that results for mesh types C and D have very small difference. Results of mesh types A and B have significant difference from those of C and D below 5000 total number of unknowns or dof. These results were presented in [8].

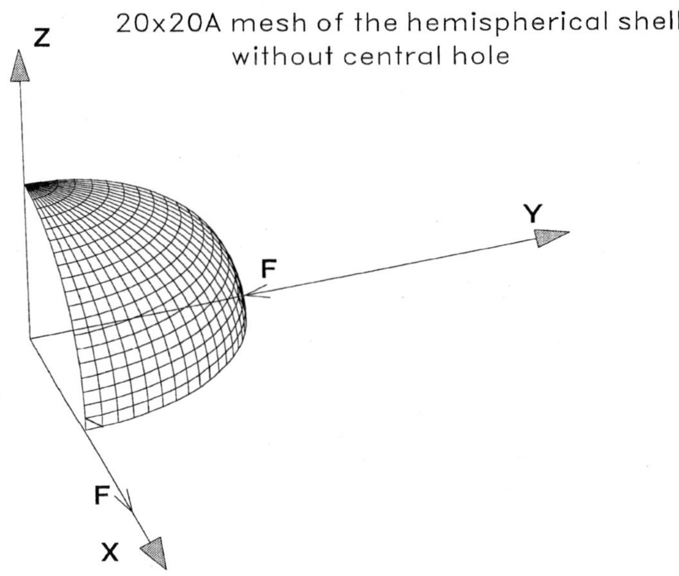

Figure 4(a). Hemispherical shell with alternating equatorial loads.

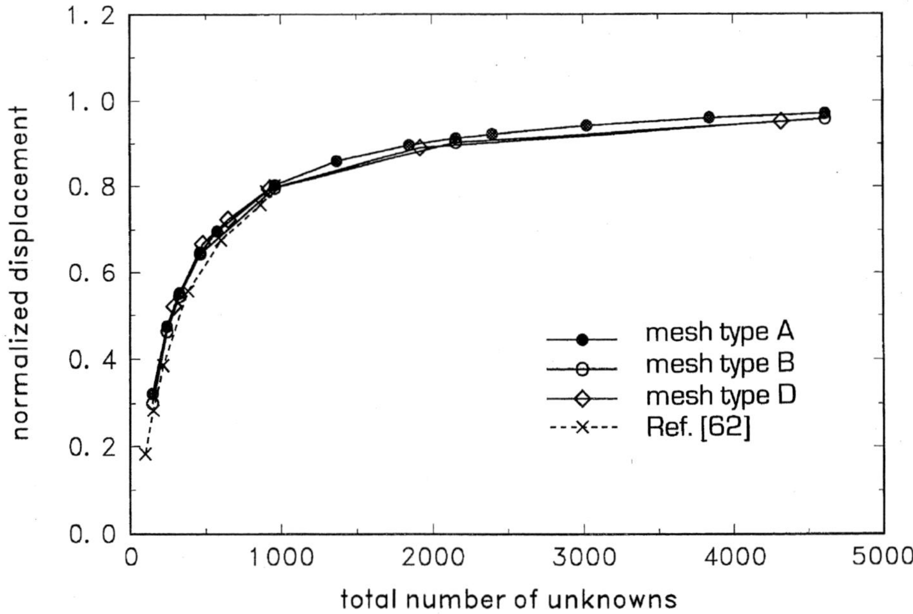

Figure 4(b). Hemispherical shell with alternating equatorial loads.

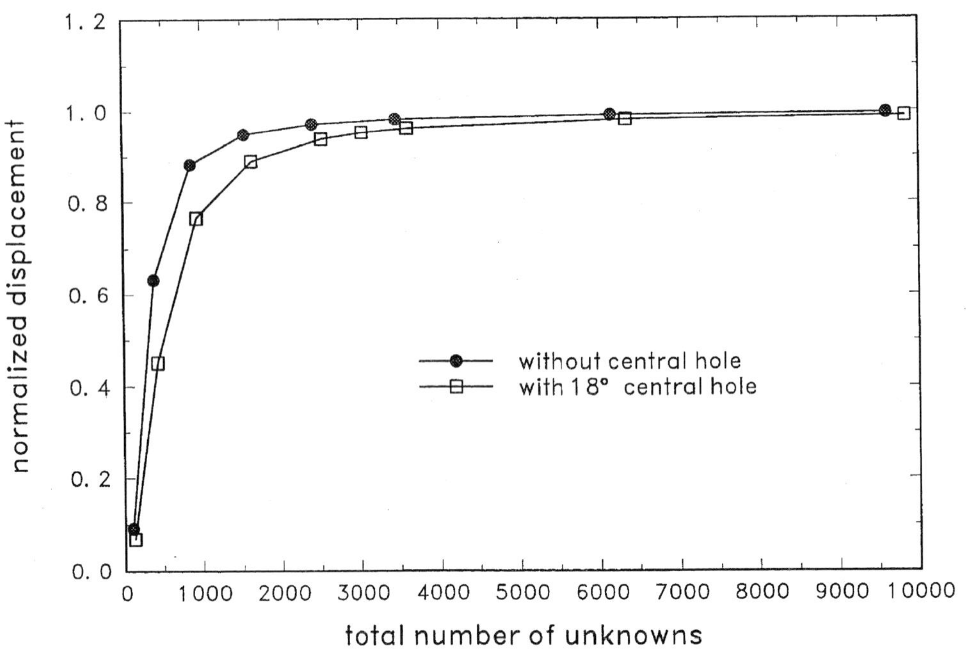

Figure 4(c). Convergence study of hemispherical shells.

6. Spline Function-Based Flat Triangular Shell Elements

Recently, spline function-based triangular shell elements have been proposed. In *Wang et al.* [85] the dynamics of thin-shell simulation through the application of triangular B-spline finite elements has been investigated. The classical Kirchhoff-Love shell theory was used and therefore transverse shear is not included. While the technique provides a powerful tool for geometrical modeling no numerical results compared with those of the benchmark suite cases was provided.

Cho *et al.* [86] implemented a framework that directly links a general tensor-based shell finite element to non-uniform rational B-spline (NURBS) geometric modeling. In addition, a six-node triangular shell element for automatic mesh generation of complex shape was developed.

7. Concluding Remarks

Many triangular shell finite elements for linear and nonlinear analysis of shell structures have been reviewed and the mixed formulation based optimal lower order flat triangular shell finite elements have been introduced in the foregoing. The optimal flat triangular shell finite elements developed by this author and his associate [62,63,8] have satisfied the *ellipticity, consistency* and *inf-sup conditions* outlined in [60]. These three conditions amount to the convergence test that consists of the patch test and eigenvalue solution. As pointed out by this author "The patch test and eigenvalue solution, however, are relatively much more easy to perform and understand conceptually. The *inf-sup* condition, on the other hand, is much more difficult to understand conceptually and establish analytically without over simplification for shell finite elements based on mixed and hybrid formulations. Numerically, it is also relatively more difficult to operate"[64].

It is believed that, in order to satisfy the patch test and eigenvalue solution in which any mixed formulation based lower order flat triangular shell finite element has to be able to provide six rigid body modes correctly when they are not constrained, the number of strain modes must satisfy the Tong, Pian and Chen condition [87,88]: $m \geq n - r$, where m is the number of assumed strain modes, n the number of generalized displacements, and r the zero-eigenvalue or rigid body modes of the element. The optimal elements introduced, of course, have satisfied this condition.

Finally, a related question is: Can the presented optimal shell finite elements be improved? With reference to the finding by Liu and To [8], the answer to this question is resoundingly negative. In other words, any attempt to improve on the optimal elements developed by the author and his associate seems to be unnecessary.

APPENDIX A. DISPLACEMENT AND ROTATION FIELDS

The two optimal shell elements have the following displacement relations. The first choice has the displacements defined by

$$\begin{aligned} u &= u_1\xi_1 + u_2\xi_2 + u_3\xi_3 + \bar{p}_1\theta_{t1} + \bar{p}_2\theta_{t2} + \bar{p}_3\theta_{t3}, \\ v &= v_1\xi_1 + v_2\xi_2 + v_3\xi_3 + \bar{q}_1\theta_{t1} + \bar{q}_2\theta_{t2} + \bar{q}_3\theta_{t3}, \\ w &= w_1\xi_1 + w_2\xi_2 + w_3\xi_3, \end{aligned} \tag{A1}$$

the second choice

$$\begin{aligned} u &= u_1\xi_1 + u_2\xi_2 + u_3\xi_3 + \bar{p}_1\theta_{t1} + \bar{p}_2\theta_{t2} + \bar{p}_3\theta_{t3}, \\ v &= v_1\xi_1 + v_2\xi_2 + v_3\xi_3 + \bar{q}_1\theta_{t1} + \bar{q}_2\theta_{t2} + \bar{q}_3\theta_{t3}, \\ w &= w_1\xi_1 + w_2\xi_2 + w_3\xi_3 - \bar{p}_1\theta_{r1} - \bar{p}_2\theta_{r2} - \bar{p}_3\theta_{r3} \\ &\quad - \bar{q}_1\theta_{s1} - \bar{q}_2\theta_{s2} - \bar{q}_3\theta_{s3}, \end{aligned} \tag{A2}$$

and rotations are linear. They are defined by

$$\begin{aligned} \theta_r &= \theta_{r1}\xi_1 + \theta_{r2}\xi_2 + \theta_{r2}\xi_3, \\ \theta_s &= \theta_{s1}\xi_1 + \theta_{s2}\xi_2 + \theta_{s3}\xi_3, \\ \theta_t &= \theta_{t1}\xi_1 + \theta_{t2}\xi_2 + \theta_{t3}\xi_3, \end{aligned} \tag{A3}$$

where

$$\bar{p}_1 = (a_{31}\xi_3 - a_{12}\xi_2)\xi_1, \quad \bar{q}_1 = (b_{31}\xi_3 - b_{12}\xi_2)\xi_1,$$
$$\bar{p}_2 = (a_{12}\xi_1 - a_{23}\xi_3)\xi_2, \quad \bar{q}_2 = (b_{12}\xi_1 - b_{23}\xi_3)\xi_2, \tag{A4}$$
$$\bar{p}_3 = (a_{23}\xi_2 - a_{31}\xi_1)\xi_3, \quad \bar{q}_3 = (b_{23}\xi_2 - b_{31}\xi_1)\xi_3,$$

in which ξ_1, ξ_2 and ξ_3 are the area co-ordinates, a_{ij} and b_{ij} are element geometric constants.

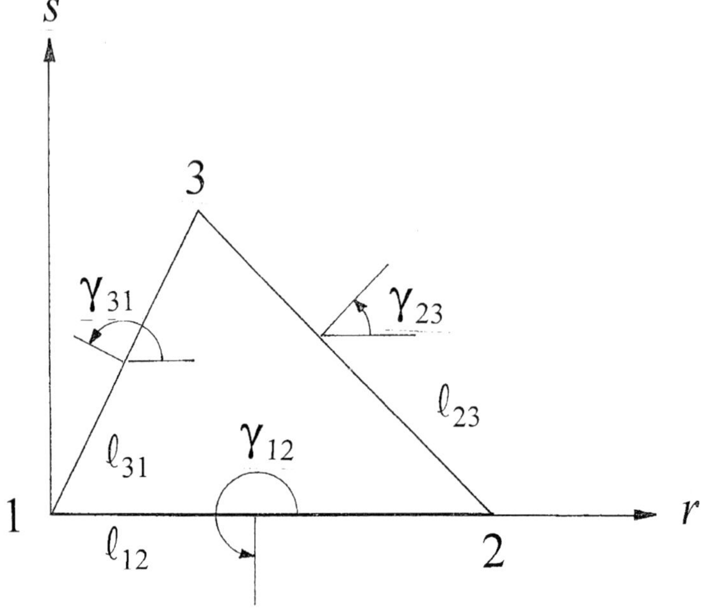

Figure A1. Geometry and local coordinates of triangular shell finite element.

With reference to Figure A1, these geometric constants are given by

$$a_{12} = \frac{1}{2}\ell_{12}\cos\gamma_{12}, \quad b_{12} = \frac{1}{2}\ell_{12}\sin\gamma_{12},$$

$$a_{23} = \frac{1}{2}\ell_{23}\cos\gamma_{23}, \quad b_{23} = \frac{1}{2}\ell_{23}\sin\gamma_{23},$$

$$a_{31} = \frac{1}{2}\ell_{31}\cos\gamma_{31}, \quad b_{31} = \frac{1}{2}\ell_{31}\sin\gamma_{31}.$$

From Eqs. (A2) through (A4) one can see that in the second optimal element the assumed translational displacement fields are quadratic when the $\bar{p}$ and $\bar{q}$ terms are included in the formulation. The selection of these terms is based on the strategy introduced by Allman [23]. The foregoing choices of the polynomials for translational and rotational dof were applied by

To and Liu [62,63,66,67] for linear and nonlinear static and dynamic analyses of isotropic plate and shell structures, and To and Wang [68-70] for static and dynamic, and linear and geometrically nonlinear analyses of laminated composite plate and shell structures.

REFERENCES

[1] Noor, A.K., Belytschko, T. and Simo, J.C. (Editors), 1989. *Analytical and Computational Models of Shells*, CED-Vol.3, A.S.M.E., New York.

[2] Yang, H.T.Y., Saigal, S. and Liaw, D.G., 1990. Advances of thin shell finite elements and some applications-Version I. *Comput. Struct.*, 35(No.4), 481-504.

[3] Kang, D.G., 1990. Present finite element technology from a hybrid formulation perspective. *Computers and Structures*, 35(No.4), 321-329.

[4] Bucalem, M. and Bathe, K.J., 1997. Finite element analysis of shell structures. *Archives of Computational Methods in Engineering*, 4, 3-61.

[5] MacNeal, R., 1998. Perspective on finite elements for shell analysis. *Finite Elements in Analysis and Design*, 30, 175-186.

[6] Mackerle, J., 2002. Finite- and boundary-element linear and nonlinear analyses of shells and shell-like structures: A bibliography (1999-2001). *Finite Elements in Analysis and Design*, 38(No.8), 765-782.

[7] Zienkiewicz, O.C., 1977. *The Finite Element Method.* 3rd ed., McGraw-Hill, New York.

[8] Liu, M.L. and To, C.W.S., 1998. A further study of hybrid strain-based three-node flat triangular shell elements. *Finite Elements in Analysis and Design*, 31, 135-152.

[9] Idelsohn, S., 1981. On the use of deep, shallow or flat shell finite elements for analysis of thin shell structures. *Comput. Meth. Appl. Mech. and Eng.*, 26, 321-330.

[10] Morley, L.S.D., 1984. A facet-like shell theory. *International Journal of Engineering Science*, 22 (No. 11/12), 1315-1327.

[11] Clough, R.W. and Johnson, C.P., 1968. A finite element approximation for the analysis of thin shells. *International Journal of Solids and Structures*, 4, 43-60.

[12] Zienkiewicz, O.C., Parikh, C.J., and King, I.P., 1968. Arch dam analysis by a linear finite element shell solution program. In *Proc. of Symp. on Arch Dams*, pp.19-22, I.C.E.

[13] Dawe, D.J., 1972. Shell analysis using a facet element. *J. Strain Anal.*, 6, 266-270.

[14] Argyris, J.H., Dunne, P.C., Malejannakis, G.A. and Schelke, E., 1977. A simple triangular facet shell element with application to linear and nonlinear equilibrium and elastic stability problems. *Comput. Meth. Appl. Mech. Eng.*, 10, 371-403; 11, 97-131.

[15] Olson, M.D. and Bearden, T.W., 1979, A simple flat triangular shell element revisited. *International Journal for Numerical Methods in Engineering*, 14, 51-68.

[16] Meek, J.L. and Tan, H.S., 1986. A faceted shell element with Loof nodes. International Journal for Numerical Methods in Engineering, 23, 49-67.

[17] Allman, D.J., 1991. Analysis of general shells by flat facet finite element approximation. *American Institute of Aeronautics and Astronautics J.*, 95, 194-203.

[18] Allman, D.J., 1994. A basic flat facet finite element for analysis of general shells. *International Journal for Numerical Methods in Engineering*, 37, 19-35.

[19] Cook, R.D., 1993. Further development of a three-node triangular shell element. *International Journal for Numerical Methods in Engineering*, 36, 1413-1425.

[20] Kabir, H.R.H., 1994. A shear locking free robust isoparametric three node triangular element for general shells. *Computers and Structures*, 51, 425-436.

[21] Onate, E., Zarate, F. and Flores, F., 1994. A simple triangular element for thick and thin plate and shell analysis. *Int. J. for Num. Meth. in Engineering*, 37, 2569-2582.

[22] Chen, W.J. and Cheung, Y.K., 1999. Refined non-conforming triangular elements for analysis of shell structures. *Int. J. for Num. Meth. in Engineering*, 46, 433-455.

[23] Allman, D.J., 1984. A compatible triangular element including vertex rotations for plane elasticity. *Computers and Structures*, 19, 1-8.

[24] Bletzinger, K.U., Bishoff, M. and Ramm, E., 2000. A unified approach for shear-locking-free triangular and rectangular shell finite elements. *Computers and Structures*, 75, 321-334.

[25] Kuznetsov, V.V. and Levyakov, S.V., 2007. Refined geometrically nonlinear formulation of a thin-shell triangular finite element. *Journal of Applied Mechanics and Technical Physics*, 48, 755-765.

[26] Strickland, G. and Loden, W., 1968. A doubly-curved triangular shell element. In *Proc. of the 2nd Conf. on Matrix Methods in Structural Mechanics*, Wright-Patterson Air Force Base, Dayton, Ohio, 15-17 October (AFFDL-TR-68-150, pp. 641-666).

[27] Bonnes, G., Dhatt, G., Giroux, Y. and Robichaud, L., 1968. Curved triangular elements for analysis of shells. In *Proceedings of the Second Conference on Matrix Methods in Structural Mechanics*, Wright-Patterson Air Force Base, Dayton, Ohio, 15-17 October. AFFDL-TR-68-150, pp. 617-640.

[28] Argyris, J.H. and Scharpf, D., 1968. The SHEBA family of shell elements for the matrix displacement method. *Aeron. Journal*, 72, 873-883; 73, 423-426 (1969).

[29] Ford, R., 1969. A triangular finite element for shell structures. CEGB Report, RD/C/N359(1969).

[30] Megard, G., 1969. Planar and curved shell elements. In *Finite Element Method in Stress Analysis*. Tapir, Trondheim.

[31] Cowper, G., Lindberg, G.M. and Olson, M.D., 1970. A shallow shell finite element of triangular shape. *International Journal of Solids and Structures*, 6, 1133-1156.

[32] Morin, N., 1970. Nonlinear analysis of thin shells. Ph.D. Thesis, Department of Civil Engineering, MIT, Cambridge, Massachussetts.

[33] Dupuis, G. and Goël, J.J., 1970. A curved finite element for thin elastic shells. *International Journal of Solids and Structures*, 6, 1413-1428.

[34] Brebbia, C.A., Sabanathan, S., Tahibilda, U. and Tottenham, H., 1971. Statics and dynamics of hydrostatically loaded shells by finite element method. In *Proceedings of the IASS Symposium on Hydromechanically Loaded Shells*, Hawaii, October.

[35] Dawe,D.J., 1975. Higher-order triangular finite element for shell analysis. *International Journal of Solids and Structures*, 11, 1097-1110.

[36] Thomas, G.R. and Gallagher, R.H., 1976. A triangular element based on generalized potential energy concepts. In *Finite Elements for Thin Shells and Curved Members*, edited by Gallagher, R.H. and Ashwell, D.G., pp.155-169. John Wiley, New York.

[37] Wempner, G.A., Oden, J.T. and Kross, D.A., 1968. Finite element analysis of thin shells. *Journal of Engineering Mechanics Division, A.S.C.E.*, 9(No.6), 1273-1294.

[38] Dhatt, G., 1970. An efficient triangular shell element. *American Institute of Aeronautics and Astronautics Journal*, 8, 2100-2102.

[39] Batoz, J.L. and Dhatt, G., 1972. Development of two simple shell elements. *American Institute of Aeronautics and Astronautics Journal*, 10, 237-248.

[40] Batoz, J.L., Cattopadhyay, A. and Dhatt, G., 1976. Finite element large deflection analysis of shallow shells. *Int. J. for Numerical Methods in Engineering*, 10, 39-58.

[41] Bathe, K.J. and Ho, L.W., 1981. A simple and effective element for analysis of general shell structures. *Computers and Structures*, 13, 673-681.

[42] Murphy, S.S. and Gallagher, R.H., 1983. Anisotropic cylindrical shell element based on discrete Kirchhoff theory. *Int. J. for Num. Meth. in Engineering*, 19, 1805-1823.

[43] Dhatt, G., Marcotte, L. and Matte, Y., 1986. A new triangular discrete Kirchhoff plate/shell element. *Int. J. for Numerical Methods in Engineering*, 23, 453-470.

[44] Carpenter, N., Stolarski, H., and Belytschko, T., 1985. A flat triangular shell element with improved membrane interpolation. *Communications in Applied Numerical Methods*, 1, 161-168.

[45] Carpenter, N., Stolarski, H., and Belytschko, T., 1986. Improvements in 3-node triangular shell elements. *Int. J. for Num. Meth. in Engineering*, 23, 1643-1667.

[46] Levy, R. and Gal, E., 2001. Geometrically nonlinear three-noded flat triangular shell elements. *Computers and Structures*, 79, 2349-2355.

[47] Ramm, E., 1977. A plate/shell element for large deflections and rotations. In *Proceedings of the Symposium on Formulations and Computational Algorithms in Finite Element Analysis*. MIT Press, Cambridge, MA.

[48] Dungar, R., Severn, R.T., and Taylor, P.R., 1967. Vibration of plate and shell structures using triangular finite elements. *J. of Strain Analysis*, 2(No.1), 73-83.

[49] Dungar, R., and Severn, R.T., 1969. Triangular finite elements of variable thickness and their application to plate and shell problems. *J. of Strain Anal.*, 4(No.1), 10-23.

[50] Pian, T.H.H., 1964. Derivation of element stiffness matrices by assumed stress distributions. *American Inst.of Aeron. and Astron. Journal*, 2, 1333-1336.

[51] Edwards, G. and Webster, J.J., 1976. Hybrid cylindrical shell finite elements. In Finite Elements for Thin Shells and Curved Members, edited by Gallagher, R.H. and Ashwell, D.G., John Wiley, New York, pp. 171-195.

[52] Rigby, F.N., Webster, J.J. and Henshell, R.D., 1983. Hybrid and Hellinger-Reissner plate and shell finite elements. In *Hybrid and Mixed Finite Element Methods* (Edited by Atluri, S.N., Gallagher, R.H. and Zienkiewicz, O.C.), pp. 73-92, John Wiley, New York.

[53] Lee, S.W., Dai, C.C. and Yeom, C.H., 1985. A triangular finite element for thin plates and shells. *Int. J. for Numerical Methods in Engineering*, 21, 1813-1831.

[54] Saleeb, A.F., Chang, T.Y., and Yingyuengyong, S., 1988. A mixed formulation of C^0-linear triangular plate/shell element, the role of edge shear constraints. *International Journal for Numerical Methods in Engineering*, 26, 1101-1128.

[55] Fish, J. and Belytschko, T., 1992. Stabilized rapidly convergent 18-degrees-of-freedom flat shell triangular element. *Int. J. for Num. Meth. in Eng.*, 33, 149-162.

[56] Simo, J.C. and Hughes, T.J.R., 1986. On the variational foundations of assumed strain methods. *A.S.M.E. Trans., Journal of Applied Mechanics*, 53, 51-53.

[57] Boisse, P., Daniel, J.L. and Gelin, J.C., 1994. A C^0 three-node shell element for nonlinear structural analysis. *Int. J. for Num. Meth. in Engineering*, 37, 2339-2364.

[58] Sze, K.Y. and Zhu, D., 1999. A quadratic assumed natural strain curved triangular shell element. *Comput. Meth. in Applied Mechanics and Engineering*, 174, 57-71.

[59] Kim, J.H. and Kim, Y.H., 2002. Three-node macro triangular shell element based on the assumed natural strains. *Computational Mechanics*, 29, 441-458.

[60] Lee, P.S. and Bathe, K.J., 2004. Development of MITC isotropic triangular shell finite elements. *Computers and Structures*, 82, 945-962.

[61] Lee, P.S., Noh, H.C. and Bathe, K.J., 2007. Insight into 3-node triangular shell finite elements: the effects of element isotropy and mesh patterns. *Computers and Structures*, 85, 404-418.

[62] To, C.W.S. and Liu, M.L., 1994. Hybrid strain based three-node flat triangular shell elements. *Finite Elements in Analysis and Design*, 17, 169-203.

[63] Liu, M.L. and To, C.W.S., 1995. Vibration of structures by hybrid strain based three-node flat triangular shell elements. *J. of Sound and Vibration*, 184(No.5), 801-821.

[64] To, C.W.S., 2002. Observations in incompatibility between finite elements and shell theory. *Proceedings of the International Conference on Computational Engineering & Sciences* (edited by Atluri, S.N. and Pepper, D.W.), July 31-August 2, Reno, NV, Paper Number 217, pp.1-6. (Also available in: *Advances in Computational Engineering & Sciences*, Tech Science Press, http://www.techscience.com)

[65] Bathe, K.J., 1996. *Finite Element Procedures*, Prentice-Hall, Englewood Cliffs, New Jersey.

[66] Liu, M.L. and To, C.W.S., 1995. Hybrid strain based three-node flat triangular shell elements I: Nonlinear theory and incremental formulation. *Computers and Structures*, 54(No. 6), 1031-1056.

[67] To, C.W.S. and Liu, M.L., 1995. Hybrid strain based three-node flat triangular shell elements II: Numerical investigation of nonlinear problems. *Computers and Structures*, 54(No. 6), 1057-1076.

[68] To, C.W.S. and Wang, B., 1998. Hybrid strain-based three-node flat triangular laminated composite shell elements for vibration analysis. *Journal of Sound and Vibration*, 211(No.2), 277-291.

[69] To, C.W.S. and Wang, B., 1998. Transient responses of geometrically nonlinear laminated composite shell structures. *Finite Elem. in Anal. and Design*, 31, 117-134.

[70] To, C.W.S. and Wang, B., 1999. Hybrid strain based geometrically nonlinear laminated composite triangular shell finite elements. *Finite Elements in Analysis and Design*, 33, 83-124.

[71] To, C.W.S. and Liu, M.L., 2001. Geometrically nonlinear analysis of layerwise anisotropic shell structures by hybrid strain based lower order elements. *Finite Elements in Analysis and Design*, 37, 1-34.

[72] Liu, M.L. and To, C.W.S., 2003. Free vibration analysis of laminated composite shell structures using hybrid strain based layerwise finite elements. *Finite Elements in Analysis and Design*, 40, 83-120.

[73] To, C.W.S. and Liu, M.L., 2003. Non-conservative and conservative loads in geometrically nonlinear shells. *Proc. of Design Eng. Technical Conferences and Computers and Information in Engineering Conference*, September 2-6, Chicago, Illinois, DETC2003/CIE-48214, pp.1-13.

[74] To, C.W.S. and Wang, B., 1996. Nonstationary random response of laminated composite structures by a hybrid strain-based laminated flat triangular shell finite element. *Finite Elements in Analysis and Design*, 23, 23-35.
[75] To, C.W.S. and Liu, M.L., 2000. Large nonstationary random responses of shell structures with geometrical and material nonlinearities. *Finite Elements in Analysis and Design*, 35, 59-77.
[76] To, C.W.S., 2008. Responses of stochastic shell structures to non-Gaussian random excitations, *Proc. of NCAD2008 NoiseCon2008-ASME NCAD*, July 28-30, Dearborn, Michigan, NCAD2008-73030, pp.1-8.
[77] To, C.W.S., 2002. Nonlinear random responses of shell structures with spatial uncertainties. *Proc. of Design Eng. Tech. Conferences and Computers and Information in Eng. Conference*, Sept. 29-Oct. 2, Montreal, Canada, DETC2002/CIE-34472, pp.1-11.
[78] To, C.W.S., 2006. Large nonlinear random responses of spatially non-homogeneous stochastic shell structures, *Proc. of the ASME 2006 Design Eng. Techn. Conf. and Computers and Information in Eng. Conf.*, September 10-13, Philadelphia, PA, DETC2006-99261, pp.1-8.
[79] To, C.W.S. and Liu, W., 2003. Analysis of laminated composite shell structures with piezoelectric components. *Mechanics of Electromagnetic Solids*, 228-250, Kluwer Academic Publishers, Norwell, MA.
[80] To, C.W.S. and Chen, T., 2007. Optimal control of random vibration in plate and shell structures with distributed piezoelectric components. *International Journal of Mechanical Sciences*, 49, 1389-1398.
[81] To, C.W.S. and Liu, M.L., 1994. Shell analysis from a hybrid strain based flat triangular finite element perspective. *A World of Shells: A Conference in Honour of Peter G. Glockner*, August 31 - September 2, Banff, Alberta, Canada, pp.315-335.
[82] Mazurkiewicz, Z.E. and Nagorski, R.T., 1991. *Shells of Revolution*, Elsevier Science Publishing Co. Inc., New York.
[83] Timoshenko, S. and Woinowsky-Krieger, S., 1956. *Theory of Plates and Shells*, 2nd ed., McGraw-Hill, New York.
[84] MacNeal, R.H. and Harder, R.L., 1985. A proposed standard set of problems to test finite element accuracy. *Finite Elements in Analysis and Design*, 1, 3-20.
[85] Wang, K.X., He, Y., Guo, X.H. and Qin, H., 2006. Spline thin-shell simulation of manifold surfaces *In Proceedings of Computer Graphics International (Lecture Notes in Computer Science,* Springer, 4035, 570-577).
[86] Cho, M.H., Choi, J.B. and Roh, H.Y., 2007. Shell finite element analysis with exact geometric quantities from NURBS surface. *The 48th AIAA/ASME/ASCE/AHS/ASC Structures, Structural Dynamics, and Material Conference*, Paper No. AIAA 2007-2386, 23-26 April, Honolulu, Hawaii.
[87] Tong, P., and Pian, T.H.H., 1969. A variational principle and the convergence of a finite element method based on assumed stress distribution. *International Journal for Numerical Methods in Engineering*, 5, 463-472.
[88] Pian, T.H.H., and Chen, D.P., 1983. On the suppression of zero energy deformation modes. *Int. Journal for Numerical Methods in Engineering*, 19, 1741-1752.

In: Continuum Mechanics
Editors: Andrus Koppel and Jaak Oja, pp.85-103
ISBN: 978-1-60741-585-5

Chapter 3

TRANSPORT CONTROL OF FLUID AND SOLUTES IN MICROCHANNELS USING AC FIELD AND SEMICONDUCTOR DIODES

Dimiter N. Petsev[1] and Orlin D. Velev[2]
Center for Biomedical Engineering, Department for Chemical and Nuclear Engineering, University of New Mexico, Albuquerque, NM, USA1[1]
Department of Chemical and Biomolecular Engineering, North Carolina State University, Raleigh, NC, USA

ABSTRACT

This chapter presents an overview of recent results on the electric field control and manipulation of fluids in microfluidic devices. The newer approaches are based on using alternating or a combination of alternating and direct current fields. The alternating field can be locally converted to direct by semiconductor diodes that may be placed at key locations where an electroosmotic force has to be applied to the fluid. Such techniques allow to design and fabricate small micrometer sized pumps and mixers. The latter are important because of the inherent low Reynolds characteristics of the flow in microchannels. The diode mixers are simple to fabricate and can be turned on and off depending on the operational requirements. Combining alternate and direct current fields and diode pumps makes possible the decoupling of the electroosmotic fluid flow from the electrophoretic particle or macromolecular mass flux. This can be exploited for precise analyte focusing, preconcentration and separation.

INTRODUCTION

Micro and nanofluidics have great potential as core technologies in a variety of devices and applications. Examples include chemical[1-14] and biomolecular[15-44] sensing, separation,[4,7,15,17,23,24,26,45-70] DNA molecule manipulation,[54,71-78] sample pre-concentration and focusing,[21,49,79-86] conducting small scale liquid reactions and manipulations,[87-89] and formation and handling of monodisperse droplets and particles.[90-108] The fabrication of microfluidic structures and patterns for these applications is based on the advancement of micro and nanofabrication techniques such as soft lithography,[109-114] bulk[115-120] and surface[1,74,121-127] machining, and chemical and thermal oxidation.[128] Other methods employed in the making of microfluidic devices include electron beam technology,[129] interferometric lithography[130-132] and patterning by self-assembly.[133-135] As the channel dimensions become smaller, the Reynolds numbers of the flows becomes lower and viscous effects dominate.[136-139] This fact leads to difficulties in the rapid processing of fluids and solutes. A typical example is mixing, which for micro and nanofluidics requires completely new engineering approaches.[47,140-162] Manipulating the fluid pressure and flow and analytes distribution in fluidic systems also requires the development of pumps and valves to direct and transport the liquid and dissolved species.[113,114,163-204]

The most facile way to manipulate the liquid flow and the distribution of charged molecular species is to use electric fields. The first widely used type of actuation relies on the use of direct current (DC) fields to induce electrophoresis and electroosmosis (Figure 1a). More recently, various phenomena based on alternating current (AC) fields have been investigated as means for pumping and manipulating particles in microfluidic devices. Here we focus on new techniques that combine both DC and AC fields in more sophisticated ways with the goal of achieving new functionality and levels of performance. Recently we developed a new method for fluid transport control and manipulation, which is based on AC field actuation in combination with semi-conductor diodes placed at specific locations along microfluidic channels.[142,205] The diodes locally rectify the AC field into DC field which drives the fluid near the diodes by electroosmosis. In this Chapter we present an overview of the fluid transport driven by AC powered semiconductor diodes in the context of the earlier developments in DC and AC electrokinetics. In the next section we discuss the governing equations describing the fluid transport in microchannels. Section 3 presents some examples that illustrate prospective useful applications of the proposed approach. The potential for the development of these techniques is discussed briefly in the concluding remarks.

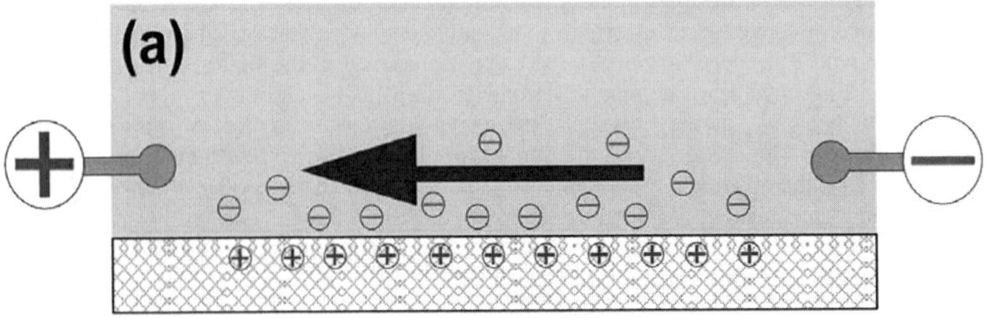

Figure 1. Schematics of the geometry and electric field actuation modes of the three types of microfluidic pumps discussed here. (a) Electroosmotic DC pumping – the ions in the double layers move towards the electrode of opposite charge dragging the liquid.

GOVERNING EQUATIONS FOR FLUID AND ANALYTE FLOW IN MICROCHANNELS

We consider here the fundamentals relations between DC and/or AC external fields and fluid flow. The fluid flux in the microchannels is given by the momentum balance equation at low Reynolds[206] flow regime[207,208]

$$\eta\nabla^2\mathbf{v} = \nabla p - \rho_e\mathbf{E} = \nabla p + \left(\varepsilon\varepsilon_0\nabla^2\Psi\right)\mathbf{E} \tag{1}$$

where $\mathbf{v}$ is the velocity field, η is the fluid viscosity, p is the pressure, ρ_e is the local charge density (due to the dissolved ionic species), $\mathbf{E}$ is the externally applied DC electric field, $\varepsilon_0 = 8.854\times10^{-12}$ F m^{-1} is the dielectric constant for vacuum, ε is the relative dielectric permittivity for the solvent (~ 80 for water at room temperature) and Ψ is the potential of the electric double layer that is formed at wall-solution interface. For most cases $\mathbf{E}$ and Ψ can be assumed to be uncoupled.[207,209] The double layer potential is related to the local charge density via the Poisson-Boltzmann equation[207,210-214]

$$\nabla^2\Psi = -\frac{\rho_e}{\varepsilon\varepsilon_0} = -\frac{e}{\varepsilon\varepsilon_0}\sum_i z_i n_i^0 \exp\left(-\frac{z_i e\Psi}{kT}\right) \tag{2}$$

where $e = 1.602\times10^{-19}$ C is the elementary charge, z_i and n_i^0 are the charge number and bulk number concentration of ionic species i. The thermal energy is included in these equations by the kT product where $k = 1.381\times10^{-23}$ J K^{-1} is the Boltzmann constant and T is the temperature. The electric double layer potential decays with the distance from the wall and its typical range is usually expressed by the so-called double layer thickness or Debye screening length $1/\kappa$ which is defined by the equation

$$\kappa^2 = \frac{e^2}{\varepsilon\varepsilon_0 kT}\sum_{i=1}^{2} z_i^2 n_i^0 \tag{3}$$

The quantity κ has the dimension of inverse length. Therefore its product with the typical characteristic length-scale is a dimensionless number that gives a qualitative measure of the relative importance of the double layer region. The double layer thickness κ^{-1} depends on the background electrolyte concentration. For example, aqueous solution of monovalent symmetric electrolyte at room temperature with concentrations ranging from 10^{-6} to 10^{-3} M results in κ^{-1} lengths varying between 300 and 10 nm respectively. These dimensions are often much smaller than the width of microchannels and wider capillaries and it is reasonable to assume that the electric double layer thickness is effectively zero (extremely thin double layer approximation).[207,208,213] Then the term that is proportional to the electric field can be omitted from Eq. (1) and the obtained result is simply

$$\eta\nabla^2\mathbf{v} = \nabla p \tag{4}$$

However, the external field still applies a force on the liquid through the infinitely thin double layer at the channel wall. This effect can be accounted for by defining a boundary condition at the wall which reads

$$\mathbf{v}\big|_{wall} = - \frac{ee_0z}{h}\mathbf{E} \tag{5}$$

The quantity ζ is historically known as the electrokinetic zeta potential. It is defined at the "shear" surface near the wall where the fluid starts slipping.[207,208,213]

In fluidic devices with narrow channels (in the nanometers range) the electrical double layer thickness might be comparable to the channel width (or radius). The infinitely thin double layer assumption may not be applicable for such small channels and Eqs. (4) and (5) do not adequately describe the fluid flow. Eq. (1) has to be used for such systems where the opposing ionic electric double layers overlap. The body force term can be calculated by using Eq. (2) and the problem becomes substantially more complex even for conventional DC field electroosmosis. This is overcome by introducing certain approximations like low ζ-potential,[215] assuming simple channel geometry (plane parallel slit) or weak double layer overlap.[216-220] Recently the potential distribution and electroosmotic flow in a cylindrical capillary with arbitrarily high ζ-potential were also calculated using the methods of matched asymptotic expansions.[221] When the electric double layer is of the order of the channel width and the surface potential is high ($e\zeta / kT > 1$) a numerical analysis is most convenient.[64]

The problem becomes even more difficult to analytically solve if nanoparticles or macromolecules with sizes that are comparable to that of the channel are also transported. The overall electrokinetic problem presents considerable mathematical difficulties. However, there is an analytical analysis of this case offered by Brenner.[222-224] It also employs the matched asymptotic expansion and the results are valid for relatively thin electric double layers around the particles and at the channel walls.

Uniform AC fields applied normally or tangentially to a charged wall do not engender unidirectional fluid flows as do the DC fields. Fluid flows, however, are generated in areas where a strong gradient of a non-uniform electric field exists across a solid-liquid interface. These flows, which have been referred to as AC electrohydrodynamics[225,226] or Induced Charge Electroosmosis (ICEO),[176,227] originate as a result of the interaction of the field with the field-induced component of the electric double layer near interfaces. The external applied voltage at these interfaces modifies the native charge on the surface thereby leading to an "induced" zeta potential added to the intrinsic zeta potential. The induced component is strongly dependent on the field frequency and electrolyte concentration. If the AC frequency is low enough, the induced double layer charge changes sign synchronously with the electric field. The counterions in the double layer move in and out of the layer during the half-cycles of the field. The induced component of the zeta potential is of the same sign as that of the applied field.[228] When the field has a component tangential to the surface a net flow towards the higher intensity area is induced. The ions in the double layer then react to tangential electric fields leading to bulk unidirectional liquid flow along the interface. Even though an AC field is applied, the bulk flow in different half cycles is always in the same direction along the field gradient.

The AC electrohydrodynamics flow velocity is given by

$$v_{AC} = - \frac{ee_0 z_{ind}}{h} E_t \tag{6}$$

where, ζ_{ind} is the induced zeta potential due to the applied external field and E_t is the tangential component of the electric field.[229,230] The induced potential is usually proportional the field that generates it in a linear manner $\zeta_{ind} \sim E_t$. Thus, the ICEO velocity is usually proportional to the squared potential, $v_{AC} \sim E_t^2$, a "signature" relation for AC electrohydrodynamic flows.[176,231]

The AC electrokinetic flows can be used in microfluidic pumping in devices with asymmetric electrodes[178,185-188,232,233] (Figure 1b) and in techniques for on-chip manipulation and collection of particles.[142,234] The AC fields do not lead to electrophoretic motion of homogeneous particles or biological macromolecules. Thus, the use of alternating fields in microfluidic pumping and mixing avoids some of the problems arising from the joint action of electroosmosis and electrophoresis in microfluidic devices manipulating biomolecules. The AC field, however, can lead to Induced Charge Electrophoresis (IECP) of Janus and other anisotropic particles[235,236] and can lead to particle attraction or repulsion to electrodes by dielectrophoresis (DEP).[237]

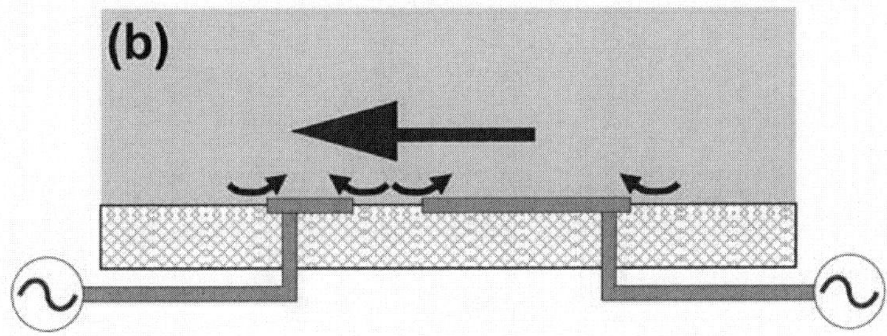

Figure 1. Schematics of the geometry and electric field actuation modes of the three types of microfluidic pumps discussed here. (b) Induced charge electroosmosis by AC field – the induced charges and the liquid near the surfaces move in the direction of the field gradient. The asymmetric configuration of the energized electrodes leads to unbalanced liquid flow.

The use of AC fields in devices operating by ICEO, ICEP and DEP is problematic for a number of reasons. The design of devices operating on field gradients is complex and in many cases requires high-end microfabrication. One factor that increases the complexity is the frequency dependence of the AC electrohydrodynamic flows occurring because the polarization effects in the electric double layer are time-dependent.[209] The upper frequency limit of ICEO pumping is ~ 5-10 kHz. This prevents the use of fields of higher frequencies, which can penetrate more materials and lead to low power dissipation. One additional problem with ICEO is that the presence of even millimolar electrolyte concentrations in combination with the low efficiency of most AC electrode pump configurations leads to liquid heating and potential chemical decomposition of the solution. Thus, a method combining some of the advantages of the DC and AC liquid handling with a simple straightforward design and fabrication can be of high value in many microfluidic devices. We present in the next section a new method based on semiconductor diode actuation bridging the DC and AC field electrokinetics techniques.

Semiconductor Diode Pumps and Mixers Powered by Alternate Current Electric Field

The major advantage of the AC fields is the ability to transmit a relatively large power flux through the chip without engendering electrokinetic flows in the channels where the field is uniform. One way to use this power in a simple way is to rectify the AC field into DC voltage and use it in local pumping without applying DC field across the whole chip. We reported how such a process can be realized by using semiconductor diodes.[142,205] Consider a microchannel where semiconductor diodes are embedded in the walls while their electrodes are in contact with the fluid (Figure 1c). If alternate current (AC) electric field is applied it is locally converted into direct current (DC) field. Hence, half of the field is available to apply electroosmotic unidirectional force on the fluid at the diode surface.[142,205] This force is periodic in time with frequency that corresponds to that of the AC field. However, for observation time scales that are longer than the inverse AC frequency we can consider the force and the velocity to be uniform. In addition, the rectified voltage is likely to be further smoothened and averaged in time by the capacitance of the double layers, especially at high frequencies. The most important effect from practical point of view is that only half of the total AC field is utilized to directionally drive the fluid into the channel. The other half is shortened by the diode and does not drive the fluid backwards.

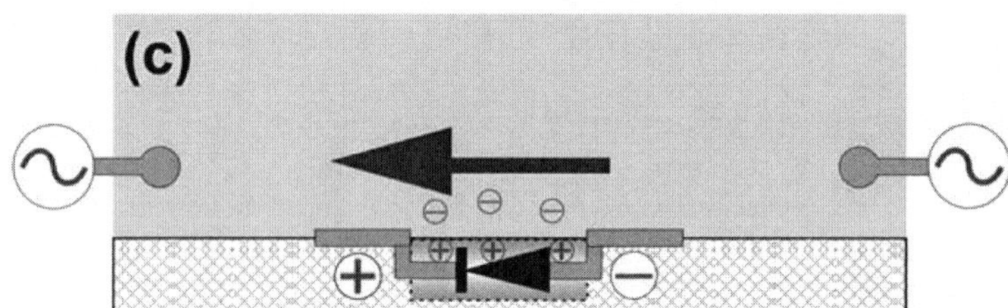

Figure 1. Schematics of the geometry and electric field actuation modes of the three types of microfluidic pumps discussed here. (c) Pumping by diodes embedded in the channel – the energy is provided by an external AC field, while the liquid moves because of DC electroosmosis localized between the diode electrodes.

The velocity of the liquid flows generated by the diodes can be estimated by relations similar to the ones derived for DC liquid actuation. The DC voltage, V_d, harvested by the diode is[142]

$$V_d = \frac{L_d}{2}(E_{ext} - E_{d0}) \tag{6}$$

where L_d is the diode length, E_{ext} is the magnitude of the external AC field and E_{d0} is the offset field that characterizes the particular pn-junction.[238] Then the electroosmotic velocity generated by the external AC field between the diode electrodes is

$$v = -\frac{ee_0 z}{2h}(E_{ext} - E_{d0}) \tag{7}$$

The factor "2" in the denominator is due to fact that only half of the total field is used; other half is "shortened" by the diode and does not lead to fluid flow.

The flow pattern in the vicinity of a diode "pump" that has been powered by an AC field in the channel is illustrated by a computer simulation in Figure 2a and an experimental visualization in Figure 2b. The pump consists of two diodes oriented in parallel at the microchannel walls. The fluid near the diode surface moves due to electroosmosis generated by the local DC field. The rest of the fluid in the channel (away from the electrodes) exerts hydrodynamic resistance, which leads to a backflow in the middle of the channel. Still, there is a net flow in one direction (right to left in the figure), which quickly obtains parabolic shape that is typical for pressure driven fluid motion.

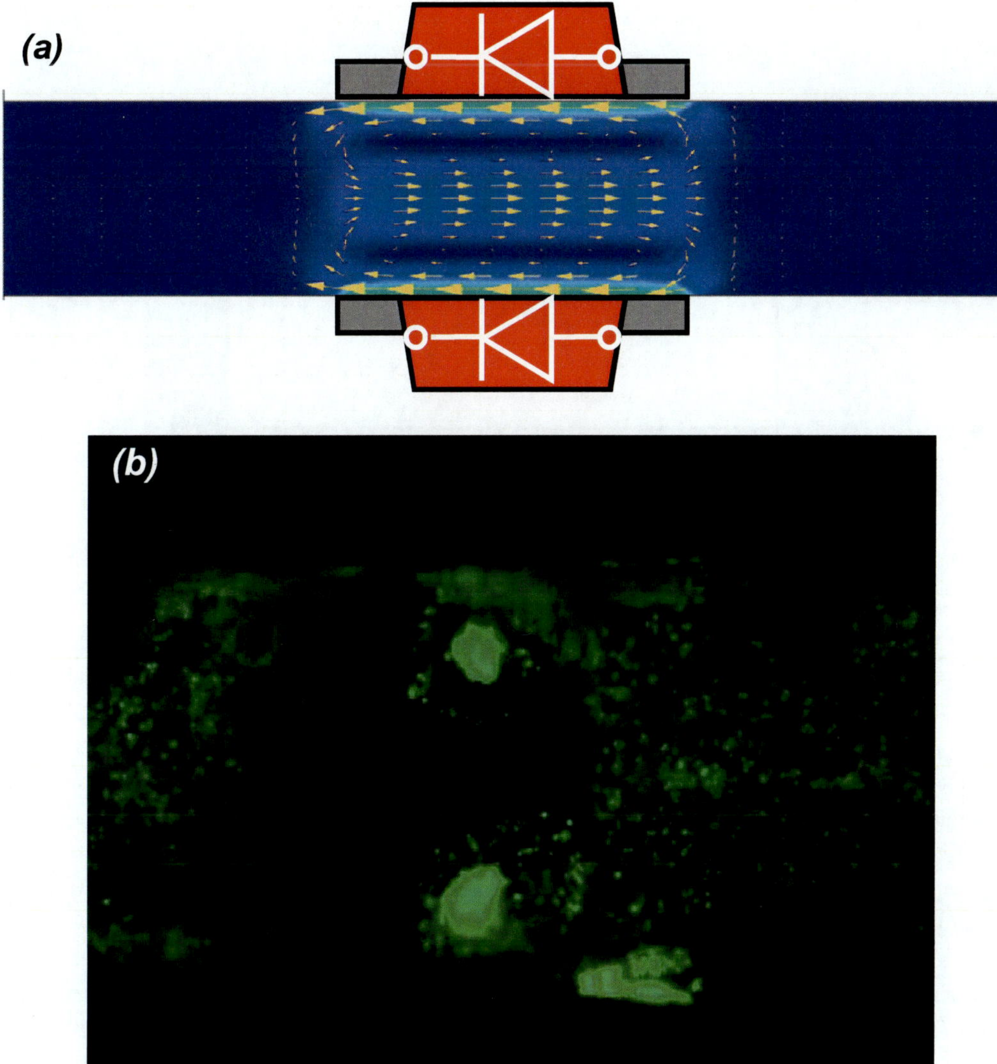

Figure 2. Schematics and flow visualization on a diode micropump. (a) A Computational Fluid Dynamics simulation of the flow field that develops in the vicinity of the AC powered diode. (b) An experimental visualization of the flow between the diodes (situated on the top and bottom but not visible in this micrograph, see Ref. [49,92,114]).

The diode pumps and a combination of AC and DC fields allows for focusing and separation of charged species. Using a loop-shaped channel (Figure 3) allows decoupling of the fluid flow from the particle motion. The applied AC field is rectified by the collinear diodes, which drive the fluid into a clock-wise circulation. The DC field does not affect the fluid flow because of the loop shape of the channel is symmetric and global DC forces counterbalance each other. It, however, gives rise to a left-to-right electrophoretic motion of the particles. In the upper part of the loop both the electrophoresis and the convective particle transport are in the same direction. In the lower part of the loop the fluid flow and the particle electrophoresis are in opposite directions and particle with lower mobility might be swept right-to-left by the fluid flow. This concept has been proven by efficiently separating two types of particles of slightly different charge.[205]

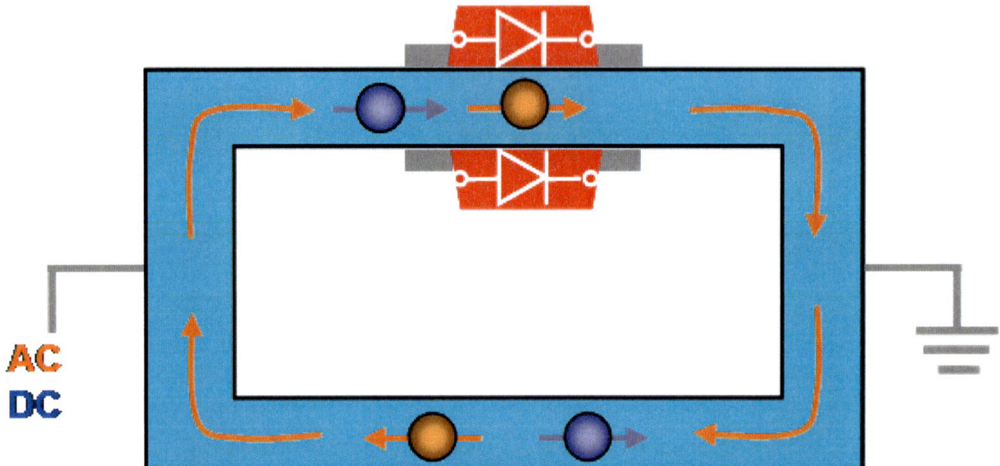

Figure 3. A sketch of the a device for particle focusing and preconcentration (see Ref. 142). This device allows decoupling of the fluid flow and electrophoretic particle motion as discussed the text.

If the diodes have an anti-parallel orientation they will generate a vortex instead of a directional flow and thus will act as an efficient microfluidic mixer. This is illustrated in Figure 4. The first one presents a computation and the second is again an experimental flow visualization of the pattern that develops at this configuration. We investigated the mixing efficiency of a diode device operating on a two concurrent laminar fluxes passing through a small chamber with oppositely facing diodes.[205] The mixing in the chamber occurs because of the vortex that is created between the oppositely aligned diodes.

The efficiency of the diode mixer was tested in Y-shaped microchannel configuration.[205] We showed that two completely separated streams (pure water and dye solution) become almost uniformly distributed 2.5 mm after the diode mixer. The fluorescence intensity averaged across the channel width obtained from the confocal micrographs confirmed that the vortex flow induced by the diodes dramatically enhanced the mixing of the two adjacent laminar streams. The degree of mixing was evaluated with a mixing index calculated as the standard deviation of the fluorescence intensity in the experimental confocal images[151]

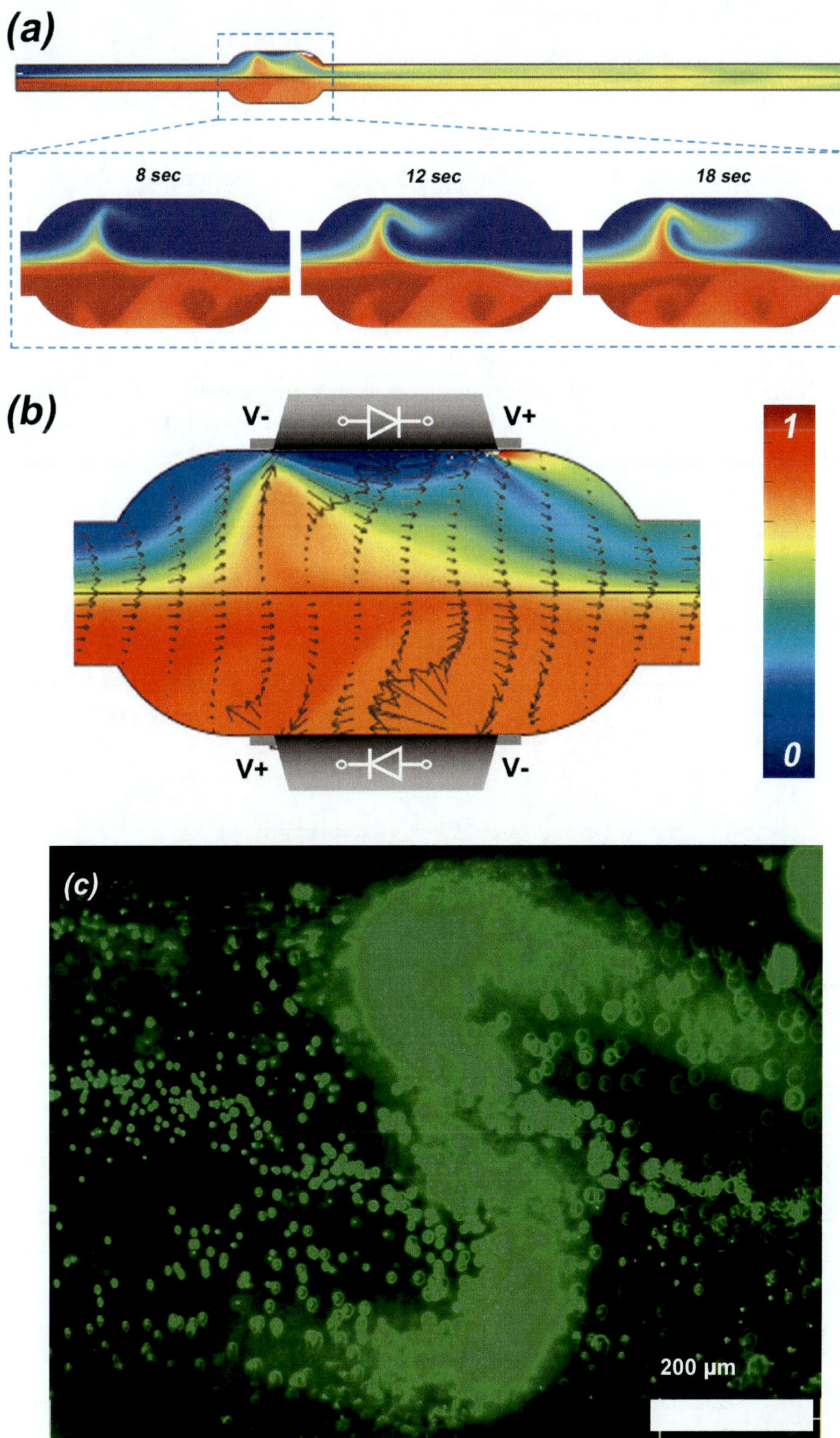

Figure 4. AC Diode mixer. (a) and (b) A Computational Fluid Dynamics simulation of the flow fields in a diode mixer. (c) Experimental flow visualization (see Ref. 142).

$$\text{Mixing Index} = \sqrt{\frac{1}{N}\sum_{k}\left(\frac{I_k - I_0}{I_0}\right)^2} \quad (8)$$

where N is the total number of pixels of the scanned image, I_k is the intensity of pixel "k" and I_0 is the average intensity over all pixels. Our experiments[205] showed that the resulting mixing index was 80% for Texas Red solution (see Figure 5). The same index would require channel length Δy_{mix} ; 68 cm if the only driving force for mixing is molecular diffusion.[205] The diode mixer is very efficient for laminar flows with low Peclet numbers ($Pe \leq 2.2\times10^3$). In contrast most passive microfluidic mixers work well only for high Peclet numbers ($Pe > 10^4$).[143,145,156] Besides the high mixing efficiency that this device could achieve it could be turned on and off remotely via the AC field and multiple mixers situated in different channels can be actuated simultaneously.

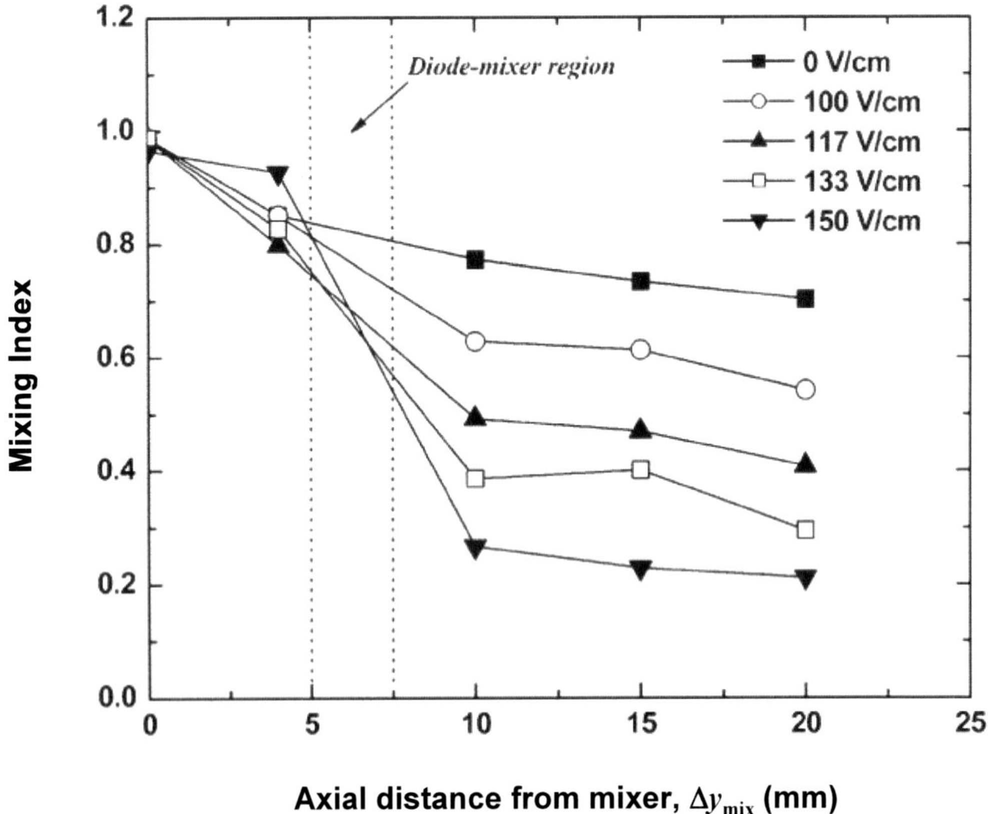

Figure 5. Mixing index at different AC fields calculated as a standard deviation of fluorescence intensity in cross-sectional confocal imaging as a function of the distance from the Y-shaped channel junction (see Ref. 205). The frequency of the AC field is 1 kHz.

CONCLUSIONS

We discuss here the origins of the DC and AC electroosmotic effects that can be used for pumping and mixing in microfluidic devices. The major effect routinely used in the present devices in DC electroosmosis, often combined with DC electrophoresis for processes of biomolecular separation. The use of AC fields in analogical applications is presently intensively investigated, but while the ICEO and ICEP effects are rather interesting from fundamental perspective, their application in engineered devices is not straightforward. We describe here a new technique combining the DC and AC effects in microfluidic devices based on semiconductor diode actuators. That technique benefits from the advantages of both other classes of electric field effects, as the power is transported through the chip by AC fields, while the actuation is carried out by DC electroosmosis between the electrodes of the diodes, which generates direct current after rectifying the AC field propagating through the chip. In the absence of the nonlinear electric element of the diodes the AC field can not lead to directional fluid motion, so it can be applied over the whole microfluidic network and be used to provide power to multiple diode pumps and mixers. This allows convenient pumping and mixing by locating diodes in all positions of the channels where such functions are needed. One of the potentially interesting future developments in this area might be the creation of a new generation of "microfluidic-electronic" chips. The microfluidic channels in such devices might directly interface the surfaces of silicon chips that would not only host the diodes and other nonlinear elements used to drive and mix the liquid, but might also include sensors and be connected to underlying electronic circuits that might provide feedback, control and logic functions. The multiditude of complex engineered structures that might provide new microfluidic functionality has only begun to be investigated.

Acknowledgments
This work was supported by the National Science Foundation grants NSF-CBET-CAREER 0844645 and NSF-CBET 0828900.

REFERENCES

[1] M. B. Stern, M.W. Geis, and J. E. Curtin, J. Vac. *Sci. Technol.* B 15, 2887 (1997).
[2] M. D. Barnes, W. B. Whitten, and J. M. Ramsey, *Anal. Chem.* 67, 418A (1995).
[3] M. D. Barnes, N. Lermer, C-Y. Kung, W. B. Whitten, J. M. Ramsey, and S. C. Hill, *Opt.* Lett. 22, 1265 (1997).
[4] J. C. Fister, S. C. Jacobson, L. M. Davis, and J. M. Ramsey, *Anal. Chem.* 70, 431 (1998).
[5] S. C. Hill, M. D. Barnes, N. Lermer, W. B. Whitten, and J. M. Ramsey, *Anal. Chem.* 70, 2964 (1998).
[6] C.-Y. Kung, M. D. Barnes, N. Lermer, W. B. Whitten, and J. M. Ramsey, *Appl. Optics* 38, 1481 (1999).
[7] J. P. Kutter, R. S. Ramsey, S. C. Jacobson, and J. M. Ramsey, *J. Microcolumn Separations* 10, 313 (1998).

[8] N. Lermer, M. D. Barnes, C-Y. Kung, W. B. Whitten, and J. M. Ramsey, *Anal. Chem.* 69, 2115 (1997).

[9] J. M. Ramsey, S. C. Jacobson, and M. R. Knapp, *Nature Medicine* 1, 1093 (1995).

[10] H. Y. Wang, R. S. Foote, S. C. Jacobson, J. H. Schneibel, and J. M. Ramsey, *Sensors and Actuators* B 45, 199 (1997).

[11] R. P. Rodgers, A. C. Lazar, P. T. A. Reilly, W. B. Whitten, and J. M. Ramsey, *Anal. Chem.* 72, 5040 (2000).

[12] C. H. Weng, W. M. Yeh, K. C. Ho, and G. B. Lee, *Sensors Actuators* B 121, 576 (2007).

[13] C. J. Huang, Y. H. Chen, C. H. Wang, T. C. Chou, and G. B. Lee, *Sensors Actuators* B 122, 461 (2007).

[14] W. Satoh, H. Hosono, H. Yokomaku, K. Morimoto, S. Upadhyay, and H. Suzuki, *Sensors* 8, 1111 (2008).

[15] R. Karnik, R. Fan, M. Yue, D. Li, P. Yang, and A. Majumdar, *Nano* Lett. 5 (5), 943 (2005).

[16] P. Fortina, J. Cheng, L. J. Kricka, L. C. Waters, S. C. Jacobson, P. Wilding, and J. M. Ramsey, *Methods in Molecular Biology* 163, 211 (2001).

[17] N. Gottschlich, C. T.Culbertson, T. E. McKnight, S. C. Jacobson, and J. M. Ramsey, *J. Chromatography* B 745, 243 (2000).

[18] N. Gottschlich, S. C. Jacobson, C. T. Culbertson, and J. M. Ramsey, *Anal. Chem.* 73, 2669 (2001).

[19] A. G. Hadd, D. E. Raymond, J. W. Halliwell, S. C. Jacobson, and J. M. Ramsey, *Anal. Chem.* 69, 3407 (1997).

[20] S. C. Jacobson, A. W. Moore, and J. M. Ramsey, *Anal. Chem.* 67, 2059 (1995).

[21] S. C. Jacobson and J. M. Ramsey, *Anal. Chem.* 69, 3212 (1997).

[22] J. Khandurina, T. E. McKnight, S. C. Jacobson, L. C. Waters, R. S. Foote, and J. M. Ramsey, *Anal. Chem.* 72, 2995 (2000).

[23] Y. Liu, R. S. Foote, S. C. Jacobson, R. S. Ramsey, and J. M. Ramsey, *Anal. Chem.* 72, 4608 (2000).

[24] Y. Liu, R. S. Foote, C. T. Culbertson, S. C. Jacobson, R. S. Ramsey, and J. M. Ramsey, *J. Microcolumn Separations* 12, 407 (2000).

[25] M. A. McClain, C. T. Culbertson, S. C. Jacobson, N. L. Allbritton, C. E. Sims, and J. M. Ramsey, *Anal. Chem.* 75, 5646 (2003).

[26] J.D. Ramsey, S.C. Jacobson, C.T. Culbertson, and J.M. Ramsey, *Anal. Chem.* 75, 3758 (2003).

[27] L. C. Waters, S.C. Jacobson, N. Kroutchinina, J. Khandurina, R.S. Foote, and J.M. Ramsey, *Anal. Chem.* 70, 158 (1998).

[28] L. C. Waters, S. C. Jacobson, N. Kroutchinina, Y. Khandurina, R. S. Foote, and J. M. Ramsey, *Anal. Chem.* 70, 5172 (1998).

[29] M. J. Anderson, B. DeLaBarre, A. Raghunathan, B. O. Palsson, A. T. Brunger, and S. R. Quake, *Biochemistry* 46, 5277 (2007).

[30] C. Hansen and S. R. Quake, *Curr Opin Struct Biol.* 13, 538 (2003).

[31] C. L. Hansen, S. Classen, J. M. Berger, and S. R. Quake, *J. Am. Chem.* Soc 128, 3142 (2006).

[32] E. P. Kartalov, J. F. Zhong, A. Scherer, S. R. Quake, C. R. Taylor, and W. F. Anderson, *Biotechniques* 40, 85 (2006).

[33] J. S. Marcus, W. F. Anderson, and S. R. Quake, *Anal. Chem.* 78, 3084 (2006).
[34] J. S. Marcus, W. F. Anderson, and S. R. Quake, *Anal. Chem.* 78, 956 (2006).
[35] J. Melin and S. R. Quake, Annu. Rev. Biophys. Biomol. *Struct.* 36, 213 (2007).
[36] E. A. Ottesen, J. W. Hong, S. R. Quake, and J. R. Leadbetter, *Science* 314, 1464 (2007).
[37] B. Kuswandi, Nuriman, J. Huskens, and W. Verboom, *Anal. Chim.* Acta 601, 141 (2007).
[38] C. J. Huang, C. C. Lu, T. Y. Lin, T. C. Chou, and G. B. Lee, J. Micromech. *Microeng.* 17, 835 (2007).
[39] X. He, D. S. Dandy, and C. S. Henry, *Sensors Actuators* B 129, 811 (2007).
[40] W. Satoh, Y. Shimizu, T. Kaneto, and H. Suzuki, *Sensors Actuators* B 123, 1153 (2007).
[41] R. R. Sathuluri, S. Yamamura, and E. Tamiya, *Biosensing for the 21st Century* 109, 285 (2008).
[42] E. Galopin, M. Beaugeois, B. Pinchemel, J. C. Camart, M. B. Ouazaoui, and V. Thomy, *Biosensors & Bioelectronics* 23, 746 (2008).
[43] P. S. Waggoner and H. G. Craighead, *Lab on a Chip* 7, 1238 (2007).
[44] C. L. Bliss, J. N. McMullin, and C. J. Backhouse, *Lab on a Chip* 7, 1280 (2007).
[45] A. Garcia, L. K. Ista, D. N. Petsev, M. J. O'Brien, P. Bisong, A. A. Mammoli, S. R. J. Brueck, and G. P. Lopez, *Lab on a Chip* 5, 1271 (2005).
[46] S. Pennathur and J.G. Santiago, *Anal. Chem.* 77, 6782 (2005).
[47] S. Vankrunkelsven, D. Clicq, D. Cabooter, and W. De Malsche, J. *Chromatography* A. 1102, 96 (2006).
[48] R.B. Schoch, A. Bertsch, and P. Renaud, *Nano* Lett. 6, 543 (2006).
[49] D. N. Petsev, G. P. Lopez, C. F. Ivory, and S. S. Sibbett, *Lab on a Chip* 5, 587 (2005).
[50] J. Fu, P. Mao, and J. Han, *Appl. Phys.* Lett. 87, 263902 (2005).
[51] H.-T. Chang, S. F. Zakharov, and A. Chrambach, *Electrophoresis* 17, 776 (1996).
[52] D. Clicq, N. Vervoort, R. Vounckx, H. Ottevaere, J. Buijs, C. Gooijer, F. Ariese, G. V. Baron, and G. Desmet, *J. Chromatography* A 979, 33 (2002).
[53] C. T. Culbertson, S. C. Jacobson, and J. M. Ramsey, *Anal. Chem.* 72, 5814 (2000).
[54] J. Han and H. G. Craighead, *Science* 288, 1026 (2000).
[55] L. K. Ista, G. P. Lopez, C. F. Ivory, M. J. Ortiz, T. A. Schifani, C. D. Schwappach, and S. S. Sibbett, *Lab on a Chip* 3, 266 (2003).
[56] B. J. Kirby and E.F. Hasselbrink, *Electrophoresis* 25, 187 (2004).
[57] T.C. Kuo and et al., *Anal. Chem.* 75, 1861 (2003).
[58] R. Levenstein, D. Hasson, and R. Semiat, *J. Membr. Sci.* 116, 77 (1996).
[59] Y. M. M. Van Lishout and D. T. Leighton, *AIChE J.* 42, 940 (1996).
[60] A. W. Moore, S. C. Jacobson, and J. M. Ramsey, *Anal. Chem.* 67, 4184 (1995).
[61] R. D. Rocklin, R. S. Ramsey, and J. M. Ramsey, *Anal. Chem.* 72, 5244 (2000).
[62] D. Ross, C. F. Ivory, L. E. Locascio, and K. E. Van Cott, *Electrophoresis* 25, 3694 (2004).
[63] H. Wang and R. H. Davis, *J. Colloid Interface Sci.* 181, 93 (1996).
[64] Z. Yuan, A. L. Garcia, G. P. Lopez, and D. N. Petsev, *Electrophoresis* 28, 595 (2007).
[65] R. Sinville and S. A. Soper, *J. Sep.* Sci. Technol. 30, 1714 (2007).
[66] Y. Zeng and D. J. Harrison, *Anal. Chem.*, 2289 (2007).
[67] A. Jain and J. D. Posner, *Anal. Chem.* 80, 1641 (2008).

[68] C. C. Striemer, T. R. Gaborski, J. L. McGrath, and P. M. Fauchet, *Nature* 445, 749 (2007).

[69] A. Mohan and P. S. Doyle, *Macromolecules* 40, 8794 (2007).

[70] G Rozing, *J. Sep.* Sci. 30, 1375 (2007).

[71] M.A. Burns, B. N. Johnson, S. N. Brahmasandra, K. Handique, J. R. Webster, M. Krishnan, T. S. Sammarco, P. M. Man, D. Jones, D. Heldsinger, C. H. Mastrangelo, and D. T. *Burke, Science* 282, 484 (1998).

[72] J. Han, S. W. Turner, and H. G. Craighead, *Phys. Rev.* Lett. 83, 1688 (1999).

[73] H. Cao, Z. Yu, J. Wang, J.O. Tegenfeldt, R.H. Austin, E. Chen, W. Wu, and S.Y. Chou, *Appl. Phys*. Lett. 81, 3058 (2002).

[74] L. J. Guo, C. X., and C. Chou, *Nano* Lett. 4, 69 (2004).

[75] S. C. Jacobson and J. M. Ramsey, *Anal. Chem*. 68, 720 (1996).

[76] M. A. Burns, C. H. Mastrangelo, T. S. Sammorco, F. P. Man, J. R. Webster, B. N. Johnson, B. Foerster, D. Jones, Y. Fields, A. R. Kaiser, and D. T. Burke, Proc. *Natl Acad.* Sci. (USA) 93, 5556 (1996).

[77] E. Pavlovic, R. Y. Lai, T. T. Wu, B. S. Ferguson, R. Sun, K. W. Plaxco, and H. T. *Soh, Langmuir* 24, 1102 (2008).

[78] C. S. J. Hou, M. Godin, K. Payer, R. Chakrabarti, and S. R. Manalis, *Lab on a Chip,* 347 (2007).

[79] S.M. Kim, M.A. Burns, and E.F. Hasselbrink, *Anal. Chem.* 78, 4779 (2006).

[80] J. Khandurina, S. C. Jacobson, L. C. Waters, R. S. Foote, and J. M. Ramsey, *Anal. Chem*. 71, 1815 (1999).

[81] H. Cui, K. Horiuchi, P. Dutta, and C. F. Ivory, *Anal. Chem.* 77, 7878 (2005).

[82] H. Cui, K. Horiuchi, P. Dutta, and C. F. Ivory, *Anal. Chem*., 1303 (2005).

[83] J. G. Shackman and D. Ross, *Anal. Chem.* 79, 6641 (2007).

[84] D. Kohlheyer, J. C. T. Eljkel, S. Schlautmann, A. van den Berg, and R. B. M. Schasfoort, *Anal. Chem.* 79, 8190 (2007).

[85] H. C. Cui, P. Dutta, and C. F. Ivory, *Electrophoresis* 28, 1138 (2007).

[86] D. Di Carlo, J. F. Edd, D. Irimia, R. G. Tompkins, and M. Toner, Anal. *Chem*. 80, 2204 (2008).

[87] T. McCreedy, Anal. *Chim.* Acta 427, 39 (2001).

[88] J. W. Hong and S. R. Quake, *Nature Biotechnology* 21, 1079 (2003).

[89] T. M. Squires and S. R. Quake, *Rev. Mod. Phys.* 7, 977 (2005).

[90] E. Villermaux, Ann. Rev. *Fluid Mech.* 39, 419 (2007).

[91] K. Ahn, C. Kerbage, T. P. Hunt, R. M. Westervelt, D. R. Link, and D. A. Weitz, *Appl. Phys.* Lett. 88, 024104 (2006).

[92] D.R. Link, E. Grasland-Mongrain, A. Duri, F. Sarrazin, Z. Cheng, G. Cristobal, M. Marquez, and D.A. Weitz, Angew. *Chem. Int.* Ed. 45, 2556 (2006).

[93] A. S. Utada, E. Lorencau, D. R. Link, P. D. Kaplan, H. A. Stone, and D. A. Weitz, *Science* 308, 537 (2005).

[94] P. Garstecki, I. Gitlin, W. DiLuzio, G.M. Whitesides, E. Kumacheva, and H. A. Stone, *Appl. Phys*. Lett. 45, 2649 (2004).

[95] P.C. Lewis, R. Graham, S. Xu, Z. Nie, M. Seo, and E. Kumacheva, *Macromolecules* 38, 4536 (2005).

[96] Z. Nie, S. Xu, M. Seo, P.C. Lewis, and E. Kumacheva, *J. Amer. Chem. Soc.* 127, 8058 (2005).

[97] M. Seo, S. Xu, Z. Nie, P.C. Lewis, R. Graham, M. Mok, and E. Kumacheva, *Langmuir* 21, 4773 (2005).

[98] H. Zhang, E. Tumarkin, R. Peerani, Z. Nie, R.M.A. Sullan, G.C. Walker, and E. Kumacheva, *J. Amer. Chem. Soc.* 128, 12205 (2006).

[99] H. Zhang, E. Tumarkin, R.M.A. Sullan, G.C. Walker, and E. Kumacheva, Macromol. *Rapid Comm.* 28, 527 (2007).

[100] J. R. Millman, K. H. Bhatt, B. G. Prevo, and O. D. Velev, *Nature Mate*r. 4, 98 (2005).

[101] O. Cayre, V. N. Paunov, and O. D. Velev, *Chem. Commun.* 18, 2296 (2003).

[102] S. Schaht, Q. Huo, I. G. Voigt-Martin, G. D. Stucky, and F. Schuth, *Science* 273, 768 (1996).

[103] J.-H. Park, C. Oh, S.-I. Shin, S.-K. Moon, and S.-G Oh, J. Colloid *Interface Sci.* 266, 107 (2003).

[104] G. Fornasieri, S. Badaire, R. Backov, O. Mondain-Monval, C. Zakri, and P. Poulin, *Adv. Mater.* 16, 1094 (2004).

[105] P. Xu, H. Wang, R. Tong, Q. Du, and W. Zhang, *Colloid Polym. Sci.* 284, 755 (2006).

[106] N. Andersson, B. Kronberg, R. Corkey, and P. Alberius, Langmuir 23, 1459 (2007).

[107] T. Jesionowski, J. Dispers. *Sci. Technol.* 22, 363 (2001).

[108] J. Collins and A. P. Lee, *Microfluidics and Nanofluidics* 3, 19 (2007).

[109] D. C. Duffy, J. C. McDonald, O. J. A. Schueller, and G. M. Whitesides, *Anal. Chem.* 70, 4974 (1998).

[110] J.C. McDonald, D.C. Duffy, J.R. Anderson, D.T. Chiu, H. Wu, O.J.A. Schueller, and G.M. Whitesides, *Electrophoresis* 21, 27 (2000).

[111] J. C. Love, J. R. Anderson, and G. M. Whitesides, *MRS Bulletin* 26, 523 (2001).

[112] S. R. Quake and A. Scherer, *Science* 290, 1536 (2000).

[113] M. A. Unger, H. P. Chou, T. Thorsen, A. Scherer, and S. R. *Quake, Science* 288, 113 (2000).

[114] M. L. Adams, M. L. Johnston, A. Scherer, and S. R. Quake, *J. Micromech. Microeng.* 15, 1517 (2005).

[115] H. Becker, L. Lowack, and A. Manz, *J. Micromech. Microeng*. 8, 24 (1998).

[116] K. Matsumoto, K. T., M. Nakao, Y. Hatamura, T. Kitamori, and T. Sawada, presented at the Proceddings of the IEEE micro electro mechanical systems, 1998 (unpublished).

[117] J. O. Tegenfeldt, O. Bakajin, and C.F. Chou, *Phys. Rev*. Lett. 86, 1378 (2001).

[118] M. Foquet, J. Korlach, and W. Zipfel, *Analytical Chemistry* 74, 1415 (2002).

[119] S. O. Kim, H. H. Solak, P. Stoykovich, N.J. Ferrier, J.J. de Pablo, and P.F. *Nealy, Nature* 424, 411 (2003).

[120] P.M. Sinha, S. Sharma, X. Liu, and M. Ferrari, *Nanotechnology* 15, S585 (2004).

[121] P. T. N. Mela, A. van den Berg, and J.E. ten Elshof, Vol., in *Encyclopedia of Nanoscience and Nanotechnology*, edited by H. S. Nalwa (2004), Vol. 6.

[122] C. K. Harnett, G.W. Coates, and H.G. Craighead, *J. Vac. Sci. Technol.* B 19, 2842 (2001).

[123] W. Li, J. O. Tegenfeldt, L. Chen, R.H. Austin, S.Y. Chou, P.A. Kohl, J. Krotine, and J.C. Sturm, *Nanotechnology* 14, 578 (2003).

[124] S. W. P. Turner, A.M. Perez, A. Lopez, and H.G. Craighead, *J. Vac. Sci. Technol.* B 16, 3835 (1998).

[125] N. R. Tas, J.W. Berenschot, P. Mela, H.V. Jansen, M. Elwenspoek, and A. van den Berg, *Nano* Lett. 2, 1031 (2002).

[126] J. C. T. Eijkel, J. Bomer, N.R. Tas, and A. van den Berg, *Lab on a Chip* 4, 161 (2004).
[127] [127]R. Jindal, J. L. Plawsky, and S. M. Cramer, *Langmuir* 21, 4458 (2005).
[128] C. Lee, E. H. Yang, N.V. Myung, and T. George, *Nano* Lett. 3, 1339 (2003).
[129] G. M. Whitesides and J.C. Love, *Sci. Am.* 285, 38 (2001).
[130] S. H. Zaidi, S.R.J. Brueck, F. M. Schellenberg, R. S. Mackay, K. Uekert, and J. J. *Persoff, presented at the SPIE*, San Jose USA, 1999 (unpublished).
[131] S. R. J. Brueck, in *International Commission on Optics*, edited by A. Guenther (SPIE Press, 2002).
[132] M.J. O'Brien, P. Bisong, L.K. Ista, E.M. Rabinovich, A.L Garcia, S.S. Sibbett, G.P. Lopez, and S.R.J. Brueck, *J. Vac. Sci. Technol.* B 21, 2941 (2003).
[133] M. Akeson, D. Branton, J.J. Kasianowicz, E. Brandin, and D.W. Deamer, *Biophys. J.* 77, 3227 (1999).
[134] E. L. Chandler, A.L. Smith, J.J. Kasianowicz, and D.L. Burden, *Langmuir* 20, 898 (2004).
[135] M. Seo, Z. Nie, S. Xu, P.C. Lewis, and E. Kumacheva, *Langmuir* 21, 4773 (2005).
[136] H. A. Stone and S. Kim, *AIChE J.* 47, 1250 (2001).
[137] H. A. Stone, A. D. Stroock, and A. Ajdari, Ann. Rev. *Fluid Mech.* 36, 381 (2004).
[138] A. Groisman and S. R. Quake, *Phys. Rev.* Lett. 92 094501, 094501 (2004).
[139] D. G. Yan, C. Yang, N. T. Nguyen, and X. Y. Huang, *Phys. Fluids 1*9, 017114 (2007).
[140] C. Tsouris, C. T. Culbertson, D. W. DePaoli, S. C. Jacobson, V. F. De Almeida, and J. M. Ramsey, *AIChE J.* 49, 2181 (2003).
[141] S. C. Jacobson, T. E. McKnight, and J. M. Ramsey, *Anal. Chem.* 71, 4455 (1999).
[142] S. T Chang, V. N. Paunov, D. N. Petsev, and O. V. Velev, *Nature Materials* 6, 235 (2007).
[143] A. D. Stroock, S. K. W. Dertinger, A. Ajdari, I. Mezi, H. A. *Stone, and G. M. Whitesides, Science* 295, 647 (2002).
[144] A. D. Stroock, S. K. W. Dertinger, G. M. Whitesides, and A. Ajdari, *Anal. Chem.* 74, 5306 (2002).
[145] A. P. Sudarsan and V. M. Ugaz, Proc. *Natl Acad. Sci.* (USA) 103, 7228 (2006).
[146] H. Andersson, W. Wijingaart, P. Nilsson, P. Enoksson, and G. Stemme, *Sensors and Actuators* B 72, 259 (2001).
[147] N. Schwesinger, T. Frank, and H. Wurmus, *J. Micromech. Microeng.* 6, 92 (1996).
[148] C-C. Hong, J-W. Choi, and C. H. Ahn, *Lab on a Chip* 4, 109 (2004).
[149] R. H. Liu, M. A. Stremler, K. V. Sharp, M. G. Olsen, J. G. Santiago, R. J. Adrian, H. Aref, and D. J. Beebe, *J. Microelectromech. Syst.* 9, 190 (2000).
[150] D-S. Kim, S-H. Lee, T-H. Kwon, and C. H. Ahn, *Lab on a Chip* 5, 739 (2005).
[151] K. S. Ryu, K. Shaikh, E. Goluch, Z. Fan, and C. Liu, *Lab on a Chip* 4, 608 (2004).
[152] A. P. Sudarsan and V. M. Ugaz, *Lab on a Chip* 6, 74 (2006).
[153] J. C. Rife, M. I. Bell, J. S. Horwitz, M. N. Kabler, R. C. Y. Auyeung, and W. J. Kim, *Sensors and Actuators* A 86, 135 (2000).
[154] P. Paik, V. K. Pamula, and R. B. Fair, *Lab on a Chip* 3, 253 (2003).
[155] F. Mugele, J. C. Baret, and D. Steinhauser, *Appl. Phys.* Lett. 88, 204106 (2006).
[156] N. Sasaki, T. Kitamori, and H-B. Kim, *Lab on a Chip* 6, 550 (2006).
[157] M. H. Oddy, J. G. Santiago, and J. C. Mikkelsen, *Anal. Chem.* 73, 5822 (2001).
[158] A. O. E. Moctar, N. Aubry, and J. Batton, *Lab on a Chip* 3, 273 (2003).

[159] M. H. Wu, J. B. Wang, T. Taha, Z. F. Cui, J. P. G. Urban, and Z. Cui, *Biomedical Microdevices* 9, 167 (2007).

[160] Y. B. Ma, C. P. Sun, M. Fields, Y. Li, D. A. Haake, B. M. Churchill, and C. M. Ho, *J. Micromech. Microeng.* 18, 045015 (2008).

[161] J. R. Pacheco, K. P. Chen, A. Pacheco-Vega, B. S. Chen, and M. A. Hayes, *Phys.* Lett. A 372, 1001 (2008).

[162] E. Villermaux, A. D. Stroock, and H. A. Stone, *Phys. Rev.* E 77, 015301 (2008).

[163] V. Studer, G. Hang, A. Pandolfi, M. Ortiz, W. F. Anderson, and S. R. Quake, *J. Appl. Phys*. 95, 393 (2005).

[164] P. Apel, Y.E. Korchev, Z. Siwy, R. Spohr, and M. Yoshida, *Nucl. Instr. Meth.* B 184, 337 (2001).

[165] H. Daiguji, Y. Oka, and K. Shirono, *Nano* Lett. 5, 2274 (2005).

[166] T.-C. Kuo, Jr. D.M. Cannon, W. Feng, M.A. Shannon, J.V. Sweedler, and P.W. Bohn, in *Micro Total Analysis Systems 2001*, edited by J.M. Ramsey and A. van den Berg (Kluwer, Dordrecht, 2001), pp. 60.

[167] Z. Siwy and A. Fulinski, *Phys. Rev.* Lett. 89, 198103 (2002).

[168] Z. Siwy, D. Dobrev, R. Neumann, C. Trautmann, and K. Voss, *Appl. Phys*. A 76, 781 (2003).

[169] Z. Siwy, P. Apel, D. Baur, D.D. Dobrev, Y.E. Korchev, R. Neumann, R. Spohr, C. Trautmann, and K. Voss, *Surface Science* 532-535, 1061 (2003).

[170] Z. Siwy, E. Heins, C.C. Harrell, P. Kohli, and C.R Martin, *J. Amer. Chem. Soc.* 126, 10850 (2004).

[171] Z. Siwy and A. Fulinski, *Am. J. Phys*. 72, 567 (2004).

[172] Z. Siwy, *Adv. Funct*. Mater. 16, 735 (2006).

[173] J. Cervera, B. Schiedt, and P. Ramirez, *Europhys.* Lett. 71, 35 (2005).

[174] D. Woermann, Phys. Chem. Chem. *Phys.* 5, 1853 (2003).

[175] D. Woermann, Nucl. Instr. Methods *Phys. Res*. B 194, 458 (2002).

[176] M. Z. Bazant and T. M. Squires, *Phys. Rev.* Lett. 92, 066101 (2004).

[177] M. Z. Bazant and Y. Ben, *Lab on a Chip* 6, 1455–1461 (2006).

[178] S. Debesset, C. J. Hayden, C. Dalton, J. C. T. Eijkel, and A. Manz, *Lab on a Chip* 4, 396–400 (2004).

[179] L.X. Chen, J.P. Ma, and Y. F. Guan, J. *Chromatogr.* A 1028, 219–226 (2004).

[180] V. Pretorius, B. J. Hopkins, and J.D. Schieke, J. *Chromatography* 99, 23–30 (1974).

[181] S. Zeng, C. H. Chen, J. C. Mikkelsen, and J. G. Santiago, *Sensors and Actuators* B 79, 107–114 (2001).

[182] L.X. Chen, J.P. Ma, F. Tan, and Y.F. Guan, *Sensors and Actuators* B 88, 260–265 (2003).

[183] L.X. Chen, J.P. Ma, and Y.F. Guan, *Microchem. J.* 75, 15 (2003).

[184] M. E Piyasena, G. P. Lopez, and D. N. Petsev, *Sensors and Actuators* B 113, 461 (2006).

[185] A. B. D. Brown, C. G. Smith, and A. R. Rennie, *Phys. Rev.* E 63, 016305 (2000).

[186] V. Studer, A. Pépin, Y. Chen, and A. Ajdari, *Microelectronic Engineering* 61/62, 915 (2002).

[187] M. Mpholo, C. G. Smith, and A. B. D. Brown, *Sensors and Actuators* B 92, 262 (2003).

[188] V. Studer, A. Pepin, Y. Chen, and A. Ajdari, *Analyst* 129, 944 (2004).

[189] A. D. Stroock, D. T. Weck, D. T. Chiu, P. J. Huck, P. J. A. Kenis, R. F. Ismagilov, and G. M. Whitesides, *Phys. Rev.* Lett. 84, 3314 (2000).
[190] N. Green, A. Ramos, A. Gonzalez, H. Morgan, and A. Castellanos, *Phys. Rev.* E 61, 4011 (2000).
[191] C. Meinhart, D. Wang, and K. Turner, *Biomedical Microdevices* 5 (2), 141 (2003).
[192] G. M. Walker and D. J. Beebe, *Lab on a Chip* 2, 131 (2002).
[193] D. E. Kataoka and S. M. Troian, *Nature* 402, 794 (1999).
[194] A. Terray, J. Oakey, and D. W. M. Marr, *Science* 296, 1841 (2002).
[195] J. Leach, H. Mushfique, R. Leonardo, M. Padgett, and J. Cooper, *Lab on a Chip* 6, 735 (2006).
[196] T. R. Kline, W. F. Paxton, Y. Wang, D. Velegol, T. E. Mallouk, and A. Sen, *J. Am. Chem.* Soc 127, 17150 (2005).
[197] W. F. Paxton, P. T. Baker, T. R. Kline, Y. Wang, T. E. Mallouk, and A. Sen, *J. Am. Chem.* Soc 128, 14881 (2006).
[198] J. Loverich, I. Kanno, and H. Kotera, *Microfluidics and Nanofluidics* 3, 427 (2007).
[199] S. K. R. S. Sankaranarayanan and V. R. Bhethanabotla, *J. Appl. Phys*. 103, 064518 (2007).
[200] I. Etchart, H. Chen, P. Dryden, J. Jundt, C. Harrison, K. Hsu, F. Marty, and B. Mercier, *Sensors and Actuators* A 141, 266 (2008).
[201] B. Bourlon, J. Wong, C. Miko, L. Forro, and M. Bockrath, *Nature Nanotechnology* 2, 104 (2007).
[202] E. Destandau, J. P. Lefevre, A. C. F. Eddine, S. Desportes, M. C. Jullien, R. Hierle, I. Leray, B. Valeur, and J. A. Delaire, *Anal. Bioanal. Chem.* 387, 2627 (2007).
[203] D. B. Weibel, A. C. Siegel, A. P. Lee, A. H. George, and G. M. Whitesides, *Lab on a Chip* 12, 1832 (2007).
[204] H. C. Cui, Z. Huang, P. Dutta, and C. F. *Ivory, Anal. Chem*. 79, 1456 (2007).
[205] S. T Chang, E. Beaumont, D. N. Petsev, and O. D. Velev, *Lab on a Chip*, 117 (2008).
[206] J Happel and H. Brenner, *Low Reynolds Number Hydrodynamics*. (Kluwer, Boston, 1983).
[207] S. S. Dukhin and B. V. Derjaguin, in *Surface and Colloid Science*, edited by E. Matijevic (Wiley Interscience, New York, 1974), Vol. 7, pp. 49.
[208] R. J. Hunter, *Zeta Potential in Colloid Science*. (Academic Press, New York, 1981).
[209] B. V. Derjaguin and S. S. Dukhin, in *Surface and Colloid Science*, edited by E. Matijevic (Wiley Interscience, New York, 1974), Vol. 7, pp. 273.
[210] S. S. Dukhin, in *Surface and Colloid Science*, edited by E. Matijevic (Wiley Interscience, New York, 1974), Vol. 7, pp. 1.
[211] G. Gouy, J. *Physique 9,* 457 (1910).
[212] G. Gouy, Ann. *Phys 7*, 129 (1917).
[213] W. B. Russel, D. A. Saville, and W. R. Schowalter, *Colloidal Dispersions*. (Cambridge University Press, Cambridge, 1989).
[214] E. J. W. Verwey and J. Th. G. Overbeek, *Theory and Stability of Lyophobic Colloids*. (Elsevier, Amsterdam, 1948).
[215] C.L. Rice and R. Whitehead, *J. Phys. Chem.* 69, 4017 (1965).
[216] D. Hildreth, *J. Phys. Chem.* 74, 2006 (1970).
[217] D. Burgreen and F.R. Nakache, *J. Phys. Chem.* 68, 1084 (1964).
[218] S. Levine, J. R. Marriot, and K. Robinson, *J. Chem. Soc. Faraday Trans II* 71, 1 (1975).

[219] S. Levine, J. R. Marriot, G Neale, and N Epstein, *J. Colloid Interface Sci.* 52, 136 (1975).

[220] D. N. Petsev, *J. Chem. Phys.* 123, 244907 (2005).

[221] D. N. Petsev and G. P. Lopez, J. *Colloid Interface Sci.* 294, 492 (2006).

[222] E. Yariv and H. Brenner, *Phys. Fluids* 14, 3354 (2002).

[223] E. Yariv and H. Brenner, *J. Fluid Mech.* 484, 85 (2003).

[224] E. Yariv and H. Brenner, SIAM *J. Appl. Math.* 64, 423 (2003).

[225] M. Trau, D. A. Saville, and I. A. Aksay, *Science* 272, 706 (1996).

[226] W. D. Ristenpart, I. A. Aksay, and D. A. Saville, *J. Fluid Mech.* 575, 83 (2007).

[227] T. M. Squires and M. Z. Bazant, *J. Fluid Mech.* 509, 217 (2004).

[228] A. Ramos, H. Morgan, N. G. Green, and A. Castellanos, *J. Phys.* D: Appl. Phys. 31, 2338 (1998).

[229] J. Lyklema, *Fundmentals of Interface and Colloid Science 2: Solid-Liquid Interfaces*. (Academic Press Inc., San Diego, 1995).

[230] H. Morgan and N. G. Green, *AC Electrokinetics: colloids and nanoparticles*. (Research Studies Press, Baldock, 2001).

[231] S. S. Dukhin and N. A. Mishchuk, Kolloidn. *Zh. 52*, 452–456 (1990).

[232] M. S. Kilic, M. Z. Bazant, and A. Ajdari, *Phys. Rev*. E 75, 021503 (2007).

[233] M. S. Kilic, M. Z. Bazant, and A. Ajdari, *Phys. Rev*. E 75, 021502 (2007).

[234] K. H. Bhatt, S Grego, and O. D. Velev, *Langmuir 21*, 6603 (2005).

[235] N. I. Gamayunov, V. A. Murtsovkin, and A. S. Dukhin, Colloid J. *USSR* (English Translation of Kolloidnyi Zhurnal) 48, 197 (1986).

[236] S. Gangwal, O. J. Cayre, M. Z. Bazant, and O. D. Velev, *Phys. Rev*. Lett. 100, 058302 (2008).

[237] O. D. Velev and K. H. Bhatt, *Soft Mater 2,* 738 (2006).

[238] B. G. Streetman, *Solid State Electronic Devices*, 3rd ed. (Prentice Hall, New Jersey, 1990).

In: Continuum Mechanics
Editors: Andrus Koppel and Jaak Oja, pp.105-146
ISBN: 978-1-60741-585-5

Chapter 4

CONTINUUM DESCRIPTION OF FLOW-LIKE LANDSLIDE DYNAMICS

M. Pirulli
Politecnico di Torino, Department of Structural and Geotechnical Engineering, Torino, Italy

ABSTRACT

Landslide run-out is a complex phenomenon, much more difficult to simulate by models than flow of fluids. The main complicating aspects concern that landslide material is often heterogeneous and its characteristics may change during the landslide movement due to drainage, hydraulic interaction between fluid and grains, comminution of grains or mixing with surface water or partly or fully liquefied superficial material entrained from the path.

The continuum mechanical theory, treating the heterogeneous and multiphase moving mass as a continuum, has emerged in the last years as a useful tool for describing the evolving geometry and the velocity distribution of a mass flowing down a surface. A hypothetical material, "equivalent fluid", whose rheology is controlled by a small number of parameters is, in fact, introduced to represent the bulk behaviour of a landslide.

After a brief introduction on landslide characteristics and dynamics, new advances in the continuum mechanical description of flow-like landslides are discussed in dedicated sections. Each section deals with one of the main aspects that characterize the physical behaviour of a landslide and presents the simplifying, but nevertheless realistic, assumptions made to streamline their mathematical formulation.

The mathematical formulation is then implemented in a numerical code (RASH3D) to test the capability of each mathematical assumption in allowing the modelling of real phenomenon dynamics. Results of numerical simulations of laboratory tests and real events are discussed in this chapter to this aim.

INTRODUCTION

Flow-like landslides such as rock avalanches, debris avalanches, debris flows and flow slides represent a significant threat to population, structures and infrastructures worldwide.

When the source of a potential landslide is identified, stabilization is not always a practical option and the consequences of failure must be considered (McDougall, 2006).

In this regard, reliable predictions of landslide runout (which includes the downslope movement and stopping of the landslide mass) can help to estimate the extent of potential impact areas and therefore contribute in reducing losses and avoiding exceedingly conservative decisions as for the urban development of a territory.

Even if the comprehension of the triggering mechanisms of a slope instability is a primarily and not completely solved problem, it is outside the aim of this chapter. It is here assumed that the mass has lost its static equilibrium and attention is focused on the analysis of the runout phase.

In the past, most practical runout prediction methods were empirical and made use of correlations based on observed events (e.g. Scheidegger, 1973; Davies, 1982; Corominas, 1996; Rickenmann, 1999). Though all of these methods are easy to use, they should only be applied to conditions strictly similar to those on which their development was based (Rickenmann, 2005).

A different and powerful approach which has emerged in recent years is that based on the continuum numerical modelling, whose models allow the flow parameters and deformation of the mass along the entire path, including deposition, to be determined (e.g. O'Brien et al., 1993; Chen and Lee, 2000; Iverson and Denlinger, 2001; McDougall and Hungr, 2005).

These models are largely based on the principles of hydrodynamics but they have gradually progressed from simple extensions of established hydrodynamic models to more complex models with important landslide-specific features. In fact, landslide motion is fundamentally different from open channel water flow. Flow-like landslides are made of a heterogeneous mass, they travel across steep and irregular soils, they contain earth materials that can resist internal strain, they can entrain substantial volumes of path material and their rheology can change during the motion (McDougall, 2006). A comprehensive model should integrate all these features together. Even if many goals have been attained in this direction, there is still room for improvement, as the most important capabilities have not yet been incorporated into a single model.

Moving from a brief introduction on flow-like landslide characteristics and dynamics, this chapter presents new advances in the continuum mechanical description of flow-like landslides.

Each of the main aspects that characterize the physical behaviour of a landslide is treated discussing the simplifying, but presumed realistic, assumptions made to streamline their mathematical formulation. In this regard, the main faced aspects concern the possible influence of the geometry discretization on the numerical flow, the importance of simulating the behaviour of a mass on a real three dimensional topography, so that aspects as run up, lateral spreading are not ignored, and the advantage of a set of implemented rheologies among which to select the more appropriate as a function of the characteristics of the event/site to be investigated.

The continuum numerical model RASH3D (Mangeney et al., 2003; Pirulli, 2005) is then presented as an example of coupling between the new advances in the continuum description of landslide dynamics and the development of an integrated continuum numerical code. This numerical implementation is intended to test the capability of each made mathematical assumption in facing the features typical of a flow-like landslide and allowing the modelling of real phenomenon dynamics. Results of numerical simulations of laboratory tests and real events are presented to this aim.

Characteristics of Flow-Like Landslides

The discussion focuses on landslides of the flow type. The reason for this is firstly that this group of mass movements is very important as a natural hazard and secondly that the continuum mechanics approach well becomes the dynamics of flows.

In 1932 Heim described the landslide of Elm (Switzerland, 1881) as "a large mass broken into thousand pieces, falling at the same time along the same course, whose debris had to flow as a single stream. The uppermost block at the very rear of the stream would attempt to get ahead. It hurried but struck the block slightly ahead, which was in the way".

Later, in 1948, Sander defined a flow as "a continuously arranged relative movement, carried out by sufficiently small (compared to the system under consideration) parts".

These considerations evidence that a flow movement is usually characterized by a high, and sometimes chaotic, degree of interaction among moving particles and that the size of each moving particle is such that the single particle is negligible (i.e. sufficiently small) respect to the whole movement to which it belongs.

The main characteristics of flow-like landslides can then be resumed as follows (Picarelli et al., 2005):

- diffuse and apparently non-localized large deformations, often giving rise to movements very similar to those exhibited by viscous fluids;
- a high mobility and capability to spread over the land, covering large distances, much larger than by other types of landslides;
- ability to adapt themselves to the slope morphology, entering and running within natural tracks or spreading laterally over flat slopes.

These characteristics depend on the mechanics of rupture and post-rupture deformation that is affected by the nature and state of the materials involved, but also by other parameters as slope morphology, initial state of stress, etc.

Since landslides can change the type of movement and the type of material as they progress, it follows that the classification of a landslide as flow can meet with great difficulties and require different discriminating factors, sometimes very subjective. As a consequence, a universal classification does not exist.

As an example the concept of flow is here discussed with reference to the widely accepted Hutchinson's (1968, 1988) or the Varnes' (1978) classifications.

As morphology is the principal factor in Hutchinson's classification mudslides, flow slides, debris flows and sturzstroms (rock avalanches) are classed together as being of flow-like form (Fig. 1). While, according to Varnes' classification they differ markedly in mechanism: mudslides predominantly slide rather than flow and are in that sense types of translational slides; flow slides and debris flows probably exhibit varying degree of sliding and flowing; while struzstroms are essentially flows (Hutchinson, 1988).

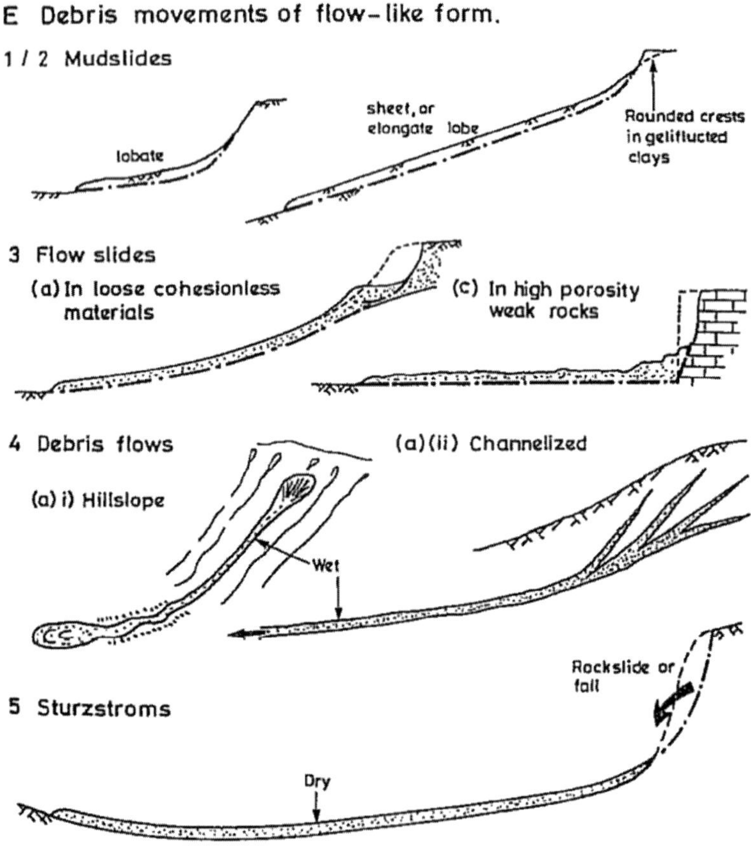

Figure 1. Main types of debris movement of flow-like form (after Hutchinson, 1988).

In the scheme of Varnes the knowledge of the mechanism assumes a fundamental value: only slope movements involving significant internal distortion of a moving mass would be classed as flows. On the contrary, as morphology is the principal factor in Hutchinson's classification, phenomena with different mechanism are classed together as a function of their overall behaviour. From this point of view, the above mentioned phenomena are all defined flow-like movements, and any commitment to a specific kinematic model is avoided.

In a way Varnes' scheme is perhaps easier to apply and requires less expertise to use but it is often difficult both to determine whether internal distortion or boundary sliding is dominant in a given case and to estimate aspects like the average grain-size distribution of a deposit. On the other way, Hutchinson's classification probably has particular appeal to the engineer contemplating stability analysis.

In the following the Hutchinson's approach will be mainly adopted.

Objectives for Continuum Model Development

Continuum dynamics models are largely based on the principles of hydrodynamics. Unlike fluids, flow-like movements sustain strain-dependent internal stresses that may be non-hydrostatic. In three-dimensions, anisotropic strain may give rise to anisotropic internal stress (McDougall and Hungr, 2004).

Volume change due to entrainment of material along the slide path is also an important characteristic of many rapid landslides. In particular, debris flows and debris avalanches can significantly increase their volume in the course of movement. Entrained and overridden material may alter the internal and basal rheologies of the slide mass (Sassa, 1985).

The rheologies may also vary in the course of movement due to comminution (Sassa, 2002) and pore pressure changes (Hutchinson, 1986; Iverson and Delinger, 2001).

Furthermore, mobile landslides travel large distances across steep and irregular terrain, often branching in several directions, separating (decoupling) longitudinally and spreading significantly.

As a consequence, a comprehensive model should not neglect the following objectives:

1) The model should permit motion across complex 3D morphology, large displacements and possible branching of the moving mass;
2) It should allow for non-hydrostatic, anisotropic internal stress;
3) It should allow for entrainment of material along the slide path;
4) It should allow the selection of a variety of material rheologies, which can vary along the slide path.

Continuum Mechanics Formulation

Models based on continuum mechanics treat the heterogeneous and multiphase moving mass of a flow-like landslide as a continuum. This implies that both the characteristic depth (*H)* and length (*L)* of a flowing mass are assumed to be large compared with the characteristic dimensions of the particles involved in the movement (Fig. 2a). By adopting this hypothesis, the dynamics of the analysed phenomena can be modelled using an "equivalent fluid", whose rheological properties are such that the bulk behaviour of the "equivalent fluid" simulates the bulk behaviour of the real mixture of the solid and fluid phases (Hungr, 1995) (Fig. 2b).

Under these conditions the mass and momentum conservation laws, written in local form and consistent with traditional soil mechanics practice, describe the behaviour of the moving mass:

$$\frac{\partial \rho}{\partial t} + \nabla \cdot (\rho \mathbf{v}) = 0 \tag{1}$$

$$\rho \frac{d\mathbf{v}}{dt} = -\nabla \cdot \boldsymbol{\sigma} + \rho \mathbf{g} \tag{2}$$

where $\mathbf{v}$ is the velocity vector, $\boldsymbol{\sigma}$ is the Cauchy stress tensor, ρ is the mass density, $\mathbf{g}$ is the gravitational acceleration vector and t is the time.

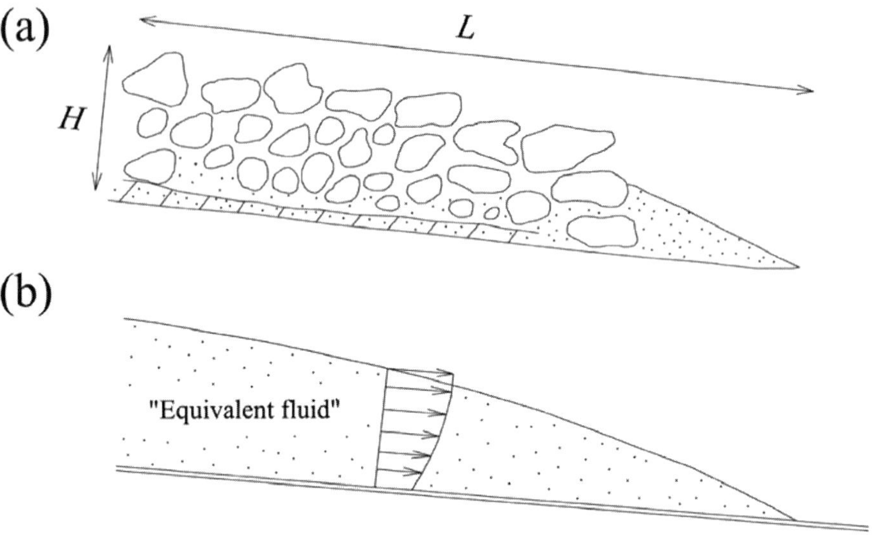

Figure 2. (a) Prototype of a heterogeneous and complex moving mass; (b) A homogeneous "apparent fluid" replaces the slide mass (after Hungr, 1995).

BOUNDARY CONDITIONS

The position of the free surface and the base of the flow may be expressed in implicit form in terms of a function ψ_s and ψ_b, respectively:

$$\psi_s(x, y, t) = [b(x, y, t) + h(x, y, t)] - z = 0 \tag{3}$$

$$\psi_b(x, y, t) = b(x, y, t) - z = 0 \tag{4}$$

where h is the depth of the flow, and b the depth of the bed measured in the z direction of an (x, y, z) right-handed Cartesian coordinate system. The z-axis is normal to the topography. (Fig. 3).

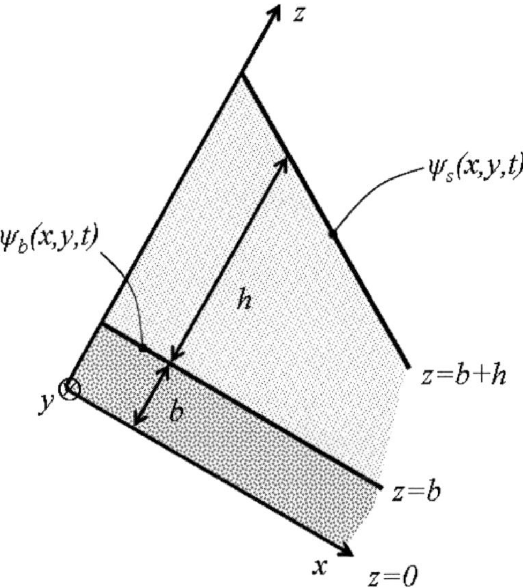

Figure 3. Orientation of the Cartesian coordinate system. ψ_s and ψ_b are the free surface and the base of the flow, respectively.

For each of these surfaces kinematic and dynamic boundary conditions are defined as follows.

The free surface (3) is assumed to be stress free (5) and it is assumed that material does not enter or leave the landslide at the free surface (i.e. particles on the free surface always remain part of the free surface) (6).

$$\boldsymbol{\sigma}\cdot \mathbf{n}\Big|_{z=b+h} = 0 \quad (5)$$

where $\mathbf{n}$ is the exterior unit normal vector.

$$\frac{d\psi_s}{dt} = \frac{d(b+h)}{dt} - \frac{dz}{dt} = \frac{\partial(b+h)}{\partial t} + v_{x(z=b+h)}\frac{\partial(b+h)}{\partial x} + v_{y(z=b+h)}\frac{\partial(b+h)}{\partial y} - v_{z(z=b+h)} = 0 \quad (6)$$

At the base of the flow (4) the component of the velocity vector normal to the slope is assumed equal to zero (i.e. there is no flow through the surface but there can be a flow tangential to it) (7) and it is allowed entrainment of material in the moving mass through erosion at the base (8)

$$\mathbf{v}\cdot \mathbf{n}\Big|_{z=b} = 0 \quad (7)$$

$$\frac{d\psi_b}{dt} = \frac{db}{dt} - \frac{dz}{dt} = \frac{\partial b}{\partial t} + v_{x(z=b)}\frac{\partial b}{\partial x} + v_{y(z=b)}\frac{\partial b}{\partial y} - v_{z(z=b)} = -E_t \quad (8)$$

where E_t is the "erosion velocity".

The shear stress acting at the base of the flow can be expressed using alternative constitutive relations that can take into account the complexity of the analysed phenomenon. A detailed description is deferred to the section "Rheological Laws".

Depth Integration

Observing field data from very large landslides evidence suggests sliding on a thin basal layer (Melosh, 1986). The basal zone is the active zone where the shear rates are high and nearly all the shear takes place. Thus the velocity profile is expected to be quite blunt and the depth-averaged velocity is quite close to the actual velocity everywhere except at the very base. Furthermore, since the vigorous shearing, fluidization and significant density variations are confined to a thin basal layer, it is sufficiently accurate to use a constant depth-averaged value for the density in the computations (Savage and Hutter, 1989).

The free flow of a granular material can then be modelled assuming that the material density (ρ) is spatially and temporally constant (i.e. incompressible). The governing mass and momentum conservation laws (1) and (2) modify as follows:

$$\nabla \cdot \mathbf{v} = 0 \tag{9}$$

$$\rho\left(\frac{\partial \mathbf{v}}{\partial t} + \mathbf{v} \cdot \nabla \mathbf{v}\right) = -\nabla \cdot \boldsymbol{\sigma} + \rho \mathbf{g} \tag{10}$$

Expanding equations (9) and (10) gives, in a reference frame linked to the topography, the following mass and momentum conservation equations:

$$\frac{\partial v_x}{\partial x} + \frac{\partial v_y}{\partial y} + \frac{\partial v_z}{\partial z} = 0 \tag{11}$$

$$\rho\left(\frac{\partial v_x}{\partial t} + v_x \frac{\partial v_x}{\partial x} + v_y \frac{\partial v_x}{\partial y} + v_z \frac{\partial v_x}{\partial z}\right) = -\left(\frac{\partial \sigma_{xx}}{\partial x} + \frac{\partial \sigma_{xy}}{\partial y} + \frac{\partial \sigma_{xz}}{\partial z}\right) + \rho g \gamma_x \tag{12}$$

$$\rho\left(\frac{\partial v_y}{\partial t} + v_x \frac{\partial v_y}{\partial x} + v_y \frac{\partial v_y}{\partial y} + v_z \frac{\partial v_y}{\partial z}\right) = -\left(\frac{\partial \sigma_{yx}}{\partial x} + \frac{\partial \sigma_{yy}}{\partial y} + \frac{\partial \sigma_{zy}}{\partial z}\right) + \rho g \gamma_y \tag{13}$$

$$\rho\left(\frac{\partial v_z}{\partial t} + v_x \frac{\partial v_z}{\partial x} + v_y \frac{\partial v_z}{\partial y} + v_z \frac{\partial v_z}{\partial z}\right) = -\left(\frac{\partial \sigma_{xz}}{\partial x} + \frac{\partial \sigma_{yz}}{\partial y} + \frac{\partial \sigma_{zz}}{\partial z}\right) + \rho g \gamma_z \tag{14}$$

where g is the constant of gravity and γ_i are coefficients, function of the local slope, defining the projection of the gravity vector along the i direction.

In the limit as $H/L \rightarrow 0$ (i.e. shallowness or long wave limit), it is also possible to integrate the balance equations (11)-(14) in depth (i.e. from the bed, $z = b$, to the free surface of the moving mass, $z = b + h$).

Depth averaging allows a complete three dimensional description of the flow to be avoided: changes in the mechanical behaviour in the flow depth are ignored and the rheology of the flowing material is incorporated into a single term that describes the frictional stress which develops at the interface between the flowing material and the rough bed surface.

Experimental measurements, both on steep slopes in two and three dimensions (e.g. Savage and Hutter, 1989: Gray et al., 1999; Wieland et al., 1999) and on curved beds (e.g. Greve and Hutter, 1993; Greve et al., 1994; Koch et al., 1994), agree relatively well with the prediction of motion and spreading of the mass given by a depth averaged model. As a consequence, starting from the introduction of depth averaged equations in the context of granular flows by Savage and Hutter (1989), many numerical models have been implemented (e.g. Chen and Lee, 2000; McDougall and Hungr, 2004; Denlinger and Iverson, 2004).

Integrating the equations (11)-(14) in depth, imposing the assumed boundary conditions, substituting appropriate depth-averaged values (e.g. $\overline{v_x} = \frac{1}{h}\int_{z=b}^{z=b+h} v_x dz$) and applying the Leibniz's rule (e.g. $\int_{z=b}^{z=b+h} \frac{\partial v_x}{\partial x} dz = \frac{\partial}{\partial x}\int_{z=b}^{z=b+h} v_x dz - v_{x(z=b+h)}\frac{\partial(b+h)}{\partial x} + v_{x(z=b)}\frac{\partial b}{\partial x}$), the initial system of equations results in the so-called depth averaged Saint Venant equations (Savage and Hutter, 1989):

$$\frac{\partial h}{\partial t} + \frac{\partial(\overline{v_x}h)}{\partial x} + \frac{\partial(\overline{v_y}h)}{\partial y} = E_t \tag{15}$$

$$\rho\left(\frac{\partial(\overline{v_x}h)}{\partial t} + \frac{\partial(\overline{v_x^2}h)}{\partial x} + \frac{\partial(\overline{v_x v_y}h)}{\partial y} - v_{x(z=b)}E_t\right) =$$
$$= -\frac{\partial(\overline{\sigma_{xx}}h)}{\partial x} - \frac{\partial(\overline{\sigma_{xy}}h)}{\partial y} - \overbrace{\left(\sigma_{xx(z=b)}\frac{\partial b}{\partial x} + \sigma_{xy(z=b)}\frac{\partial b}{\partial y} - \sigma_{zx(z=b)}\right)}^{T_x} + \rho g \gamma_x h \tag{16}$$

$$\rho\left(\frac{\partial(\overline{v_y}h)}{\partial t} + \frac{\partial(\overline{v_y v_x}h)}{\partial x} + \frac{\partial(\overline{v_y^2}h)}{\partial y} - v_{y(z=b)}E_t\right) =$$
$$= -\frac{\partial(\overline{\sigma_{xy}}h)}{\partial x} - \frac{\partial(\overline{\sigma_{yy}}h)}{\partial y} - \overbrace{\left(\sigma_{xy(z=b)}\frac{\partial b}{\partial x} + \sigma_{yy(z=b)}\frac{\partial b}{\partial y} - \sigma_{zy(z=b)}\right)}^{T_y} + \rho g \gamma_y h \tag{17}$$

$$\rho\left(\frac{\partial\left(\overline{v_z h}\right)}{\partial t}+\frac{\partial\left(\overline{v_z v_x h}\right)}{\partial x}+\frac{\partial\left(\overline{v_z v_y h}\right)}{\partial y}-v_{z(z=b)}E_t\right)= \\ =-\frac{\partial\left(\overline{\sigma_{xz}}h\right)}{\partial x}-\frac{\partial\left(\overline{\sigma_{yz}}h\right)}{\partial y}-\overbrace{\left(\sigma_{xz(z=b)}\frac{\partial b}{\partial x}+\sigma_{yz(z=b)}\frac{\partial b}{\partial y}-\sigma_{zz(z=b)}\right)}^{T_z}+\rho g\gamma_z h \tag{18}$$

where $\overline{\mathbf{v}}=\left(\overline{v_x},\overline{v_y},\overline{v_z}\right)$ denotes the depth-averaged flow velocity, $\overline{\boldsymbol{\sigma}_{ij}}$ is the depth-averaged stress tensor, and $\mathbf{T}=\left(T_x,T_y.T_z\right)=\boldsymbol{\sigma}\cdot\mathbf{n}\big|_{z=b}$ is the traction vector ($\mathbf{n}$ is the unit vector normal to the bed).

Because the depth-wise velocity profile may be non uniform, a momentum correction ζ must be applied to relate $\overline{v_x^2}$ to $\overline{v_x}^2$ and $\overline{v_x v_y}$ to $\overline{v_x}\,\overline{v_y}$. To give some idea of numerical values of ζ note that for a parabolic velocity profile (corresponding to no sliding, all differential shear) with vanishing basal velocity ζ = 6/5, whereas for a uniform profile (all sliding and no differential shear) ζ = 1. Since it is likely that sliding is present, the active shear zone is confined to a thin basal layer and the velocity profile is blunt (Melosh, 1986), it may, without introducing a large error, choose ζ = 1 (Savage and Hutter, 1989). This allows the derivation to proceed with a further separation of terms:

$$h\left(\frac{\partial\overline{v_x}}{\partial t}+\overline{v_x}\frac{\partial\overline{v_x}}{\partial x}+\overline{v_y}\frac{\partial\overline{v_x}}{\partial y}\right)+\left(\overline{v_x}-v_{x(z=b)}\right)E_t=-\frac{1}{\rho}\frac{\partial\left(\overline{\sigma_{xx}}h\right)}{\partial x}-\frac{1}{\rho}\frac{\partial\left(\overline{\sigma_{xy}}h\right)}{\partial y}+\frac{1}{\rho}T_x+g\gamma_x h \tag{19}$$

$$h\left(\frac{\partial\overline{v_y}}{\partial t}+\overline{v_x}\frac{\partial\overline{v_y}}{\partial x}+\overline{v_y}\frac{\partial\overline{v_y}}{\partial x}\right)+\left(\overline{v_y}-v_{y(z=b)}\right)E_t=-\frac{1}{\rho}\frac{\partial\left(\overline{\sigma_{xy}}h\right)}{\partial x}-\frac{1}{\rho}\frac{\partial\left(\overline{\sigma_{yy}}h\right)}{\partial y}+\frac{1}{\rho}T_y+g\gamma_y h \tag{20}$$

$$h\left(\frac{\partial\overline{v_z}}{\partial t}+\overline{v_x}\frac{\partial\overline{v_z}}{\partial x}+\overline{v_y}\frac{\partial\overline{v_z}}{\partial y}\right)+\left(\overline{v_z}-v_{z(z=b)}\right)E_t=-\frac{1}{\rho}\frac{\partial\left(\overline{\sigma_{xz}}h\right)}{\partial x}-\frac{1}{\rho}\frac{\partial\left(\overline{\sigma_{yz}}h\right)}{\partial y}+\frac{1}{\rho}T_z+g\gamma_z h \tag{21}$$

Equations (19)–(21) are the most general form of the Eulerian depth-averaged governing equations, obtained without any approximation. In the following, some assumptions will lead to simplify and close the equations.

FURTHER APPROXIMATIONS

According to Gray et al. (1999), the shallowness or long-wave limit (i.e. $H/L \rightarrow 0$) leads to neglect:

1) the acceleration normal to the base of the flow (i.e. topography) and the horizontal gradient of the stress in the non depth-averaged z equation, as a consequence equation (14) can be rewritten as:

$$\sigma_{zz} = \rho g \gamma_z (b + h - z) \tag{22}$$

for the above mentioned reasons plus condition (7), the depth-averaged z equation (21) can be rewritten as:

$$T_{z(z=b)} = -\rho g \gamma_z h \tag{23}$$

2) the horizontal gradient of the stress in the depth-averaged x and y equations, as a consequence equations (19)-(20) can be rewritten as:

$$h\left(\frac{\partial \overline{v_x}}{\partial t} + \overline{v_x}\frac{\partial \overline{v_x}}{\partial x} + \overline{v_y}\frac{\partial \overline{v_y}}{\partial y}\right) + \left(\overline{v_x} - v_{x(z=b)}\right)E_t = -\frac{1}{\rho}\frac{\partial\left(\overline{\sigma_{xx}}h\right)}{\partial x} + \frac{1}{\rho}T_x + g\gamma_x h \tag{24}$$

$$h\left(\frac{\partial \overline{v_y}}{\partial t} + \overline{v_x}\frac{\partial \overline{v_y}}{\partial x} + \overline{v_y}\frac{\partial \overline{v_y}}{\partial x}\right) + \left(\overline{v_y} - v_{y(z=b)}\right)E_t = -\frac{1}{\rho}\frac{\partial\left(\overline{\sigma_{yy}}h\right)}{\partial y} + \frac{1}{\rho}T_y + g\gamma_y h \tag{25}$$

The simplified model would assume isotropy of normal stresses, i.e. $\sigma_{xx} = \sigma_{yy} = \sigma_{zz}$, but flow-like movements sustain strain-dependent internal stresses that may be non-hydrostatic. As a consequence, the most general model should define the ratio of the longitudinal stresses to the normal stress equal to the earth pressure coefficients, K (e.g. $K_x = \frac{\sigma_{xx}}{\sigma_{zz}}$ and $K_y = \frac{\sigma_{yy}}{\sigma_{zz}}$).

In its turn the K coefficient can be defined as a function of the dynamic basal friction angle, φ, and the internal friction angle, δ (Gray et al., 1999).

Depth-averaging the σ_{xx}, σ_{yy} and σ_{zz} values and substituting the obtained terms into equations (24)-(25) gives the following:

$$h\left(\frac{\partial\overline{v_x}}{\partial t}+\overline{v_x}\frac{\partial\overline{v_x}}{\partial x}+\overline{v_y}\frac{\partial\overline{v_x}}{\partial y}\right)+\frac{\partial\left(K_x g\gamma_z\frac{h^2}{2}\right)}{\partial x}=\frac{1}{\rho}T_x+g\gamma_x h-\left(\overline{v_x}-v_{x(z=b)}\right)E_t \quad (26)$$

$$h\left(\frac{\partial\overline{v_y}}{\partial t}+\overline{v_x}\frac{\partial\overline{v_y}}{\partial x}+\overline{v_y}\frac{\partial\overline{v_y}}{\partial x}\right)+\frac{\partial\left(K_y g\gamma_z\frac{h^2}{2}\right)}{\partial y}=\frac{1}{\rho}T_y+g\gamma_y h-\left(\overline{v_y}-v_{y(z=b)}\right)E_t \quad (27)$$

DISCRETIZATION

The governing equations introduced in the previous section represent a system of partial differential equations. Only for very simple problems does a closed form analytical solution exist. In general a numerical solution is the only practical way to solve the problem.

In particular, four steps are necessary to replace the continuous problem through a finite set of discrete values:

- geometry definition;
- mesh generation;
- space discretization;
- time discretization.

Because of the discretization, there is only a finite number of discrete values in the representation, and each values is only specified to finite precision. The information about the detailed solution structure within the discrete spatial cells and time steps is inevitably lost and the result of the computational solution is given only with some approximation.

In the space discretization, the flow variables and their derivatives are discretized with reference to the *x*, *y*, *z* axes in a definite time point *t*. Condition for carrying out this procedure is the definition of a mesh. To perform the space discretization, different methods are possible, and the most common for engineering purposes are:

- Finite Difference Methods
- Finite Volume Methods
- Finite Element Methods

In the time discretization, necessary condition for the stability of an explicit scheme is the CFL condition (Courant-Friedrichs-Lewi condition) which limits the time step size. The CFL condition is based on the physical principle that flow disturbances can travel no faster than the local propagation speed of waves of different types throughout the domain, being the speed of sound and the flow speed combined, and therefore information should advance by no more than one cell spacing in a single time step (Audusse et al., 2000)

Implicit methods do not have time step restrictions but require solving matrix equations at each time step and the cost per time step can be quite high compared with the explicit methods.

In the following section, the RASH3D code (Pirulli, 2005) that is based on the finite volume method and a kinetic scheme will be analysed in detail.

RASH3D Description

The RASH3D code (Pirulli, 2005) originates from a pre-existing model (SHWCIN) based on the classical finite volume approach for solving a hyperbolic system of equations using the concept of cell centred conservative quantities, developed by Audusse et al. (2000) and Bristeau et al. (2001) to compute the Saint-Venant equation in hydraulic problems.

An extension of SHWCIN to simulate dry granular flows using a kinetic scheme was initially introduced by Mangeney et al. (2003).

Pirulli (2005) introduced further extensions of SHWCIN to prevent the mesh-dependency problems, enable simulation of motion across irregular 3D terrains and run analyses with different rheologies. RASH3D is the new upgraded code.

Numerical Implementation

The system of equations (15), (26), (27) is discretized in RASH3D on a triangular finite element mesh through a kinetic scheme based on a finite volume approach.

Finite Volume Method

The basic idea of the finite volume methods is to divide the spatial domain into control volumes, C_i, (*finite volumes,* also called *grid cells* or *dual cells*) and to apply the transport equation, in integral form, to each volume. Dual cells C_i are obtained by joining the centres of mass of the triangles surrounding each vertex P_i of the mesh (Fig. 4).

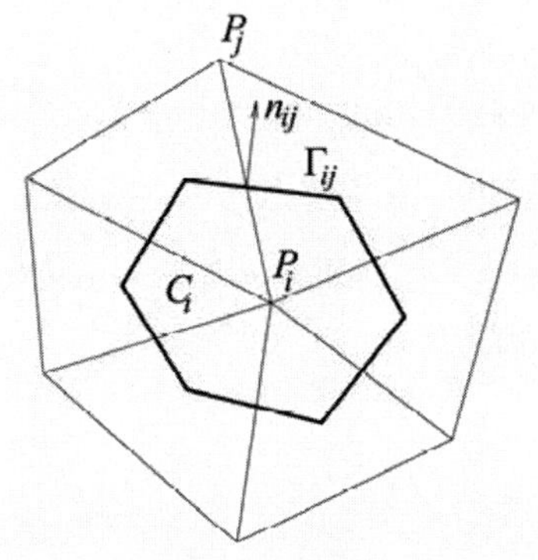

Figure 4. Triangular finite element mesh for dual cell Ci (after Mangeney et al., 2003).

Assuming $E_t = 0$ and isotropy of stresses, the system of equations (15), (26) and (27) reads:

$$\frac{\partial \mathbf{U}}{\partial t} + \nabla \cdot \mathbf{F}(\mathbf{U}) = \mathbf{B}(\mathbf{U}) \tag{28}$$

with

$$\mathbf{U} = \begin{pmatrix} h \\ h\overline{v_x} \\ h\overline{v_y} \end{pmatrix}; \quad \mathbf{F}(\mathbf{U}) = \begin{pmatrix} h\overline{v_x} & h\overline{v_y} \\ h\overline{v_x^2} + \frac{1}{2}gh^2 & h\overline{v_x v_y} \\ h\overline{v_x v_y} & h\overline{v_y^2} + \frac{1}{2}gh^2 \end{pmatrix}; \quad \mathbf{B}(\mathbf{U}) = \begin{pmatrix} 0 \\ \frac{1}{\rho}T_x + g\gamma_x h \\ \frac{1}{\rho}T_y + g\gamma_y h \end{pmatrix} \tag{29}$$

Integrating equation (28) in space and time on the set C_i X (t^n, t^{n+1}),

$$\int_{t^n}^{t^{n+1}} \int_{C_i} \frac{\partial \mathbf{U}}{\partial t} dC dt + \int_{t^n}^{t^{n+1}} \int_{C_i} \nabla \cdot \mathbf{F}(\mathbf{U}) dC dt = \int_{t^n}^{t^{n+1}} \int_{C_i} \mathbf{B}(\mathbf{U}) dC dt \tag{30}$$

and applying the Gauss theorem to the second term:

$$\int_{t^n}^{t^{n+1}} \int_{C_i} \frac{\partial \mathbf{U}}{\partial t} dC dt + \int_{t^n}^{t^{n+1}} \int_{\partial C_i} \mathbf{F}(\mathbf{U}) \cdot \mathbf{n} dC dt = \int_{t^n}^{t^{n+1}} \int_{C_i} \mathbf{B}(\mathbf{U}) dC dt, \tag{31}$$

where $\mathbf{n} = (n_x, n_y)$ denotes the unit outward normal to the boundary ∂C_i of the element C_i, the finite volume scheme is then obtained from equation (31) as follows

$$\boldsymbol{U}_i^{n+1} = \boldsymbol{U}_i^n - \sum_{j \in N_i} \frac{\Delta t L_{ij}}{|C_i|} \boldsymbol{F}\left(\boldsymbol{U}_i^n, \boldsymbol{U}_j^n, \mathbf{n}_{ij}\right) + \frac{\Delta t}{|C_i|} \int_{C_i} \mathbf{B}(\mathbf{U}) dC \tag{32}$$

where N_i is the set of nodes P_j surrounding the node P_i, $|C_i|$ is the area of C_i, $\mathbf{n}_{ij}$ is the unit normal to the boundary edge Γ_{ij} that divides cell C_i from C_j and outward to C_i, $\boldsymbol{U}_i^{n+1}$ and $\boldsymbol{U}_i^n$ are the approximation for the ith cell average of the exact solution $\mathbf{U}$ at times t^n and t^{n+1}, respectively (Mangeney et al., 2003).

$$\boldsymbol{U}_i \approx \frac{1}{|C_i|} \int_{C_i} \mathbf{U} dC \tag{33}$$

$F\left(U_i^n, U_j^n, \mathbf{n}_{ij}\right)$ denotes an interpolation of the normal component of the flux $\mathbf{F}(\mathbf{U})\cdot\mathbf{n}_{ij}$ along the boundary edge Γ_{ij}, L_{ij} is the length of Γ_{ij}, Δt is the time step ($t^n = n\Delta t$).

The source term **B(U)** can be split in $\mathbf{B}(\mathbf{U})_s$ and $\mathbf{B}(\mathbf{U})_f$ that are the bottom slope and friction slope, respectively:

$$\mathbf{B}(\mathbf{U})_s = \begin{pmatrix} 0 \\ g\gamma_x h \\ g\gamma_y h \end{pmatrix}, \tag{34a}$$

$$\mathbf{B}(\mathbf{U})_f = \begin{pmatrix} 0 \\ \frac{1}{\rho}T_x \\ \frac{1}{\rho}T_y \end{pmatrix} \tag{34b}$$

$\mathbf{B}(\mathbf{U})_s$ and $\mathbf{B}(\mathbf{U})_f$ will betreated in the sections "Source Term: Bottom Slope" in an explicit way and "Source Term: Friction Slope" in a semi-implicit way, respectively.

The summation sign in the second term of the right hand side of equation (32) indicates that the computation here includes all the boundary edges of the considered *i*th cell.

The cell-averaged values are modified in each time-step by the flux through the edges of the grid cells, and the primary problem is to determine good numerical flux functions (Audusse et al., 2000). With regard to this, the kinetic scheme is here proposed to define fluxes at the control volume interfaces through a fictitious description of the microscopic behaviour of the system at the interfaces.

Triangular Mesh Selection

A numerical problem which can occur when solving topology optimization is the fact that different solutions can be obtained just by choosing different element types and/or element numbers. This is the mesh-dependency problem. Ignorance of mesh dependency can lead to wrong numerical calculations. Mesh structures need to be developed to minimize mesh dependency without compromising the finite computing resource and/or incurring large computational expense.

The triangular finite element mesh implemented in a finite volume approach gives mesh-dependency problems when, a triangular regular mesh (Fig. 5a), *structured mesh,* is adopted. Fitting the mesh structure could be a way of reducing the problem but a considerable increase of the cpu-time would occur, without a complete removal of the problem.

As a consequence, a triangular irregular mesh, *unstructured mesh* (Fig. 5b), built up using the Delaunay triangulation technique and the Voronoi diagram, has to be preferred.

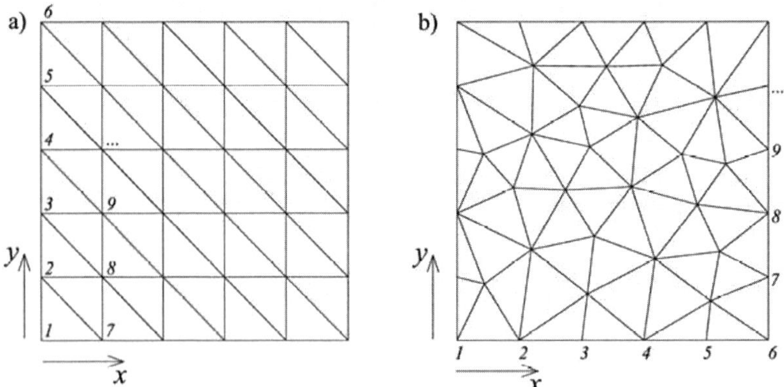

Figure 5. (a) A structured mesh. (b) Propagation of hemi-spherical mass on horizontal plane at t = 0.3s using a structured mesh (after Pirulli et al., 2007).

Kinetic Scheme

The kinetic approach consists in describing the temporal evolution of a system made of a large number of particles through 1) a distribution function of fictitious particles $M(t, x, y, \xi)$ with velocity ξ, and 2) a Boltzmann-type kinetic equation describing the temporal evolution of the density M of the particles.

The scheme is here discussed by omitting the source term which is treated in dedicated sections.

In this frame the functions $(h, \overline{\mathbf{v}})$ are solutions to (28)-(29) if, and only if, $M(t, x, y, \xi)$ satisfies the Boltzmann-type kinetic equation (Audusse et al., 2000):

$$\frac{\partial M}{\partial t} + \xi \cdot \nabla_{\mathbf{x}} M - g \cdot \nabla_{\xi} M = Q(t, x, y, \xi) \tag{35}$$

where $Q(t, x, y, \xi)$, which is the "collision term", is usually neglected in numerical schemes.

The microscopic density M of the particles that are present at time t in the neighbourhood $\Delta x \Delta y$ of a position (x, y) and with a velocity ξ can be defined as a so called Gibbs equilibrium (Audusse et al., 2000).

$$M(t, x, y, \xi) = \frac{h(t, x, y)}{c^2} \chi(\boldsymbol{\omega}) \qquad \text{with} \qquad \boldsymbol{\omega} = \frac{\xi - \overline{\mathbf{v}}(t, x, y)}{c} \tag{36}$$

where c is defined by $c^2 = gh/2$ and $\chi(\omega)$ be a positive and even function defined on $\Re^2$ (i.e. $\chi(\omega) = \chi(-\omega) \geq 0$) and satisfying

$$\int_{\Re^2} \chi(\boldsymbol{\omega}) d\boldsymbol{\omega} = 1 \tag{37a}$$

$$\int_{\Re^2} \omega_i \omega_j \chi(\boldsymbol{\omega}) d\boldsymbol{\omega} = \delta_{ij} \tag{37b}$$

with δ_{ij} the Kronecker symbol and $\boldsymbol{\omega} = (\omega_x, \omega_y)$.

Due to the properties of χ, it is proved that the Saint-Venant equations (28)-(29) are equivalent to equation (36) integrated in ξ against 1, ξ, ξ^2.

$$\begin{pmatrix} h \\ h\bar{\mathbf{v}} \\ h\bar{\mathbf{v}}^2 + \dfrac{gh^2}{2} \end{pmatrix} = \int_{\Re^2} \begin{pmatrix} 1 \\ \xi \\ \xi^2 \end{pmatrix} M(t, x, y, \xi) d\xi \tag{38}$$

It can be finally stated that the integration of (35) in space and time on the set C_i X (t^n, t^{n+1}) is equivalent to equation (32).

The fundamental consequence of the obtained relations is that the non-linear system (28)-(29) can be viewed as a linear transport equation (35) on a non-linear quantity M (36), for which it is easier to find a simple numerical scheme with good properties (Audusse et al., 2000).

The application of a simple upwind scheme to microscopic equation (35) allows the fluxes defined in equation (32) to be evaluated as

$$\boldsymbol{F}(\boldsymbol{U}_i, \boldsymbol{U}_j, \mathbf{n}_{ij}) = \boldsymbol{F}^+(\boldsymbol{U}_i, \mathbf{n}_{ij}) + \boldsymbol{F}^-(\boldsymbol{U}_j, \mathbf{n}_{ij}) \tag{39}$$

$$\boldsymbol{F}^+(\mathbf{U}_i, \mathbf{n}_{ij}) = \int_{\xi \cdot \mathbf{n}_{ij} \geq 0} \xi \cdot \mathbf{n}_{ij} \begin{pmatrix} 1 \\ \xi \end{pmatrix} M_i(\xi) d\xi = \begin{pmatrix} \boldsymbol{F}_h^+ \\ \boldsymbol{F}_q^+ \end{pmatrix} \tag{40}$$

$$\boldsymbol{F}^-(\mathbf{U}_j, \mathbf{n}_{ij}) = \int_{\xi \cdot \mathbf{n}_{ij} \leq 0} \xi \cdot \mathbf{n}_{ij} \begin{pmatrix} 1 \\ \xi \end{pmatrix} M_j(\xi) d\xi = \begin{pmatrix} \boldsymbol{F}_h^- \\ \boldsymbol{F}_q^- \end{pmatrix} \tag{41}$$

An example of function χ satisfying the above described properties is a simple rectangular function

$$\chi(\boldsymbol{\omega}) = \begin{cases} \dfrac{1}{12} & \text{for } |\omega_i| \leq \sqrt{3},\ i = x, y \\ 0 & \text{otherwise} \end{cases} \tag{42}$$

Source Term: Bottom Slope

Working on the simulation of real world geometry, the strong irregularity of the bottom means that the numerical treatment of the source terms plays an important role on the final accuracy of the obtainable results.

The numerical pointwise treatment of $\mathbf{B}(\mathbf{U})_S$ (34a) is not difficult if the bottom slope in the x ant y directions can be easily determined. However, generally the six vertices of each dual cell do not lie on the same plane (Fig. 6); therefore the slope of the cell is not trivially computable (Valiani et al., 2002).

To face this problem, the following steps are here followed:

- the first step consists of a division of the dual cell into six triangular elements (dual triangles), with a common vertex coincident with P_i (Fig. 6);

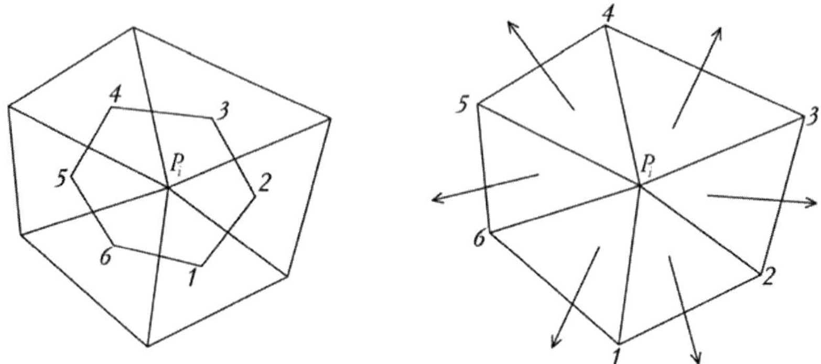

Figure 6.(a) Dual cell for node Pi; (b) Subdivision of the dual cell in six dual triangles.

- the second step is the determination of the equation of the plane passing through the three vertices of each triangle. As an example, if the $1P_i2$ (Fig. 6) triangle is chosen such an equation is

$$\det\begin{vmatrix} x & y & z_b & 1 \\ x_1 & y_1 & z_1 & 1 \\ x_2 & y_2 & z_2 & 1 \\ x_{Pi} & y_{Pi} & z_{Pi} & 1 \end{vmatrix} = 0 \qquad (43)$$

or $z_b = ax + by + c$ in compact form.

Now, the slopes of each dual triangle are simply calculable. For the triangle $1P_i2$ such slopes are:

$$\gamma_x = -\frac{\dfrac{\partial z_b}{\partial x}}{\sqrt{1+\left(\dfrac{\partial z_b}{\partial x}\right)^2}} = -\frac{a}{\sqrt{1+a^2}} \qquad (44)$$

$$\gamma_y = -\frac{\frac{\partial z_b}{\partial y}}{\sqrt{1+\left(\frac{\partial z_b}{\partial y}\right)^2}} = -\frac{b}{\sqrt{1+b^2}} \tag{45}$$

where

$$a = \frac{(y_2 - y_{Pi})z_1 + (y_{Pi} - y_1)z_2 + (y_1 - y_2)z_{Pi}}{x_{Pi}y_1 - x_2y_1 + x_1y_2 - x_{Pi}y_2 + x_2y_{Pi} - x_1y_{Pi}} \tag{46}$$

$$b = \frac{(x_2 - x_{Pi})z_1 + (x_{Pi} - x_1)z_2 + (x_1 - x_2)z_{Pi}}{x_{Pi}y_1 - x_2y_1 + x_1y_2 - x_{Pi}y_2 + x_2y_{Pi} - x_1y_{Pi}} \tag{47}$$

The surface integral on each cell is now written as a sum of six surface integrals on the six triangular elements (dual triangles). The constancy of bottom slope of each element allows one to write the corresponding integral as a constant times an area. As an example, for the dual cell i with A,B,C,D,E,F the six dual triangles, one obtains at time n:

$$\int_{C_i} \mathbf{B}(\mathbf{U}_i^n)_s = \begin{pmatrix} 0 \\ gh_i^n \sum_\nu \gamma_{x\nu} C_\nu \\ gh_i^n \sum_\nu \gamma_{y\nu} C_\nu \end{pmatrix}, \nu = A,B,C,D,E,F \tag{48}$$

it is then obtained:

$$\boldsymbol{B}(\boldsymbol{U}_i^n)_s \approx \frac{1}{|C_i|}\int_{C_i} \mathbf{B}(\mathbf{U}_i^n)_s dC \approx \frac{1}{|C_i|}\begin{pmatrix} 0 \\ gh_i^n \sum_\nu \gamma_{x\nu} C_\nu \\ gh_i^n \sum_\nu \gamma_{y\nu} C_\nu \end{pmatrix} \tag{49}$$

where h_i is the water level assumed constant over each cell, C_ν are the areas of each dual triangle, and $\gamma_{x\nu}$ and $\gamma_{y\nu}$ are the slopes in the x and y directions, respectively, of each dual triangle.

Equation (32) can be then rewritten as:

$$\boldsymbol{U}_i^{n+1} = \boldsymbol{U}_i^n - \sum_{j\in N_i} \frac{\Delta t L_{ij}}{|C_i|} \boldsymbol{F}(\boldsymbol{U}_i^n, \boldsymbol{U}_j^n, \mathbf{n}_{ij}) + \boldsymbol{B}(\boldsymbol{U}_i^n)_s + \frac{\Delta t}{|C_i|}\int_{C_i} \mathbf{B}(\mathbf{U})_f dC \tag{50}$$

Source Term: Friction Slope

Depth averaging allows to avoid a complete three dimensional description of the flow: the complex rheology of the granular material is incorporated in a basal shear term $\mathbf{T_t} = \left(T_x, T_y\right)$ that develops at the interface between the flowing material and the rough surface and opposes motion.

Once the flux and the bottom slope are defined, the friction term $\left(\frac{1}{\rho}\mathbf{T}_{ti}^{n+1}\Delta t\right)$ is introduced using a semi-implicit scheme.

Equation (50) without the friction term corresponds to (Mangeney et al., 2003):

$$h_i^{n+1} = h_i^n - \sum_{j \in K_i} \frac{\Delta t L_{ij}}{A_i} \boldsymbol{F}_h\left(\boldsymbol{U}_i^n, \boldsymbol{U}_j^n, \mathbf{n}_{ij}\right) \tag{51}$$

$$\mathbf{q}_i^{n+1} = \mathbf{q}_i^n - \sum_{j \in K_i} \frac{\Delta t L_{ij}}{A_i} \boldsymbol{F}_{\mathbf{q}}\left(\boldsymbol{U}_i^n, \boldsymbol{U}_j^n, \mathbf{n}_{ij}\right) + \Delta t\, \boldsymbol{B}\left(\boldsymbol{U}_i^n\right)_s \tag{52}$$

where $\mathbf{q} = h\overline{\mathbf{v}_t}$.

Introducing the friction term, equation (51) is unchanged while equation (52) modifies as follows:

$$\tilde{\mathbf{q}}_i^{n+1} = \mathbf{q}_i^{n+1} + \frac{1}{\rho}\mathbf{T}_{ti}^{n+1}\Delta t \tag{53}$$

(i.e: $\boldsymbol{B}\left(\boldsymbol{U}_i^n\right)_f \approx \frac{1}{|C_i|}\int_{C_i} \mathbf{B}\left(\mathbf{U}_i^n\right)_f dC \approx \begin{pmatrix} 0 & \frac{1}{\rho}T_{xi}^n & \frac{1}{\rho}T_{yi}^n \end{pmatrix}^T$)

The γ_z value to be used in defining the traction $\mathbf{T}_t$ is initially analysed. Due to the possible spatial variation of the slope, the dual triangles and surfaces defined for each cell in the previous section are also used for the present purpose. As an example, for the dual cell C_i with A,B,C,D,E,F the six dual triangles, $\boldsymbol{\gamma}_z$ is defined as follows:

$$\boldsymbol{\gamma}_z = \frac{1}{|C_i|}\begin{pmatrix} \sum_\nu \gamma_{zx\nu} C_\nu \\ \sum_\nu \gamma_{zy\nu} C_\nu \end{pmatrix}, \nu = A, B, C, D, E, F \tag{54}$$

where $|C_i|$ is the area of C_i and C_ν is the area of a dual triangle of C_i, and $\gamma_{zx\nu}$ and $\gamma_{zy\nu}$ are the normal to the slopes in the x and y directions, respectively, of each dual triangle:

$$\gamma_{zx} = -\frac{1}{\sqrt{1+\left(\frac{\partial z_b}{\partial x}\right)^2}} = -\frac{1}{\sqrt{1+a^2}} \tag{55}$$

$$\gamma_{zy} = -\frac{1}{\sqrt{1+\left(\frac{\partial z_b}{\partial y}\right)^2}} = -\frac{1}{\sqrt{1+b^2}} \tag{56}$$

a and b are defined in equations (46) and (47), respectively.

Moving to the traction term $\mathbf{T}_t$, equation (53) shows the linear variation of the traction $\mathbf{T}_{ti}^{n+1}$ as a function of $\tilde{\mathbf{q}}_i^{n+1}$:

$$\mathbf{T}_{ti}^{n+1} = \frac{\rho}{\Delta t}\tilde{\mathbf{q}}_i^{n+1} - \frac{\rho}{\Delta t}\mathbf{q}_i^{n+1} \tag{57}$$

Equation (57) reduces in the direction of the flow to a scalar equation and the representation on the $\left(\tilde{q}_i^{n+1}, T_{ti}^{n+1}\right)$ graph gives a family of lines with slope $\frac{\rho}{\Delta t}$ (Fig. 7):

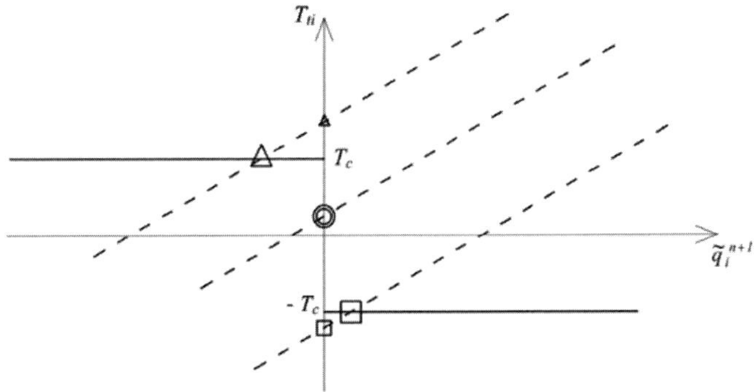

Figure 7. Resolution of the tangential traction by projection on the admissible state imposed by the Coulomb friction law. Solid lines represent the domain of admissible state of traction, dashed lines represent the family of line obtained from the momentum conservation equation. Small triangle, circle and square represent the possible values of T_{ti} when $\tilde{\mathbf{q}}_i^{n+1} = 0$. Large triangle, circle and square represent the projection of T_{ti} on the admissible state of traction (after Mangeney et al., 2003).

According to the Coulomb friction law:

- the mass stops if the acting shear stress ($\mathbf{T}_t$) is lower than the resisting shear stress (T_c) at the base of the moving mass (circles in Fig. 5).
- the mass moves if the acting shear stress ($\mathbf{T}_t$) is larger or equal to the resisting shear stress (T_c) at the base of the moving mass (squares and triangles in Fig. 7)

The resisting shear stress (T_c) is given by:

$$T_c = |T_z|\tan\varphi = |\sigma_{zz|b}|\tan\varphi = \rho g \gamma_z h \tan\varphi \,. \tag{58}$$

It is constant at a given time and defines the admissible state of the traction, T_{ti} (Fig. 6).

Since the $\mathbf{q}_i^{n+1}$ value is known (see Section "Kinetic Scheme"), the intercept of (57) on the ordinate axis is known and consequently the line of the family is identified.

For a given node i and a given time $n+1$ at $\tilde{q}_i^{n+1} = 0$ the reciprocal position of T_{ti} and T_c can be:

$$\left\|\mathbf{T}_{ti}^{n+1}\right\| < T_c = \rho g \gamma_{zi} h_i^{n+1} \tan\varphi \rightarrow \mathbf{v}_{ti}^{n+1} = 0 \tag{59}$$

$$\left\|\mathbf{T}_{ti}^{n+1}\right\| \geq T_c = \rho g \gamma_{zi} h_i^{n+1} \tan\varphi \quad \rightarrow \quad \left\|\mathbf{T}_{ti}^{n+1}\right\| = T_c \quad \rightarrow \quad \mathbf{T}_{ti}^{n+1} = -\rho g \gamma_{zi} h_i^{n+1} \tan\varphi \,\mathrm{sgn}\left(\overline{\mathbf{v}}_{ti}^{n+1}\right) \tag{60}$$

The flux with friction results the following:

$$\tilde{\mathbf{q}}_i^{n+1} = \mathbf{q}_i^{n+1} - \Delta t g \gamma_{zi} h_i^{n+1} \tan\varphi \,\mathrm{sgn}\left(\overline{\mathbf{v}}_{ti}^{n+1}\right) \tag{61}$$

Since $\mathrm{sgn}\left(\overline{\mathbf{v}}_{ti}^{n+1}\right) = \mathrm{sgn}\left(\mathbf{q}_i^{n+1}\right) = \mathrm{sgn}\left(\tilde{\mathbf{q}}_i^{n+1}\right)$ it can be finally stated:

$$\tilde{\mathbf{q}}_i^{n+1} = \left(\left\|\mathbf{q}_i^{n+1}\right\| - \Delta t g \gamma_{zi} h_i^{n+1} \tan\varphi\right) \frac{\mathbf{q}_i^{n+1}}{\left\|\mathbf{q}_i^{n+1}\right\|} \tag{62}$$

Rheological Laws

Different rheologies exist to describe the basal shear term $\mathbf{T_t} = \left(T_x, T_y\right)$ that develops at the interface between the flowing material and the rough surface. The following are here implemented:

- Frictional rheology, the resisting shear forces at the base of the flowing mass are assumed to depend on the normal stress, but not on velocity

$$\mathbf{T}_t = -\rho g \gamma_z h \tan\varphi \,\mathrm{sgn}\left(\overline{\mathbf{v}}_t\right) \tag{63}$$

where φ is the dynamic basal friction angle and $\overline{\mathbf{v}}_t = \left(\overline{v_x}, \overline{v_y}\right)$.

- Voellmy rheology, which consists of a turbulent term, v^2/ξ, accounting for velocity-dependent energy losses, and a friction term for describing the stopping mechanism.

$$\mathbf{T}_t = -\left(\rho g \gamma_z h \mu + \rho g \frac{\overline{\mathbf{v}}_t^2}{\xi}\right) \operatorname{sgn}\left(\overline{\mathbf{v}}_t\right) \tag{64}$$

where ξ is the turbulence coefficient (ξ = C^2, where C is the Chézy coefficient), μ (=tan φ) is the friction coefficient; the others terms are similar as in equation (63).

- Quadratic rheology, in which the shear resistance stress is provided by the following expression:

$$\mathbf{T}_t = -\left(\tau_y + \frac{k\eta}{8h}\left|\overline{\mathbf{v}}_t\right| + \rho g \frac{n_{td}^2 \overline{\mathbf{v}}_t^2}{h^{\frac{1}{3}}}\right) \operatorname{sgn}\left(\overline{\mathbf{v}}_t\right) \tag{65}$$

where τ_y is the Bingham yield stress, η is the Bingham viscosity, n_{td} is the equivalent Manning coefficient for turbulent and dispersive shear stress components and k is the flow resistance parameter. The first and the second terms on the right hand side of equation (65) are the yield term and the viscous term as defined in the Bingham equation, respectively. The last term represents the turbulence contribution (O'Brien et al. 1993).

Model Evaluation

Results of controlled laboratory experiments are a useful tool for validating a code once a new aspect is numerically implemented, but laboratory cannot reproduce on site events due to the well-known scale effects. Further, although it is scientifically appealing to be able to measure the input parameters independently, no standard tests are available to measure, for example, the properties of coarse rock avalanche debris travelling at extremely rapid velocities. Such properties, even if measurable, may change significantly during the course of motion, along with the rheology itself, and may be scale-dependent.

Once a code is validated, the back-analysis of large-scale occurred events can be the way to test the capability of the approach in simulating the complexity of flow-like landslides and to calibrate rheological parameters for forward-analyses. Rheological parameters are constrained by systematic adjustment during trial-and-error back-analysis of full-scale prototype events. Simulation is typically achieved by matching the simulated travel distance, velocities and extent and depth of the deposit to those of the prototype. Reasonable confidence limits have obviously to be applied in using these parameters for prediction purposes (McDougall et al., 2008).

Parameter selection for the purposes of prediction can also be difficult because, while simulation of a single event can be performed quite efficiently, back-analysis of a significant number of similar prototype events is required for the calibrated parameters to be physically and/or statistically justifiable.

In the following sections the influence of the geometry discretization on the obtained numerical flow (§ "The Mesh-Dependency Problem"), the importance of a tool able of simulating the behaviour of the mass on a real three dimensional topography (§ "Motion Across Complex 3D Morphology") and the advantage of a set of implemented rheologies among which to select the more appropriate as a function of the characteristics of the event/site (§ "Calibration of Rheological Parameters") are discussed.

The Mesh-Dependency Problem

Structured and unstructured meshes are here compared to test their influence on numerical results. To this aim, a simple numerical simulation in which a mass is released from a hemi-spherical cap on a horizontal plane is carried out (Fig. 8).

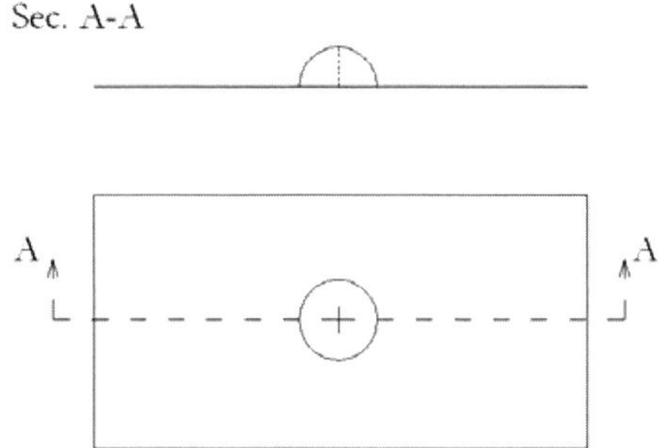

Figure 8. Hemi-spherical mass on a horizontal plane (Pirulli, 2005).

The limits of a structured mesh immediately emerged. Instead of a symmetric propagation in all the directions, the mass is widely influenced by the geometry of the mesh (Fig. 9a).

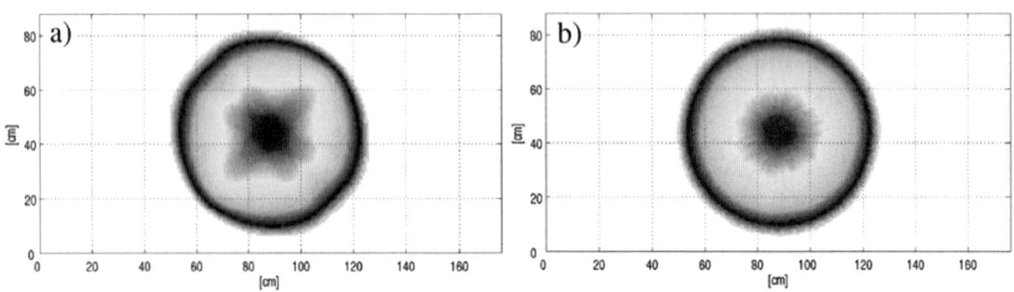

Figure 9. Propagation of hemi-spherical mass on horizontal plane at t = 0.3s using (a) a structured mesh, (b) an unstructured mesh (after Pirulli et al., 2007).

Moving to an unstructured mesh allowed to reduce the asymmetric effects (Fig. 9b) keeping unchanged the cpu-time.

Motion Across Complex 3D Morphology

The importance of studying the motion of a mass across a 3D terrain is here evidenced analysing a laboratory deflected run out experiment and the dynamics of the Val Pola rock avalanche (Italy, 1987).

These two examples allow to verify the robustness of the approach in simulating the behaviour of a moving mass on bend and in case of run up and lateral spreading. The importance of a 3D model clearly emerges.

Deflected Run out Experiment

A laboratory experiment was conducted at the University of British Columbia with dry polystyrene beads by McDougall and Hungr (2004). The material was released from a box onto a chute with variable slope (to control the approach velocity), ran out onto a 20° approach slope and was deflected by a dike oriented obliquely to the flow direction. The deflection angle, λ (plan angle between the initial direction of motion and the intersection of the dyke and approach planes), and the dike dip angle, ψ_d, were variable. A photograph of the laboratory apparatus is shown in Fig. 10.

The box used to contain and release the material at start up could not be replicated by a digital 3D sliding surface, due to its infinitely sloping sidewalls. Therefore, an imaginary release chute and initial distribution of material were used (McDougall, 2006). By trial and error, the position, width and velocity of the simulated flow front were synchronized with the experiment at the start of the 20° approach slope.

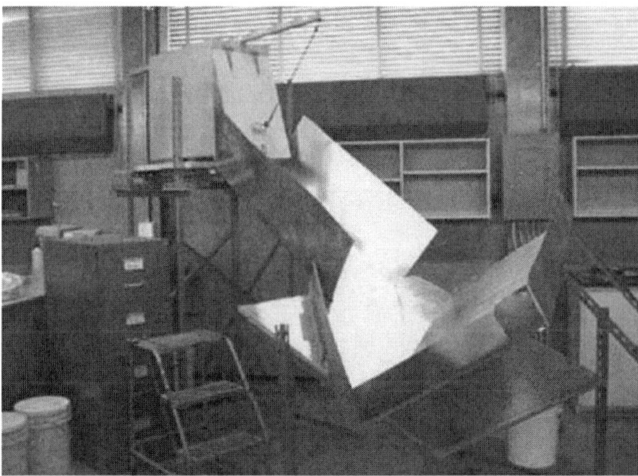

Figure 10. Photograph of laboratory apparatus used for deflected runout experiments (Image courtesy of S. McDougall, University of British Columbia).

A simulation of an experiment configured with $\lambda = 60°$ and $\psi_d = 33°$ is shown in Fig. 11, were laboratory results and RASH3D numerical analyses are compared.

Numerical results were obtained with a frictional rheology assuming a basal friction angle, φ, equal to 20° and an internal friction angle, δ, equal to 25°. These friction angles are within a small range of values measured in separate laboratory tests by McDougall, placing a conical pile of beads on a sheet metal plane and, respectively: 1) measuring the tilt angle that initiates basal sliding, and 2) measuring the angle of repose of the material itself.

With rheological parameters calibrated on the basis of the previous laboratory test (φ =20° and δ=25°) the model RASH3D produces rather good predictions of maximum runup distance, as well as the position and distribution of slide material in time.

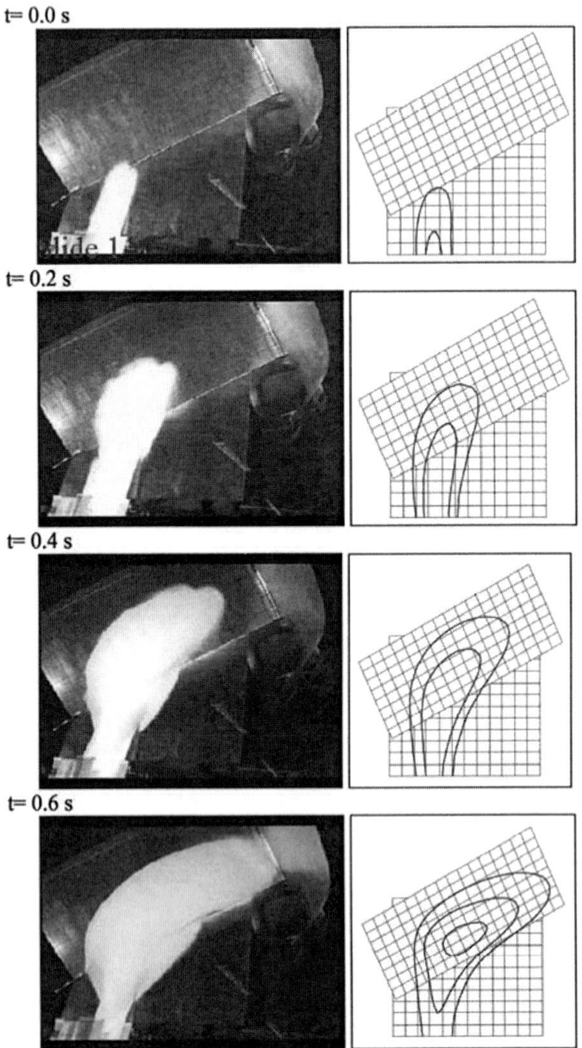

Figure 11.Comparison between the deflected runout experiment using λ=60° and ψ_d=33° (left side) and RASH3D results (right side). The planes are marked with a 10cm square grid (Photographs courtesy of S. McDougall, University of British Columbia).

Val Pola Rock Avalanche (Italy, 1987)

Valtellina is a valley of glacial origin located in Lombardia Region, Northern Italy, near the Swiss border (Fig. 12). The Val Pola rock avalanche moved down from the east flank of

Mount Zandila, a 2936 m peak on the right side of Valtellina, and took its name after the canyon which marked the northern limit of the rock mass (Govi et al., 2002).

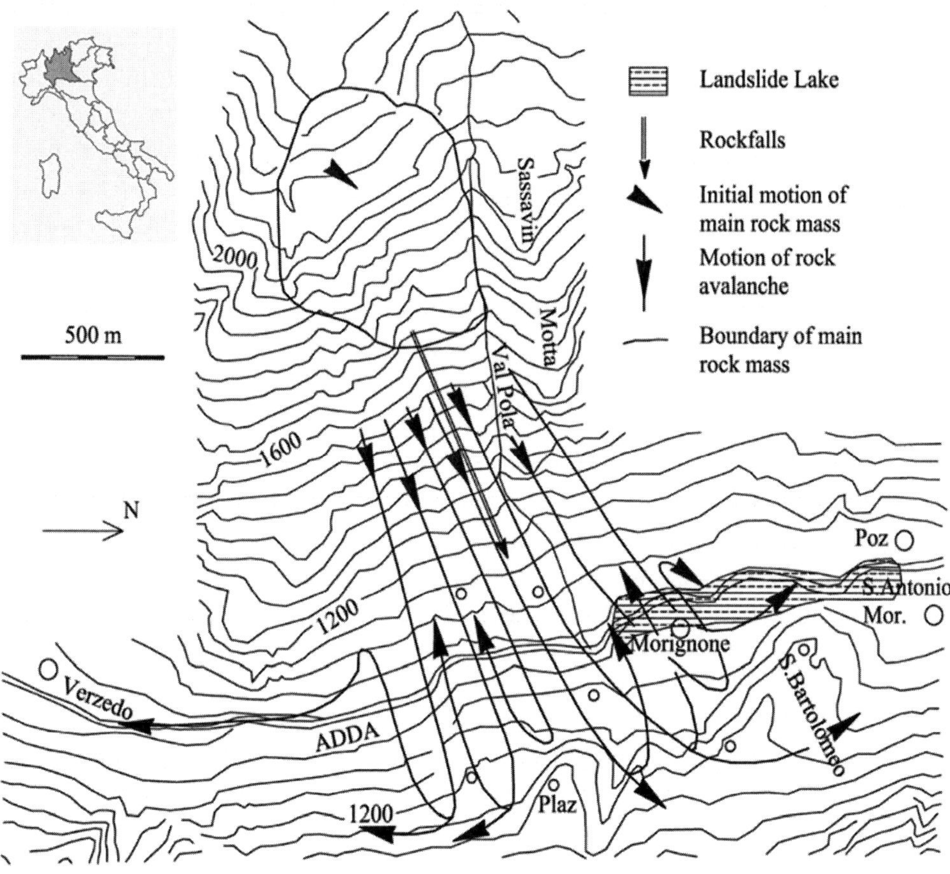

Figure 12. State of the site in the early morning of July 28, before the rock avalanche and rock avalanche kinematics (modified after Govi et al., 2002).

The period between July 15 and 28, 1987 was of particularly heavy precipitation in the Alps. A great number of flood disasters and gravitational mass movements were consequences of this meteorological coincidence.

The Val Pola rock avalanche began at about 7:25 a.m. of July 28 by the detachment of an estimated rock volume of about $30 \cdot 10^6$ m^3.

The slide mechanism was fully determined by major structures: a fault plane dipping at 40° formed the back scarp of the slide, another fault, coinciding with the heavily eroded Val Pola, formed the northern margin and a set of bedding joints dipping 32° traversed the slope. The resulting shape was a remarkably regular compound wedge (Smith & Hungr, 1992).

The rock mass initially shifted slowly towards the north, parallel with the dip direction of the bedding joints. Subsequently, it collapsed toward the valley. Seven men disappeared in the stream of rock fragments. A village and six hamlets, all previously evacuated, were destroyed.

The debris mass crossed the bottom of the valley, ran up the opposite slope and then parted into two arms. Part of the material in the southern arm came to rest while the other fell back, again crossed the valley and ran up the source slope (Fig. 12). Finally, a part of the mass was channelled southwards along the valley bottom. Similarly, part of the debris in the northern arm stopped, while another part ran back across the valley bottom and back up the source slope. But in this second case, a considerable part of the debris mass plunged into the small landslide lake formed in the previous days, and raised a huge wave of debris-water swept north along the valley bottom for more than 2 km, inundating some not evacuated villages and killing 22 people (Govi et al., 2002). When the process stopped, this arm presented a marshy area on its north side, which was followed by a new greater lake northern (Fig. 13) that can be considered as a second phase of the event much more similar to the behaviour of a debris flow.

Exposed debris consisted mainly of diorite fragments; gabbro and paragneiss were locally present (Chiesa and Azzoni, 1988). Although blocks up to 10 m and over were fairly common (some of which consisted of a conglomerate of ice and rock), the grain size of the debris ranged mostly between 0.5 m and sand size. Since about 3000 trees had been destroyed, abundant wood was scattered throughout. The maximum thickness of the accumulation was measured of about 90 m.

Taking into account that the mass descended a slope dipping about 32° along a path practically free of obstacles, it must be concluded that the initial energetic input was great. In this frame the high speed calculated by Costa (1991) of about 76-108 m/s can be considered reasonable.

To calibrate the parameters for an assumed rheology, information about the run out area shape are necessary.

If the path followed by the mass along the source slope, the maximum run up of the mass along the opposite slope and the south boundary of the spreading area can be defined in an enough detailed way from available data, the north boundary of the spreading area is a little uncertain due to the behaviour assumed by the mass when it plunged into the landslide lake. It follows that the rock avalanche run out area to be compared with numerical results neglects both the long wave and the marshy area (Fig. 13a).

A second element of comparison is the longitudinal profile of the final deposit (Fig. 13b). As the distal point of the final deposit does not correspond to the maximum run up, it is fundamental to combine values in order to obtain a good agreement in term of run out area shape, maximum run up and final deposit profile.

Results obtained calibrating both a Frictional and a Voellmy rheology are here discussed.

The simple frictional model was initially considered. The tentative value of 16°, obtained from literature (Hungr and Evans, 1996), gave rather good results. Changing this value, run out area results are acceptable in the range of ±1deg while they got worse if ±2deg is assumed (Pirulli and Mangeney, 2008).

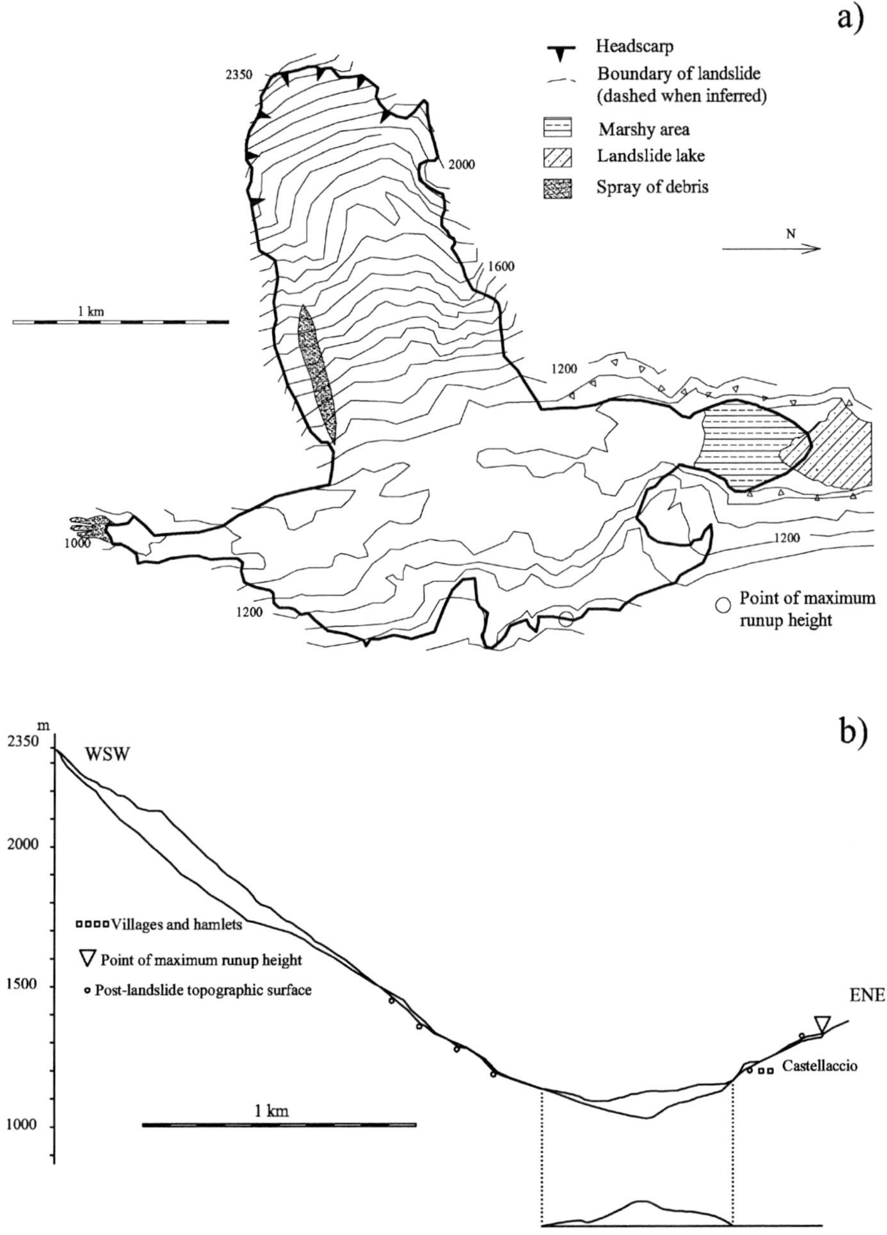

Figure 13. (a) Rock avalanche morphology. In the inset, subdivision by sector. Dotted line gives the position of the profile considered in (b). (b) Profile of the rock avalanche with- pre- and post-landslide topographic surfaces. Small circles mark the post- landslide surface in the positions where uncertainties might arise (modified after Govi et al., 2002).

From a comparison in terms of final deposit profile, it emerges that the profile obtained with 15° is the best in term of distal point position even if the maximum runup is overestimated (Fig. 14a-14b).

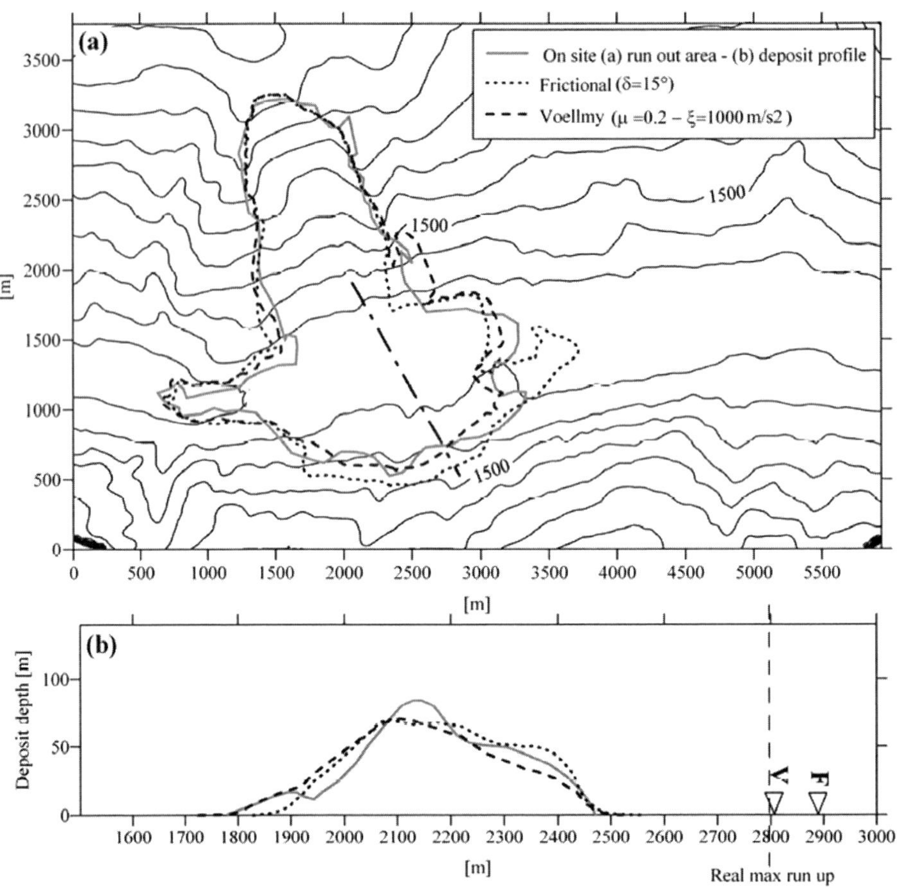

Figure 14. Results of parametric analyses carried out assuming a frictional and a Voellmy rheologies. The symbol ∇ gives the position of the maximum run up as a function of the assumed rheology (modified after Pirulli and Mangeney, 2008).

In case of frictional rheology a dynamic friction angle of 15°-16° is considered as able to give satisfactory results in terms of run out area shape (e.g. runup and lateral spreading of the mass) and final deposit profile. A good agreement with the value of friction angle assumed by Hungr and Evans (1996) was obtained.

In case of a Voellmy rheology, the influence of each rheological parameter was investigated making three sets of parametric analyses starting again from values proposed by Hungr and Evans (1996):

[1] friction coefficient constant (μ = 0.1) and turbulence coefficient variable (ξ)
[2] friction coefficient variable (μ) and turbulence coefficient constant (ξ=500m/s^2)
[3] friction coefficient constant (μ = 0.2) and turbulence coefficient variable (ξ)

The results of the first set of analyses [1] shows that in terms of run out area shape the more satisfactory results are obtained using ξ=350m/s^2 but with this combination of parameters the deposit profile largely differs from the on site profile.

This lack of good results gave reason of running a second set of analyses [2] in which a constant turbulence coefficient is assumed (ξ=500m/s^2) with an increasing friction angle. The more satisfactory results are obtained using μ = 0.14 but the obtained deposit profile still does not correspond to the on site profile.

To obtain a good agreement both in term of runout area and deposit profile, it was tried to move to an higher value of friction coefficient assuming μ = 0.2 and changing the turbulence coefficient [3]. This assumption gave the best results with a turbulence coefficient equal to ξ=1000m/s^2 (Fig. 14a-14b).

It can be then underlined that using a Voellmy rheology, a rather good correspondence in terms of run out area shape, maximum run up and final deposit profile required an higher value of friction coefficient (μ = 0.2) than that proposed by Hungr & Evans (1996). Assuming lower values of friction angle any change of the turbulence coefficient was not able to give a good deposit profile.

It can be finally stated that the RASH3D code, for the two assumed rheologies, was able to reproduce the runout area and the final deposit of the Val Pola event in a rather satisfactory way. Nevertheless, the precision in approximating the run up and the lateral spreading of the mass was very satisfactory in case of Voellmy rheology, while an overestimation of the propagation is evidenced in case of the Frictional rheology (Fig. 14b).

Calibration of Rheological Parameters

In this Section are analysed two real cases: the Thurwieser rock avalanche, to evidence the importance of the rheology selection procedure, and the Tate's Cairn debris flow, to remark the difficulty in choosing the appropriate combination of rheological values for carrying out a forward-analysis.

The Thurwieser Rock Avalanche (Italy, 2004)

The Thurwieser rock avalanche, involving a rock volume of about $3 \cdot 10^6$ m^3, detached on September 18th, 2004 from the South-East flank of the Thurwieser Peak (3657 m a.s.l.), in the Central Italian Alps (Fig. 15). Permafrost degradation was supposed to be the most probable triggering factor of the event (Cola, 2005). Similar phenomena, even with smaller volumes, have occurred since the summer of 2003 in the Alps (e.g., Bernina, the Matterhorn, Mont Blanc) following periods of exceptionally high temperatures.

The moving mass initially ran on a pre-existing sliding surface located between 3280 m and 3630 m a.s.l., then fell down on a flat portion of the Vedretta dello Zebrù glacier below (2900 m a.s.l.), crossed the area on which the 5th Alpini Refuge is built and channelized along the right branch of the Rio Marè stream below, a tributary of the Zebrù stream, where it stopped at about 2220 m a.s.l., not far from the Baita del Pastore area (Fig. 15).

The steep slope and the existing lobe determined the absence of deposits upper-level. A film of debris partially covered the glacier, while the main deposit was concentrated, as above mentioned, along the Rio Marè stream.

Due to the characteristics of both the moving material (i.e. rock, debris, ice, snow) and the superficial layer of the propagation path (e.g. the presence of a glacier), a distance of at least 2600m was covered by the mass. The mean mass velocity can be estimated in about 30-40 m/s, as usually recorded in this type of events (Dei Cas et al., 2004).

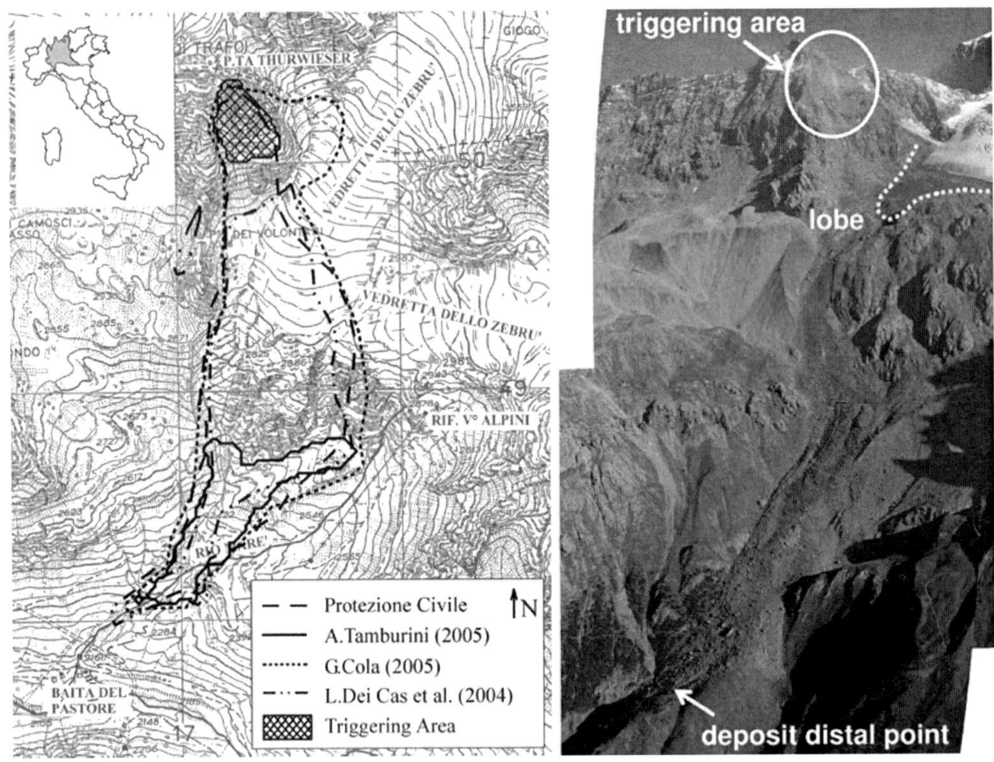

Figure 15. The Thurwieser rock avalanche. Location of the study area and boundary of the runout area on detailed cartography, as surveyed on site by different authors. Image courtesy of Regione Lombardia (Italy).

The on site boundaries of the Thurwieser runout area are not uniquely defined. Aerial photos and surveys of different authors have been then superimposed and numerical results were systematically compared with the obtained envelope. Three of the four considered interpretations (Protezione Civile, Cola 2005 and Dei Cas et al. 2004) sketch the boundary of the whole runout, propagation plus deposit, while Tamburini's (2005) interpretation is focused only on the delimitation of the main deposit in the Rio Marè stream (Fig. 15).

Three subsequent sets of analyses, based on the assumption of different rheological laws, were necessary to correctly back-analyse the Thurwieser rock avalanche (Pirulli, 2008): Hp.1) a Frictional rheology with a unique value of the friction angle (φ) for the whole slope; Hp.2) a Frictional rheology with a friction angle changing as a function of the characteristics of the slope: φ_1, the value of the friction angle outside the glacier (zone 1), and φ_2, the value of the friction angle on the glacier (zone 2); and Hp.3) a Voellmy rheology with two parameters to be set for each zone of the slope: the friction angle (φ) and the turbulence coefficient (ξ).

In the Hp.1, the presence of the glacier on the propagation of the mass is not expressly taken into account and an average friction angle, which includes the characteristics of the whole path of propagation, is looked for. But, parametric analyses gave unsatisfactory results in terms of both propagation path and deposit shape.

The value of the friction angle that was able to reproduce the correct travel distance (φ = 25°) determined, contrarily to reality, the deposition of large quantities of material on the glacier (Fig. 16a). Reducing the value of the friction angle (φ=20°), an excessive travel distance of the front was obtained with unchanged deposition on the glacier.

In the Hp.2, the value of the friction angle to be assumed on the glacier (φ_2) was initially selected from the literature. Assuming φ_2 = 0.6° (Bottino et al. 2002) and calibrating φ_1, a good approximation of the travel distance of the Thurwieser rock avalanche was obtained with φ_1 = 30°, but an excessive spreading of the mass along the runout path was observed (Fig. 16b). A further calibration of φ_1 with a different value for φ_2 did not change the results.

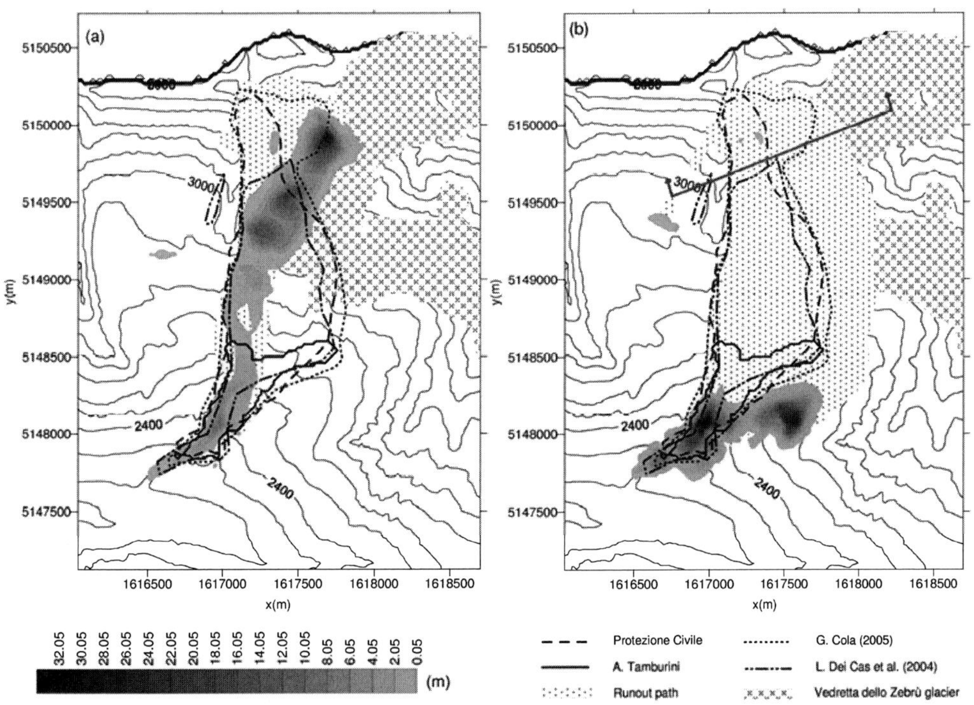

Figure 16. Comparison between the boundaries of the Thurwieser rock avalanche runout area, as surveyed on site by different authors and the depth contours of sliding debris, as simulated by RASH3D, assuming a Frictional rheology with (a) one friction angle value along the whole path of propagation (φ = 25°), and (b) two friction angle values (φ_1=30°, φ_2=0.6°) (modified after Pirulli, 2008).

It can therefore be stated that Frictional rheology is not able to accurately simulate the Thurwieser rock avalanche dynamics. Transition to a more complex rheology was then necessary. This change implies the calibration of a larger number of rheological parameters.

In Hp.3, two different values of friction angle were considered for zone 1 and zone 2 (as in Hp.2) but a common turbulence coefficient was assumed for both zones. The best approximation of the real event was obtained assuming φ_1=20°, φ_2=0.6°, ξ=100m/s^2. For these best fit data, the calculated sequence of entire landslide process in terms of depths of the sliding mass is visualized in Fig. 17.

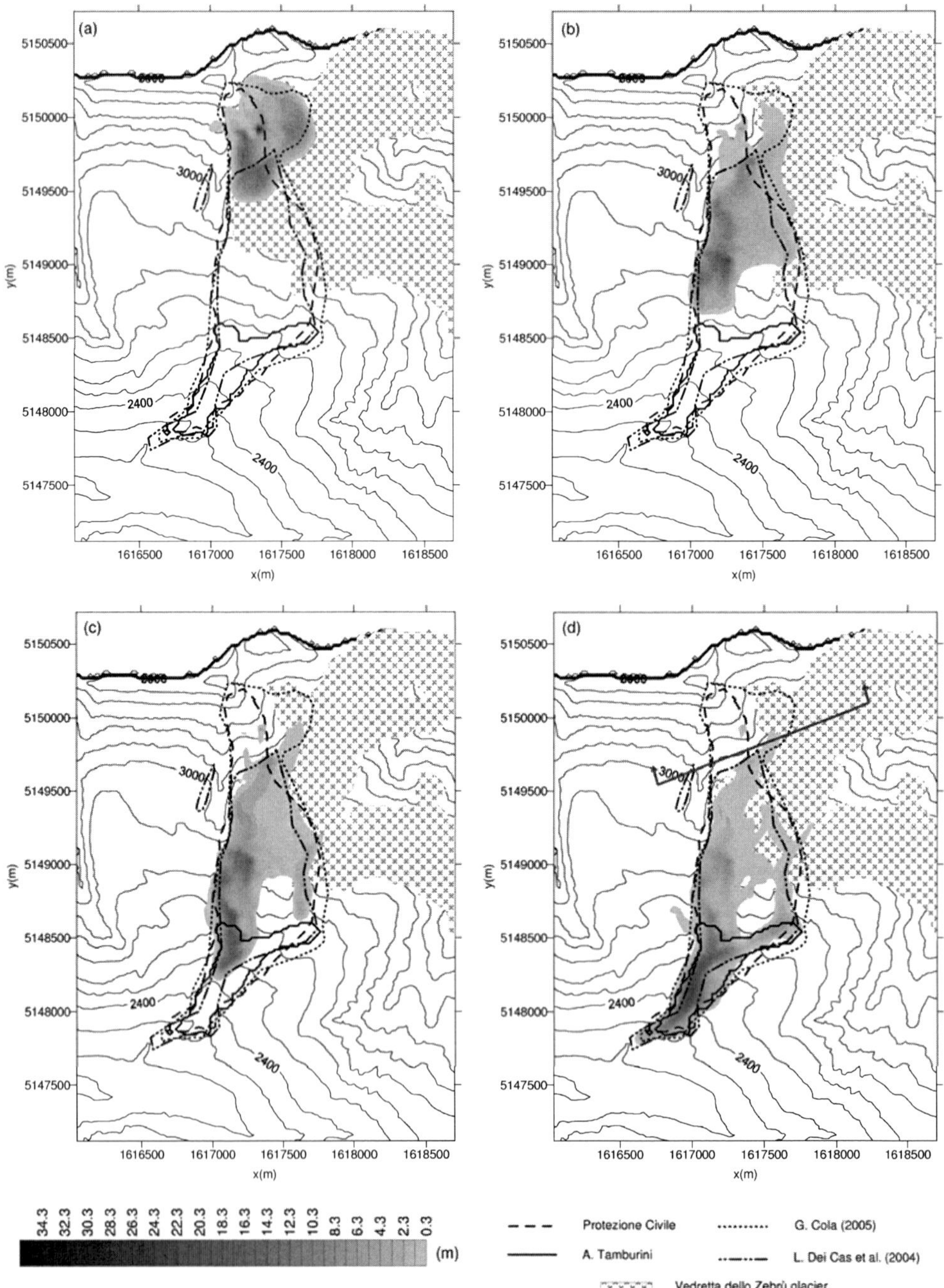

Figure 17. Comparison between the boundaries of the Thurwieser rock avalanche runout area, as surveyed on site by different authors and the depth contours of sliding mass, as simulated by RASH3D, assuming a Voellmy rheology (combination n° 3: φ_1=20°, φ_2=0.6°, ξ=100m/s2) (modified after Pirulli, 2008).

In comparison with the Frictional results, it immediately emerges that the Voellmy rheology allows the spreading of the mass along the path of propagation to be controlled in a satisfactory way.

A comparison between the Voellmy best simulation (φ_1=20°, φ_2=0.6°, ξ=100m/s^2) and the Frictional best simulation (φ_1=30°, φ_2=0.6°) was carried out in an attempt to evaluate which factors allow the Voellmy rheology to simulate the Thurwieser dynamics better than the Frictional rheology. A transversal section placed in the upper part of the glacier was therefore chosen to analyse at which velocity the moving mass crosses the glacier (indicated in Figs. 16b and 17d). The calculated sequence of profiles is represented in Fig. 18.

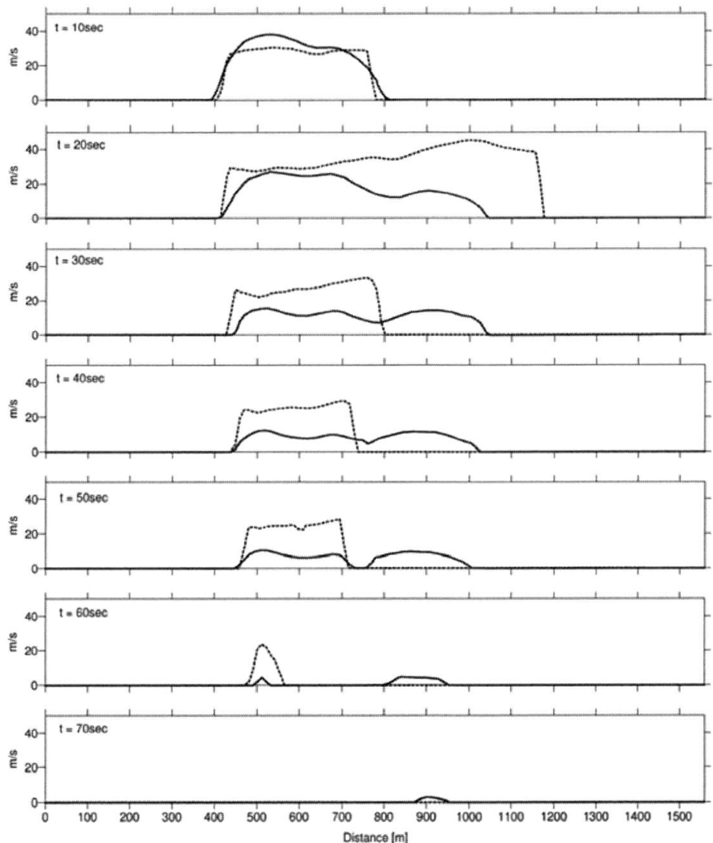

Figure 18. Velocity of the mass along a transversal section of the propagation path at different times. Dotted line: simulated Frictional velocity; Continuous line: simulated Voellmy velocity. The position of the considered section is indicated in the plane in Fig. 16b and Fig. 17d (modified after Pirulli, 2008).

At t = 10s, the Voellmy and Frictional velocity profiles are similar and a velocity of about 30m/s is reached.

At t = 20s, 30s it can be observed that:

1) in the case of the Frictional rheology, the high velocity at which the mass crosses the considered section does not allow a correct control of the propagation on the glacier (the mass spreads excessively);

2) in the case of the Voellmy rheology, on the contrary, the lower velocity of the mass allows a better control of the propagation on the glacier and a better fit of the onsite surveyed runout path.

The remaining profiles (t = 40s, 50s, 60s, 70s) further stress that the Voellmy velocities are lower than the Frictional velocities.

These differences in velocities can be considered the element that has allowed, in case of the Voellmy rheology, a reduced spreading of the mass along the whole propagation path and the deposition of a layer of material on the glacier.

The Tate's Cairn Debris Flow (Hong Kong, 2005)

In the early morning of 22 August 2005, following heavy rainfall on 19 and 20 August 2005, a debris flow was reported to have occurred on a natural hillside about 200 m to the north of Tate's Ridge and about 500 m south of Kwun Ping Road, Kwun Yam Shan, Hong Kong (Fig. 19).

The location of the August 2005 debris flow is probably associated with the geomorphological setting of the site, within, and adjacent to, a densely vegetated, linear topographical depression (about 60 m wide by 100 m in length) between two rounded spurlines and above a convex break-in-slope at the head of an ephemeral drainage line (Fig. 19).

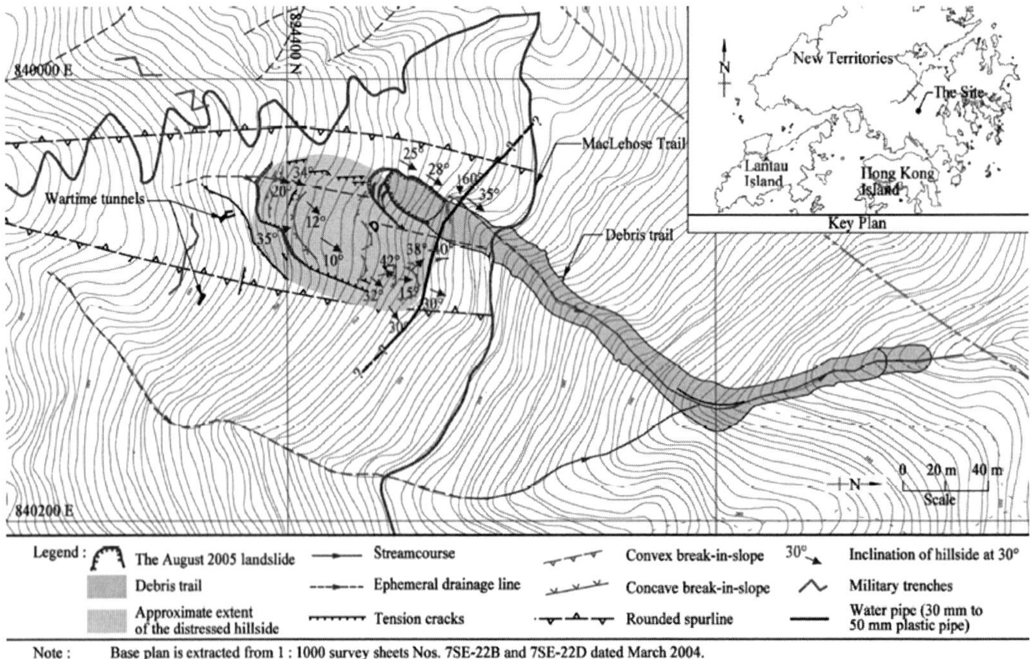

Figure 19. The August 2005 debris flow – Location plan (after MGS, 2007).

The aerial photographic records indicate that the August 2005 debris flow occurred at about the same location as two smaller pre-1956 landslides. The previous instability and progressive degradation of the hillside appear typical of a hillside retreat process. Both surface runoff and subsurface flow would be directed towards the landslide site at the head of the drainage line and the presence of an interconnected subsurface drainage network would have promoted rapid groundwater flow to the landslide site (MGS, 2007).

The crown of the debris flow source area is located at an elevation of about 448 m asl with the toe located at about 430 m asl. The failure involved up to 5 m depth of colluvium with a total volume of about 2350 m³. A large portion of the displaced material (about 1350 m³) remained within the source area as intact rafts separated by a series of stepped tension cracks. The rest of the detached mass (about 1000 m³) entered the ephemeral drainage line below, and developed into a channelised debris flow (Fig. 19).

The landslide debris traveled a total distance of about 330 m down the drainage line and came to rest at two distinct boulder dams within the drainage line.

The difference in elevation between the landslide source and the end of the debris trail was approximately 138 m, with a travel angle of about 24° (Wong and Ho 1996).

As a consequence of the event an approximately 10 m section of the MacLehose Trail, that is about 30 m to the north of the toe of the August 2005 debris flow source area, was severed but no casualties were reported.

A detailed inspection of the hillside above the August 2005 landslide in March 2006 revealed an extensive system of tension cracks (Fig. 19). These tension cracks define an area of distressed hillside with an estimated volume of 10000 m^3 located on the southeast side of the August 2005 landslide source area. The possible toe of this distressed hillside lies at an elevation between 430m asl and 440m asl and has an average inclination of about 20°-25° (Fig. 19).

The hillside above the distressed hillside is vegetated and is inclined at about 30°, gradually reducing to about 15° near Tate's Ridge located about 100 m to the south. The hillside below the study area is densely vegetated and is inclined at about 30° to 40° that gradually reduces to 10° to 20° along the streamcourse.

Given that the majority of the tension crack faces appeared fresh, the largest portion of the identified movement was postulated to have likely triggered by the August 2005 rainstorm. The possibility of the fresh tension cracks being a reactivation of the distressed hillside due to the severe rainstorm in August 2005 cannot be ruled out (MGS, 2007).

Theoretical stability analyses and groundwater monitoring data to date suggest that the development of a perched water table above the colluvium could initiate a failure potentially leading to the formation of the main scarps (MGS, 2007).

The back-analysis of the August 2005 debris flow was carried comparing the results obtainable with the Frictional, the Voellmy and the Quadratic rheologies. Once the appropriate rheology is identified, calibrated rheological parameters can be useful for running the forward-analysis of the distressed hillside. In the present chapter, only the influence of the rheology on the quality of the back analysis is discussed.

Starting from the simple Frictional model a calibration of the rheological parameters has given good back-analysis results setting the basal friction angle equal to 27° (onsite estimated travel angle was of about 23°). In case of Quadratic rheology it was possible to obtain a satisfactory representation of the mass runout by assuming τ_y = 1.2kPa – η = 40Pa•s – n = 0.03, while the Voellmy rheology gives the best results with φ = 25° and ξ = 1000m/s^2.

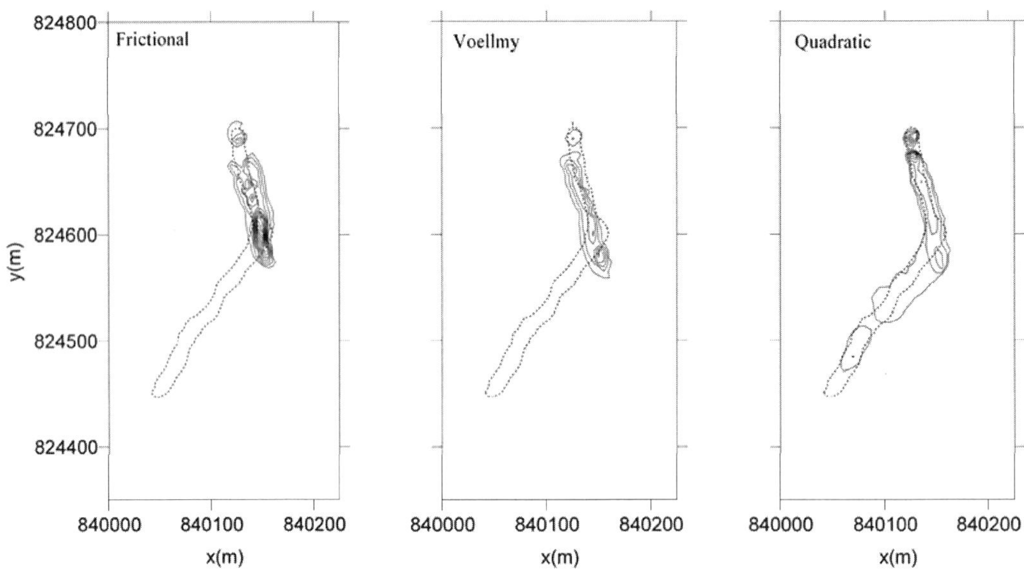

Figure 20. The August 2005 debris flow – Results of parametric analyses carried out assuming a frictional, a Voellmy and a Quadratic rheologies.). Dashed line: on site run out path.

In comparison to the Frictional rheology, the Voellmy rheology introduces an additional parameter to be calibrated but, as the Frictional rheology, it was not able to simulate the deposition of material along the path of propagation, which at the contrary was observed on field (MGS, 2007). In both the cases, the final deposit of the mass is concentrated in the distal part of the propagation path (Fig. 20).

It is also observed that the value of the friction angle calibrated for the Frictional rheology (27°) and that calibrated for the Voellmy rheology (25°) are very close. This small difference in the value of the friction angle coupled with a high value of the turbulence coefficient (1000m/s^2), which, in its turn, reduce the contribution of the turbulent part of the Voellmy rheology, leads to very similar results in both the Frictional and the Voellmy rheologies.

Respect to the previous analyses, the Quadratic rheology adds a further parameter to be calibrated but it allowed a better approximation of the deposition of material along the path of propagation (Fig. 20).

A splitting of the mass in the final part of the deposit emerges with all the rheologies due to the existence of two boulder dams that were probably formed in previous landslide incident(s) (MGS, 2007).

CONCLUSION

Flow-like landslides are among the most destructive and difficult to prevent types of landslide phenomena. Their impact is becoming stronger and stronger due to increasing tourism and the construction of new roads and railways in mountainous areas (Bonnard et al., 2004).

With prediction losses could be reduced, as they could provide means to define the hazardous areas, estimate the intensity of the hazard and work out the parameters for the identification of appropriate protective measures. At the same time, reliable predictions of run out could help to avoid exceedingly conservative decisions regarding the development of hazardous areas.

Dealing with the run out problem, it has been observed that a continuum mechanics approach is flexible enough to allow that many of the aspects characterising propagation of a mass on a complex topography can be taken into account.

As an example, the system of governing equations used in RASH3D has been derived from first principles using a straightforward series of explicit and justifiable assumptions and approximations. The generality of the system, in accordance with the equivalent fluid approach, makes it applicable in a wide variety of situations.

Results of some carried out numerical analyses have evidenced the possible influence of the geometry discretization on the obtained numerical flow, the importance of a tool able of simulating the behaviour of a mass on a real three dimensional topography, so that aspects as run up, lateral spreading are not ignored, and the advantage of a set of implemented rheologies among which to select the more appropriate as a function of the characteristics of the event/site to be analysed.

Further developments of the undertaken research work are needed from both a theoretical and a numerical point of view. These have to be done by keeping in mind the final aim, which is to provide a tool whose application could give useful information for investigating, within realistic geological contexts, the dynamics of flows and of their arrest phase. In particular, as regards the RASH3D code a fundamental aim that is being pursued concerns the numerical implementation of entrainment of material along the path of propagation, which is a main aspect when events like debris flows are investigated.

ACKNOWLEDGMENTS

The author wishes to thank Dr. Giovanni Mortara (CNR-IRPI Torino, Italy) and Dr. Andrea Tamburini (CESI S.p.A, Milano, Italy) for the data concerning the Thurwieser rock avalanche; Dr. Franco Godone (CNR-IRPI, Torino, Italy) and Dr. Luca Mallen (ARPA Piemonte, Torino, Italy) for their contribution to the development of the digital terrain model of Val Pola; the Geotechnical Engineering Office, Civil Engineering and Development Department of the Government of the Hong Kong SAR for the provision of the digital terrain model of the Hong Kong landslide; Dr. Anne Mangeney (IPGP, France) and Dr. Marie-Odile Bristeau (INRIA, France) for having offered the use of the SHWCIN code and for having helped solve some fundamental numerical problems; Prof. Renato Lancellotta and Prof. Claudio Scavia for having reviewed this manuscript.

Paper reviewed by: Prof. Renato Lancellotta and Prof. Claudio Scavia - Politecnico di Torino, Department of Structural and Geotechnical Engineering, Italy.

REFERENCES

Audusse, E. Bristeau, M.O. & Perthame B. (2000). *Kinetic schemes for Saint-Venant equations with source terms on unstructured grids,* INRIA Rep. 3989, Natl.Inst.for Res.in Comput. Sci. and Control, Le Chesnay, France.

Bonnard, C. Forlati, F. & Scavia, C. (Eds.) (2004). *Identification and mitigation of large landslide risks in Europe, Advances in risk assessment,* 5th Framework Program: Imiriland Project, A.A. Balkema.

Bottino, G. Chiarle, M. Joly, A. & Mortara, G. (2002). *Modelling rock avalanches and their relation to permafrost degradation in glacial environments,* Permafrost and Periglacial Processes, 13, 283-288.

Bristeau, M.O. & Coussin, B. (2001). *Boundary conditions for the shallow water equations solved by kinetic schemes,* INRIA Rep. 4282, Natl.Inst.for Res.in Comput. Sci. and Control, Le Chesnay, France.

Chen, H. & Lee, C.F. (2000). Numerical simulation of debris flows, *Canadian Geotechnical Journal,* 37, 146-160.

Chiesa, S. & Azzoni, A. (1988). Esecuzione di rilievi geologici e geostrutturali sulla frana di Val Pola [A geological and geostructural field survey of the Val Pola landslide]. Bergamo, *ISMES Report PROG/ASP/4284, RTF-DGM-02150,* p. 57.

Cola, G. (2005). *La grande frana della cresta sud-est della Punta Thurwieser* (Thurwieser-Spitze) 3658 m (Alta Valtellina, Italia), Terra Glacialis, Anno VIII, 9-37.

Corominas, J. (1996). The angle of reach as a mobility index for small and large landslides, *Can. Geotech. J.,* 33, 260-271.

Costa, J.E. (1991). *Nature, mechanics and mitigation of the Val Pola landslide,* Valtellina, Italy, 1987-1988. Zeirschrift fur Geomorphologie N.F. 35, 15-38.

Davies, T.R. (1982). Spreading of rock avalanches by mechanical fluidization, *Rock Mechanics,* 15, 9-24.

Dei Cas, L. Mannucci, G. & Tropeano, D. (2004). *Thurwieser, 18 settembre 2004, frana la cima in Alta Val Zebrù, SLM Sopra il Livello del Mare,* 17, 16-21.

Denlinger, R.P. & Iverson, R.M. (2004). Granular avalanches across irregular three-dimensional terrain: 1. Theory and computation. *Journal of Geophysical Research,* 109, F01014.

Govi, M. Gullà, G. & Nicoletti, P.G. (2002). Val Pola rock avalanche of July 28, 1987, in Valtellina (Central Italian Alps), In: Evans, S.G. & DeGraff, J.V. (Eds.), Catastrophic Landslides: Effects, Occurrence, and Mechanisms, *Review in Engineering Geology Volume XV,* The Geological Society of America.

Gray, J.M.N.T. Wieland, M. & Hutter, K. (1999). *Gravity-driven free surface flow of granular avalanches over complex basal topography.* Proceedings of the Royal Society of London, A, 455, 1841-1874.

Greve, R. & Hutter, K. (1993). Motion of a granular avalanche in a convex and concave curved chute: experiments and theoretical predictions. *Philosophical Transactions of the Royal Society of London, Physical Sciences and Engineering,* 342(1666), 573-600.

Greve, R. Koch, T. & Hutter, K. (1994). *Unconfined flow of granular avalanches along a partly curved surface,* Part 1: Theory. Proceedings of the Royal Society of London, A, 445: 399-413.

Heim, A. (1932). *Landslides and human lives* (Bergsturz und Menschenleben), In: Skermer N. (Ed.), Bi-Tech Publishers, Vancouver, p. 196.

Hungr, O. (1995). A model for the runout analysis of rapid flow slides, debris flows, and avalanches. *Canadian Geotechnical Journal*, 32, 610-623.

Hungr, O. & Evans, S.G. (1996). *Rock avalanche run out prediction using a dynamic model,* Proceeding 7th International Symposium on Landslides, Trondheim, Norway, 1, 233-238.

Hutchinson, J.N. (1968). Mass movement, In: Fairbridge R.W. (ed.), Encyclopedia of Geomorphology, Reinhold Publishers, New York, 688-695.

Hutchinson, J.N. (1986). A sliding-consolidation model for flow-slides. *Canadian Geotechnical Journal*, 23, 115-126.

Hutchinson, J.N. (1988). General Report: Morphological and geotechnical parameters of landslides in relation to geology and hydrogeology, In: Bonnard C. (ed), Procs. 5th *International Symposium on Landslides*, A.A. Balkema, Rotterdam, Netherlands, 1, 3-35.

Iverson, R.M. Denlinger, R.P. (2001). Flow of variably fluidized granular masses across three-dimensional terrain. 1: Coulomb mixture theory, *Journal of Geophysical Research,* 106(B1), 537-552.

Koch, T. Greve, R. & Hutter, K. (1994). *Unconfined flow of granular avalanches along a partly curved surface*, Part 2: Experiments and numerical computations. Proceedings of the Royal Society of London, A, 445, 415-435.

Mangeney, A. Vilotte, J.P. Bristeau, M.O. Perthame, B. Bouchut, F. Simeoni, C. & Yerneni, S. (2003). Numerical modelling of avalanche based on Saint Venant equations using a kinetic scheme, *J. Geophiys. Research*, 108, B11.

McDougall, S. Hungr, O. (2004). A model for the analysis of rapid landslide runout motion across three-dimensional terrain, *Canadian Geotechnical Journal*, 41(6), 1084-1097.

McDougall, S. Hungr, O. (2005). Dynamic modelling of entrainment in rapid landslides, *Can. Geotech. J.*, 42(5), 1437-1448.

McDougall, S. (2006). *A new continuum dynamic model for the analysis of extremely rapid landslide motion across complex* 3D terrain, Ph.D. thesis, University of British Columbia, Canada, pp.253.

McDougall, S. Pirulli, M. Hungr, O. & Scavia, C. (2008). *Progress in landslide continuum dynamic modelling*, Special Lecture, In: Chen et al. (Eds.), Procs. 10th International Symposium on Landslides and Engineered Slopes, Xi'an, China, 30 Giugno - 4 Luglio 2008, 1, 145-157.

MGS (2007). Detailed Study of the 22nd August 2005 *Landslide and Distress on the Natural* Hillside above Kwun Ping Road, Kwun Yam Shan, Shatin. Landslide Study Report 5/2007, Geotechnical Engineering, Maunsell Geotechnical Services Ltd.

Melosh, J. (1986). *The physics of very large landslides,* Acta Mech., 64, 89-99.

O'Brien, J.S. Julien, P.Y. & Fullerton W.T. (1993). Two-dimensional water flood and mudflow simulation, *J. Hydrol. Eng.,* 119(2), 244-261.

Picarelli, L. Oboni, F. Evans, S.G. Mostyn, G. & Fell, R. (2005). *Hazard characterization and quantification,* In: Hungr, O. Fell, R. Couture, R. Eberhardt E. (Eds.), Landslide risk management, Taylor & Francis Group, London, 27-61.

Pirulli, M. (2005) *Numerical modelling of landslide runout, a continuum mechanics approach,* Ph.D. thesis, Politecnico di Torino, Italy.

Pirulli, M. Bristeau, M.O. Mangeney, A. & Scavia C. (2007). The effect of the earth pressure coefficients on the runout of granular material. *Environmental modelling & software,* 22, 1437-1454.

Pirulli, M. & Mangeney, A. (2008). Results of back-analysis of the propagation of rock avalanches as a function of the assumed rheology, *Rock Mechanics and Rock Engineering,* 41(1), 59-84.

Pirulli, M. (2008). The Thurwieser rock avalanche (Italiana Alps): Description and dynamic analysis, *Engineering Geology*, doi:10.1016/j.engggeo.2008.10.007.

Pirulli, M. & Sorbino, G. (2008). Assessing potential debris flow runout: a comparison ot two simulation models, *Natural Hazards and Earth System Sciences*, 8, 961-971.

Rickenmann, D. (1999). Empirical relationships for debris flows, *Natl. Hazards*, 19(1), 47-77.

Rickenmann, D. (2005). Runout prediction methods, In: Jakob, M. & Hungr, O. (Eds.), *Debris-flow Hazards and Related Phenomena*, Springer Praxis, 305-324.

Sander, B. (1948). *Einfűhrung in die Gefűgekunde der geologischen kőrper,* Wien, Springer.

Sassa, K. (1985). *The mechanism of debris flows*, In: Procs. XI International Conference on Soil Mechanics and Foundation Engineering, San Francisco, 1, 37-56.

Sassa, K. 2002. *Mechanism of rapid and long travelling flow phenomena in granular soils.* In: Sassa K editor, Proceedings of the UNESCO/IGCP International Symposium on Landslide Mitigation and Protection of Cultural and Natural Heritage, Kyoto. Disaster Prevention Research Institute, Kyoto University. 11-30.

Savage, S.B. & Hutter, K. (1989). The motion of a finite mass of granular material down a rough incline. *Journal of Fluid Mechanics*, 199, 177-215.

Scheidegger, A.E. (1973). *On the prediction of the reach and velocity of catastrophic landslides, Rock Mechanics,* 5, 231-236.

Smith, D. & Hungr, O. (1992). *Failure behaviour of large rockslide*, Report to the Geological Survey of Canada and B.C.Hydro and Power Authority, DSS Contract Number 23397-9-0749/01-SZ), Thurber Engineering Ltd, Vancouver, B.C.

Tamburini, A. 2005. *Personal communication.*

Valiani, A. Caleffi, V. & Zanni, A. (2002). Case study: Malpasset dam-break simulation using a 2D finite-volume method, *Journal of Hydraulic Engineering*, 128(5), 460-472.

Varnes, D.J. (1978). *Slope movements types and processes*. In: Schuster, R.L. & Krizek, R.J. (Eds.), Landslides, Analysis and Control. Transportation Research Board, National Academy of Sciences, Washington, DC., Special Report 176, 11-33.

Wieland, M. Gray, J.M.N.T. & Hutter, K. (1999). Channelized free-surface flow of cohesionless granular avalanches in a chute with shallow lateral curvature, *Journal of Fluid Mechanics,* 392, 73-100.

Wong, H.N. & Ho, K.K.S. (1996). *Travel distance of landslide debris.* In Proceedings of the 7th International Symposium on Landslides, Trondheim, Norway, 1, 417-422.

In: Continuum Mechanics
Editors: Andrus Koppel and Jaak Oja, pp.147-171

ISBN: 978-1-60741-585-5

Chapter 5

EXAMINATION OF CRACKS BASED ON CONTINUUM-MECHANICS

Agnes Horvath[*]
Department of Mechanics, University of Miskolc, H-3515, Miskolc, Egyetemváros, Hungary

ABSTRACT

The phenomenon of failure by catastrophic crack propagation in structural materials poses problems of design and analysis in many fields of engineering. Cracks are present to some degree in all structures. They may exist as basic defects in the constituent materials or they may be induced in construction or during service life. The continuum-mechanics can be applied for macro cracks.

Over the past decades the finite element technique has become firmly established as a useful tool for numerical solution of engineering problems. In order to be able to apply the finite element method to the efficient solution of fracture problems, adaptations or further developments must be made. Using the finite element method, a lot of papers deal with the calculation of stress intensity factors for two- and three-dimensional geometries containing cracks of different shapes under various loadings to elastic bodies. In order to increase the accuracy of the results, special singular and transition elements have been used. They are described together with methods for calculating the stress intensity factors from the computed results. These include the displacement substitution method, *J*-integral and the virtual crack extension technique. Despite of the large number of published finite element stress intensity factor calculations there are not so many papers published on *J*-integral to elastic-plastic bodies.

At the vicinity of a crack tip the strains are not always small, but they may be large ones, too. In this case the *J*-integral can also be applied to characterise the cracks in elastic or elastic-plastic bodies.

This chapter describes the computation of the two dimensional *J*-integral in the case of small and large strains to elastic and elastic-plastic bodies and represents some numerical examples, too.

[*] Telephone number: 36-46-565-162, Fax number: 36-46-565-163, E-mail: mechva@uni-miskolc.hu

INTRODUCTION

Different fracture mechanical parameters can be used to characterize the cracks. One of them is the *J*-integral. In the literature there are relatively few papers dealing with the *J*-integral for large strains. Lau and his co-workers [1], [2] presented a revised *J*-estimation method under large plastic deformation. May and Kobayashi [3] investigated plane stress stable crack growth and the *J*-integral using Moiré interferometry to determine the two orthogonal displacements in a single edge crack specimen. Boothman and his co-workers [4] developed the *J*- and *Q*-estimation schemes for homogeneous plates. Jackiewicz [5] applied a hybrid model of steel cracking. The hybrid model uses a finite element simulation combined with an experimental test realised in the macro scale. Bouchard and his co-workers [6] demonstrated their two-dimensional local approach finite element study compared with the conventional *J*-estimation schemes and cracked body *J*-integral analysis. Saczuk and co-workers [7] presented a continuum model with inelastic material behaviour and a generalisation of the *J*-integral.

The aim of this chapter is a further development of the two-dimensional *J*-integral on the base of the continuum-mechanics for large strains in the case of elastic or elastic-plastic material behaviour, the computation of this *J*-integral using the finite element method and the presentation of some numerical examples.

FUNDAMENTAL CONCEPTS AND NOTATIONS

The continuum-mechanics is the part of the mechanics, it deals with the mechanical motion of the bodies using continuum models. The general theory of the continuum-mechanics applies to 3-dimensional models. The tensors have important role. The total peculiarities of a continuum are called the initial configuration in the $t = t_0 = 0$ moment and they are called the present configuration in a t arbitrary moment. It is supposed the continuum has a natural state without any deformation in the initial configuration.

The mechanical motion of a continuum is analysed in a reference co-ordinate system which is usually the Descartes rectangular (xyz) co-ordinate system. An arbitrary point of the moving continuum is denoted by $\hat{P}$. The position of this point is P^0 in the initial configuration and P in the present configuration. In the (xyz) reference co-ordinate system the coordinates of the point P^0 are x^0, y^0, z^0 and in this point the basic vectors of the local coordinate system $\mathbf{e}_x^0, \mathbf{e}_y^0, \mathbf{e}_z^0$. The coordinates of the point P are x, y, z and in this point the basic vectors of the local co-ordinate system are $\mathbf{e}_x, \mathbf{e}_y, \mathbf{e}_z$. In the initial configuration all quantities are denoted by superscript ^{0}.

The scalars are denoted by thin italic letters, e.g. U, s. The notations of the vectors are standing bold letters, e.g. $\mathbf{u}$, $\mathbf{A}$. The tensors are denoted by slanting bold letters, e.g. $\boldsymbol{T}$, $\boldsymbol{F}$.

Using the index notation the tensors are denoted by the same thin slanting character with index as in the invariant notation. In the initial configuration all quantities are denoted by superscript ^{0}, too. The indexes may be superscripts, subscripts or they may be in mixed position. Two same indexes are silent, the others are free. The summarizing convention is valid except for x, y, z.

The inverse of a tensor is denoted by superscript $^{-1}$, the notation of a transposed tensor is superscript ^{T}. The material derivative of a quantity according to time is denoted by · above it.

The notation of a scalar multiplication is ·, the double scalar multiplication is denoted by ··.

DEVELOPMENT OF *J*-INTEGRAL

Figure 1 shows a line integral path, which encloses the crack tip and has initial and end points, which lie, on the two crack faces. It has been shown independently by Rice [8] and Cherepanov [9] that the following integral quantity is path independent when taken along any path, which satisfies the above conditions:

$$J = \int_{\Gamma} \left(U n_1 - T_i \, \partial u_i / \partial x_1 \right) \mathrm{d}s \tag{1}$$

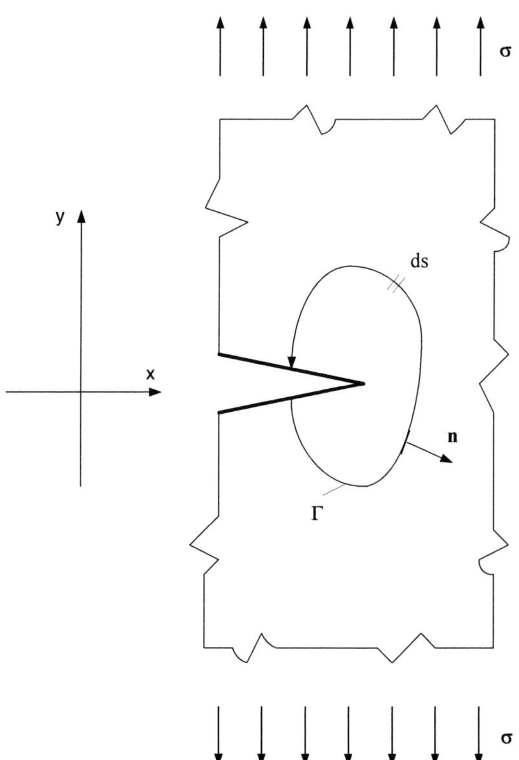

Figure 1. Contour path for *J* - integral evaluation

In this expression U is the strain energy density, T_i is the traction vector on a plane defined by the outward drawn normal, n_i and u_i is the displacement vector, ds is the element of arc along the path, Γ. For a closed path not containing the crack tip, $J = 0$.

Knowles and Sternberg [10] noted that this expression could be considered as the first component of a vector:

$$J_k = \int_{\Gamma} \left(U\, n_k - T_i\, \partial u_i / \partial x_k \right) \mathrm{d}\, s, \quad k = 1,2 \tag{2}$$

This integral is also path - independent provided the contour touches each surface of the crack at the tip. For elastic - plastic applications it is necessary to employ the appropriate definition of the strain energy density:

$$U = U_e + U_p \tag{3}$$

U_e is given by

$$U_e = \frac{1}{2}\, \sigma_{ij} \left(\varepsilon_{ij} \right)_e \tag{4}$$

where $\left(\varepsilon_{ij} \right)_e$ denotes the elastic components of strain. The plastic work contribution is given by

$$U_p = \int_0^{\bar{\varepsilon}_p} \bar{\sigma}\, \mathrm{d}\, \bar{\varepsilon}_p \tag{5}$$

In this expression $\bar{\sigma}$ and $\bar{\varepsilon}_p$ are the effective stress and effective plastic strain respectively.

Figure 2 represents the motion of a continuum with the initial and present configurations.

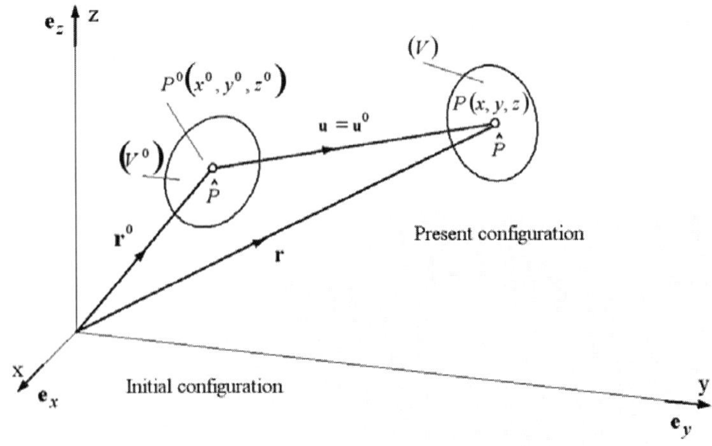

Figure 2. Motion of the continuum in the (xyz) reference coordinate-system.

Let us suppose that equation (2) is valid in the present configuration for large strains. As the initial configuration is known it is necessary to express the quantities in the integrand by means of strain and stress tensors attached to this configuration. Such tensors are the Green-

Lagrange ($\boldsymbol{E}^0$) strain and II. Piola-Kirchhoff ($\boldsymbol{T}^0$) stress tensors. For elastic applications it can be proved that instead of the strain energy density U we can write the next formula:

$$U^0 = \frac{1}{2} \boldsymbol{E}^0 \cdot\cdot \boldsymbol{T}^0 \tag{6}$$

Let us suppose that there is no heat effect, therefore the following expression can be written for the material derivative of the specific inner energy:

$$\rho \dot{e}_B = \boldsymbol{T} \cdot\cdot \boldsymbol{D} = \frac{\mathrm{d}V^0}{\mathrm{d}V} \boldsymbol{T}^0 \cdot\cdot \dot{\boldsymbol{E}}^0, \tag{7}$$

where

$\boldsymbol{T}$ Cauchy stress tensor,

$\boldsymbol{D} = \frac{1}{2}(\mathbf{v} \circ \nabla + \nabla \circ \mathbf{v})$ strain velocity tensor,

$\mathbf{v}$ velocity field,

ρ density.

The material derivative of the inner energy for the whole continuum is the next:

$$\dot{E}_B = \int_{(V)} \dot{e}_B \,\rho \,\mathrm{d}V = \int_{(V)} \boldsymbol{T} \cdot\cdot \boldsymbol{D} \,\mathrm{d}V = \int_{(V^0)} \boldsymbol{T}^0 \cdot\cdot \dot{\boldsymbol{E}}^0 \,\mathrm{d}V^0 \tag{8}$$

Let us suppose that the material equation for an elastic body is the following:

$$T^{0k\ell} = C^{k\ell pq} E^0_{pq}, \tag{9}$$

where $C^{k\ell pq}$ is the forth order tensor of the material constants.

Using equation (9) the integrand in equation (8) can be reformed:

$$\boldsymbol{T}^0 \cdot\cdot \dot{\boldsymbol{E}}^0 = \dot{\boldsymbol{E}}^0 \cdot\cdot \boldsymbol{T}^0 = \dot{E}^0_{k\ell} C^{k\ell pq} E^0_{pq} = \frac{1}{2}\left(E^0_{k\ell} C^{k\ell pq} E^0_{pq} \right)^{\cdot}. \tag{10}$$

Substituting (10) into equation (8) we obtain the next formula:

$$\dot{E}_B = \int_{(V^0)} \frac{1}{2}\left(E^0_{k\ell} C^{k\ell pq} E^0_{pq}\right)^{\cdot} \mathrm{d}V^0. \tag{11}$$

The integration of equation (11) between $t_0 = 0$ and t results the following expression:

$$E_B = \int_{(V^0)} \frac{1}{2}\left(E^0_{k\ell} C^{k\ell pq} E^0_{pq}\right) \mathrm{d}V^0 \tag{12}$$

In equation (12) the integrand is the strain energy density. In other form we can write:

$$E_B = \int_{(V^0)} \frac{1}{2} \boldsymbol{E}^0 \cdot\cdot \boldsymbol{T}^0 \mathrm{d}V^0 = \int_{(V^0)} U^0 \mathrm{d}V^0 \tag{13}$$

In two dimensional cases the formula for U^0 is the next:

$$U^0 = \frac{1}{2}\left(E^0_{xx} T^0_{xx} + 2E^0_{xy} T^0_{xy} + E^0_{yy} T^0_{yy}\right). \tag{14}$$

The element of arc is

$$\mathrm{d}s = \mathrm{d}s^0 \, \lambda_s = \mathrm{d}s^0 \sqrt{1 + 2\mathbf{e}^0 \cdot \boldsymbol{E}^0 \cdot \mathbf{e}^0} \tag{15}$$

where $\mathrm{d}s^0$ is the element of arc, $\mathbf{e}^0$ is the tangent vector along the curve Γ in the initial configuration and λ_s is the stretch. The following expression can be used for the reforming of the $\mathbf{t} = \boldsymbol{T} \cdot \mathbf{n}$ traction vector:

$$\boldsymbol{T} = \frac{1}{\delta} \boldsymbol{F} \cdot \boldsymbol{T}^0 \cdot \boldsymbol{F}^T \tag{16}$$

where

$\boldsymbol{T}$ Chauchy stress tensor,

$\boldsymbol{T}^0$ II. Piola –Kirchhoff stress tensor,

$\boldsymbol{F}$ strain gradient tensor,

$\mathbf{n}$ the outward drawn normal in the present configuration,

$\delta = Det\left|\boldsymbol{F}\right|$ Jacobian determinant.

Applying $\mathrm{d}\,\mathbf{A} = \delta\left(\boldsymbol{F}^{-1}\right)^{T}\cdot\mathrm{d}\,\mathbf{A}^{0}$ formula between the vectorial surface elements in the initial and present configuration, the traction vector can be expressed in the next form:

$$\begin{aligned}\mathbf{t} &= \boldsymbol{F}\cdot\boldsymbol{T}^{0}\cdot\frac{\mathrm{d}\,\mathbf{A}^{0}}{\mathrm{d}\,A} = \boldsymbol{F}\cdot\boldsymbol{T}^{0}\cdot\mathbf{n}^{0}\frac{\mathrm{d}\,A^{0}}{\mathrm{d}\,A} = \frac{1}{\lambda_{A}}\boldsymbol{F}\cdot\boldsymbol{T}^{0}\cdot\mathbf{n}^{0} = \\ &= \frac{1}{\delta\sqrt{\mathbf{n}^{0}\cdot\left(2\boldsymbol{E}^{0}+\boldsymbol{I}\right)^{-1}\cdot\mathbf{n}^{0}}}\boldsymbol{F}\cdot\boldsymbol{T}^{0}\cdot\mathbf{n}^{0}\end{aligned} \tag{17}$$

where

$\boldsymbol{E}^{0}$ Green-Lagrange strain tensor,

$\boldsymbol{I}$ unit tensor,

$\mathbf{n}^{0}$ the outward drawn normal in the initial configuration,

It can be seen in Figure 2 that $\mathbf{r} = \mathbf{r}^{0} + \mathbf{u}^{0}$, therefore we can write the next expressions:

$$\mathrm{d}y = \frac{\partial y}{\partial x^{0}}\mathrm{d}x^{0} + \frac{\partial y}{\partial y^{0}}\mathrm{d}y^{0} = \mathrm{d}y^{0} + \frac{\partial u_{y}^{0}}{\partial x^{0}}\mathrm{d}x^{0} + \frac{\partial u_{y}^{0}}{\partial y^{0}}\mathrm{d}y^{0} \tag{18}$$

$$\mathrm{d}x = \frac{\partial x}{\partial x^{0}}\mathrm{d}x^{0} + \frac{\partial x}{\partial y^{0}}\mathrm{d}y^{0} = \mathrm{d}x^{0} + \frac{\partial u_{x}^{0}}{\partial x^{0}}\mathrm{d}x^{0} + \frac{\partial u_{x}^{0}}{\partial y^{0}}\mathrm{d}y^{0} \tag{19}$$

As $\mathbf{u} = \mathbf{u}^{0}$, the derivatives of the displacement vector are the next:

$$\frac{\partial \mathbf{u}^{0}}{\partial x} = \frac{\partial u_{x}^{0}}{\partial x}\mathbf{e}_{x}^{0} + \frac{\partial u_{y}^{0}}{\partial x}\mathbf{e}_{y}^{0} = \left(\frac{\partial u_{x}^{0}}{\partial x^{0}}\frac{\partial x^{0}}{\partial x} + \frac{\partial u_{x}^{0}}{\partial y^{0}}\frac{\partial y^{0}}{\partial x}\right)\mathbf{e}_{x}^{0} + \left(\frac{\partial u_{y}^{0}}{\partial x^{0}}\frac{\partial x^{0}}{\partial x} + \frac{\partial u_{y}^{0}}{\partial y^{0}}\frac{\partial y^{0}}{\partial x}\right)\mathbf{e}_{y}^{0} \tag{20}$$

$$\frac{\partial \mathbf{u}^{0}}{\partial y} = \frac{\partial u_{x}^{0}}{\partial y}\mathbf{e}_{x}^{0} + \frac{\partial u_{y}^{0}}{\partial y}\mathbf{e}_{y}^{0} = \left(\frac{\partial u_{x}^{0}}{\partial x^{0}}\frac{\partial x^{0}}{\partial y} + \frac{\partial u_{x}^{0}}{\partial y^{0}}\frac{\partial y^{0}}{\partial y}\right)\mathbf{e}_{x}^{0} + \left(\frac{\partial u_{y}^{0}}{\partial x^{0}}\frac{\partial x^{0}}{\partial y} + \frac{\partial u_{y}^{0}}{\partial y^{0}}\frac{\partial y^{0}}{\partial y}\right)\mathbf{e}_{y}^{0} \tag{21}$$

As $\mathbf{r} = \mathbf{r}^{0} + \mathbf{u}^{0}$, in the *(xyz)* reference co-ordinate system the formulas below are valid for two dimensional problems:

$$x = x^0 + u_x^0,\; y = y^0 + u_y^0 . \tag{22}$$

Applying (22) we can write the $\boldsymbol{F}$ strain gradient, the $\boldsymbol{F}^{-1}$ inverse strain gradient and the δ Jacobian determinant:

$$[\boldsymbol{F}] = \begin{bmatrix} \dfrac{\partial x}{\partial x^0} & \dfrac{\partial x}{\partial y^0} \\ \dfrac{\partial y}{\partial x^0} & \dfrac{\partial y}{\partial y^0} \end{bmatrix} = \begin{bmatrix} 1+\dfrac{\partial u_x^0}{\partial x^0} & \dfrac{\partial u_x^0}{\partial y^0} \\ \dfrac{\partial u_y^0}{\partial x^0} & 1+\dfrac{\partial u_y^0}{\partial y^0} \end{bmatrix}, \tag{23}$$

$$\left[\boldsymbol{F}^{-1}\right] = \begin{bmatrix} \dfrac{\partial x^0}{\partial x} & \dfrac{\partial x^0}{\partial y} \\ \dfrac{\partial y^0}{\partial x} & \dfrac{\partial y^0}{\partial y} \end{bmatrix} = \frac{1}{\delta} \begin{bmatrix} 1+\dfrac{\partial u_y^0}{\partial y^0} & -\dfrac{\partial u_x^0}{\partial y^0} \\ -\dfrac{\partial u_y^0}{\partial x^0} & 1+\dfrac{\partial u_x^0}{\partial x^0} \end{bmatrix}, \tag{24}$$

$$\delta = Det\,|\,\boldsymbol{F}\,| = \left(1+\frac{\partial u_x^0}{\partial x^0}\right)\left(1+\frac{\partial u_y^0}{\partial y^0}\right) - \frac{\partial u_x^0}{\partial y^0}\frac{\partial u_y^0}{\partial x^0} . \tag{25}$$

Using (22) - (25), formulas (20) and (21) can be expressed in an other form:

$$\frac{\partial \mathbf{u}^0}{\partial x} = \frac{1}{\delta}\left[\left(1+\frac{\partial u_y^0}{\partial y^0}\right)\frac{\partial u_x^0}{\partial x^0} - \frac{\partial u_x^0}{\partial y^0}\frac{\partial u_y^0}{\partial x^0}\right]\mathbf{e}_x^0 + \frac{1}{\delta}\left[\left(1+\frac{\partial u_y^0}{\partial y^0}\right)\frac{\partial u_y^0}{\partial x^0} - \frac{\partial u_y^0}{\partial y^0}\frac{\partial u_y^0}{\partial x^0}\right]\mathbf{e}_y^0 \tag{26}$$

$$\frac{\partial \mathbf{u}^0}{\partial y} = \frac{1}{\delta}\left[\left(1+\frac{\partial u_x^0}{\partial x^0}\right)\frac{\partial u_x^0}{\partial y^0} - \frac{\partial u_x^0}{\partial x^0}\frac{\partial u_x^0}{\partial y^0}\right]\mathbf{e}_x^0 + \frac{1}{\delta}\left[\left(1+\frac{\partial u_x^0}{\partial x^0}\right)\frac{\partial u_y^0}{\partial y^0} - \frac{\partial u_y^0}{\partial x^0}\frac{\partial u_x^0}{\partial y^0}\right]\mathbf{e}_y^0 \tag{27}$$

Substituting (6) - (27) into equation (2), we can obtain the components of J in two dimensions:

$$J_x = \int_{(\Gamma)} \left[U^0 \left(\mathrm{d}y^0 + \frac{\partial u_y^0}{\partial x^0}\mathrm{d}x^0 + \frac{\partial u_y^0}{\partial y^0}\mathrm{d}y^0 \right) - \mathbf{t}\frac{\partial \mathbf{u}^0}{\partial x}\mathrm{d}s^0 \lambda_s \right] \tag{28}$$

$$J_y = \int_{(\Gamma)} \left[-U^0 \left(\mathrm{d}x^0 + \frac{\partial u_x^0}{\partial x^0}\mathrm{d}x^0 + \frac{\partial u_x^0}{\partial y^0}\mathrm{d}y^0 \right) - \mathbf{t}\frac{\partial \mathbf{u}^0}{\partial y}\mathrm{d}s^0 \lambda_s \right] \tag{29}$$

For elastic - plastic applications it is necessary to employ the appropriate definition of the strain energy density:

$$U^0 = U_e^0 + U_{pl}^0 \tag{30}$$

where U_e^0 is given in equation (6) and U_{pl}^0 is similar to equation (5):

$$U_{pl}^0 = \int_0^{\overline{E}_{pl}^0} \overline{T}^0 \, \mathrm{d}\overline{E}_{pl}^0 \tag{31}$$

In this expression $\overline{T}^0$ and $\overline{E}_{pl}^0$ are the effective stress and effective plastic strain in the initial configuration. Using the finite element method the integration in equation (28) and equation (29) must be undertaken numerically.

In the case of inclined cracks two co-ordinate systems are necessary in the initial configuration using the appropriate transformation:

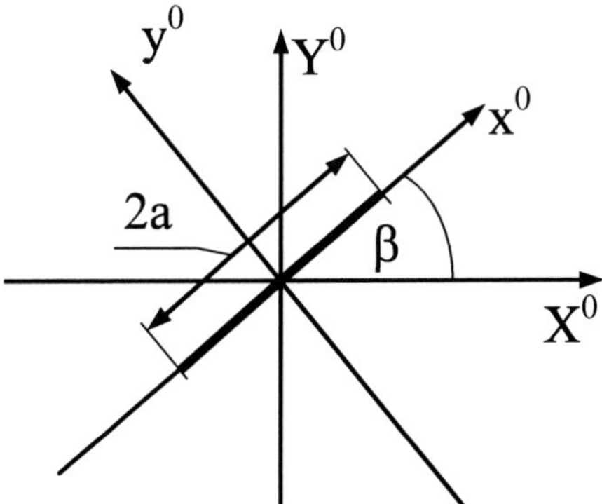

Figure 3. Co-ordinate systems in the initial configuration.

Path-Independence of the J-Integral

Rice already investigated the problem of path-independence [8]. Except him other researchers also examined this question, e.g. Atluri [11], Brocks and Scheider [12], Wang and his co-worker [13].

Henceforth the path-independence of (28) and (29) are proved for large strains in two dimensional cases. The following assumptions have been made:

- The material of the body is homogeneous.
- There are no body forces.
- The stress and strain fields depend on two co-ordinates (x^0, y^0).
- The crack is straight.

- The stress-free crack borders parallel to x^0.

Figure 4 demonstrates a closed integral path not containing the crack tip.

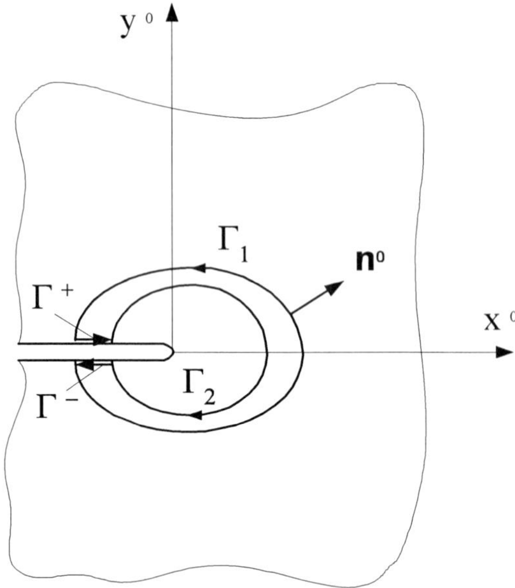

Figure 4. Closed contour for J-integral evaluation.

The closed contour Γ - does not include a singularity- is the next:

$$\Gamma = \Gamma_1 \cup \Gamma^+ \cup \Gamma_2 \cup \Gamma^- . \tag{32}$$

Then *J = 0* along a closed contour Γ for large strains, too. Let us examine the component J_x along the path Γ.

$$\begin{aligned} J_x = 0 = & \int\limits_{(\Gamma_2)}^{\rightarrow} [U^0(\mathrm{d}y^0 + \frac{\partial u_y^0}{\partial x^0}\mathrm{d}x^\circ + \frac{\partial u_y^0}{\partial y^0}\mathrm{d}y^0) - \mathbf{t}\cdot\frac{\partial \mathbf{u}^0}{\partial x}\lambda_s \,\mathrm{d}s^0] + \\ & + \int\limits_{(\Gamma_1)}^{\leftarrow} [U^0(\mathrm{d}y^0 + \frac{\partial u_y^0}{\partial x^0}\mathrm{d}x^\circ + \frac{\partial u_y^0}{\partial y^0}\mathrm{d}y^0) - \mathbf{t}\cdot\frac{\partial \mathbf{u}^0}{\partial x}\lambda_s \,\mathrm{d}s^0] + \\ & + \int\limits_{(\Gamma^+)} [U^0(\mathrm{d}y^0 + \frac{\partial u_y^0}{\partial x^0}\mathrm{d}x^\circ + \frac{\partial u_y^0}{\partial y^0}\mathrm{d}y^0) - \mathbf{t}\cdot\frac{\partial \mathbf{u}^0}{\partial x}\lambda_s \,\mathrm{d}s^0] + \\ & + \int\limits_{(\Gamma^-)} [U^0(\mathrm{d}y^0 + \frac{\partial u_y^0}{\partial x^0}\mathrm{d}x^\circ + \frac{\partial u_y^0}{\partial y^0}\mathrm{d}y^0) - \mathbf{t}\cdot\frac{\partial \mathbf{u}^0}{\partial x}\lambda_s \,\mathrm{d}s^0] . \end{aligned} \tag{33}$$

The sign → above the integral means that the integration contour runs clockwise. The sign ← means that the integration contour runs counter clockwise. The values of the integrals along Γ^+ and Γ^- are the same, they differs from each others in indication, therefore

$$\begin{aligned} 0 = & \overrightarrow{\int_{(\Gamma_2)}} [U^0(\mathrm{d}y^0 + \frac{\partial u_y^0}{\partial x^0}\mathrm{d}x^o + \frac{\partial u_y^0}{\partial y^0}\mathrm{d}y^0) - \mathbf{t}\cdot\frac{\partial \mathbf{u}^0}{\partial x}\lambda_s\,\mathrm{d}s^0] + \\ & + \overleftarrow{\int_{(\Gamma_1)}} [U^0(\mathrm{d}y^0 + \frac{\partial u_y^0}{\partial x^0}\mathrm{d}x^o + \frac{\partial u_y^0}{\partial y^0}\mathrm{d}y^0) - \mathbf{t}\cdot\frac{\partial \mathbf{u}^0}{\partial x}\lambda_s\,\mathrm{d}s^0]. \end{aligned} \tag{34}$$

We obtain the next expression when the integration contour Γ_2 runs counter clockwise:

$$\begin{aligned} & \overrightarrow{\int_{(\Gamma_2)}} [U^0(\mathrm{d}y^0 + \frac{\partial u_y^0}{\partial x^0}\mathrm{d}x^o + \frac{\partial u_y^0}{\partial y^0}\mathrm{d}y^0) - \mathbf{t}\cdot\frac{\partial \mathbf{u}^0}{\partial x}\lambda_s\,\mathrm{d}s^0] = \\ & = - \overleftarrow{\int_{(\Gamma_2)}} [U^0(\mathrm{d}y^0 + \frac{\partial u_y^0}{\partial x^0}\mathrm{d}x^o + \frac{\partial u_y^0}{\partial y^0}\mathrm{d}y^0) - \mathbf{t}\cdot\frac{\partial \mathbf{u}^0}{\partial x}\lambda_s\,\mathrm{d}s^0]. \end{aligned} \tag{35}$$

Substituting (35) into equation (34) we obtain the following formula:

$$\begin{aligned} 0 = & - \overleftarrow{\int_{(\Gamma_2)}} [U^0(\mathrm{d}y^0 + \frac{\partial u_y^0}{\partial x^0}\mathrm{d}x^o + \frac{\partial u_y^0}{\partial y^0}\mathrm{d}y^0) - \mathbf{t}\cdot\frac{\partial \mathbf{u}^0}{\partial x}\lambda_s\,\mathrm{d}s^0] + \\ & + \overleftarrow{\int_{(\Gamma_1)}} [U^0(\mathrm{d}y^0 + \frac{\partial u_y^0}{\partial x^0}\mathrm{d}x^o + \frac{\partial u_y^0}{\partial y^0}\mathrm{d}y^0) - \mathbf{t}\cdot\frac{\partial \mathbf{u}^0}{\partial x}\lambda_s\,\mathrm{d}s^0]. \end{aligned} \tag{36}$$

Rearrangement of (36) results:

$$\begin{aligned} & \overleftarrow{\int_{(\Gamma_2)}} [U^0(\mathrm{d}y^0 + \frac{\partial u_y^0}{\partial x^0}\mathrm{d}x^o + \frac{\partial u_y^0}{\partial y^0}\mathrm{d}y^0) - \mathbf{t}\cdot\frac{\partial \mathbf{u}^0}{\partial x}\lambda_s\,\mathrm{d}s^0] = \\ & \overleftarrow{\int_{(\Gamma_1)}} [U^0(\mathrm{d}y^0 + \frac{\partial u_y^0}{\partial x^0}\mathrm{d}x^o + \frac{\partial u_y^0}{\partial y^0}\mathrm{d}y^0) - \mathbf{t}\cdot\frac{\partial \mathbf{u}^0}{\partial x}\lambda_s\,\mathrm{d}s^0]. \end{aligned} \tag{37}$$

So we obtain the path independence of the first component of the *J* vector. This holds for the other component and in the case of inclined cracks, too.

APPLICABILITY OF SPECIAL ISOPARAMETRIC ELEMENTS

The applicability of the special isoparametric elements is well-known for small strains. The aim of this section is the examination of these elements for large strains.

A. Singular Isoparametric Elements

To create an element possessing a singularity of order $r^{(1-m)/m}$, consider a one-dimensional element which may form a side of a 2D or 3D nth order isoparametric element (see Figure 5).

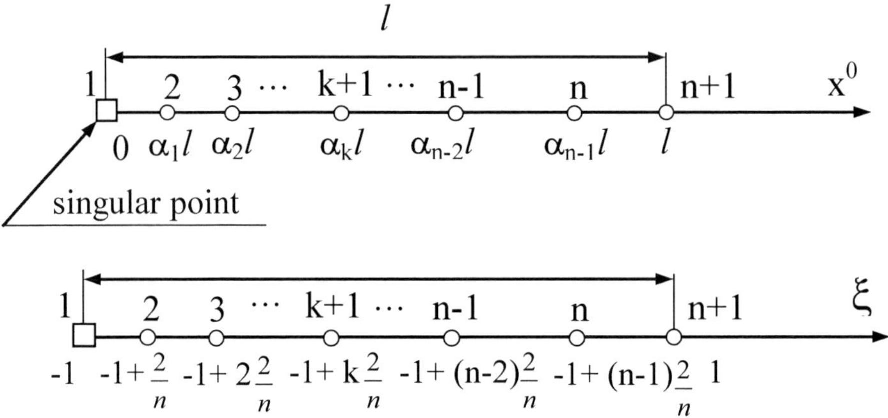

Figure 5. Element co-ordinate mapping.

This transformation is accomplished by means of the usual isoparametric mapping technique [14]. We obtain the next expressions without any details:

$$x^0 = \frac{\ell}{2^m}(1+\xi)^m \tag{38}$$

$$\xi = -1 + 2\left(\frac{x^0}{\ell}\right)^{1/m} \tag{39}$$

The exponent m determines the order of the strain singularity, and n is the order of the element. Between m and n the following relation is valid: $m \leq n$. The displacement u_x^0 can be written in the next form in the case of an isoparametric element:

$$u_x^0 = b_0 + b_1\xi + b_2\xi^2 + \cdots + b_n\xi^n, \quad n \geq 2. \tag{40}$$

From (40) we have the displacement derivative:

$$\frac{\partial u_x^0}{\partial \xi} = b_1 + 2b_2\xi + 3b_3\xi^2 + \cdots + n\,b_n\xi^{(n-1)} \tag{41}$$

This expression has an other form substituting (39) into (41):

$$\frac{\partial u_x^0}{\partial \xi} = b_1 + 2b_2\left[-1+2\left(\frac{x^0}{\ell}\right)^{1/m}\right] + 3b_3\left[-1+2\left(\frac{x^0}{\ell}\right)^{1/m}\right]^2 + \cdots + n\,b_n\left[-1+2\left(\frac{x^0}{\ell}\right)^{1/m}\right]^{(n-1)} \tag{42}$$

In the x^0 - direction the strain is then

$$E_{xx}^0 = \frac{d\,u_x^0}{d\,x^0} + \frac{1}{2}\left(\frac{d\,u_x^0}{d\,x^0}\right)^2 \tag{43}$$

Equation (43) obtains the next form by means of (40):

$$E_{xx}^0 = \frac{d\,u_x^0}{d\,\xi}\frac{d\,\xi}{d\,x^0} + \frac{1}{2}\left(\frac{d\,u_x^0}{d\,\xi}\frac{d\,\xi}{d\,x^0}\right)^2 \tag{44}$$

Substituting (42) and the derivative of (39) into (44) the following formula can be obtained:

$$\begin{aligned} E_{xx}^0 = {} & A_1\left(x^0\right)^{\frac{1-m}{m}} + A_2\left(x^0\right)^{\frac{2-m}{m}} + A_3\left(x^0\right)^{\frac{3-m}{m}} + \cdots + A_n\left(x^0\right)^{\frac{n-m}{m}} + \\ & + A_{n+1}\left(x^0\right)^{\frac{2(1-m)}{m}} + A_{n+2}\left(x^0\right)^{\frac{2(1-m)+1}{m}} + \cdots + A_{3n-1}\left(x^0\right)^{\frac{2(n-m)}{m}}, \end{aligned} \tag{45}$$

where

$$A_1 = C\left(b_1 - 2b_2 + 3b_3 - \cdots \pm nb_n\right),$$

$$A_2 = \frac{C}{\ell^{1/m}}2\left[2b_2 - 6b_3 + 12b_4 - \cdots \pm (n-1)nb_n\right],$$

$$\vdots$$

$$A_n = \frac{C}{\ell^{(n-1)/m}}2^{(n-1)}nb_n,$$

$$\begin{aligned} A_{n+1} = {} & \frac{C^2}{2}[\ b_1^2 + 4b_2^2 + \cdots + n^2b_n^2 - 4b_1b_2 + 6b_1b_3 - \cdots \pm 2nb_1b_n \pm \cdots \\ & \pm 2(n-1)nb_{(n-1)}b_n\ ], \end{aligned}$$

$$A_{n+2} = \frac{C^2}{2\,\ell^{1/m}} 2^2 \left[\ 2b_1b_2 - 6b_1b_3 + 18b_2b_3 - 4b_2^2 - 18b_3^2 - \cdots \right.$$

$$\left. \pm n^2(n-1)b_{(n-1)}b_n - (n-1)(nb_n)^2\ \right],$$

$$\vdots$$

$$A_{3n-1} = \frac{C^2}{2\,\ell^{2(n-1)/m}} 2^{2(n-1)} (nb_n)^2,$$

$$C = \frac{2}{m}\frac{1}{\ell^{1/m}}.$$

Equation (45) demonstrates that the strain is singular at $x^0 = 0\,(\xi = -1)$. The leading strain term is of order $\left(x^0\right)^{\frac{1-m}{m}}$, as $x^0 \to 0$, and this is the required strain singularity. That means, the type of the strain singularity can vary between $\left(x^0\right)^{-1/2}$ and $\left(x^0\right)^{-1}$. Rice [15] has shown that the singularity in strain at a crack tip is of order $\left(x^0\right)^{-1/(1+N)}$, where N is a hardening exponent varying between 1 and 0 from purely elastic to perfectly plastic response. Between N and m the following relation is valid:

$$N = -\frac{1}{1-m}, \qquad m = \frac{1+N}{N}. \tag{46}$$

As $m \geq 2$ is an integer number, the value of N is definable using m. The relation is valid in the opposite direction, too.

B. Isoparametric Transition Elements

A generalisation of the $r^{(1-m)/m}$ strain singularity of higher-order isoparametric elements is presented. By variable placement of the side nodes between their original and singular positions, the point of singularity sensed by element can be controlled. The transition elements have a strain singularity outside their domain. The singular and non-singular elements are special cases of the general mapping. The transition elements, together with the singular isoparametric elements, can be used for solving crack problems.

Pu, Hussain and Lorensen [16] have demonstrated the 12-node quadrilateral isoparametric elements. It has been demonstrated that the inverse square root singularity of the strain field at the crack tip could be obtained by placing the two side nodes at 1/9 and 4/9 of the length of the side from the crack tip. Lynn and Ingraffea [17] have developed a transition element possessing a singularity of order $r^{-1/2}$ outside the element. In practical application of this element the principal parameter affecting the accuracy of the solution is the ratio of the singular element length to the crack length. If this ratio approaches a small number, modelling of singular behaviour is lost because of the non-singular behaviour of neighbouring elements. Better modelling by replacing the non-singular neighbouring elements with transition elements possessing the same order of singularity at the crack tip is the cause of their development. If we move the singular point from its original position (see Fig. 5.) outside the element, we obtain the transition element. Let us consider that the position of the singular point is at $x = -q\ell$. We obtain the next expressions [18] without any details:

$$\alpha_k = \left[\frac{(n-k)q^{1/m} + k(1+q)^{1/m}}{n} \right]^m - q, \qquad k = 1,2,3,\ldots,n-1 \tag{47}$$

Using equation (47) we can determine the position of the side nodes and the desired singularity occurs outside the element. This formula is valid in the case of small and large strains, too.

If we take into consideration the values of $k = 0$ and $k = n$, they will be the equivalent of $\alpha_0 = 0$ and $\alpha_n = 1$. The exponent m determines the order of the strain singularity, and n is the order of the element. Between m and n next relation is also valid: $m \leq n$.

Equation (47) relates the location of the element side nodes α_k to the singular point q outside the element domain, at which the $r^{(1-m)/m}$ singularity is sensed. It becomes apparent that such a transition element can be easily constructed by simply shifting the side nodes to some properly calculated points.

It can be shown easily that the singular and non-singular elements are special cases of the general mapping.

Case 1. Non-Singular Isoparametric Elements

In order to obtain these elements, in equation (47) we must choose $m = 1$ and $q = 0$. Then we have the next expression:

$$\alpha_k = \frac{k}{n} \tag{48}$$

Case 2. Singular Isoparametric Elements

Choosing $q = 0$ and $m \geq 2$ in equation (47), then

$$\alpha_k = \left(\frac{k}{n} \right)^m \tag{49}$$

Equation (48) is the upper, and equation (49) the lower limit of equation (47) for the same values of m and n.

NUMERICAL EXAMPLES

The author made FORTRAN programs by means of Microsoft Developer Studio 97 to compute the *J*-integral numerically for small and large strains. At a real physical problem the strains are either small or large depending on the loading of the body. At the demonstrated example both kind of strains were calculated in order to come to know that is there any difference or not between the two kind of strains. At the computation of the elastic-plastic problems the Von Mises yield criterion, the Newton-Raphson iteration techniques and the Euler-Cauchy incremental method was applied.

1. Model

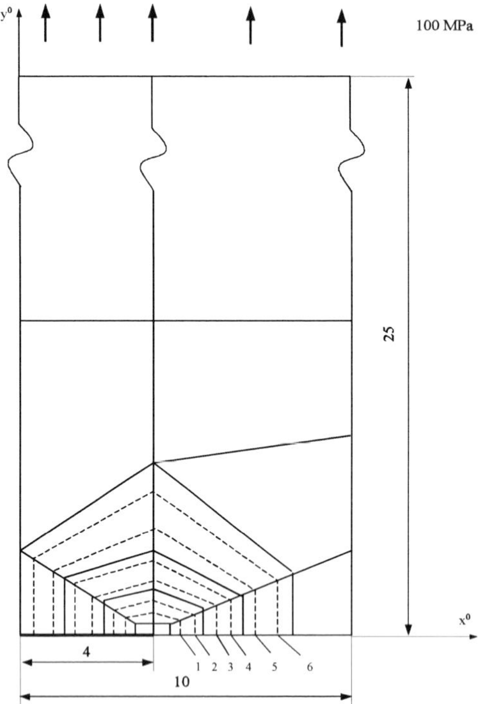

Figure 6. Finite element mesh.

The first example is a central crack problem (the crack was on the axe x^0). The example considered is that of a plate under tension which contains a crack of length *8 mm* perpendicular to the direction of loading. The width of the plate is *20 mm* and the thickness assumed to be unity. The length of the plate is *50 mm*. Two computations were made: one for small and the other for large strains. In these calculations the material was linear elastic with the properties $E = 10000\ MPa$ and $\nu = 0{,}3$. The applied tensile traction was $p = 100\ MPa$. The finite element mesh represented only one quarter of the plate because of the symmetrical properties of the body (Figure 6). The finite element mesh doesn't contain singular and transition elements. The dashed lines denote the integral paths going through the Gauss points of the elements.

Theoretically J_y is zero for this problem. Figure 7 represents the calculated J-integral values.

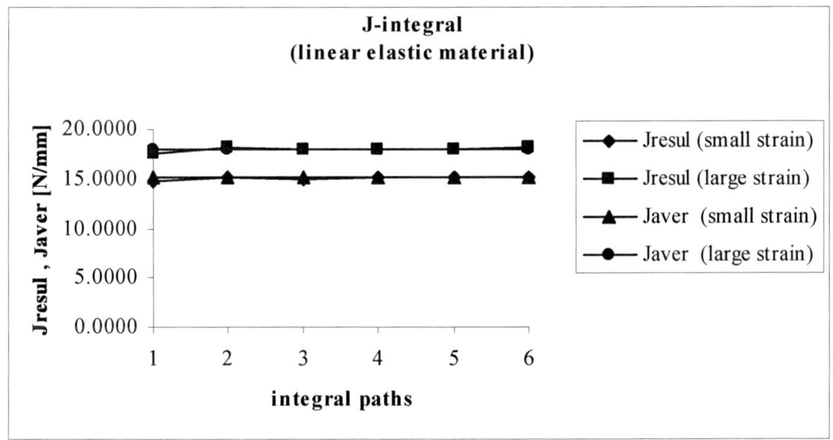

Figure 7. Calculated J-integral values for elastic case.

These figures demonstrate J_{resul} which means the next expression: $J_{resul} = \sqrt{\left(J_x^2 + J_y^2\right)}$. J_{aver} is the average of J_{resul} along the integral paths.

2. Model

Two calculations were made for the same finite element mesh (see Figure 6). In the second computations the material was linear elastic - linear hardening with the properties $E = 10000\ MPa$, $\nu = 0,3$ and $H'=0,1E$. The yield stress was $\sigma_F = 350\ MPa$. The loading was applied in incremental steps. The increments were: $1.0\ p = 100\ MPa$, $3 \times 0.1p = 10\ MPa$. The calculated J-integral values can be seen in Figure 8 for small and in Figure 9 for large strains.

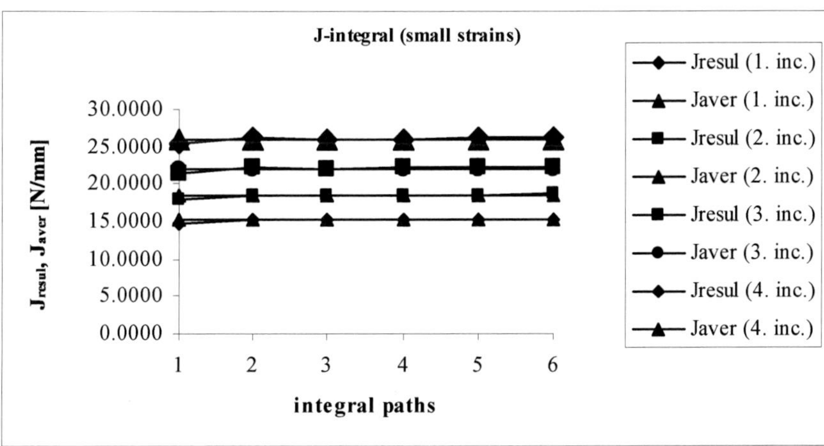

Figure 8. J-integral values for small strains.

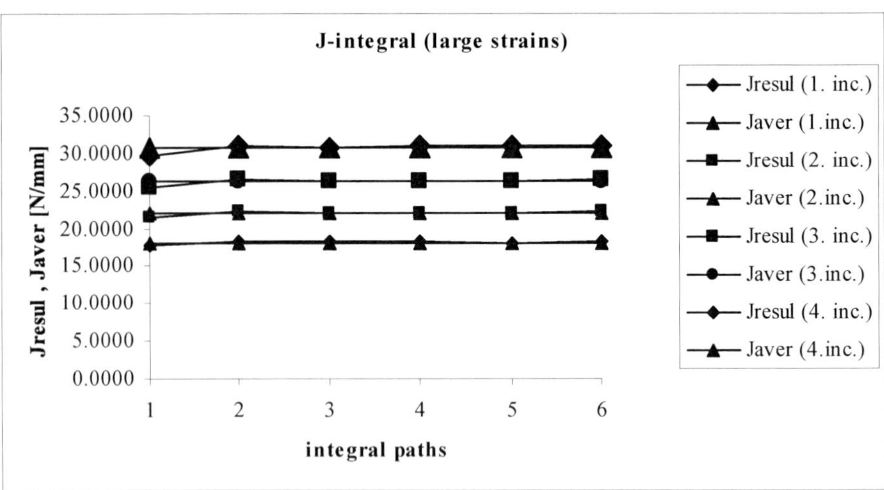

Figure 9. J-integral values for large strains.

3. Model

Two calculations were made for the same finite element mesh (see Figure 6). Figure 10 demonstrates the applied material model.

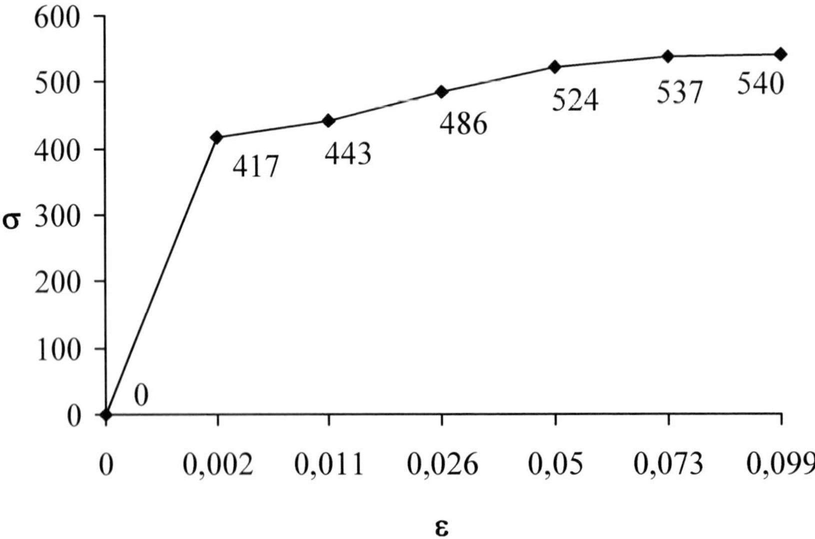

Figure 10. Applied material model.

In the third computations the applied material model was elastic – plastic with the properties $E = 211000\ MPa$ and $\nu = 0{,}3$. The yield stress was $\sigma_F = 417\ MPa$. The loading was applied in incremental steps. The increments were: $1.0\,p = 100\ MPa$, $3 \times 0.1p = 10\ MPa$. The calculated J-integral values can be seen in Figure 11 for small and in Figure 12 for large strains.

Convergence of the Solution

We investigate the convergence of the solution using 1. MODEL. The analytical solution is known [19] in the case of small strains. Further computations were made with the refinements of the finite element mesh. Figure 13-18 demonstrate these finite element meshes.

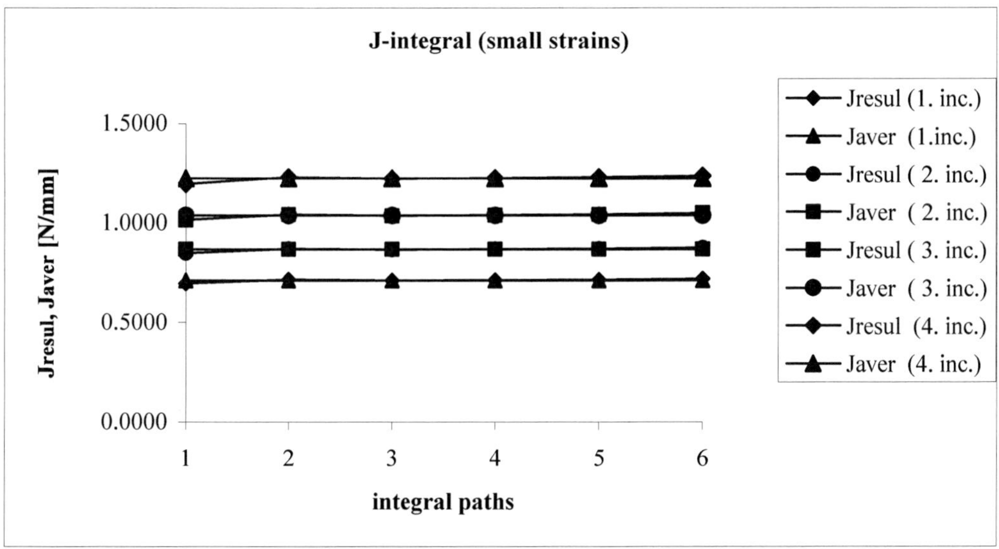

Figure 11. J-integral values for small strains.

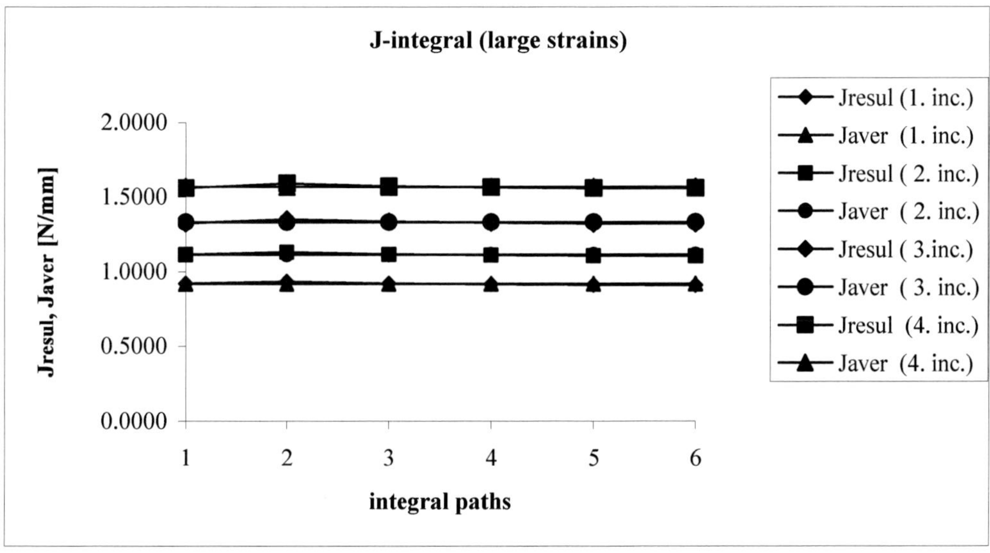

Figure 12. J-integral values for large strains.

Figure 13. 42 elements.

Figure 14. 50 elements.

Figure 15. 58 elements.

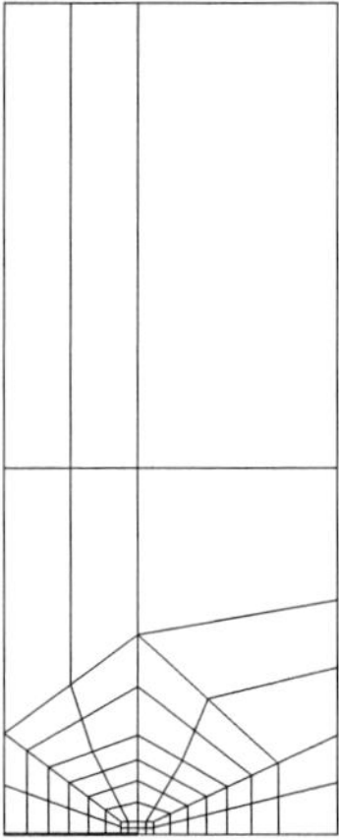

Figure 16. 66 elements.

Figure 17. 72 elements.

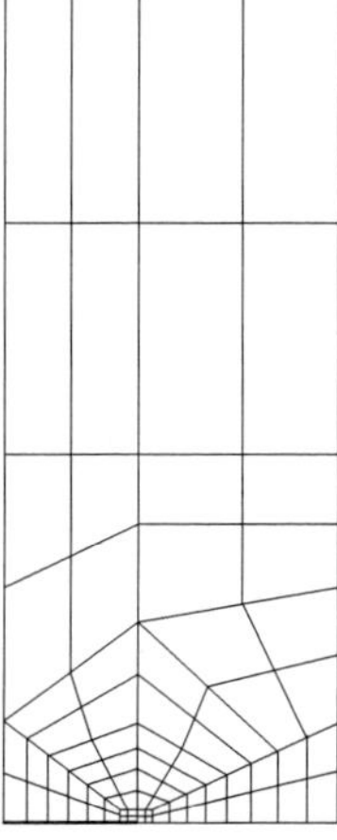

Figure 18. 80 elements.

Two other calculations were made for the original finite element mesh (Figure 6). In the first case the mesh contained two singular isoparametric elements at the crack tip. In the second case there were singular elements and their neighboring elements were transition isoparametric elements in the finite element mesh. One more computation was made for the mesh contained 42 elements using singular elements at the crack tip (Figure 13). One more calculation was made for the mesh contained 50 elements using singular elements at the tip (Figure 14).

Table 1 summarizes the special features of these finite element meshes. Table 2 demonstrates the calculated J_x integral values, the analytical solution and the relative error. The relative error applies to the analytical solution and can be computed in the next form:

$$h = \frac{J_{xcal} - J_{xanal}}{J_{xanal}} \cdot 100\,(\%) \tag{50}$$

Table 1. Special features of the meshes.

Small strains, linear elastic material					
Sign of calculation	Number of elements	Number of nodes	DOF	Singular element	Transition element
1	20	79	158	no	no
2	42	151	302	no	no
3	50	177	354	no	no
4	58	203	406	no	no
5	66	229	458	no	no
6	72	249	498	no	no
7	80	277	564	no	no
8	20	79	158	yes	no
9	20	79	158	yes	yes
10	42	151	302	yes	no
11	50	177	354	yes	no

Table 2. J_x values and the relative error (small strains).

Small strains , linear elastic material			
Sign of calculation	J_{xcal} [N/mm]	J_{xanal} [N/mm]	Relative error [%]
1	15,2159	15,4842	-1,7327
2	15,3145	15,4842	-1,0960
3	15,3189	15,4842	-1,0675
4	15,3441	15,4842	-0,9048
5	15,3461	15,4842	-0,8919
6	15,3471	15,4842	-0,8854
7	15,3526	15,4842	-0,8499
8	15,3519	15,4842	-0,8544
9	15,4310	15,4842	-0,3436
10	15,3984	15,4842	-0,5541
11	15,4280	15,4842	-0,3629

Figure 19 demonstrates the analytical and calculated J_x integral values.

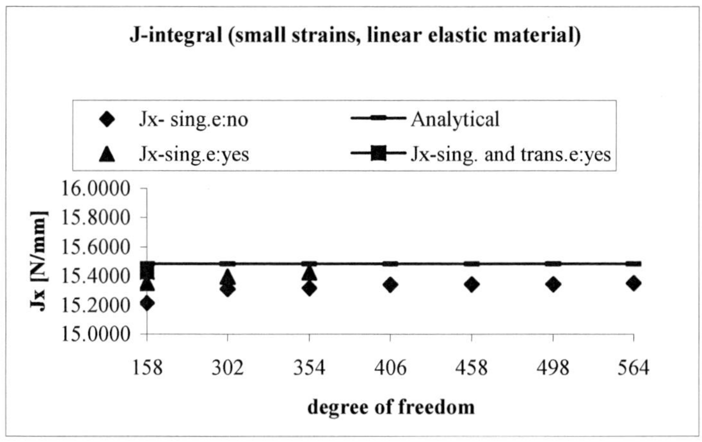

Figure 19. Analytical and calculated J_x values (small strains).

In the case of large strains - as the author knows - there are no data in the literature which are suitable for the comparison. Computations were made for the same finite element meshes. The calculated results can be seen in Table 3 and in Figure 20.

Table 3. Special features of the meshes and the calculated J_x values (large strains).

Large strains, linear elastic material					
Number of elements	Number of nodes	Degree of freedom	J_x [N/mm]	Singular element	Transition element
20	79	158	18,0459	no	no
42	151	302	18,0605	no	no
50	177	354	18,0770	no	no
58	203	406	18,0952	no	no
66	229	458	18,1057	no	no
72	249	498	18,1186	no	no
80	277	564	18,1219	no	no
20	79	158	18,1317	yes	no
42	151	302	18,1480	yes	no
50	177	354	18,1568	yes	no
20	79	158	18,1634	yes	yes

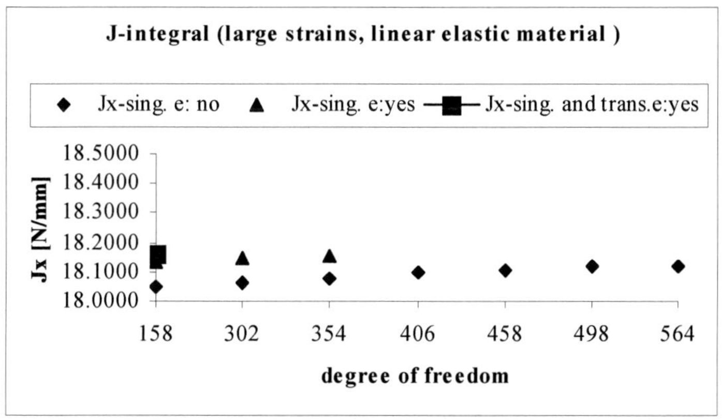

Figure 20. Calculated J-integral values (large strains).

CONCLUSION

This chapter presented the formulations and applicability of the *J*-integral for large strains in the case of elastic and elastic-plastic materials at the solution of crack problems based on continuum-mechanics. The calculated *J*-integral values are higher for large strains, as it was demonstrated by means of the numerical example. But this fact means that the safety of the cracked body increases from the aspect of service life. The characteristics of the diagrams are very similar to those which were obtained for small strains.

It is well-known that this integral quantity is path independent for small strains. The path independence of the two dimensional *J*-integral was also proved for large strains, which were supported by the computed results.

The chapter presented the mapping and applicability of special isoparametric elements in the finite element meshes. The results support that the special isoparametric elements can be applied in the case of large strains, too. The investigation of the convergence was also demonstrated.

REFERENCES

[1] C.L. Lau, M.M.K. Lee and A.R Luxmoore, Methodologies for predicting *J*-integrals under large plastic deformation-I. Further developments for tension loading, *Eng. Frac. Mech.* 1994, 49, No. 3, 337-354.

[2] C.L. Lau, M.M.K. Lee and A.R. Luxmoore, Methodologies for predicting *J*-integrals under large plastic deformation-II. Single edge notch specimens in pure bending, *Eng. Frac. Mech.* 1994, 49, No. 3, 355-369.

[3] G.B. May and A.S. Kobayashi, Plane stress stable crack growth and *J*-integral/HRR field, *Int. J. Solids Structures*, 1995, 32, No. 6/7, 857-881.

[4] D.P. Boothman, M.M.K. Lee, A.R. Luxmoore, The effects of weld mismatch on *J*-integrals and *Q*-values for semi-elliptical surface flaws, *Eng. Frac. Mech.* 1999, 64, 433-458.

[5] J. Jackiewicz, Numerical aspects of non-local modelling of the damage evolution in elastic-plastic materials, *Comp. Mat. Science,* 2000, 19, No. 1-4, 235-251.

[6] P.J. Bouchard, M.R. Goldthorpe, P. Prottey, *J*-integral and local damage fracture analyses for a pump casing containing large weld repairs, *Int. J. Pressure Vessels*, 2001, 78, 295-305.

[7] J. Saczuk, H. Stumpf, C. Vallee, A continuum model accounting for defect and mass densities in solids with inelastic material behaviour, *Int. J. of Solids and Structures,* 2001, 38, No. 52, 9545-9568.

[8] J. R. Rice, A path independent integral and the approximate analysis of strain concentration by notches and cracks, *J. App. Mech.* 1968, 34, 379-386.

[9] G. P. Cherepanov Cracks in solids, *Prikl. Mat. Mekh.* 1967, 25, 476-488.

[10] J. K. Knowles, E. Sternberg, On a class of conservation laws in linearized and finite elastostatics, *Arch. Rat. Mech. Anal.* 1972, 44, 187-211.

[11] S. N. Atluri, Path-independent integrals in finite elasticity and inelasticity, with body forces, inertia, and arbitrary crack-face conditions, *Eng. Frac. Mech.* 1982, 16, No.3, 341-364.

[12] W. Brocks and I. Scheider, Numerical Aspects of the Path-Dependence of the J-Integral in Incremental Plasticity, *GKSS Forschungszentrum, Geesthacht,* 2001, 1-33.

[13] X.-M. Wang and Y.-P. Shen, The conservation laws and path-independent integrals with an application for linear electro-magneto-elastic media, *Int. J. Solids Structures,* 1966, 33, No.6, 865-878.

[14] Á. Horváth, Higher-order singular isoparametric elements for crack problems, *Comm. in Num. Meth. Eng.* 1994, 10, 73-80.

[15] J. R. Rice, in *Fracture*, 1968, Vol. II, 191-311, ed. Liebowitz, Academic press.

[16] S. L. Pu, M. A. Hussain and W: E: Lorensen, The collapsed cubic isoparametric element as a singular element for crack problems, *Int. J. Num. Meth. Eng.* 1978, 12, 1727-1742.

[17] P. P. Lynn and A. R. Ingraffea, Transition elements to be used with quarter-point crack tip elements, *Int. J. Num. Meth. Eng.* 1978, 12, 1031-1036.

[18] Á. Horváth, General forming of transition elements, *Comm. in Num. Meth. Eng.* 1994, 10, 267-273.

[19] T. K. Hellen, On the method of virtual crack extensions, *Int. J. Num. Meth. Eng.* 1975, 9, 187-207.

In: Continuum Mechanics
Editors: Andrus Koppel and Jaak Oja, pp.173-191
ISBN: 978-1-60741-585-5

Chapter 6

REVIEW ON METHODOLOGIES OF PROGRESSIVE FAILURE ANALYSIS OF COMPOSITE LAMINATES

P. F. Liu [*] ***and J.Y. Zheng***
Zhejiang University, Hangzhou, China

ABSTRACT

Stiffness degradation for laminated composites such as carbon fiber/epoxy composites is an important physical response to the damage and failure evolution under continuous or cyclic loads. The ability to predict the initial and subsequent evolution process of such damage phenomenon is essential to explore the mechanical properties of laminated composites. This chapter gives a general review on the popular methodologies which deal with the damage initiation, stiffness degradation and final failure strength of composite laminates. These methodologies include the linear/nonlinear stress calculations, the failure criteria for initial microcracking, the stiffness degradation models and solution algorithms in the progressive failure analysis. It should be pointed out that the assumption of constant damage variable which is introduced into the constitutive equations of laminated composites to simulate the stiffness degradation properties is less effective and practical than that of changed damage variable with loads in the framework of continuum damage mechanics (CDM). Also, different damage evolution laws using CDM should be assumed to describe three failure modes: fiber breakage, matrix cracking and interfacial debonding, respectively.

Keywords: composite laminates; progressive failure analysis; continuum damage mechanics (CDM)

* Corresponding author. Tel.: +86 571 87953393; fax: +86 571 87952110. E-mail address: pfliu1980@yahoo.com (P.F. Liu)

1. INTRODUCTION

The laminated composites such as carbon fiber/epoxy composites have been increasingly employed in the wide fields of engineering application ranging from sport, mechanics, automobile, aerospace to ocean in order to develop the lightweight composite components due to their advantages such as high strength/stiffness-to-weight ratio, excellent resistance to fatigue and corrosion as well as satisfactory durability [1,2].

The design of the lightweight composite laminated structures relates the physical and mechanical properties of materials to the geometry shapes of structures. Before the structure optimization which aims to achieve the minimum weight, an initial exploration of the structural load-carrying capacity which is represented by the structural failure and damage behaviors is required for the reliable and economical design of composite laminates. Various damage modes may be provoked under complex loading environment and these damage modes continue to evolve with external loads. Therefore, the damage evolution and accumulation of laminated composites largely affect the mechanical strength and physical properties of composites. In this case, the knowledge of the damage tolerance of composite laminates plays an important role in the practical design of load-bearing composite structures [3-5].

The stiffness degradation phenomenon of fiber-reinforced composite (FRC) laminates under continuous external loads is an important response in the process of progressive failure of composites, which can be considered as the macroscopic representation of the microcrack propagation [6-8]. Currently, various theories and failure criteria have been proposed to predict the failure initiation, the progressive failure and the final damage properties of composite laminates. Over the past five decades, there have been continuous efforts in developing failure criteria to predict the initial failure and microcracking of composite laminates such as the well-known maximum stress/strain, Hashin, Hoffman, Tsai-Hill and Tsai-Wu criteria [9-16]. These failure criteria specify the different failure modes with respect to various failure mechanisms such as the fiber breakage, matrix cracking and interfacial debonding. In terms of the progressive failure analysis of composite laminates, the methodologies that accounts for the continuous stiffness degradation and predicts the final failure strength of composites is significant. Since Kachanov [17] first applied the CDM to study the creep rupture of metals, progressive failure analysis using CDM has been proved to be an excellent method to acquire the damage initiation and accumulation information of composite laminates. The CDM replaces the mechanical properties of the damaged materials with those of the homogenous materials by associating the damage mechanisms with their effects on the elastic constants of materials.

Already, a large number of theoretical models for fiber-reinforced composite laminates have been established by defining the damage tensors and their damage evolution laws based on the CDM theory. The refs.[18-26] introduced internal state variables to simulate the damage evolution in FRCs. These variables were related to the mechanical aspects of damage mechanisms and the dissipation energy required for the evolution of the damaged states. Thus, different damage mechanisms may have distinct internal variables to track the damage evolution. The refs.[27-30] established the thermodynamic models to simulate the damage evolution of composite laminates by relating the CDM theory with the finite element method. However, these thermodynamic models were all limited to the plane structures. For 3D composite cylindrical laminates, Perreux *et al.*[31] and Ferry *et al.*[32] derived the constitutive relationships of the damaged materials by using the CDM and the fracture mechanics. However, these models cannot explain different failure modes of FRCs because they defined only the individual thermodynamic conjugate force and damage variable. Recently, Perreux *et al.*[33], Liu and Zheng [34] described the damage evolution process by defining three independent damage variables to explain three failure modes above with respect to the carbon fiber/epoxy composite hydrogen storage vessels in area of hydrogen fuel cell vehicles.

With the rapid development of the calculation performance of computer, the finite element analysis has become a powerful tool to deal with the complex numerical problems. It is crucial to develop effective methodologies to implement the progressive failure analysis for the laminated composite structures. Zhang *et al.*[35] and Orifici *et al.*[36] conducted comprehensive reviews respectively on the damage constitutive relationships, the failure criteria and failure modes, but did not gave no information on the damage evolution using CDM theory and the effective solution algorithms of the finite element analysis. After introducing the various failure criteria and the thermodynamic theory (CDM theory), this chapter aims to explore the finite element implementation of the stiffness degradation and failure strength of composite laminates by discussing several practicable finite element algorithms. Finally, a case for the progressive failure analysis of the carbon fiber-reinforced composite hydrogen storage vessel is given and the numerical results are compared with those obtained by experiments and other existing models.

2. Failure Criteria of Composite Laminates

The failure of composite laminates is a complex problem and the modes of failure depend on the loading, the geometry size, the physical and mechanical properties of materials as well as the specimen defect. The composite laminates may fail in the different forms of fiber breakage, matrix cracking and interfacial debonding. The failure of the first ply represents the initiation of the failure and damage evolution of the whole composite laminates, but the failure of the first ply does not represent the final catastrophic damage of structures since the residual bearing ability of composite laminates still remains, which can be reflected by continuous ply failure and macroscopic stiffness degradation [37,38]. As the local/global buckling or unstability in the composite laminates occurs which is represented by the appearance of many singular plies with a lot of extremely small values approaching zero in the stiffness matrices, the composite laminates start to enter into a so-called strain softening stage. At this stage, the convergence problem about how to accurately calculate the final failure strength of composite laminates is a challenging task which requires some special manipulation mathematically.

Several phenomenological failure criteria on the initial failure of laminates with fiber orientations have been developed in terms of their properties with regard to the complicated nature of laminates over the past five decades. These failure criteria which appear representative of most of those are the maximum stress/strain, Hashin, Hoffman, Tsai-Hill and Tsai-Wu criteria [9-16]. Failure criteria for composite laminates can also be classified into two types: the non-interactive failure criteria and interactive failure criteria. The non-interactive failure criteria, sometimes also called independent failure criteria, assume no interactions between the stress or strain components and compare the individual stress or strain components with the corresponding material allowable strength values. The maximum stress/strain criteria belong to this category. The failure surfaces for these criteria are rectangular in stress and strain space, respectively [37]. In contrast, the interactive failure criteria such as the Hoffman, Tsai-Hill, Tsai-Wu criteria involve interactions between stress and strain components. The interactive failure criteria generally use the polynomial tensor expressions based on the material strengths to describe a failure surface and to separate each mode of failure [39].

Among all interactive failure criteria, the most general polynomial failure criterion for composite materials is the tensor polynomial criterion proposed by Tsai and Wu [9]. For anisotropic composite laminates, the generalized Tsai-Wu failure criterion can be written as the following form

$$F_i\sigma_i + F_{ij}\sigma_i\sigma_j + F_{ijk}\sigma_i\sigma_j\sigma_k \geq 1, \quad i,j,k = 1,2,...,6 \tag{1}$$

In the following, taking the carbon fiber-reinforced composite cylindrical laminates for example, the expression for the Tsai-Wu failure criterion is discussed. Fig.1 shows the cylindrical part of the developed composite hydrogen storage vessel, where a cylindrical coordinate system is defined and r, θ, z denote the radial, hoop and axial directions, respectively. The cylindrical part is composed of an aluminum liner and several carbon fiber/epoxy composite layers. Fig.2 describes a representative volume element (RVE) taken from Fig.1 and indicates the principle direction (1,2,3) of a composite layer under the cylindrical coordinate (r,θ, z). The carbon fiber/epoxy composites are considered to be linear-elastic and transversely isotropic. The same mechanical properties on the plane (2-3) perpendicular to the longitudinal direction (1) are assumed.

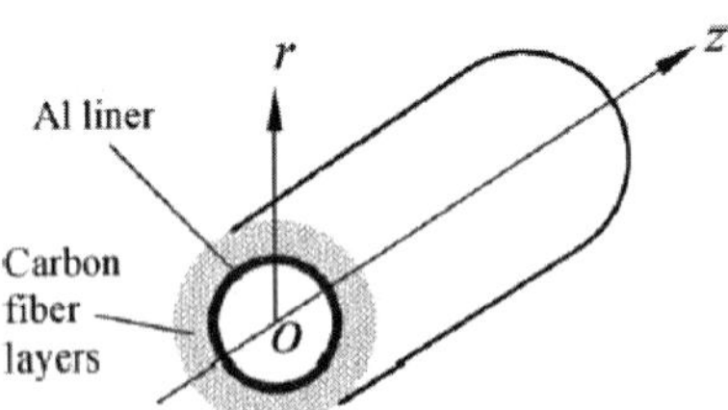

Figure 1. Fiber winding hydrogen storage vessel under the cylindrical coordinate (r, θ, z).

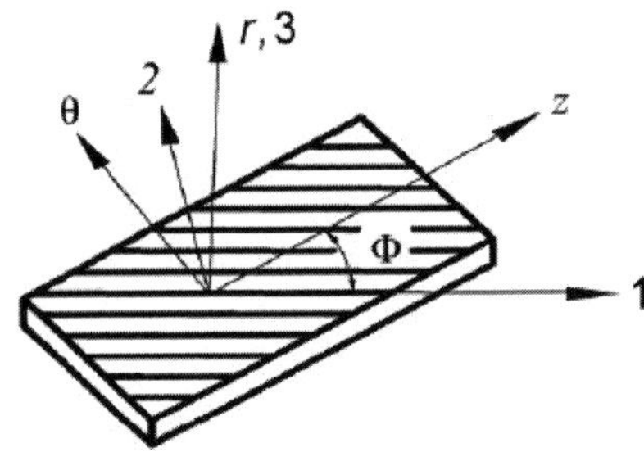

Figure 2. Schematic illustration of the on-axis coordinate (1, 2, 3) and the off-axis coordinate (r, θ, z).

The on-axis stress-strain relationships of the *k*th ($k=1,2,\cdots,1+n_s$) layer under the defined cylindrical coordinate system are expressed as

$$\begin{bmatrix} \sigma_1 \\ \sigma_2 \\ \sigma_3 \\ \sigma_6 \end{bmatrix}^{(k)} = \begin{bmatrix} C_{11} & C_{12} & C_{13} & C_{16} \\ C_{21} & C_{22} & C_{23} & C_{26} \\ C_{13} & C_{23} & C_{33} & C_{36} \\ C_{16} & C_{26} & C_{36} & C_{66} \end{bmatrix}^{(k)} \begin{bmatrix} \varepsilon_1 \\ \varepsilon_2 \\ \varepsilon_3 \\ \varepsilon_6 \end{bmatrix}^{(k)} \tag{2}$$

where $C_{ij}(i,j=1,2,3,6)$ are the on-axis elastic constants of materials. $\sigma_i(i=1,2,3,6)$ and $\varepsilon_i(i=1,2,3,6)$ are on-axis stresses and strains, respectively [34,38]. $\sigma_i(i=4,5)=0$ and $\varepsilon_i(i=4,5)=0$. n_s is the number of composite layers.

The expanded form for the Tsai-Wu failure criterion is expressed as [34,38]

$$F_{11}\sigma_1^2 + F_{22}\sigma_2^2 + F_{66}\sigma_6^2 + F_1\sigma_1 + F_2\sigma_2 + 2F_{12}\sigma_1\sigma_2 \geq 1 \tag{3}$$

where σ_1 and σ_2 are the on-axis stresses in the longitudinal and transverse directions and σ_6 is the on-axis in-plane shear stress.

The strength parameters F_{11}, F_{22}, F_{66}, F_1, F_2 and F_{12} are given by

$$F_{11} = \frac{1}{X_t X_c}, F_{22} = \frac{1}{Y_t Y_c}, F_{66} = \frac{1}{S^2}, F_1 = \frac{1}{X_t} - \frac{1}{X_c}, F_2 = \frac{1}{Y_t} - \frac{1}{Y_c}, F_{12} = -\frac{1}{2}\sqrt{F_{11}F_{22}} \tag{4}$$

where X_t and X_c are the longitudinal tensile and compressive strengths, respectively. Y_t and Y_c are those for the transverse direction. S is the in-plane shear strength.

The Tsai-Wu failure criterion identifies a ply failure, but it can not identify the modes of failure. Three stress components $H_i(i=1,2,6)$ in Eq.(3) are separated to identify the failure modes of the failed composite plies [34,40]

$$\left.\begin{aligned} H_1 &= F_1\sigma_1 + F_{11}\sigma_1^2 \\ H_2 &= F_2\sigma_2 + F_{22}\sigma_2^2 \\ H_6 &= F_{66}\sigma_6^2 \end{aligned}\right\} \tag{5}$$

When the Tsai-Wu failure criterion for each ply is satisfied, the maximum one of $H_i(i=1,2,6)$ can be found. If H_1 is the maximum, the fiber breakage is the dominating damage mode; If H_2 is the maximum, the matrix cracking is the dominating one; If H_6 is the maximum, the shear failure is the dominating one.

Tay *et al.* [41], Davila *et al.* [42] and Sleight [43] made an extensive literature survey of existing failure criteria for composites laminates. They pointed out that the maximum stress/ strain, Tsai-Hill, and Hashin criteria are generalized for either tensile or compressive stresses and the corresponding (tensile or compressive) strength value must be chosen based on the sign of the applied stress. The Tsai-Wu criterion is designed for use in all quadrants of the stress plane and 3D problems and may be directly used without modification for different stress signs. The Tsai-Wu criterion requires a biaxial test to experimentally determine the interaction term F_{12}, which is the only difference between the Tsai-Wu and Hoffman criteria and is found to be insignificant for the most part by Narayanaswami and Adelman [44]. Both the Tsai-Hill and Tsai-Wu criteria allow quadratic stress interactions, but Tsai-Hill is a purely quadratic criterion without linear stress term included by the Tsai-Wu criterion.

In addition, several new failure theories are developing such as the fracture energy based failure criteria [45,46], the strain invariant failure theory proposed by Tay *et al.* and Li *et al.* [41,47,48], the multicontinuum theory proposed by Mayes and Hansen [49,50] and the micromechanics-based failure theory proposed by Zhu and Ha [51,52].

3. CONTINUUM DAMAGE MECHANICS THEORY

In the continuum damage mechanics (CDM) theory, the loss of stiffness can be physically considered a consequence of distributed microcracks and microvoids. Over the past five decades, the CDM theory has been widely used to predict the isotropic/anisotropic stiffness degradation of composite laminates by introducing a phenomenological damage variable D associated with various damage modes such as the fiber breakage, matrix cracking and interfacial debonding [18-34]. According to Lemaitre and Chaboche [53,54], Ristinmaa and Ottosen [23,55], a brief review of the progressive failure analysis using CDM is given below.

For the damaged composite laminates, the Cauchy stress tensor σ can be substituted by the nominal stress tensor $\bar{\sigma}$ [17,23]

$$\sigma = (1-D)\bar{\sigma} \tag{6}$$

where $0 \le D < 1$. $D = 0$ represents the perfect materials and $D = 1$ denotes the completely damaged materials. According to the damage mechanics concept, the stresses in Eq.(6) are effective stresses.

If the damage and plasticity is uncoupled, the Helmholtz free energy per unit mass ψ for elastic-plastic materials under isothermal conditions is written as [19-22]

$$\rho\psi = \rho\psi^{e}(\varepsilon_{ij} - \varepsilon_{ij}^{p}, D_{ij}) + \rho\psi^{p}(\kappa) + \rho\psi^{d}(\kappa) \tag{7}$$

where ψ^{e}, ψ^{p} and ψ^{d} are the free energy which represents the elastic, plastic deformations and damage hardening, respectively. κ is an internal variable and ρ is the density.

The thermodynamic conjugate forces $\left(Y_{ij}, B, R\right)$ corresponding to the internal variables $\left(D_{ij}, \kappa, \alpha\right)$ are expressed as

$$\sigma_{ij} = \rho \frac{\partial \psi}{\partial \varepsilon_{ij}}, \quad Y_{ij} = -\rho \frac{\partial \psi}{\partial D_{ij}}, \quad B = \rho \frac{\partial \psi}{\partial \kappa}, \quad R = \rho \frac{\partial \psi}{\partial \alpha} \tag{8}$$

The thermodynamic formulation gives the thermodynamic forces and the dissipation inequality equation but no information about the evolution laws for internal variables. The only restriction imposed by the continuum thermodynamics on the evolution laws is that the Clausius-Duhem dissipation inequality must be fulfilled, which takes the following form under isothermal conditions [21-24]

$$\gamma = -\rho \dot{\psi} + \sigma_{ij} \dot{\varepsilon}_{ij} \geq 0 \tag{9}$$

where γ is the power of dissipation due to damage and a dot denotes rate with respect to time.

From Eqs.(7)-(9), the dissipation inequality is expressed as

$$\gamma = \sigma_{ij} \varepsilon_{ij}^{p} + Y_{ij} \dot{D}_{ij} - B\dot{\kappa} - R\dot{\alpha} \geq 0 \,(10)$$

If the dissipation γ reaches the maximum, the damage evolution law is given by

$$\dot{D}_{ij} = \dot{\lambda}^{d} \frac{\partial F^{d}}{\partial Y_{ij}}, \quad \dot{\kappa} = -\dot{\lambda}^{d} \frac{\partial F^{d}}{\partial B}, \quad \dot{\varepsilon}_{ij}^{p} = \dot{\lambda}^{p} \frac{\partial F^{p}}{\partial \sigma_{ij}}, \quad \alpha^{p} = -\dot{\lambda}^{p} \frac{\partial F^{p}}{\partial R} \tag{11}$$

where $F^{d}(Y_{ij}, D, B)$ and $F^{p}(\sigma_{ij}, D, R)$ are the damage potential functions. $\dot{\lambda}^{d} \geq 0$ and $\dot{\lambda}^{p} \geq 0$ are called the consistency parameters and they are assumed to obey the Kuhn-Tucker consistency requirements [23]

$$\begin{aligned} &\dot{\lambda}^{d} \geq 0, \quad F^{d} \leq 0, \quad \dot{F}^{d} = 0 \\ &\dot{\lambda}^{p} \geq 0, \quad F^{p} \leq 0, \quad \dot{F}^{p} = 0 \end{aligned} \tag{12}$$

In the associative plastic flow criterion, the functions F^{d} and F^{p} in Eq.(11) can be generally replaced by the corresponding plastic potential function G^{d} and G^{p}. Based on the CDM theory, the refs.(18)-(26) proposed the isotropic/anisotropic stiffness degradation and damage evolution models by introducing a two-order or four-order damage tensor. In these models, the relationships between the damage dissipation potential *F*, the conjugate force *Y* and the damage variable *D* are addressed, which are explained by different damage evolution laws. Besides, the micromechanics theory which assumes a representative volume element (RVE) is associated with the CDM theory to describe the internal damage properties.

Furthermore, the damage/plasticity coupling model is developed to describe the effect of plastic deformations on the damage properties of materials. Based on the CDM framework, the refs.[27-34] established the thermodynamic models to describe the progressive failure properties and to interpret the stiffness degradation of composite laminates. In these models, the relationships between the damage variables, conjugate forces and internal stresses/strains are further formulated and the three failure modes above are assumed. Specially, for 3D elastic carbon fiber/epoxy composite cylindrical laminates, Liu and Zheng [34] proposed an energy-based anisotropic stiffness degradation model including three failure modes above to derive the damage constitutive relationship, which is written as

$$\begin{bmatrix} \sigma_1 \\ \sigma_2 \\ \sigma_3 \\ \sigma_6 \end{bmatrix}^{(k)} = \begin{bmatrix} (1-D_1)C_{11} & & & \\ C_{12} & (1-D_2)C_{22} & Symmetric & \\ C_{13} & C_{23} & C_{33} & \\ C_{16} & C_{26} & C_{36} & (1-D_6)C_{66} \end{bmatrix}^{(k)} \begin{bmatrix} \varepsilon_1 \\ \varepsilon_2 \\ \varepsilon_3 \\ \varepsilon_6 \end{bmatrix}^{(k)} \tag{13}$$

where $D_i(i=1,2,6)$ represent the scalar damage variables for fiber breakage, matrix cracking and fiber/matrix interface failure, respectively. The detailed damage evolution laws for three damage variables $D_i(i=1,2,6)$ proposed by us see the ref. [34].

4. Finite Element Implementation of Progressive Failure Analysis

In recent years, the exploration for the methodologies of progressive failure analysis of composite laminates using the finite element method has attracted much attention around the world. In order to interpret the damage constitutive relationship and stiffness degradation properties of composite laminates under continuous loading, a set of effective finite element methods to be developed is significant. How to implement the initiation and damage evolution for three different failure modes above using the finite element analysis is an important research focus. The refs.[18-34] proposed different finite element methods respectively to perform the progressive failure analysis of composite laminates. Specially, Ochoa and Reddy [56] presented an excellent overview of the basic steps for performing the progressive failure analysis. However, the discussion on the calculation algorithms available for the convergence problem in the finite element analysis has not been involved. In the following, the finite element implementation of the progressive failure analysis of composite laminates is particularly addressed in detail.

A typical finite element equation which accounts for the geometrically nonlinear behavior in the progressive failure analysis is given by

$$K_T(u)u = F \tag{14}$$

where u is the displacement vector, K_T is the tangent stiffness matrix which depends on the material properties and the unknown displacement u and F is the load vector. Generally, the well-known Newton-Raphson iterative algorithm can be widely used to solve the non-linear equations due to its rapid convergence speed.

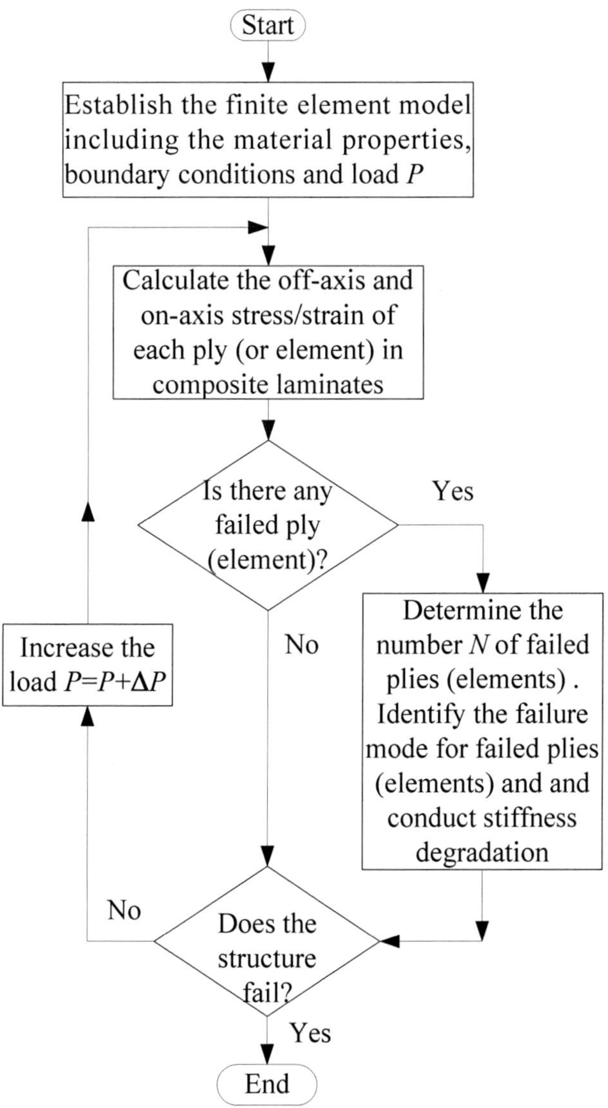

Figure 3. Flow chart of the progressive failure analysis.

By using these nonlinear equations, the flow chart for the progressive failure analysis of composite laminates using the finite element method is illustrated in Fig.3, which is summarized by four points: 1. At each load step, the finite element analysis is performed and the on-axis stresses/strains at each ply (or each element) are obtained. 2. The stresses/strains at each ply (or each element) are compared with the material allowable values and used to determine whether some plies (elements) have failed according to a certain failure criteria. If no failure is detected, then the applied load is increased and the analysis continues. 3. If some plies (or elements) fail, the stiffness constants are degraded according to the proposed stiffness degradation model based on the CDM theory. This is practically realized by multiplying a small value within 0 and 1 on the elastic constants of the failed plies (or elements). At this time, as the initial nonlinear solution no longer corresponds to an equilibrium state, the equilibrium of composite structure requires to be re-established using the modified mechanical properties for the failed plies (or elements) while maintaining the current load level. This adjustment accounts for the material nonlinearity due to local stiffness degradation in the finite element analysis. 4. this iterative process continues which repeatedly obtains nonlinear equilibrium solutions after each load step until a sudden catastrophic failure appears, which indicates the advent of the strain softening stage.

However, as the incremental load increases, the convergence problem will appear in the nonlinear finite element solution, which is represented by two points: 1. After some plies (or elements) repeatedly fail, the corresponding elastic constants degraded and ultimately may become the values approaching zero, leading to the appearance of many "ill-conditioned" finite element equations; 2. The sudden catastrophic failure of composite laminates indicates the final failure of composite laminates. In this case, the conventional Newton-Raphson algorithm can not further track the load path and becomes incapable because the integrated structural stiffness matrices at the final failure point are singular, as shown in Fig.4.

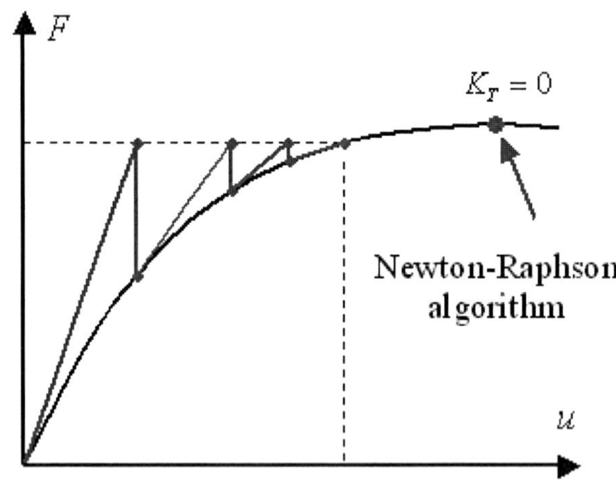

Figure 4. Newton-Raphson iterative algorithm.

In order to solve this problem, Tay *et al.* [41,48,57,58] proposed an element failure method for the progressive failure analysis of composite laminates so that more rapid convergence and computational robustness are achieved. This method essentially manipulates the nodal forces of finite elements directly to simulate the effect of damage while leaving the material stiffness values unchanged. However, this method has not been widely used since it cannot effectively acquire the information on stiffness degradation.

Geometrically nonlinear static problems sometimes involve buckling or collapse behavior, where the load-displacement response shows negative stiffnesses and negative eigenvalues. At this time, the structure must release strain energy to remain in equilibrium. Currently, three applicable methods are proposed to solve the softening problem. First, the quasi-static analysis can be substituted by a slow dynamic analysis, which is actually realized by modeling the response with inertia effects. Second, associated with the Newton-Raphson iterative method, the adaptive damping matrices are introduced into the nonlinear finite element equations to ensure positive eigenvalues at the structural softening stage. Third, the load matrices are multiplied by a variable arc-length factor to circumvent instability, but the loading must be proportional where the load magnitudes are governed by this factor. This method is also called as the "modified riks method". In the following, three algorithms above for the progressive failure analysis are discussed in detail to solve the softening problem.

[1] The finite element equation in the **dynamic algorithm** is given by [59,60]

$$K_T u + M\ddot{u} = F \tag{15}$$

where M is the mass matrix. When K_T becomes singular or non-positive after the failure point, the adaptable mass matrix M is introduced to prevent singularity in the iterative solution. For stable nonlinear calculations, the inertia effect represented by the matrix M should be small enough not to affect the calculation accuracy.

[2] (2) The finite element equation in the **nonlinear stabilization algorithm** is expressed as [59,60]

$$K_T u + C\dot{u} = F \tag{16}$$

where C is the damping matrix. As K_T becomes singular or non-positive after the failure point, the elements in C increase to obtain the solutions for u in such a way that the viscous forces introduced are sufficiently large to prevent instantaneous collapse but small enough not to affect the behavior significantly while the problem is stable. It is more efficient and accurate in most cases to relate the nonlinear stabilization algorithm with the multiframe restart analysis in the finite element analysis.

[3] (3) The finite element equation in the **arc-length algorithm** is written as [61-65]

$$K_T u = \lambda F \tag{17}$$

where λ is a load factor within -1 and 1 which changes with the stiffness matrix K_T to ensure an accurate solution of u. The arc-length algorithm imposes another constraint, which is stated as

$$\sqrt{\Delta u^2 + \lambda^2} = R \tag{18}$$

where Δu is the displacement increment and R is the arc-length radius.

It can be found from numerical calculations that the arc-length algorithm can ensure highest solution precision among three algorithms though the calculations are relatively time-consuming. In contrast, the dynamic algorithm may lead to large errors since the differentiation in the finite element method is approximately substituted by the finite difference though its convergence velocity is more rapid than other two algorithms. Besides, the prediction could be inaccurate when the problem is highly nonlinear for the stabilization algorithm. Although it can be competent for both local and global instability problem with few limitations, the nonlinear stabilization algorithm cannot capture the negative-slope portion of the load-displacement curve for global instability. Therefore, the three algorithms above should be appropriately chosen in such a way that the theoretical results are in good agreement with the experimental results.

Generally, the progressive failure analysis can be implemented using one of three algorithms associated with the finite element software ABAQUS or ANSYS [59,60]. In general, for the damage and failure problems of composite laminates, the eigenvalue buckling analysis is initially performed to provide prelimilary information about a structural catastrophic failure, followed by the subsequent progressive failure analysis. For complex cases for damage and failure of composite laminates, parallel calculations adopted by using high-performance computer are necessary to improve calculation velocity and efficiency [34].

5. Progressive Failure Analysis of Carbon Fiber/Epoxy Composite Cylindrical Laminates

Based on the CDM theory, an energy-based stiffness degradation model is proposed to predict the progressive failure properties of the carbon fiber/epoxy composite cylindrical laminates, that is, the high pressure hydrogen storage vessel in the field of hydrogen cell vehicle. The Tsai-Wu failure criterion is employed and three failure modes above are included in this model and three damage evolution laws are proposed by Liu and Zheng [34]. A 3D finite element model is established by the finite element software ANSYS, as shown in Fig.5 and the progressive failure analysis is implemented by using the multiframe restart analysis and the arc-length algorithm is used to solve the sudden catastrophic failure problem (softening problem). For the developed composite hydrogen storage vessel with 40MPa working pressure, the inner radius and thickness of the liner are 44mm and 1.8mm, respectively. The thickness of each composite layer is 0.42mm. The number of the composite winding layers is $n_s = 10$. Thus, the radius of the outmost layer is 50mm. The fiber volume fraction is V_f=70% in carbon fiber/epoxy composites. The length of the cylindrical part is 160 mm. The 3-D eight-node solid element SOLID95 and 3-D eight-node anisotropic solid element SOLID64 are adopted to mesh the liner layer and carbon fiber/epoxy composite layers, respectively. The mesh model includes 1000 Al liner elements and 5000 composite elements.The winding angles of composite layers from the inner layer to the outer layer are ϕ=90^0,-90^0,18.9^0,-18.9^0, 90^0,-90^0, 28.6^0, -28.6^0, 90^0,-90^0, respectively. Perfect bonding between the liner layer and fiber layers is assumed. The symmetric constraints are applied on the symmetric sides and the increasing pressure P is loaded on the inner surface of the liner layer.

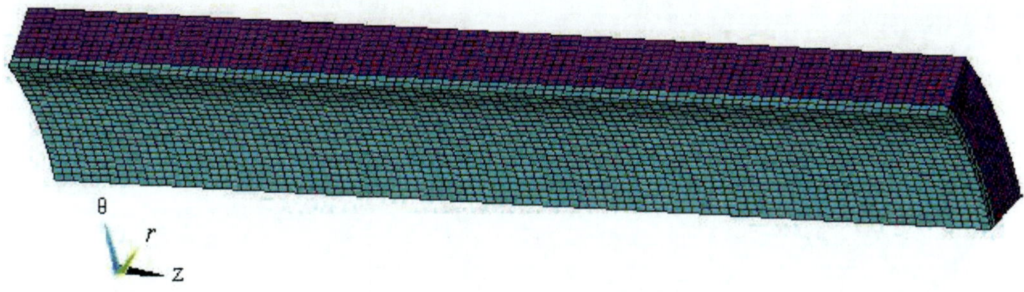

Figure 5. A 30^0 axisymmetric finite element model of Al-carbon fiber/epoxy composite cylindrical laminates.

The mechanical parameters of materials are listed in Table.1. σ_s and σ_b represent the yielding and tensile strengths, respectively. τ is the shear modulus in the bilinear hardening model. The strength parameters of T700 fiber/epoxy composites are listed in Table 2. Table 3 gives the damage parameters of composite laminates determined by performing the pressure tests on the developed composite hydrogen storage vessels.

A subroutine is coded using the ANSYS-APDL (ANSYS Parametric Design Language) and integrated into the main program. Initially, the pressure value P is specified to ensure no element failure. 5000 material numbers, 5000 damage variables D_i (i=1,2,6) and 5000 conjugate forces Y_i (i=1,2,6) for 5000 composite elements are defined.

Table 1. Mechanical properties of materials [34,38]

E_1(GPa) E_2(GPa) v_{12} v_{23} σ_s(MPa) σ_b(MPa) τ(MPa) ρ(Kg/m^3)
6061 Al 70 70 0.3 0.3 246 324 600 2700
T700/epoxy 181 10.3 0.28 0.49 □ □ □ 1550

Table 2. Strength parameters for T700/epoxy composites [34,38]

X_t(MPa) X_c(MPa) Y_t(MPa) Y_c(MPa) S(MPa)
2150 2150 298 298 778

Fig.6 shows the distributions of the Mises equivalent stress under the internal pressure 40MPa, 80MPa and 120MPa, respectively. The increase of the Mises stress indicates the damage and failure evolution process of structure. The nonlinear relationships between the Mises stress and the internal pressure arise mainly from the stiffness degradation after element failure. Besides, it is revealed that matrix cracking accounts for the dominating role in the process of the damage evolution of carbon fiber/epoxy composite laminates.

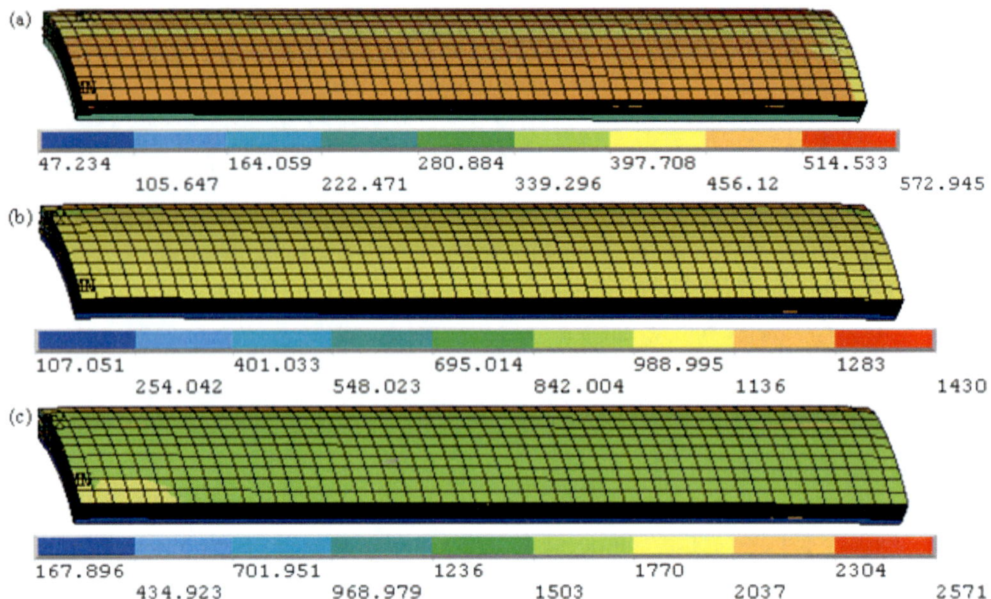

Figure 6. Distributions of Mises equivalent stress under internal pressure (a) 40MPa, (b) 80MPa and (c) 120MPa.

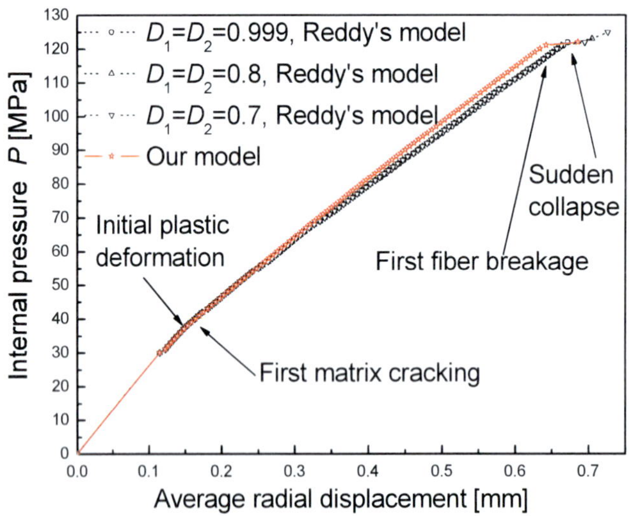

Figure 7. Internal pressure-average radial displacement curve.

Fig.7 shows the internal pressure-average radial displacement curve on the inner surface of the liner. After the structure enters into the plastic deformation stage, the non-linear damage initiation is marked by matrix cracking. Within a quite long area of the curve, the structure experiences a relatively steady failure stage. After conducting the stiffness degradation on the failed elements, the stresses redistribute and the stresses of the failed elements are concentrated on the neighbouring elements. When the stress concentrations develop to some extent, the composite structures completely lose their load-carrying capacity and ultimately collapse. This is represented by a small abrupt turn in the last portion of the curve. From the finite element analysis, the final failure pressure of the composite structure is about 122MPa, which is in good agreement with the experimental value 125-126MPa by the burst test of hydrogen storage vessel developed by us, as shown in Fig.8. Reddy *et al.* [66], Akhras and Li [67] also conducted stiffness degradation on the anisotropic elastic constants of composite materials after identifying the failure modes. In their work, the elastic modulus and Poisson's ratio of composite materials are multiplied by an average value D within 0 and 1 regardless of the damage evolution laws, which yet neglects the evolution properties of damage variables D with increasing internal pressure. For different type of failure modes and composite structures, how to accurately determine the true damage evolution laws which stem essentially from the evolvement of microcracks and microvoids is worthy of further discussion and research.

Figure 8. Hydrogen storage vessel after burst.

6. Conclusion

This chapter discusses the methodologies of progressive failure analysis of composite laminates including the initial failure criteria, the damage evolution model based on the CDM theory and the corresponding finite element method on how to implement the variable-stiffness finite element solution with increasing loads. It should be emphasized that the assumption of the changed damage variable with loads is more rational and practical than the constant damage variable, and different failure modes should be associated with different damage evolution laws. However, the damage evolution laws should be chosen in such a way that the theoretical objective value approaches the experimental results. In addition, the appropriate selection and evaluation of calculation performance for different algorithms in the finite element analysis are also crucial problems which deserve further research.

Acknowledgments

This research is supported by the Chinese postdoc science funding project (Number: 20070411175), the high-technology research and development program (863 program) of China (Numbers: 2006AA05Z143, 2006AA11A188 and 2006AA11A187) and the key project of national programs for fundamental research and development (973 program) of China (Number: 2007CB209706).

References

[1] Young, K. S. (1992). Advanced composites storage containment for hydrogen. *International Journal of Hydrogen Energy*, 17 (7), 505-507.

[2] Chapelle, D. D., & Perreux, D. (2006). Optimal design of a Type 3 hydrogen vessel: Part I-Analytic modelling of the cylindrical section. *International Journal of Hydrogen Energy*, 31 (5), 627-638.

[3] Effendi, R. R., Barrau, J. J., & Guedra-Degeorges, D. (1995). Failure mechanism analysis under compression loading of unidirectional carbon/epoxy composites using micromechanical modelling. *Composite Structures,* 31 (2), 87-98.

[4] Phillips, E. A., Herakovich, C. T., & Graham, L. L. (2001). Damage development in composites with large stress gradients, *Composites Science and Technology,* 61 (15), 2169-2182.

[5] Bussiba, A., Kupiec, M., Ifergane, S., Piat, R., & Böhlke,T. (2008). Damage evolution and fracture events sequence in various composites by acoustic emission technique. *Composites Science and Technology*, 68 (5), 1144-1155.

[6] Zhang, D. X., Ye, J.Q., & Lam, D. (2006). Ply cracking and stiffness degradation in cross-ply laminates under biaxial extension, bending and thermal loading. *Composite Structures,* 75 (1-4), 121-131.

[7] Bouazza, M., Tounsi, A., Benzair, A., & Adda-bedia, E. A. (2007). Effect of transverse cracking on stiffness reduction of hygrothermal aged cross-ply laminates. *Materials & Design*, 28 (4), 1116-1123.

[8] Spootswood, M.S., & Palazotto, A. N. (2001). Progressive failure analysis of a composite shell. *Composite Structures*, 53 (1), 117-131.

[9] Tsai, S. W., & Wu, E. M. (1971). A general theory of strength for anisotropic materials. *Journal of Composite Material,* 5 (1), 58-80.

[10] Cui, W. C., Wisnom, M.R., & Jones, M. (1992). A comparison of failure criteria to predict delamination of unidirectional glass/epoxy specimens waisted through the thickness. *Composites*, 23 (3), 158-166.

[11] Hinton, M. J., & Soden, P. D. (1998). Predicting failure in composite laminates: the background to the exercise. *Composite Science Technology*, 58 (7), 1001-1010.

[12] Soden, P. D., Hinton, M. J., & Kaddour, A. S. (1998). A comparison of the predictive capabilities of current failure theories for composite laminates. *Composites Science and Technology,* 58 (7), 1225-1254.

[13] Theocaris, P. S. (1992). Weighing failure tensor polynomial criteria for composites. *International Journal of Damage Mechanics*, 1 (1), 4-46.

[14] Rotem, A., & Hashin, Z. (1975). Failure modes of angle ply laminates. *Journal of Composite Materials*, 9 (2), 191-206.

[15] Narayanaswami, R., & Adelman, H. M. (1977). Evaluation of the tensor polynomial and hoffman strength theories for composite materials. *Journal of Composite Materials*, 11(4), 366-377.

[16] Hashin, Z. (1980). Failure criteria for unidirectional fiber composites. *Journal of Applied Mechanics*, 47, 329-334.

[17] Kachanov, L. M. (1958). *Time of the rupture process under creep conditions.* IVZ Akad. Nauk S.S.R. Otd Tech Nauk, 8, 26-31. (In Russian)

[18] Schapery, R. A. (1990). A theory of mechanical behavior of elastic media with growing damage and other changes in structure. *Journal of the Mechanics and Physics of Solids,* 38 (2), 215-253.

[19] Murakami, S., & Kamiya, K. (1997). Constitutive and damage evolution equations of elastic-brittle materials based on irreversible thermodynamics. *International Journal Mechanical Science*, 39(4), 473-486.

[20] Hayakawa, K., Murakami, S., & Liu, Y. (1998). An irreversible thermodynamics theory for elastic-plastic- damage materials. *European Journal of Mechanics-A/Solids,* 17 (1), 13-32.

[21] Tang, X. S., Jiang, C. P., & Zheng, J. L. (2002). Anisotropic elastic constitutive relations for damaged materials by application of irreversible thermodynamics. *Theoretical and Applied Fracture Mechanics*, 38 (3), 211-220.

[22] Brünig, M. (2003). An anisotropic ductile damage model based on irreversible thermodynamics. *International Journal of Plasticity*, 19 (10), 1679-1713.

[23] Olsson, M., & Ristinmaa, M. (2003). Damage evolution in elasto-plastic materials response due to different concepts. *International Journal of Damage Mechanics,* 12, 115-139.

[24] Basu, S., Waas, A. M., & Ambur, R. D. (2007). Prediction of progressive failure in multidirectional composite laminated panels. *International Journal of Solids and Structures,* 44 (9), 2648-2676.

[25] Maimí, P., Camanho, P. P., & Mayugo, J. A. (2007). A continuum damage model for composite laminates: Part I- constitutive model. *Mechanics of Materials,* 39 (10), 897-908.

[26] Maimí, P., Camanho, P. P., & Mayugo, J. A. (2007). A continuum damage model for composite laminates: Part II - computational implementation and validation. *Mechanics of Materials,* 2007, 39 (10), 909-919.

[27] Kwon,Y. W., & Liu, C. T. (1997). Study of damage evolution in composites using damage mechanics and micromechanics. *Composite Structures*, 38 (1-4), 133-139.

[28] Schipperen, J. H. A. (2001). An anisotropic damage model for the description of transverse matrix cracking in a graphite–epoxy laminate. *Composite Structures,* 53 (3), 295-299.

[29] Maa, R. H., & Cheng, J. H. (2002). A CDM-based failure model for predicting strength of notched composite laminates. *Composites Part B: Engineering*, 33 (6), 479-489.

[30] Camanho, P. P., Maimí, P., & Dávila, C. G. (2007). Prediction of size effects in notched laminates using continuum damage mechanics. *Composites Science and Technology*, 67 (13), 2715-2727.

[31] Perreux, D., Robinet, D., & Chapelle, D. (2006). The effect of internal stress on the identification of the mechanical behaviour of composite pipes. Composites Part A: *Applied Science and Manufacturing*, 37 (4), 630-635.

[32] Ferry, L., Perreux, D., Rousseau, J., & Richard, F. (1998). Interaction between plasticity and damage in the behaviour of [+φ,−φ] fibre reinforced composite pipes in

biaxial loading (internal pressure and tension). *Composites Part B: Engineering,* 29 (6), 715-723.

[33] Perreux, D., & Thiebaud, F. (1995). Damaged elasto-plastic behaviour of [+φ,−φ] fibre-reinforced composite laminates in biaxial loading. *Composites Science and Technology*, 54 (3), 275-285.

[34] Liu, P. F., & Zheng, J.Y. (2008). Progressive failure analysis of carbon fiber/epoxy composite laminates using continuum damage mechanics. *Materials Science and Engineering*:A, 485 (1-2), 711-717.

[35] Zhang, Y. X., & Yang, C. H. (2008). Recent developments in finite element analysis for laminated composite plates. *Composite Structures* (In press).

[36] Orifici, A.C., Herszberg, I., & Thomson, R. S. (2008). Review of methodologies for composite material modelling incorporating failure. *Composite Structures*, 86 (1-3), 194-210.

[37] Reddy, Y. S. N., & Pandey, A. K. (1987). A first-ply failure analysis of composite laminates. *Computers and Structures*, 25 (3), 371-393.

[38] Zheng, J. Y., & Liu, P. F. (2008). Elasto-plastic stress analysis and burst strength evaluation of Al-carbon fiber/epoxy composite cylindrical laminates. *Computational Materials Science,* 42 (4), 453-461.

[39] Cuntze, R. G., & Freund, A. (2004). The predictive capability of failure mode concept-based strength criteria for multidirectional laminates. *Composites Science and Technology*, 64 (3-4), 343-377.

[40] Padhi, G.S., Shenoi, R.A., Moy, S.S.J., & Hawkins, G.L. (1997). Progressive failure and ultimate collapse of laminated composite plates in bending. *Composite Structures*, 40 (3-4), 277-291.

[41] Tay, T.E., Liu, G., Tan, V. B.C., Sun, X.S., & Pham, D.C. (2008). Progressive failure analysis of composites. *Journal of Composite Materials,* 42 (18), 1921-1966.

[42] Davila, C. G., Camanho, P. P., & Rose, C. A. (2005). Failure criteria for FRP laminates. *Journal of Composite Materials*, 39 (4), 323-345.

[43] Sleight, D. W. (1999). *Progressive failure analysis methodology for laminated composite structures.* NASA TP-209107, 48-63.

[44] Wu, R.Y., Stachurski, Z. (1984). Evaluation of the normal stress interaction parameter in the tensor polynomial strength theory for anisotropic materials. *Journal of Composite Materials,* 18 (5), 456-463.

[45] Hwu, C., Kao, C. J., & Chang, L. E. (1995). Delamination fracture criteria for composite laminates. *Journal of Composite Materials*, 29 (15), 1962-1987.

[46] Huang, H. S., Springer, G. S., & Christensen, R. M. (2003). Predicting failure in composite laminates using dissipated energy. *Journal of Composite Materials,* 37 (23), 2073-2099.

[47] Li, R., Kelly, D., & Ness, R. (2003). Application of a first invariant strain criterion for matrix failure in composite materials. *Journal of Composite Materials*, 37 (22), 1977-2000.

[48] Tay, T.E., Liu, G., Yudhanto, A., & Tan, V. B.C. (2008). A micro-macro approach to modeling progressive damage in composite structures. *International Journal of Damage Mechanics,* 17, 5-28.

[49] Mayes, J.S., & Hansen, A.C. (2004). Composite laminate failure analysis using multicontinuum theory. *Composites Science and Technology,* 64 (3-4), 379-394.

[50] Mayes, J.S., & Hansen, A.C. (2004). A comparison of multicontinuum theory based failure simulation with experimental results. *Composites Science and Technology*, 64 (3-4), 517-527.

[51] Zhu, H., Sankar, B.V., & Marrey, R.V. (1998). Evaluation of failure criteria for fiber composites using finite element micromechanics. *Journal of Composite Materials,* 32 (8), 766-782.

[52] Ha, S.K., Jin, K.K., & Huang, Y. (2008). Micro-mechanics of failure (MMF) for continuous fiber reinforced composites. *Journal of Composite Materials*, 42 (18), 1873-1895.

[53] Lemaitre, J., & Chaboche, J. L. (1990). *Mechanics of solid materials*. Cambridge, UK: Cambridge University Press.

[54] Lemaitre, J. (1992). *A course on damage mechanics*. New York, USA: Springer-Verlag.

[55] Ristinmaa, M. & Ottosen, N. S. (1998). Viscoplasticity based on an additive split of the conjugated forces. *European Journal of Mechanics-A/Solids*, 17 (2), 207-235.

[56] Ochoa, O., & Reddy, J.N. (1992). *Finite element analysis of composite laminates.* Dordrecht, Netherlands: Kluwer Academic Publishers.

[57] Tay, T.E., Tan, S.H.N., Tan, V.B.C. & Gosse, J. H. (2005). Damage progression by the element-failure method (EFM) and strain invariant failure theory (SIFT). *Composites Science and Technology*, 65 (6), 935-944.

[58] Tay, T.E., Tan, V.B.C., & Deng, M. (2003). Element-failure concepts for dynamic fracture and delamination in low-velocity impact of composites. *International Journal of Solids and Structures,* 40 (3), 555-571.

[59] Hibbit, Karlsson, & Sorensen. (2007). *ABAQUS User's Manual.* version 6.8.

[60] Swanson, J. (2007). *ANSYS User's Manual*. version 11.0.

[61] Riks, E. (1979). An incremental approach to the solution of snapping and buckling problems. *International Journal of Solids and Structures*, 15, 529-551.

[62] Ramm, E. (1981). Strategies for tracing the nonlinear response near limit points. *In: nonlinear finite element analysis in structural mechanics* (68-89). New York, USA: Springer-Verlag.

[63] Crisfield, M. A. (1983). An arc-length method including line searches and accelerations. *International Journal for Numerical Methods in Engineering,* 19, 1269-1289.

[64] Liu, P. F., Zheng, J.Y., Ma, L., Miao, C. J., & Wu, L. L. (2008). Calculations of plastic collapse load of pressure vessel using FEA. *Journal of Zhejiang University Science A*, 9 (7), 900-906.

[65] Liu, P. F., Xu, P., Han, S.X., & Zheng, J.Y. (2008). Optimal design of pressure vessel using an improved genetic algorithm. *Journal of Zhejiang University Science* A, 9 (9), 1264-1269.

[66] Reddy, Y. S. N. (1995). Non-linear progressive failure analysis of laminated composite plates. *International Journal Non-Linear Mechanics,* 30 (5), 629-649.

[67] Akhras, G., & Li, W.C. (2007). Progressive failure analysis of thick composite plates using the spline finite strip method. *Composite Structures,* 79 (1), 34-43.

In: Continuum Mechanics ISBN: 978-1-60741-585-5
Editors: Andrus Koppel and Jaak Oja, pp. 193-221

Chapter 7

DIFFERENTIAL FORM OF CONTINUUM MECHANICS: OPERATORS AND EQUATIONS

Kazuhito Yamasaki
Kobe University, Kobe, Japan

Abstract

The continuum mechanics in terms of the differential forms is proposed. We introduce the dual material space-time which consists of the strain space-time and the stress space-time. In this case, there is a one-to-one correspondence between the kinds of the basic equations in the continuum mechanics and the kinds of the basic operators in the differential forms. That is, the kinematic and constitutive equations can be derived by the exterior differential operator and the Hodge star operator, respectively. Other compound equations such as the Navier equation, Laplace (wave) equation and the incompatibility equation can be derived by the combination of the basic operators. This systematic approach allows us to find (i) the anti-exact solution of the Navier equation and (ii) the J-integral in fracture mechanics. The result (ii) means that the continuum mechanics in terms of the differential forms describes a partial aspect of the fracture mechanics. Moreover, the differential form approach allows us to link the deformation field with the non-deformation field such as the electromagnetic field. As an example, we take up the piezoelectric and Villari effects and derive the constitutive equations for these effects. These constitutive equations can be interpreted geometrically as the interaction among the geometrical objets of the space-time.

1 Introduction

The continuum mechanics is concerned with the interaction of the deformation field with the non-deformation field such as electromagnetic field and thermodynamic field, because many interesting physical phenomena which possess important technological applications lie at this interaction [1]. The non-deformation field of the deformation field is not the sum of these two fields, but a nonlinear composition of them [1], so the mathematical description of the interaction is complicated. Now, the differential forms have been known as the geometric language which facilitates the mathematical description of physical fields. Therefore, recently, this mathematical tool becomes a central method for modern mathematical physics (e.g., [2, 3]). In fact, the geometry of electromagnetic field (the gauge field) and the thermodynamic field have already been described by the differential forms (e.g., [4, 5, 6]).

On the other hand, the differential forms have not yet been widely applied to the deformation field except for strain space-time. Thus, the purpose of this chapter is to express the deformation field in terms of the differential forms. This mathematical description can be straight-forward and relatively simple. Moreover, this geometrical approach shall allow us to demonstrate the geometrical interpretation of the interaction of deformation field with the non-deformation field. Now, there are various equations to describe the deformation fields. In this paper, we take up the following equations, which are absolutely necessary to describe the field, in the following sections.

(a) Constitutive equations (e.g., Hooke's law) in the section 3.1.

(b) Kinematic equations (e.g., conservation law of momentum) in the sections 3.2 and 3.3.

(c) Navier equations in the section 4.

(d) Laplace (wave) equations in the section 5.

(e) Incompatibility equations (extended St. Venant compatibility equations) in the section 6.1.

(f) Stress functions (e.g., Airy stress functions) in the section 6.2.

The problem of how to express these equations in terms of the differential forms can be solved by noting the kinds of operators. In the differential forms, there are three basic operators: an exterior differential operator, a Hodge star operator and a homotopy operator. As we will see, the equations (a)$\sim$(f) can be derived in the systematic way by the combination of these basic operators.

The structure of this chapter is as follows.

In the section 2, we review the differential forms, because it is not the standard language in the continuum mechanics. As a good example to understand the calculation techniques of the differential forms, we take up electromagnetism.

In the section 3, we present the dual material space-time with defect filed: one is the strain space-time, and the other is stress space-time[7, 8]. By using the exterior differential operator, we derive all the kinematic and the continuity equations in the dual material space-time. Moreover, by using the Hodge star operator, we derive all the constitutive equations that link the stress space-time with the strain space-time.

In the section 4, we derive the Navier equation by the combination of the basic operators. Moreover, by applying the Homotopy operator to the Navier equation, we suggest an anti-exact solution of the equation.

In the section 5, we derive the wave equation in the dual material space-time by using the adjoint operator. Moreover, as a Hodge duality of the Navier equation in the section 4, we derive the Laplace equation for the defects field.

In the section 6, we derive the incompatibility equation by using a new defined exterior differential operator. Moreover, as a Hodge duality of the incompatibility equation, we derive the generalized equation for stress functions.

In the section 7, we define the deformation energy in terms of the differential forms, and show that this corresponds to the J-integral in fracture mechanics. This means that the continuum mechanics in terms of the differential forms describes a partial aspect of the fracture mechanics.

In the section 8, we reconsider the results in the section 3 form the viewpoint of the differential geometry.

In the section 9, we take up two well-known phenomena: piezoelectric and Villari effects. These are good examples to consider the electrodynamics of continua in terms of the differential forms.

The sections 10 and 11 are devoted to discussion and conclusion, respectively.

2 Differential Forms

As we will see, differential forms unable us to simplify the formulas of deformation fields. However, the differential forms are not the standard languages of the continuum mechanics. Then, in this section, the techniques of calculations in the differential forms are briefly summarized. We do not mathematically deal with the differential forms, but use it just as a basis for calculation, paying attention to the application of it to the continuum mechanics. For this purpose, we take up electromagnetism, because this is a good example to understand the calculation techniques of the differential forms (e.g., [4]).

In some cases mathematical details are omitted, so see the references [2, 6] for investigating the matter in more detail.

2.1 The definition of differential forms and wedge product

Let x^i be an arbitrary orthogonal. Let Λ^p be the set consists of the differential forms of rank p. In this case, $\omega \in \Lambda^p$ is defined as

$$\omega = \frac{1}{p!}\omega_{i_1 \cdots i_p} dx^{i_1} \wedge \cdots \wedge dx^{i_p}, \tag{1}$$

where the symbol $\wedge$ is the wedge (or exterior) product. The wedge product $\wedge$ satisfies the following laws of association, distribution and commutation:

$$(\alpha \wedge \beta) \wedge \gamma = \alpha \wedge (\beta \wedge \gamma), \tag{2}$$

$$(\alpha + \beta) \wedge \gamma = \alpha \wedge \gamma + \beta \wedge \gamma, \tag{3}$$

$$\alpha_p \wedge \beta_q = (-1)^{pq} \beta_q \wedge \alpha_p, \tag{4}$$

where $\alpha_p \in \Lambda^p$ and $\beta_q \in \Lambda^q$. For instance, the following commutation is often used:

$$dx^i \wedge dx^j = -dx^j \wedge dx^i \ (i \neq j), \tag{5}$$

$$dx^i \wedge dx^i = 0. \tag{6}$$

In this section, we consider the (3+1)-dimensional space-time: $dx^1, dx^2, dx^3, dx^0 (= c_L dt$, where c_L is the velocity of light and t is the time). Take the electromagnetic field for example. The scalar potential ϕ is 0-form:

$$\phi = \phi, \tag{7}$$

the electric field $\mathcal{E}$ is 1-form

$$\mathcal{E} = \mathcal{E}_1 dx^1 + \mathcal{E}_2 dx^2 + \mathcal{E}_3 dx^3 = \mathcal{E}_A dx^A, \tag{8}$$

the magnetic flux density $\mathcal{B}$ is 2-form B =$\mathrm{B}_{23} dx^2 \wedge dx^3 + \mathcal{B}_{31} dx^3 \wedge dx^1 + \mathcal{B}_{12} dx^1 \wedge dx^2$ $= \mathcal{B}^1 ds_1 + \mathcal{B}^2 ds_2 + \mathcal{B}^3 ds_3 = \mathcal{B}^A ds_A$, and the charge density ρ_c is 3-form:

$$\rho_c = \tilde{\rho}_c dx^1 \wedge dx^2 \wedge dx^3 = \tilde{\rho}_c dV, \tag{9}$$

where ds_A is the oriented surface element and dV is the oriented volume element.

By the combination of $\mathcal{E}$ and $\mathcal{B}$, we can obtain the electromagnetic field 2-form $\mathcal{F}$:

$$\mathcal{F} = \mathcal{B} + \mathcal{E} \wedge dt. \tag{10}$$

The electromagnetic potential 1-form $\mathcal{A}$ is given by the combination

$$\mathcal{A} = \tilde{\mathcal{A}} - \phi dt, \tag{11}$$

where $\tilde{\mathcal{A}}$ is the vector potential 1-form. The potential $\mathcal{A}$ is related to the field $\mathcal{F}$ though the equation

$$\mathcal{F} = d\mathcal{A}, \tag{12}$$

where d is the exterior differential operator, which will be taken up below.

2.2 Exterior differential operator

The exterior differential operator d is defined by

$$d\omega = \frac{1}{p!} \partial_j \omega_{i_1 \cdots i_p} dx^j \wedge dx^{i_1} \wedge \cdots \wedge dx^{i_p}, \tag{13}$$

where $\partial_j(\cdots) = \partial(\cdots)/\partial x^j$. The comparison between (1) and (13) shows that d: $\Lambda^p \rightarrow \Lambda^{p+1}$. In this paper, we take up the (3+1)-dimensional exterior derivative operator:

$$d\omega_p = d_s \omega_p + (-1)^p \partial_t \omega_p \wedge dt, \tag{14}$$

where $\omega_p \in \Lambda^p$, s refers to pure space-differentiation and t to time-differentiation. For instance, (11) gives $d\mathcal{A} = d_s \tilde{\mathcal{A}} - (\partial_t \tilde{\mathcal{A}} + d_s \phi) \wedge dt$. Thus, from (10) and (12), we obtain B $= \mathrm{d}_s \tilde{\mathcal{A}}$, $\mathcal{E} = -\partial_t \tilde{\mathcal{A}} - d_s \phi$. To see what these results correspond to in vector analysis, let us calculate the exterior space-derivative of (7), (8) and (2.1). By using (5) and (6), we obtain

$$d_s \phi = \partial_1 \phi dx^1 + \partial_2 \phi dx^2 + \partial_3 \phi dx^3, \tag{15}$$

$$d_s \mathcal{E} = (\partial_2 \mathcal{E}_3 - \partial_3 \mathcal{E}_2) ds_1 + (\partial_3 \mathcal{E}_1 - \partial_1 \mathcal{E}_3) ds_2 + (\partial_1 \mathcal{E}_2 - \partial_2 \mathcal{E}_1) ds_3, \tag{16}$$

$$d_s \mathcal{B} = (\partial_1 \mathcal{B}^1 + \partial_2 \mathcal{B}^2 + \partial_3 \mathcal{B}^3) dV. \tag{17}$$

These correspond to grad ϕ, div $\mathcal{B}$ and curl $\mathcal{E}$, respectively. These results are generalized as follows: $\mathrm{d}_s(\omega_0) \leftrightarrow$ grad$(\cdots)$, $d_s(\omega_1) \leftrightarrow$ curl$(\cdots)$, $d_s(\omega_2) \leftrightarrow$ div$(\cdots)$, where $\omega_p \in \Lambda^p$. Therefore, (2.2) means the usual form: $\mathcal{B} = \mathrm{curl}\,\tilde{\mathcal{A}}$ and $\mathcal{E} = -\partial_t\tilde{\mathcal{A}} - \mathrm{grad}\,\phi$.

From (13), the double application of d is given by

$$dd\omega = \frac{1}{p!}\partial_k\partial_j\omega_{i_1\cdots i_p}dx^k \wedge dx^j \wedge dx^{i_1} \wedge \cdots \wedge dx^{i_p}. \tag{18}$$

Since $\partial_k\partial_j(\cdots)dx^k \wedge dx^j = 0$, (18) means the important rule

$$dd = 0. \tag{19}$$

From (2.2), this corresponds to

$$d_s d_s(\omega_0) = 0 \leftrightarrow \text{curl grad}(\cdots) = 0, \tag{20}$$

$$d_s d_s(\omega_1) = 0 \leftrightarrow \text{div curl}(\cdots) = 0. \tag{21}$$

For instance, we take up (12): $\mathcal{F} = d\mathcal{A}$. From (19), we obtain

$$d\mathcal{F} = dd\mathcal{A} = 0. \tag{22}$$

From (10) and (14), the concrete form of (22) is given by dF $=\mathrm{d}_s\mathcal{B} + (\partial_t\mathcal{B} + d_s\mathcal{E}) \wedge dt = 0$. Thus, we have

$$d_s\mathcal{B} = 0 \text{ and } \partial_t\mathcal{B} + d_s\mathcal{E} = 0. \tag{23}$$

From (16) and (17), this corresponds to div $\mathcal{B} = 0$ and $\partial_t\mathcal{B}$+curl $\mathcal{E} = 0$. This means that (22) is the homogeneous Maxwell's equations in terms of the differential forms.

In a similar fashion, we obtain the inhomogeneous Maxwell's equations:

$$d\mathcal{G} = \mathcal{J}, \tag{24}$$

with

$$\mathcal{G} = \mathcal{D} - \mathcal{H} \wedge dt, \tag{25}$$

$$\mathcal{J} = \rho_c - j \wedge dt, \tag{26}$$

where $\mathcal{D}$ is the electric flux density 2-form, $\mathcal{H}$ is the magnetic field 1-form, ρ_c is the charge density 3-form and j is the local current density 2-form. The time-like and space-like components of (24) is given by

$$d_s\mathcal{D} = \rho_c \text{ and } d_s\mathcal{H} = j + \partial_t\mathcal{D}. \tag{27}$$

The field $\mathcal{F}$ is related to the field $\mathcal{G}$ through the constitutive equation:

$$\mathcal{G} = \sqrt{\frac{\epsilon}{\mu}} * \mathcal{F}, \tag{28}$$

where $*$ is the Hodge star operator, which will be taken up below. The constants ϵ and μ are permittivity and permeability, respectively. Since $c_L = 1/\sqrt{\epsilon\mu}$, the relation (28) can be rewritten as

$$\mathcal{G} = \epsilon c_L * \mathcal{F} \text{ or } * \mathcal{F} = \mu c_L\,\mathcal{G}. \tag{29}$$

2.3 Hodge star operator

Le us consider the Hodge duality that plays an important role in the continuum mechanics in terms of the differential forms. In the n-dimensional space-time, the Hodge star operator is defined by

$$* = \frac{(-1)^s}{p!} \epsilon_{i_1 \cdots j_{p+1} \cdots j_n} g^{i_{p+1} j_{p+1}} g^{i_{p+2} j_{p+2}} \cdots g^{i_n j_n}, \tag{30}$$

where s is the number of negative signs of the metric and $\epsilon_{i_1 \cdots j_{p+1} \cdots j_n}$ is the Levi-Civita tensor. As an example, let us calculate the dual of $dx^1 \wedge dx^2$ in the (3+1)-dimensional space-time. Since $g^{11} = g^{22} = g^{33} = 1$, $g^{00} = -1$ and $s = 1$, we obtain $*(dx^1 \wedge dx^2) = (-1)\epsilon_{1230} g^{33} g^{00} dx^3 \wedge dx^0$
$= (-1)(-1)(+1)(-1) dx^3 \wedge dx^0$
$= -dx^3 \wedge dx^0$, or

$$*ds_3 = -dx^3 \wedge dx^0. \tag{31}$$

In a similar fashion, we can obtain other components as follows.
(I) (3+1)-dimensional space-time: $g^{00} = -1$, $g^{11} = g^{22} = g^{33} = 1$ and $s = 1$. $*(ds_A) = -dx^A \wedge dx^0$, $*(dx^A \wedge dx^0) = ds_A$, $*(dx^A) = -ds_A \wedge dx^0$, $*(ds_A \wedge dx^0) = -dx^A$, $*(dx^0) = -dV$, $*(dV) = -dx^0$, $*(dV \wedge dx^0) = 1$, $*(1) = -dV \wedge dx^0$. (II) 3-dimensional space: $g^{11} = g^{22} = g^{33} = 1$ and $s = 0$. $*(ds_A) = dx^A$, $*(dx^A) = ds_A$, $*(dV) = 1$, $*(1) = dV$. The results above show that $*$: $\Lambda^p \to \Lambda^{n-p}$.

From $** (ds_A) = -* (dx^A \wedge dx^0) = -ds_A$ and $** (dx^A) = -* (ds_A \wedge dx^0) = dx^A$, we find that

$$** = -1 \text{ for even-form}, \tag{32}$$

$$** = 1 \text{ for odd-form}, \tag{33}$$

in the (3+1)-dimensional space-time. In the 3-dimensional space, we find that

$$** = 1 \text{ for any-form}. \tag{34}$$

These results are generalized as follows:

$$** = (-1)^{p(n-p)+s}. \tag{35}$$

As an example, let us calculate the concrete form of (28): $\sqrt{\epsilon/\mu} * \mathcal{F} = \mathcal{G}$. From (8), (2.1) and (10), the Hodge dual of $\mathcal{F}$ is given by $\sqrt{\frac{\epsilon}{\mu}} * \mathcal{F} = \sqrt{\frac{\epsilon}{\mu}} * (\mathcal{B}^A ds_A + \frac{\mathcal{E}_A}{c_L} dx^A \wedge dx^0)$
$= \sqrt{\frac{\epsilon}{\mu}} (-\mathcal{B}_A dx^A \wedge dx^0 + \frac{\mathcal{E}^A}{c_L} ds_A)$
$= \sqrt{\frac{\epsilon}{\mu}} (-\mathcal{B} c_L \wedge dt + \frac{\mathcal{E}}{c_L})$
$= \sqrt{\frac{\epsilon}{\mu}} (-\mathcal{B} \frac{1}{\sqrt{\epsilon\mu}} \wedge dt + \sqrt{\epsilon\mu}\, \mathcal{E})$
$= -\frac{1}{\mu} \mathcal{B} \wedge dt + \epsilon\, \mathcal{E} dt$
$= \mathcal{G}$
$= \mathcal{D} - \mathcal{H} \wedge dt$. Thus, we obtain the usual form of the constitutive equations:

$$\mathcal{D} = \epsilon\, \mathcal{E} \text{ and } \mathcal{B} = \mu\, \mathcal{H}. \tag{36}$$

This means that (28) is the (3+1)-dimensional expression of the previous constitutive equations.

2.4 Adjoint operator

Next, we take up the adjoint operator δ. In n-dimensional space-time, this operator is defined by

$$\delta = (-1)^{p(n-p+1)+s} * d*, \tag{37}$$

where p is the order of the differential form on which δ operates and s is the number of negative sigen of the metric. In the (3+1)-dimensional space-time ($n = 4$ and $s = 1$),

$$*d* = -\delta. \tag{38}$$

In the 3-dimensional space ($n = 3$ and $s = 0$),

$$*d* = (-1)^p \delta. \tag{39}$$

From (19): $dd = 0$, (37) shows that

$$\delta\delta = 0. \tag{40}$$

This operator is related to the Laplace operator $\triangle$ as follows:

$$-\triangle = \delta d + d\delta. \tag{41}$$

From $dd = 0$ and $\delta\delta = 0$, (41) shows that $d\triangle = \triangle d, \delta\triangle = \triangle\delta, *\triangle = \triangle *$. As an example, let us calculate the electromagnetic field in 3-dimensional space. In this case, by applying $*$ to the second equation of (27) and using the 3-dimensional version of (28): $\mathcal{D} = \epsilon * \mathcal{E}$ and $\mathcal{B} = \mu * \mathcal{H}$ and (2.2), we obtain

$$\tilde{\mathcal{A}} + d_s(\delta\tilde{\mathcal{A}} - \mu\epsilon\partial_t\phi) = -\mu * j, \tag{42}$$

with

$$= \triangle - \mu\epsilon\partial_t\partial_t = \triangle - (1/c^2)\partial_t\partial_t. \tag{43}$$

The operator is known as the wave operator (or the d'Alembertian operator).

Now, the field $\mathcal{F}(= d\mathcal{A})$ is invariant under the translation $\mathcal{A} \to \mathcal{A} + d\delta\mathcal{A}$, because $dd\delta\mathcal{A} = 0$. This translation is called the gauge translation (e.g., [4]). In the electromagnetism, there are two well-known gauge conditions: the Coulomb gauge $\delta\tilde{\mathcal{A}} = 0$ and the Lorentz gauge $\delta\tilde{\mathcal{A}} = \mu\epsilon\partial_t\phi$. The wave equation (42) becomes $\tilde{\mathcal{A}} - \mu\epsilon\partial_t d_s\phi = -\mu * j$ in the Coulomb gage, and becomes $\tilde{\mathcal{A}} = -\mu * j$ in the Lorentz gauge.

2.5 Homotopy operator

In order to express the effect of the defects field, we introduce the linear homotopy operator H based on Ref.[6]. The operator H is well defined on the starshaped region as a Riemann-Gravis integral and have the following four properties:

H1. H maps Λ^k into Λ^{k-1} for $k \geq 1$ and maps Λ^0 into zero.
H2. $dH + Hd =$ identity for $k \geq 1$ and $(Hdf)(x^i) = f(x^i) - f(x_0^i)$ for $k = 0$.
H3. $(HH\omega)(x^i) = 0$ and $(H\omega)(x_0^i) = 0$.
H4. $HdH = H$ and $dHd = d$.

From **H2**, any differential form $\omega \in \Lambda^k$ satisfies $\omega = dH\omega + Hd\omega = \omega_e + \omega_a$, where $\omega_e = dH\omega$ is the exact part of ω and $\omega_a = Hd\omega$ is the anti-exact (or non-integral) part of ω. From (19), we obtain $d\omega_e = ddH\omega = 0$, i.e., the exact part is the kernel of d. From **H3**, we obtain $H\omega_a = HHd\omega = 0$, i.e., the anti-exact part is the kernel of H. In the case of $d\omega = 0$, we obtain $\omega_a = 0$, i.e., $\omega = \omega_e$. This means that every closed form $(d\omega = 0)$ is exact $(\omega = \omega_e)$ in the starshaped region, which is called Poincareé lemma.

One of the useful applications of homotopy operator H is the invert of exterior differentiation operator d. For instance, we can rewrite the equation (12): $\mathcal{F} = d\mathcal{A}$ as

$$H\mathcal{F} = Hd\mathcal{A} = \mathcal{A}_a = \mathcal{A} - \mathcal{A}_e. \tag{44}$$

If and only if $\mathcal{A}$ is an antiexact (i.e., $\mathcal{A}_e = 0$), (44) shows $\mathcal{A} = H\mathcal{F}$. This means that the solution of the equation $\mathcal{F} = d\mathcal{A}$ can be obtained as $\mathcal{A} = H\mathcal{F}$ by using the homotopy operator.

2.6 Other useful formulas

The exterior derivative of the wedge product of $\alpha_p \in \Lambda^p$ and $\beta_q \in \Lambda^q$ is given by

$$d(\alpha_p \wedge \beta_q) = d\alpha_p \wedge \beta_q + (-1)^p \alpha_p \wedge d\beta_q. \tag{45}$$

From (2.2), it is found that this generalizes the formulations in vector analysis. For instance, $d(\alpha_0 \wedge \beta_2) = d\alpha_0 \wedge \beta_2 + \alpha_0 \wedge d\beta_2$ corresponds to $\mathrm{div}(\alpha_0\beta_2) = \mathrm{grad}(\alpha_0)\beta_2 + \alpha_0 \mathrm{div}\beta_2$.

The generalized Stokes's theorem in differential forms is given by

$$\int_M d\alpha = \int_{\partial M} \alpha, \tag{46}$$

where ∂M is the boundary of the manifold M and $\alpha \in \Lambda^{\partial M}$. From (2.2), in the case of α being 1-form or 2-form, (46) corresponds to the Stokes theorem: $\int_S \mathrm{curl}\alpha = \int_C \alpha$ and the Gauss's theorem: $\int_V \mathrm{div}\alpha = \int_S \alpha$, respectively.

3 Dual Material Space-Time

In this section, we derive the basic equations in continuum mechanics: constitutive, kinematic and continuity equations in the systematic way based on the differential forms. We consider the (3+1)-dimensional material space-time whose coordinate is given by $(x^1, x^2, x^3, x^0 = ct)$. In this case, c is not the velocity of light but that of elastic wave: $c = \sqrt{e/\rho}$, where e is the elastic modulus and ρ is the density of mass.

3.1 Constitutive equations: the usefulness of Hodge star operator

We present the dual material space-time with defect filed: one is the strain space-time, and the other is stress space-time[7, 8]. This dual structure is similar to that in electromagnetism: one is the electric field and the other is the magnetic field. The idea of the dual material space-time was evoked by the RAAG's work and the concept of the Tonti's diagram (e.g., [9, 10, 11, 12]).

The quantities in the strain space-time are summarized as follows [13, 14]:

$$\text{distortion-velocity 1-form: } B^i = \beta^i + v^i dt, \tag{47}$$

$$\text{bend-twist-spin 2-form: } K^i = \kappa^i - \omega^i \wedge dt, \tag{48}$$

$$\text{dislocation 2-form: } A^i = \alpha^i + I^i \wedge dt, \tag{49}$$

$$\text{disclination 3-form: } \Theta^i = \theta^i - J^i \wedge dt, \tag{50}$$

where distortion: $\beta^i = \beta^i_A dx^A$, velocity: v^i, bend-twist: $\kappa^i = \kappa^{iA} ds_A$, spin: $\omega^i = \omega^i_A dx^A$, dislocation density: $\alpha^i = \alpha^{iA} ds_A$, dislocation current: $I^i = I^i_A dx^A$, disclination density: $\theta^i = \vartheta^i dV$, disclination current: $J^i = J^{iA} ds_A$. It is interesting that the quantities in strain space-time are visible, i.e., we can see them with our own eyes.

On the other hand, the stress space-time consists of the invisible quantities such as stress and momentum. The quantities in the stress space-time are summarized as follows [7, 8]:

$$\text{stress function-potential 1-form: } F_i = -\phi_i dt + \gamma_i, \tag{51}$$

$$\text{couple-stress function-potential 2-form: } C_i = c_i \wedge dt - \psi_i, \tag{52}$$

$$\text{couple-stress and angular momentum 2-form: } M_i = m_i \wedge dt + a_i, \tag{53}$$

$$\text{stress and momentum 3-form: } S_i = -\sigma_i \wedge dt - P_i, \tag{54}$$

where stress function: ϕ_i, stress potential: $\gamma_i = \gamma_{iA} dx^A$, couple-stress function: $c_i = c_{iA} dx^A$, couple-stress potential: $\psi_i = \psi_i^A ds_A$, couple-stress: $m_i = m_{iA} dx^A$, angular momentum: $a_i = a_i^A ds_A$, stress: $\sigma_i = \sigma_i^A ds_A$, momentum: $P_i = p_i dV$.

Of course, we feel seeing the stress due to the strain through the Hooke's law. However, in this case, we directly observe the strain but not the stress itself. This implies that we can observe the stress space-time when the quantities in the stress space-time are mapped on to the strain space-time. In this case, the mapping is interpreted physically as the constitutive equation. As mentioned in the section 2.3, the constitutive equation is the physical expression of the Hodge duality. Then, let us link the quantities in the stress space-time with those in the strain space-time through the Hodge duality.

Recall that p-form in n-dimensional space is mapped on to $(n-p)$-form by the Hodge star operator $*$. Since S_i is a 3-form in the stress space-time (see (54)), the dual form is a 1-form in the strain space-time, i.e., a distortion 1-form (47). Then, we express this duality by the following linear relation:

$$\frac{1}{\sqrt{e\rho}} * S_i = B^i. \tag{55}$$

Since $c = \sqrt{e/\rho}$, (55) can be rewritten as

$$S_i = \rho c * B^i \text{ or } * B^i = e^{-1} c S_i. \tag{56}$$

From (2.3) and (2.3), the explicit form of (55) is given by 1$\frac{}{\sqrt{e\rho} * S_i = \frac{1}{\sqrt{e\rho}} * (-\sigma_i^A ds_A \wedge \frac{1}{c} dx^0 - p_i dV) = \frac{1}{\sqrt{e\rho}} (\frac{1}{c} \sigma^i_A dx^A + c\, p^i dt) = \frac{1}{e} \sigma^i_A dx_A + \frac{1}{\rho} p^i dt = B^i = \beta^i_A dx^A + v^i dt.}$
Therefore, the space-like and time-like components of (55) are given by

$$\sigma^i_A = e \beta^i_A, \tag{57}$$

$$p^i = \rho v^i. \tag{58}$$

(57) is the Hooke's law, and (58) is the definition of momentum. This result yields new insight to continuum mechanics: the two equations have been recognized as entirely unrelated, but in fact they are the time-like and the space-like components of the (3+1)-dimensional constitutive equation (55).

In a similar fashion, we can derive the following (3+1)-dimensional constitutive equations (and their space-like and time-like components):

$$\frac{1}{\sqrt{e\rho}} * F_i = \Theta^i \; (\phi^i = e\vartheta^i \text{ and } \gamma^{iA} = \rho J^{iA}), \tag{59}$$

$$\frac{1}{\sqrt{e\rho}} * M_i = K^i \; (m^{iA} = e\kappa^{iA} \text{ and } a^i_A = \rho\omega^i_A), \tag{60}$$

$$\frac{1}{\sqrt{e\rho}} * C_i = A^i \; (c^{iA} = e\alpha^{iA} \text{ and } \psi^i_A = \rho I^i_A). \tag{61}$$

The space-like and the time-like components of (59) correspond to the constitutive equations in the micropolar theory (or the Cosserat continuum) (e.g., [15, 16]). The space-like components of (60) and (61) have been already pointed out by Ref. [10]. On the other hand, the time-like components of (60) and (61) are often ignored in the previous paper, although Ref. [17] has pointed out the existence of these constitutive equations. In summary, the Hodge star operator $*$, which links the stress space-time and the strain space-time, enables us to derive the various constitutive equations of previous papers in a systematic way.

3.2 Stress space-time: the usefulness of exterior differential operator

In this section, we reformulate the previous stress space in a (3+1)-dimensional space-time based on the differential forms. The exterior differential operator d, which links p-form to $(p+1)$-form in the stress space-time, shall enable us to derive two kinematic equations and two continuity equations in stress space-time.

Since $d : \Lambda^p \mapsto \Lambda^{p+1}$, the exterior differential of the couple-stress function-potential 2-form (52): dC_i is related to the 3-form in the stress space-time, i.e., the stress 3-form (54):

$$S_i = dC_i. \tag{62}$$

This is a second kinematic equation in the stress space-time. From (14): $d(\cdots) = d_s(\cdots) + (-1)^p \partial_t(\cdots) \wedge dt$, the explicit form of (62) is

$$-\sigma_i \wedge dt - P_i = (-\partial_t \psi_i + d_s c_i) \wedge dt - d_s \psi_i. \tag{63}$$

Then, we have

$$-\sigma_i = -\partial_t \psi_i + d_s c_i, \tag{64}$$

$$P_i = d_s \psi_i. \tag{65}$$

From (19): $dd = 0$, (62) yields a second continuity equation in strain space-time:

$$dS_i = 0. \tag{66}$$

The explicit form of (66) is

$$(\partial_t P_i - d_s \sigma_i) \wedge dt - d_s P_i = 0, \tag{67}$$

that is,

$$\partial_t P_i = d_s \sigma_i, \tag{68}$$

$$d_s P_i = 0. \tag{69}$$

(68) is the conservation law of momentum.

The first kinematic equation in the stress space-time is given by

$$M_i = dF_i + C_i. \tag{70}$$

This yields the first continuity equation in the stress space-time:

$$S_i = dM_i, \tag{71}$$

where we use (62). The explicit forms of (70) and (71) are $\mathrm{m}_i = -\partial_t \gamma_i - d_s \phi_i + c_i$, $a_i = d_s \gamma_i - \psi_i$, $-\sigma_i = \partial_t a_i + d_s m_i$, $-P_i = d_s a_i$. The first equation of (3.2) is the conservation law of angular momentum.

As we will see in the section 8, the physical equations derived in this section can be interpreted geometrically as the basic equations in the differential geometry such as the structure equation and the Bianchi identities. That is, the first (70) and the second (62) kinematic equations correspond to the first and the second structure equations. The first (71) and the second (66) continuity equations correspond to the first and the second Bianchi identities.

3.3 Strain space-time: the usefulness of homotopy operator

In this section, we review the strain space-time based on Ref. [14]. The homotopy operator H shall enable us to introduce the defects field into the continuum mechanics.

For instance, we take up the bend-twist κ^i defined by the first definition of (3.1). Since κ^i is 2-form, $d_s \kappa^i$ is related to 3-form in strain space-time, i.e., disclination density 3-form:

$$\theta^i = d_s \kappa^i. \tag{72}$$

From the property **H2** in section 2, we can divide the bend-twist as $\kappa^i = \kappa^i_e + \kappa^i_a$, where $\kappa^i_e = dH\kappa^i$ is the exact part and $\kappa^i_a = Hd\kappa^i$ is the anti-exact part. Since the exact part is the kernel of d: $d_s \kappa^i_e = 0$, the equation (72) becomes

$$\theta^i = d_s \kappa^i_a. \tag{73}$$

(73) shows the important fact: the anti-exact part of the deformation field is accompanied by the defects field. In fact, the anti-exact part of the distortion 1-form is related to the dislocation density 2-form:

$$\alpha^i = d_s \beta^i + \kappa^i = d_s \beta^i_a + \kappa^i. \tag{74}$$

The results above can be simplified and systematized by a reformulation in a (3+1)-dimensional space-time as follows. The second kinematic equation is given by

$$\Theta^i = dK^i. \tag{75}$$

This yields the second continuity equation:

$$d\Theta^i = 0. \tag{76}$$

The space-like and the time-like components of (75) and (76) are given by $\theta^i = d_s\kappa^i$, $J^i = d_s\omega^i - \partial_t\kappa^i$, $\mathrm{d}_s\theta^i = 0, \partial_t\theta^i + d_sJ^i = 0$.

The first kinematic equation is given by

$$A^i = dB^i + K^i. \tag{77}$$

This yields the first continuity equation:

$$dA^i = \Theta^i, \tag{78}$$

where we use (75). Explicit forms of (77) and (78) are $\alpha^i = d_s\beta^i + \kappa^i$, $I^i = d_sv^i - \partial_t\beta^i - \omega^i$, $\mathrm{d}_s\alpha^i = \theta^i, -\partial_t\alpha^i - d_sI^i = J^i$.

As already pointed out by Ref. [14], the kinematic and the continuity equations correspond to the structure equation and the Bianchi identities (see also [18, 19, 20]).

4 Navier Equation and Its Antiexact Solution

4.1 The Navier equations

In this section, we derive the Navier equation in terms of the differential form. Needless to say, in the continuum mechanics, the Navier equation is the most basic equation to describe the deformation field.

The tensor expression of the Navier equation that describes the displacement field due to the body force f_i^B is given by

$$e_{ijkl}\partial_j\partial_l u_k - \rho\partial_t\partial_t u_i = f_i^B, \tag{79}$$

where e_{ijkl} is the tensor expression of the elastic modulus. This equation can be derived by combining two equations, i.e., the equation of motion:

$$\partial_j\sigma_{ij} = \partial_t p_i, \tag{80}$$

and the constitutive equations:

$$\sigma_{ij} = e_{ijkl}\beta_{kl}, \tag{81}$$

$$p_i = \rho v_i. \tag{82}$$

Moreover, to derive the usual form (79), we use geometrical relations:

$$\beta_{kl} = \partial_l u_k, \tag{83}$$

$$v_i = \partial_t u_i. \tag{84}$$

In the differential forms, the equation of motion is extended like (66): $dS_i = 0$. The constitutive equations are extended like (55): $(1/\sqrt{e\rho}) * S_i = B^i$. Moreover, the geometric relation is extended like $\mathrm{B}^i = dHB^i + HdB^i = du^i + B^i_a$
$(= d_s u^i + \partial_t u^i dt + B^i_a)$
$(= d_s u^i + v^i dt + B^i_a)$,
where $u^i = HB^i$ is a displacement 0-form and $B^i_a = HdB^i$ is the anti-exact part of B^i. As a result, the Navier in terms of the differential forms is given by combining (66), (55) and (4.1):

$$\sqrt{e\rho} d * du^i = -\sqrt{e\rho} d * B^i_a. \tag{85}$$

(85) is the Navier equation for a displacement field due to the anti-exact part of the distortion. i.e., dislocations. By applying the operator $*$ and using (37) and (41), the equation (85) can be rewritten as the wave equation:

$$\sqrt{e\rho}(\Delta u^i + d\delta u^i) = \sqrt{e\rho}\delta B^i_a. \tag{86}$$

The relation between the Navier equation and the wave equation in terms of the differential forms will be take up in more detail in the section 5.2.

In a similar fashion, we can obtain the Navier equation for a rotational displacement 1-form $r^i = HK^i$ due to the anti-exact part of the bend-twist, i.e., disclinations (and stress):

$$\sqrt{e\rho} d * dr^i = -\sqrt{e\rho} d * K^i_a - S_i, \tag{87}$$

or

$$\sqrt{e\rho}(\Delta r^i + d\delta r^i + du^i) = \sqrt{e\rho}(\delta K^i_a - B^i_a). \tag{88}$$

To derive (87), we use the equation of motion (71): $dM_i = S_i$, the constitutive equation (60): $M_i = \sqrt{e\rho} * K^i$ and the geometrical equation: $\mathrm{K}^i = dHK^i + HdK^i = dr^i + K^i_a$
$(= d_s r^i - \partial_t r^i dt + K^i_a)$
$(= d_s r^i - \omega^i dt + K^i_a)$.
By applying $*$ to (88) and using $\delta\delta = 0$ and $\delta\Delta = \Delta\delta$, we can obtain (86).

The results given above can be generalized as in the following analysis. The basic operators in differential form are exterior differential operator and Hodge star operator. As we have seen in sections 3.1, 3.2 and 3.3, the operator d derives continuity equations, and the operator $*$ derives the constitutive equations. Thus, we operate with d and $*$ on an arbitrary p-form X in n-dimensional space to derive generalized continuity and constitutive equations:

$$dX = Y, \tag{89}$$

$$*X = Z, \tag{90}$$

where Y is a $(p+1)$-form and Z is an $(n-p)$-form. Moreover, we use homotopy operators to derive generalized geometric relations:

$$Z = dHZ + HdZ = dz + Z_a, \tag{91}$$

where $z = HZ$ is a generalized displacement $(n-p-1)$-form and $HdZ = Z_a$ is the anti-exact part of Z. By combining (89), (90) and (91), we obtain

$$Nz = -d * Z_a + (-1)^{p(n-p)+s} Y, \tag{92}$$

or

$$-(\triangle z + d\delta z) = -\delta Z_a + (-1)^{p-n} * Y, \tag{93}$$

where we define the new operator $N = d{*}d : \Lambda^p \mapsto \Lambda^{n-p}$. (92) is the general form of the Navier equation. That is, when we recognize z as the ordinary and rotational displacements, (92) corresponds to (85) and (87), respectively. Thus, in this paper, we call N the Navier operator. In the section 5, we shall derive the Laplace equation for defects field by using the Navier operator N.

4.2 Anti-exact solution of the Navier equation

It is important to solve the Navier equation for displacement field. As mentioned in the section 2.5, the homotopy operator H is a useful tool to derive the equation in terms of the differential form. Then, we attempt to solve the Navier equation by using the homotopy operator. For simplicity we set $\sqrt{e\rho} = 1$.

First, we solve (85) for u^i. By operating H, we have -Hd$*B_a^i = Hd * du^i$
$= Hd(*du^i)$
$= (*du^i)_a$
$= *du^i - (*du^i)_e$. Moreover, by operating $H*$, we have -H$*Hd*B_a^i = Hdu^i - H*(*du^i)_e$
$= u_a^i - H * (*du^i)_e$
$= u^i - u_e^i - H * (*du^i)_e$.
Therefore, in the case of $u_e^i = 0$ and $(*du^i)_e = 0$, the Navier equation (85) has the solution as

$$u^i = -H * Hd * B_a^i. \tag{94}$$

In a similar fashion, we obtain the solution of (87) for r^i:

$$r^i = H * H(d * K_a^i + S_i). \tag{95}$$

under the conditions that $r_e^i = 0$ and $(*dr^i)_e = 0$. Moreover, the solution of (92) for the $(n-p-1)$-form z is given by

$$z = H * H\{(-1)^{(p-n)p-s+1} d * Z_a + Y\}, \tag{96}$$

under the conditions that $z_e = 0$ and $(*dz)_e = 0$. This is a general form of the solution of the Navier equation in the case of no exact parts.

In summary, the Navier equation and its anti-exact solution are given by the form of $d * d(\cdots)$ and $H * H(\cdots)$, respectively.

5 Wave and Laplace Equations for Defects Field

5.1 Wave equations: the usefulness of adjoint operator

In this section, we derive the wave equation in terms of the differential forms. The wave equation is one of the most basic equation to describe the dynamics of the deformation field. As we will see, the wave equation in the strain (or stress) space is derived by the basic equation in the it's Hodge dual space, i.e., the stress (or strain) space.

We consider the wave equation in the 3-dimensional space. In this case, the constitutive equations defined in the section 3.1 are re-defined as $\mathrm{e}^{-1} * \sigma_i = \beta^i, \rho^{-1} * P_i = v^i$, $\mathrm{e}^{-1} * \phi_i = \theta^i, \rho^{-1} * \gamma_i = J^i$, $\mathrm{e}^{-1} * m_i = \kappa^i, \rho^{-1} * a_i = \omega^i$, $\mathrm{e}^{-1} * c_i = \alpha^i, \rho^{-1} * \psi_i = I^i$. Moreover, the divisions of the distortion and the bend-twist: (4.1) and (4.1) are re-defined as

$$\beta^i = d_s H \beta^i + H d_s \beta^i = d_s u^i + \beta^i_a, \tag{97}$$

$$\kappa^i = d_s H \kappa^i + H d_s \kappa^i = d_s r^i + \kappa^i_a. \tag{98}$$

First, we derive a wave equation for the displacement field. By applying $*$ to the time-like component of the second continuity equation in the stress space-time (68), we obtain $\partial_t(*P_i) = *d_s\sigma_i$. From the constitutive equation (5.1), this becomes

$$\rho \partial_t \partial_t u^i = e * d_s * \beta^i = -e\delta\beta^i, \tag{99}$$

where we use the definition of the adjoint operator (37) in the last step. When β^i is exact, the Laplace operator (41) gives $\delta\beta^i = \delta d_s u^i = -\Delta u^i - d_s\delta u^i$. Therefore, (99) becomes

$$u^i + d_s\delta u^i = 0, \tag{100}$$

where $= \Delta - (1/c^2)\partial_t\partial_t$ with $c = \sqrt{e/\rho}$. This is a wave equation for the displacement field. If we choice the gauge: $\delta u^i = 0$, this becomes $u^i = 0$.

In a similar fashion, when the bend-twist is exact: $\kappa^i = d_s r^i$, the time-like component of the first continuity equation in the stress space-time (3.2) gives a wave equation for the rotational displacement field:

$$r^i + d_s\delta r^i - \beta^i = 0. \tag{101}$$

When the distortion is exact, $d_s\delta r^i - \beta^i = d_s(\delta r^i - u^i)$. In this case, if we choice the gauge: $\delta r^i = u^i$, (101) becomes $r^i = 0$.

The time-like component of the second kinematic equation (64) gives a wave equation for the distortion field:

$$\beta^i - \beta^i + \frac{1}{c^2}\partial_t(d_s v^i - \omega^i) + d_s\delta\beta^i - \delta\kappa^i = 0. \tag{102}$$

When the distortion and the bend-twist are exact, (102) can be rewritten as

$$\beta^i - d_s u^i + r^i + d_s\delta r^i - \beta^i = 0. \tag{103}$$

Moreover, (100) and (101) valid, so the equation (103) becomes $\beta^i = 0$.

Finally, the time-like component of the first kinematic equation (3.2) gives a wave equation for the bend-twist field:

$$\kappa^i + \frac{1}{c^2}\partial_t d_s\omega^i + d_s\delta\kappa^i - d_s\beta^i = 0. \tag{104}$$

When the distortion and the bend-twist are exact, this becomes $\kappa^i = 0$.

In a similar fashion, we can derive wave equations in the stress space-time by using the kinematic and the continuity equations in the strain space-time. For instance, the time-like component of the first kinematic and continuity equations in the strain space-time give the wave equation for the couple-stress potential and the stress-potential, respectively:

$$\psi_i + d_s\delta\psi_i - \psi_i + \frac{1}{c^2}\partial_t d_s c_i - a_i = 0, \tag{105}$$

$$\gamma_i + d_s\delta\gamma_i - \gamma_i - \frac{1}{c^2}\partial_t(m_i + d_s\phi_i) + \delta a_i = 0. \tag{106}$$

5.2 Laplace equation (wave equation for defects field)

In this section, we consider the Laplace equation in terms of the differential forms. From (41), the Laplace operator $\triangle$ is given by $-\triangle = d_s\delta + \delta d_s = \{d_s, \delta\}$, where $\delta = (-1)^{p(n-p+1)+s} * d_s *$. On the other hand, in the 3-dimensional space ($n = 3$ and $s = 0$), the Navier operator, defined in the section 4.1, is given by $N = d_s * d_s$. In this case, we obtain $d_s\delta = (-1)^{p(4-p)}N*$ and $\delta d_s = (-1)^{p(4-p)} * N$. This means that the Laplace operator is related to the Navier operator through the Hodge star operator:

$$-\triangle = (-1)^{p(4-p)}\{N, *\}. \tag{107}$$

For instance, we take up the disclination density 3-form and apply $-\triangle$ to it:

$$-\triangle\theta^i = -\{N, *\}\theta^i = -N * \theta^i - *N\theta^i. \tag{108}$$

Since $\mathrm{N}*\theta^i = Ne^{-1}\phi_i$
$= e^{-1}d_s * d_s\phi_i$
$= e^{-1}d_s * (-m_i - \partial_t\gamma_i + c_i)$
$= e^{-1}d_s(-e\kappa^i - \partial_t\rho J^i + e\alpha^i)$
$= e^{-1}(-e\theta^i - \rho\partial_t(-\partial_t\theta^i) + e\theta^i)$
$= e^{-1}\rho\partial_t\partial_t\theta^i$
$= (1/c^2)\partial_t\partial_t\theta^i$,
and

$$*N\theta^i = *d_s * d_s\theta^i = 0, \tag{109}$$

it is found that the Laplace equation (108) yields the wave equation for the disclination density:

$$\theta^i = 0. \tag{110}$$

It is interesting that the disclination density take this form without any additional conditions (gauge conditions).

In a similar fashion, from the Laplace equation for the dislocation density 2-form:

$$-\triangle\alpha^i = \{N, *\}\alpha^i, \tag{111}$$

we obtain the wave equation:

$$\alpha^i = \frac{2}{c^2}\partial_t J^i. \tag{112}$$

If the bend-twist is exact, $\partial_t J^i = \partial_t(d_s\omega^i - \partial_t\kappa^i) = \partial_t(d_s\partial_t\gamma^i - \partial_t d_s\gamma^i) = 0$, then (112) becomes $\alpha^i = 0$.

5.3 (3+1)-dimensional expression of wave equations

The results in the sections 5.1 and 5.2 can be easily generalized in the (3+1)-dimensional space-time by using the basic equations (62), (66), (70), (71), and the constitutive equations (55), (59), (60), (61).

The first and the second continuity equations in the stress space-time, (71) and (66) give the wave equations for the exact part in the strain space-time:

$$\sqrt{e\rho}(\triangle r^i + d\delta r^i + du^i) = \sqrt{e\rho}(\delta K^i_{\text{a}} - B^i_{\text{a}}), \tag{113}$$

$$\sqrt{e\rho}(\triangle u^i + d\delta u^i) = \sqrt{e\rho}\delta B^i_{\text{a}}. \tag{114}$$

These equations have already been derived in the section 4.1. If we choice the gauges $\delta r^i = -u^i$ and $\delta K^i_{\text{a}} = B^i_{\text{a}}$, the equation (113) becomes the "covariant" form: $\sqrt{e\rho}(\triangle r^i) = 0$. From $\delta\delta = 0$, these gauge conditions lead the another gauge conditions $\delta u^i = \delta B^i_{\text{a}} = 0$. Therefore, (114) becomes $\sqrt{e\rho}(\triangle u^i) = 0$.

The first and the second kinematic equations in the stress space-time, (70) and (62) give the wave equations for the anti-exact part in the strain space time.

$$\sqrt{e\rho}(\triangle K^i_{\text{a}} + d\delta K^i_{\text{a}} - dB^i_{\text{a}}) = 0, \tag{115}$$

$$\sqrt{e\rho}(\triangle B^i_{\text{a}} + d\delta B^i_{\text{a}}) = 0. \tag{116}$$

If we choice the gauge $\delta K^i_{\text{a}} = B^i_{\text{a}}$, (115) becomes $\sqrt{e\rho}(\triangle K^i_{\text{a}}) = 0$. Moreover, this gauge condition means $\delta B^i_{\text{a}} = 0$, so (116) becomes $\sqrt{e\rho}(\triangle B^i_{\text{a}}) = 0$.

The equations (115) and (116) can be rewritten as $\sqrt{e\rho}(\delta dK^i_{\text{a}} + dB^i_{\text{a}}) = 0$ and $\sqrt{e\rho}\delta dB^i_{\text{a}} = 0$. By applying d to these equations, we obtain the wave equations for the defects field:

$$\sqrt{e\rho}\triangle\Theta^i = 0, \tag{117}$$

$$\sqrt{e\rho}\triangle(A^i - K^i) = 0. \tag{118}$$

(117) shows that the disclination 3-from take the "covariant" form without any additional conditions.

6 Incompatibility Equations and Stress Functions

A compatibility in the strain space is necessary for strain components to have solutions of displacement components. In the continuum mechanics including defects filed, we should extend the compatibility equation to include the incompatibility tensor (e.g., [19]). Our aim in this section is to derive the incompatibility equation in the strain space by using the differential forms. Moreover, we derive the incompatibility equation in the stress space and show that this equation includes previous stress functions such as Beltrami, Morera, Maxwell and Airy stress functions.

6.1 Incompatibility equation in strain space

Let us consider the 3-dimensional space. The incompatibility equations in the strain space are given by the relation between the disclination densities and the distortions (e.g., [19]):

$$\theta^i = \theta^i(\beta^i). \tag{119}$$

A simple derivation of (119) is given by the substitution of (3.3): $\alpha^i = d_s\beta^i + \kappa^i$ into the (3.3): $\theta^i = d_s\alpha^i$ as follows

$$\theta^i = d_s d_s\beta^i + d_s\kappa^i. \tag{120}$$

(120) seems to satisfy (119), but the the square of the exterior differential operator vanishes: $d_s d_s\beta^i = 0$. Since θ^i is a 3-form and β^i is a 1-form, we should use $d_s : \Lambda^p \mapsto \Lambda^{p+1}$ twice to derive (119), though the associated term always vanishes such as (120).

Then, we define the new exterior differential operator acting from the right: $\overleftarrow{d_s}$, where the symbol $\leftarrow$ means acting from the right. For instance, the alternative expression of $\theta^i = d_s\alpha^i$ is given by

$$\theta^i = \alpha^i\overleftarrow{d_s}. \tag{121}$$

The substitution of (3.3) into (121) leads to the equation satisfies (119):

$$\eta^i = d_s\beta^i\overleftarrow{d_s}, \tag{122}$$

where $\eta^i = \theta^i - \kappa^i\overleftarrow{d_s}$. The new defined quantity η^i corresponds to the incompatibility tensor (in the case of no extra matter except for dislocations and disclinations) (e.g., [21]). Now, as we have seen in (2.2), one-to-one correspondences between operators in differential forms and in tensor analysis are given by $d_s\Lambda^0 \leftrightarrow \mathrm{grad}(\cdots)$, $d_s\Lambda^1 \leftrightarrow \mathrm{curl}(\cdots)$ and $d_s\Lambda^2 \leftrightarrow \mathrm{div}(\cdots)$. By analogy with these, we set

$$d_s\Lambda^0\overleftarrow{d_s} = \mathrm{grad}(\cdots)\overleftarrow{\mathrm{grad}}, \tag{123}$$

$$d_s\Lambda^1\overleftarrow{d_s} = \mathrm{curl}(\cdots)\overleftarrow{\mathrm{curl}}, \tag{124}$$

$$d_s\Lambda^2\overleftarrow{d_s} = \mathrm{div}(\cdots)\overleftarrow{\mathrm{div}}. \tag{125}$$

Note that acting twice from one-side is not generally equivalent to acting once from each side, that is, $\mathrm{curl}\,\mathrm{curl}(\cdots) \neq \mathrm{curl}(\cdots)\overleftarrow{\mathrm{curl}}$ and so on. Since $\beta^1 \in \Lambda^1$, (122) corresponds to

$$\eta = \mathrm{curl}\,\beta\,\overleftarrow{\mathrm{curl}}. \tag{126}$$

This is called the incompatibility equation. In the particular case of $\eta = 0$, components of (126) are given by

$$0 = \epsilon_{ikl}\epsilon_{jmn}\partial_k\partial_m\beta_{ln}. \tag{127}$$

This is the well-known St. Venant compatibility equation, which means that the topology of the continuum is invariant under elastic deformations. However, when the deformation accompanied anelastic deformations such as emergency of defects field, the topology of the continuum is not invariant, so the left-hand side of (122) is not zero.

6.2 Incompatibility equation in stress space and stress functions

Next, we drive the stress functions from the incompatibility equation. By using the constitutive equations (5.1), (5.1) and (5.1), the incompatibility equation in the strain space (122) can be rewritten as that in the stress space:

$$*\phi_i - (*m_i)\overleftarrow{d_s} = d_s(*\sigma_i)\overleftarrow{d_s}. \tag{128}$$

What is the physical meaning of the incompatibility equation in the stress space? To consider this, we introduce the dual stress space consists of a dual stress 3-form $\tilde{\sigma}^i = *\phi_i$, a dual couple-stress function 2-form $\tilde{c}^i = *m_i$ and a dual stress function 1-form $\tilde{\phi}^i = *\sigma_i$. In this case, (128) becomes

$$\tilde{\sigma}^i = d_s\tilde{\phi}^i\overleftarrow{d_s} + \tilde{c}^i\overleftarrow{d_s}, \tag{129}$$

or as the tensor analysis expression

$$\tilde{\sigma} = \operatorname{curl}\tilde{\phi}\,\overleftarrow{\operatorname{curl}} + \tilde{c}\,\overleftarrow{\operatorname{div}}. \tag{130}$$

It is interesting that (130) includes the previous stress functions as described below. In the particular case of $\tilde{c}\,\overleftarrow{\operatorname{div}} = 0$, (130) corresponds to the equation for the Beltrami stress functions:

$$\tilde{\sigma}_{ij} = \epsilon_{ikl}\epsilon_{jmn}\partial_k\partial_m\tilde{\phi}_{ln}. \tag{131}$$

The nondiagonal components of Beltrami stress functions are Morera stress functions and the diagonal components are Maxwell stress functions:

$$\tilde{\sigma}_{ij} = \delta_{ij}\partial_k\partial_k\tilde{\phi} - \partial_i\partial_j\tilde{\phi}. \tag{132}$$

where $\tilde{\phi} = \tilde{\phi}_{mm}$. Moreover, in the two-dimensional case, the Maxwell stress functions are the well-known Airy stress functions: $\tilde{\sigma}_{11} = \partial_2\partial_2\tilde{\phi}, \tilde{\sigma}_{22} = \partial_1\partial_1\tilde{\phi}, \tilde{\sigma}_{12} = \tilde{\sigma}_{21} = -\partial_1\partial_2\tilde{\phi}$.

7 Energy Integral and J-Integral

In the previous sections, we do not use the integral method in terms of the differential form. Then, in this section, we apply the generalized Stokes's theorem (46) to the continuum mechanics. As an example, we take up the energy integral in fracture mechanics called the J-integral [22, 23, 24].

7.1 Energy integral (J-integral) in fracture mechanics

In the 3-dimensional strain space, let us define the energy density 3-form due to the deformation:

$$W = \beta^i \wedge *\beta^i + \kappa^i \wedge *\kappa^i = \beta^i \wedge \sigma_i + \kappa^i \wedge m_i, \tag{133}$$

where we use (5.1): $e^{-1} * \sigma_i = \beta^i$, (5.1): $e^{-1} * m_i = \kappa^i$ and set $e = 1$ for simplicity. Therefore, the change in the energy density due to the distortion and the bend-twist is given by

$$d_sW = \frac{\partial W}{\partial \beta^i} \wedge d_s\beta^i + \frac{\partial W}{\partial \kappa^i} \wedge d_s\kappa^i = \sigma_i \wedge d_s\beta^i + m_i \wedge d_s\kappa^i. \tag{134}$$

Since the stress and the couple-stress satisfy the law of the conversation of momentum (68): $\partial_t P_i = d_s\sigma_i$ and that of angular momentum (3.2): $-\sigma_i = \partial_t a_i + d_s m_i$, we obtain

$$d_s(\sigma_i \wedge \beta^i) = \partial_t P_i \wedge \beta^i + \sigma_i \wedge d_s\beta^i, \tag{135}$$

and

$$d_s(m_i \wedge \kappa^i) = (-\sigma_i - \partial_t a_i) \wedge \kappa^i - m_i \wedge d_s\kappa^i, \tag{136}$$

where we use the formula (45): $d(\alpha_p \wedge \beta_q) = d\alpha_p \wedge \beta_q + (-1)^p \alpha_p \wedge d\beta_q$ for $\alpha_p \in \Lambda^p$ and $\beta_q \in \Lambda^q$. Therefore, the equation (134) is rewritten as

$$d_s(W - \sigma_i \wedge \beta^i + m_i \wedge \kappa^i) + \partial_t P_i \wedge \beta^i + \partial_t a_i \wedge \kappa^i + \sigma_i \wedge \kappa^i = 0. \tag{137}$$

The integral of (137) along the boundary of the M-dimensional manifold is given by

$$\int_{\partial\partial M} (W - \sigma_i \wedge \beta^i + m_i \wedge \kappa^i) + \int_{\partial M} (\partial_t P_i \wedge \beta^i + \partial_t a_i \wedge \kappa^i + \sigma_i \wedge \kappa^i) = 0, \tag{138}$$

where we use the generalized Stokes' theorem (46):

$$\int_{\partial M} d\omega = \int_{\partial\partial M} \omega, \tag{139}$$

for $\omega \in \Lambda^{M-2}$. From the divisions (97): $\beta^i = d_s u^i + \beta^i_a$ and (98): $\kappa^i = d_s r^i + \kappa^i_a$, we can divide the integral (138) into the exact part and the anti-exact part. In this case, the exact-part corresponds to the J-integral for Cosserat continuum:

$$\Pi = \int_{\partial\partial M} (W - \sigma_i \wedge d_s u^i + m_i \wedge d_s r^i) + \int_{\partial M} (\partial_t P_i \wedge d_s u^i + \partial_t a_i \wedge d_s r^i + \sigma_i \wedge d_s r^i). \tag{140}$$

Ref.[25] has obtained the J-integral from translational invariance of the deformation energy, and shown $\Pi = 0$ for the integral domain is free of singularities. This is a generalization of the correspondent result for the path independent J-integral of the standard continuum. If there is a singularity within the integral domain then $\Pi \neq 0$ in general.

Using the definition (140), the integral (138) is rewritten as

$$\Pi = \int_{\partial\partial M} (\sigma_i \wedge \beta^i_a - m_i \wedge \kappa^i_a) - \int_{\partial M} (\partial_t P_i \wedge \beta^i_a + \partial_t a_i \wedge \kappa^i_a + \sigma_i \wedge \kappa^i_a). \tag{141}$$

This shows that the anti-exact parts can be recognized as the singularities in the present paper. In the present paper, the term "singularity" is used in the sense defined in Ref.[22], that is, an extended state of internal stress, rather than a singularity in the mathematical sense. From (3.3): $\theta^i = d_s\kappa^i$ and (3.3): $\alpha^i = d_s\beta^i + \kappa^i$, the first and the second term in (141) are rewritten as

$$\int_{\partial\partial M} \sigma_i \wedge \beta^i_a = \int_{\partial M} d_s(\sigma_i \wedge \beta^i_a) = \int_{\partial M} \{\partial_t P_i \wedge \beta^i_a + \sigma_i \wedge (\alpha^i - \kappa^i)\}, \tag{142}$$

$$\int_{\partial\partial M} (-m_i \wedge \kappa^i_a) = \int_{\partial M} d_s(-m_i \wedge \kappa^i_a) = \int_{\partial M} (\partial_t a_i \wedge \kappa^i_a + \sigma_i \wedge \kappa^i_a + m_i \wedge \theta^i). \tag{143}$$

Therefore, we can rewrite (141) as the J-integral for dislocation-disclination fields:

$$\Pi = \int_{\partial M} (\sigma_i \wedge \alpha^i + m_i \wedge \theta^i - \sigma_i \wedge \kappa^i). \tag{144}$$

The J-integral (144) implies that the J-integral does not necessarily vanish due to dislocation-disclination fields. In fact, the path-independence no longer holds for a heterogeneous material with a dislocation cloud ahead of the crack tip [26].

7.2 Generalized Peach-Koehler force

The physical meaning of the usual J-integral (and the Eshelby's static energy-momentum tensor) is the force on the dislocation field. Then, let us consider the force on the dislocation-disclination field by using (144). Since the force on the dislocation field is expressed in terms of the Burgers vector, the force on the dislcoation-disclination fields should be represented by not only the term of the Burgers vector but also the term of the Frank vector. The components of Burgers vector b^i and that of the Frank vector f^i are defined by the integrals:

$$b^i = \int_{\partial S} \beta^i_a = \int_S d_s\beta^i_a = \int_S (\alpha^i - \kappa^i), \tag{145}$$

$$f^i = \int_S \kappa^i_a = \int_V d_s\kappa^i_a = \int_V \theta^i, \tag{146}$$

where we use (3.3), (3.3) and d(exact part) $= 0$. Next, we rewrite (144) in the static case. In this case, $d_s\sigma_i = \partial_t P_i = 0$, then we approximate σ_i by a constant. Moreover, $d_s m_i = -\sigma_i - \partial_t a_i = -\sigma_i$, then the J-integral (144) can be rewritten as

$$\Pi = \sigma_i \wedge \int_{\partial M} (\alpha^i - \kappa^i) - \sigma_i \wedge \int_M \theta^i, \tag{147}$$

where we use (3.3): $d_s\theta^i = 0$. Since α^i and κ^i are 2-forms and θ^i is a 3-form, we have the correspondences $\partial M \to \partial S$ and $M \to V$. Thus, from the definitions (145) and (146), the static J-integral (147) becomes

$$\Pi = \sigma_i \wedge (b^i - f^i). \tag{148}$$

This is the generalized Peach-Koehler force that extends the previous result to include the effect of the disclination field.

8 Geometrical Interpretation of Basic Equations

A deformed media including a defect field has been formulated by the various mathematical tools such as the differential geometry and the abstract algebra (e.g., [14, 20, 27, 28, 29, 30, 31]). In this wide field, we take up the well-known result: the geometrical interpretation of equations from a view point of the differential forms.

8.1 Differential geometry in terms of the differential forms

First, we review the differential geometry. In the modern geometry, the connection Γ^i_{jk} plays an important role, because this allows us to compare the local geometry at various points. From this local approach, we can estimate the global quantity of space such as a torsion T^i_{jk} and a curvature tensors R^i_{jkl}:

$$T^i_{jk} = \Gamma^i_{jk} - \Gamma^i_{kj}, \tag{149}$$

$$R^i_{jkl} = \partial_k \Gamma^i_{lj} - \partial_l \Gamma^i_{kj} + \Gamma^m_{lj}\Gamma^i_{km} - \Gamma^m_{kj}.\Gamma^i_{lm}, \tag{150}$$

where we take a natural frame of local coordinates.

In the differential forms, a generalized connection 1-form, a generalized torsion 2-form and a generalized curvature 2-form are defined as

$$\Gamma^i_j = \Gamma^i_{kj}\varphi^k, \tag{151}$$

$$T^i = \frac{1}{2}T^i_{jk}\varphi^j \wedge \varphi^k, \tag{152}$$

$$R^i_j = \frac{1}{2}R^i_{jkl}\varphi^k \wedge \varphi^l, \tag{153}$$

where φ^i is a dual basis 1-form. The geometrical objects T^i and R^i_j should satisfy the following equations:

$$T^i = D\varphi^i, \tag{154}$$

$$R^i_j = D\Gamma^i_j, \tag{155}$$

where the operator D is a covariant exterior differential operator, which gives

$$D\varphi^i = d\varphi^i + \Gamma^i_j \wedge \varphi^j, \tag{156}$$

$$D\Gamma^i_j = d\Gamma^i_j + \Gamma^i_k \wedge \Gamma^k_j. \tag{157}$$

(154) and (155) are called the first and the second Cartan structure equations, respectively.

By applying d to (154), we have

$$dT^i = d(d\varphi^i) + R^i_j \wedge \varphi^j - \Gamma^i_j \wedge T^j, \tag{158}$$

where we use (155). Therefore, the integrability condition, $d(d\varphi^i) = 0$ gives

$$DT^i = R^i_j \wedge \varphi^j, \tag{159}$$

with

$$DT^i = dT^i + \Gamma^i_j \wedge T^j. \tag{160}$$

(159) is called the first Bianchi identities (the integrability condition of a torsion). By applying d to (155), we have

$$dR^i_j = d(d\Gamma^i_j) + R^i_k \wedge \Gamma^k_j - \Gamma^i_k \wedge R^k_j. \tag{161}$$

Therefore, the integrability condition, $d(d\Gamma^i_j) = 0$ gives

$$DR^i_j = 0, \tag{162}$$

with

$$DR^i_j = dR^i_j - R^i_k \wedge \Gamma^k_j + \Gamma^i_k \wedge R^k_j. \tag{163}$$

(162) is called the second Bianchi identities (the integrability condition of a curvature). As we will see, the structure equations and the identities in the differential geometry corresponds to the kinematic and the continuity equations in the continuum mechanics (e.g., [14]).

8.2 Differential geometrical expression of the dual material space-time

It has been shown that the physical quantities in the strain space-time can be expressed in terms of geometrical objects (e.g., [14]):

$$B^i = \hat{\varphi}^i, \tag{164}$$

$$K^i = \hat{\Gamma}^i_j \wedge \hat{\varphi}^j, \tag{165}$$

$$A^i = \hat{T}^i, \tag{166}$$

$$\Theta^i = \hat{R}^i_j \wedge \hat{\varphi}^j - \hat{\Gamma}^i_j \wedge \hat{T}^j, \tag{167}$$

where the symbol $\widehat{(\cdots)}$ means the geometrical objects in the strain space-time. In this case, the first kinematic equation in strain space-time (77): $A^i = dB^i + K^i$ gives the first structure equation (154) as follows:

$$\hat{T}^i = d\hat{\varphi}^i + \hat{\Gamma}^i_j \wedge \hat{\varphi}^i. \tag{168}$$

$$\hat{T}^i = D\hat{\varphi}^i. \tag{169}$$

The second kinematic equation in strain space-time (75): $\Theta^i = dK^i$ gives the second structure equation (155) as follows: $\hat{R}^i_j \wedge \hat{\varphi}^j - \hat{\Gamma}^i_j \wedge \hat{T}^j = d(\hat{\Gamma}^i_j \wedge \hat{\varphi}^j)$
$= d\hat{\Gamma}^i_j \wedge \hat{\varphi}^j - \hat{\Gamma}^i_j \wedge (\hat{T}^j - \hat{\Gamma}^j_k \wedge \hat{\varphi}^k)$.

$$\hat{R}^i_k = D\hat{\Gamma}^i_k \text{ for } \hat{\varphi}^k \neq 0. \tag{170}$$

Moreover, the first continuity equation in strain space-time (78): $dA^i = \Theta^i$ gives the first Bianchi identities (159) as follows:

$$d\hat{T}^i = \hat{R}^i_j \wedge \hat{\varphi}^j - \hat{\Gamma}^i_j \wedge \hat{T}^j. \tag{171}$$

$$D\hat{T}^i = \hat{R}^i_j \wedge \hat{\varphi}^j. \tag{172}$$

The second continuity equation in strain space-time (76): $d\Theta^i = 0$ gives the second Bianchi identities (162) as follows: $\mathrm{d}(\hat{R}^i_j \wedge \hat{\varphi}^j - \hat{\Gamma}^i_j \wedge \hat{T}^j) = d\hat{R}^i_j \wedge \hat{\varphi}^j + \hat{R}^i_j \wedge d\hat{\varphi}^j - d\hat{\Gamma}^i_j \wedge \hat{T}^j + \hat{\Gamma}^i_j \wedge d\hat{T}^j$
$= (d\hat{R}^i_k - \hat{R}^i_j \wedge \hat{\Gamma}^j_k + \hat{\Gamma}^i_j \wedge \hat{R}^j_k) \wedge \hat{\varphi}^k$
$= 0.$

$$D\hat{R}^i_k = 0 \text{ for } \hat{\varphi}^k \neq 0. \tag{173}$$

The constitutive equations (55), (59), (60) and (61) allow us to express the physical quantities in the stress space-time by the geometrical objects:

$$S_i = *\hat{\varphi}^i, \tag{174}$$

$$M_i = -*(\hat{\Gamma}^i_j \wedge \hat{\varphi}^j), \tag{175}$$

$$C_i = -*\hat{T}^i, \tag{176}$$

$$F_i = *(\hat{R}^i_j \wedge \hat{\varphi}^j - \hat{\Gamma}^i_j \wedge \hat{T}^j), \tag{177}$$

where we set $\sqrt{e\rho} = 1$ for simplicity. In this case, the kinematic and continuity equations in the stress space-time: (62), (66), (70) and (71) give

$$*\hat{\varphi}^i = -d * \hat{T}^i, \tag{178}$$

$$d * \hat{\varphi}^i = 0, \tag{179}$$

$$- * (\hat{\Gamma}^i_j \wedge \hat{\varphi}^j) = d * (\hat{R}^i_j \wedge \hat{\varphi}^j - \hat{\Gamma}^i_j \wedge \hat{T}^j) - *\hat{T}^i, \tag{180}$$

$$*\hat{\varphi}^i = -d * (\hat{\Gamma}^i_j \wedge \hat{\varphi}^j). \tag{181}$$

By using (154), these equations can be summarized as $\mathrm{d}*\hat{\varphi}^i = 0, d * d\hat{\varphi}^i = 0$. From (41), we finally obtain

$$\Delta\hat{\varphi}^i = 0. \tag{182}$$

Note that this equation is derived under all the basic equations in strain and stress space-time and all the constitutive equations. In other words, the deformed media including a defect field should satisfy the equation (182).

9 Electrodynamics of Continua in Terms of the Differential Forms

In the sections 2 and 3, we have formulated the electromagnetism and the continuum mechanics in terms of the same mathematical language, i.e., the differential forms. It is expected that this similar approach reduces the difficulties in linking the electromagnetic field and the deformation field. Our aim in this section is to discover this link based on the Hodge duality. As an example, we take up the well-known phenomena: piezoelectric and Villari effects. State concisely, the former is to link the deformation field to the electric field, and the later is to link it to the magnetic field.

9.1 Linear constitutive equations

First, let us consider the piezoelectric effect. This is the ability of some materials to effect a change in an electric potential in response to applied mechanical stress. Converse phenomenon is also known: the mechanical deformation in response to applied electric field. In the linear analysis, these phenomenon are expressed by the following constitutive equations:

$$\mathcal{D} = \epsilon\mathcal{E} + l_1\sigma, \tag{183}$$

$$\beta = e^{-1}\sigma + l_2\mathcal{E}. \tag{184}$$

Recall that $\mathcal{D}$ is the electric flux density, $\mathcal{E}$ is the electric field, σ is the stress and β is the distortion. l_1 and l_2 are constants. Let us consider (183) and (184) from a viewpoint of the differential forms in the 3-dimensional space.

As suggested by (183), the piezoelectric effect is the interaction among $\mathcal{E}$, $\mathcal{D}$ and σ. Since $\mathcal{D}, \sigma \in \Lambda^2$ and $\mathcal{E} \in \Lambda^1$, we cannot directly link $\mathcal{E}$ to $\mathcal{D}$ and σ. Then, we use the Hodge star operator $*$: $\Lambda^p \rightarrow \Lambda^{3-p}$ as follows: $\epsilon * \mathcal{E}^i = \mathcal{D}_i + l_1(-\sigma_i)$ $= \mathcal{D}_i + l_1(\text{the time-like component of } S_i)$.

From $*\mathcal{E}^i = *(\mathcal{E}^i_A dx^A) = \mathcal{E}^A_i ds_A$, $\mathcal{D}_i = \mathcal{D}^A_i ds_A$ and $\sigma_i = \sigma^A_i ds_A$, the equation (9.1) gives $\epsilon\mathcal{E}^A_i = \mathcal{D}^A_i - l_1\sigma^A_i$, which corresponds to (183). Next, we take up the converse piezoelectric effect that is the interaction among β, $\mathcal{E}$ and σ. Since $\beta, \mathcal{E} \in \Lambda^1$ and $\sigma \in \Lambda^2$, we have $\mathrm{e}^{-1} * \sigma_i = \beta^i + l_2\mathcal{E}^i$
$= \beta^i + l_2(\text{the time-like component of } \mathcal{F}^i)$.
From $*\sigma_i = *(\sigma^A_i ds_A) = \sigma^i_A dx^A$, $\beta^i = \beta^i_A dx^A$ and $\mathcal{E}^i = \mathcal{E}^i_A dx^A$, the equation (9.1) gives $e^{-1}\sigma^i_A = \beta^i_A + l_2\mathcal{E}^i_A$, which corresponds to (184).

Second, let us consider the Villari effect: a change in magnetization in response to a stress. The converse Villari effect is sometime called the Joule effect. The linear equation for The Villari effect and the Joule effect is given by[32]

$$\mathcal{B} = \mu\mathcal{H} + l_3\sigma, \tag{185}$$

$$\beta = \mathrm{e}^{-1}\sigma + l_4\mathcal{H}, \tag{186}$$

where l_3 and l_4 are constants. Recall that $\mathcal{B}$ is the magnetic flux density, $\mathcal{H}$ is the magnetic field, and β is the distortion. As suggested by (185), the Villari effect is the interaction among $\mathcal{B}, \sigma \in \Lambda^2$ and $\mathcal{H} \in \Lambda^1$. Then, we express it by $\mu * \mathcal{H}^i = \mathcal{B}_i + l_3(-\sigma_i)$
$= \mathcal{B}_i + l_3(\text{the time-like component of } S_i)$,
that is, $\mu\mathcal{H}^A_i = \mathcal{B}^A_i - l_3\sigma^A_i$. Since the Joule effect is the interaction among $\beta, \mathcal{H} \in \Lambda^1$ and $\sigma \in \Lambda^2$, we express is by $\mathrm{e}^{-1} * \sigma_i = \beta^i + l_4(-\mathcal{H}^i)$
$= \beta^i + l_4(\text{the time-like component of } \mathcal{G}^i)$,
that is, $e^{-1}\sigma^i_A = \beta^i_A - l_4\mathcal{H}^i_A$.

Since only the constitutive equations can link the deformation field to the electromagnetic filed, the equations (9.1), (9.1), (9.1) and (9.1) can be recognized as the foundation for deriving the various equations in electrodynamics of continua. As an example, we will derive the new wave equations in the next section.

9.2 Wave equations

In dielectric crystals, it is known that mechanical vibrations are generated by electric fields through the piezoelectric effect, and conversely, mechanical vibrations produce electric wave through the converse piezoelectric effect (e.g., [1]). Then, let us derive the wave equation that can account for this coupling in terms of the differential forms.

First, we consider the electromagnetic wave generated by the mechanical vibrations. In the case of the piezoelectric effect, the constitutive equation $\mathcal{D}_i = \epsilon * \mathcal{E}^i$ is extended as (9.1). In this case, the usual wave equation (42) is extended as

$$\tilde{\mathcal{A}}^i + d_s(\delta\tilde{\mathcal{A}}^i - \mu\epsilon\partial_t\phi^i) = -\mu * j_i - \mu l_1\partial_t * \sigma_i. \tag{187}$$

On the other hand, in the case of the Villari effect, the constitutive equation $\mathcal{B}^i = \mu * \mathcal{H}_i$ is extended as (9.1), then

$$\tilde{\mathcal{A}}^i + d_s(\delta\tilde{\mathcal{A}}^i - \mu\epsilon\partial_t\phi^i) = -\mu * j_i - l_3\delta\sigma^i. \tag{188}$$

Therefore, both effects exist, we obtain

$$\tilde{\mathcal{A}}^i + d_s(\delta\tilde{\mathcal{A}}^i - \mu\epsilon\partial_t\phi^i) = -\mu * j_i - l_3(\delta\sigma^i + \mu\frac{l_1}{l_3}\partial_t * \sigma_i). \tag{189}$$

If we choice the gauge: $\delta\sigma^i = -\mu(l_1/l_3)\partial_t * \sigma_i$, we obtain the usual wave equation (42).

Next, we consider the mechanical vibrations generated by the electromagnetic field. In the case of the converse piezoelectric effect, the constitutive equation $\sigma_i = e{*}\beta^i$ is extended as (9.1). In this case, the usual wave equation (100) is extended as

$$u^i + d_s\delta u^i = l_2\delta\mathcal{E}^i. \tag{190}$$

On the other hand, in the case of the converse Villari effect, the constitutive equation $\sigma_i = e * \beta^i$ is extended as (9.1), then

$$u^i + d_s\delta u^i = -l_4\delta\mathcal{H}^i. \tag{191}$$

Therefore, both effects exist, we obtain

$$u^i + d_s\delta u^i = \delta(l_2\mathcal{E}^i - l_4\mathcal{H}^i). \tag{192}$$

In the particular case of $l_2\mathcal{E}^i = l_4\mathcal{H}^i$, we obtain the usual wave equation (100).

9.3 Electromagnetic field and defects field

In the previous sections 9.1 and 9.2, we consider a S_i-$\mathcal{F}_i$ system and a S_i-$\mathcal{G}_i$ system. As we have seen in the section 3.1, S_i is the 3-form in the stress space-time. On the other hand, in the dual space, i.e., the strain space-time, the 3-form is Θ^i. Thus, it is expected that we can build up a Θ_i-$\mathcal{F}_i$ system and a Θ_i-$\mathcal{G}_i$ system.

For instance, let us rewrite (9.1). In the Θ_i-$\mathcal{G}_i$ system, the corresponding constitutive equation is given by $\mu * \mathcal{H}_i = \mathcal{B}^i + l_5(\text{the time-like component of } \Theta^i)$
$= \mathcal{B}^i + l_5(-\mathcal{J}^i)$,
where l_5 is a constant. From (3.3), this becomes

$$\mathcal{B}^i = \mu * \mathcal{H}_i - l_5(\partial_t\alpha^i + d_s I^i). \tag{193}$$

This equation shows the well-known phenomenon: the magnitude of the magnetic flux is influenced by the moving dislocations.

10 Discussion

The purpose of this chapter is to express the continuum mechanics in terms of the differential forms. For this purpose, we introduce the dual material space-time: one is the strain space-time consists of the visible physical quantities, and the other is the stress space-time consists of the invisible physical quantities. We can observe the invisible quantities in the stress space-time when they are mapped on to the strain space-time through the Hodge star operator. This operator physically corresponds to the constitutive equation. In this paper, several physical equations can be derived by the corresponding operators. Then, first, let us summarize and discuss the relationship between the operators and the equations.

10.1 Operators and equations

There are two basic equations in the continuum mechanics: the kinematic equation and the constitutive equation. The kinematic equations are valid for all types of substances (e.g., gases, fluids, solids), and the phenomenological properties of materials are brought into play through the constitutive equations [1]. In this paper, as we have seen in the section 3, these two basic equations can be derived by the two basic operators in the differential forms: exterior differential operators: d $\leftrightarrow$ kinematic equations, Hodge star operators: $*\leftrightarrow$ constitutive equations, where we omit continuity equations, because they are another expressions of the kinematic equations. This correspondence means that not the operator d but the operator $*$ stipulates the the phenomenological properties of materials.

Moreover, as we have seen in the sections 4 and 5, certain kinds of compound operators lead certain kinds of compound equations: Navier operators: d $*d \leftrightarrow$ Navier equations, Laplace (or wave) operators: $\{ *d*, d\} \leftrightarrow$ Laplace (or wave) equations. This means that the fact that the Navier and Laplace (or wave) equations can be derived by the combination of the basic equations, corresponds to the fact that the Navier and Laplace (or wave) operators can be derived by the combination of basic operators.

Of course, the other compound equations can also be derived by the combination of d and $*$. For instance, in the section 6.1, we have derived the incompatibility equation by the operator $d_s(\cdots)\overleftarrow{d_s}$, and have derived the stress functions by the dual operator $d_s * (\cdots)\overleftarrow{d_s}$. One of the merits of using the operators is in the simplicity of obtaining the solution of the equation. In fact, in the section 4.2, we have obtained the anti-exact solution of the Navier equation by using the homotopy operator H. In summary, these results suggest the following systematic way: we can derive the equation by the combination of d and $*$, and can solve it by applying the operator H.

10.2 Other physical fields

In the section 9, we consider the relationship between the deformation field and the electromagnetic field. From a view point of the fiber bundle, the electromagnetic field can be recognized as the curvature form. In this case, the potential is the connection form. (Compare (12) with (155) in the case of the abelian group $U(1)$: $A \wedge A = 0$.) Thus, the Maxwell equation corresponds to the Bianchi identities. On the other hand, as we have seen in the section 8, the physical quantities in the continuum mechanics can also be recognized as the geometrical objects. In this case, the continuity equation corresponds to the Bianchi identities. Therefore, it is expected that the interaction between the deformation field and the electromagnetic field can be expressed in a uniform manner, i.e., the interaction among the geometrical objects of the space-time. For instance, the equation (193) can be interpreted as the interaction of the curvature (magnetic field) with the torsion (dislocations field).

Since we express the electromagnetism and the continuum mechanics in terms of the same mathematical language, i.e., the differential forms, it is relatively easy to link the deformation field with the electromagnetic field. This kind of linkage is done by the constitutive equation, i.e., the Hodge star operator $*$. This approach allows us to link one physical field with the other physical field. Now, it has been known that the thermodynamics can be also formulated by the differential forms. Therefore, we should link the deformation field with the thermodynamic field through the operator $*$. For instance, the first law of thermodynamics is expressed as the relationship among the 1-form thermodynamic quantities such as $d\tau$, where τ is the temperature 0-form. Therefore, the stress 2-form σ_i and the distortion 1-form β^i is related to the 1-form $d\tau$ as following form: $*\sigma_i = e\beta^i + l_6 d\tau^i$, where l_6 is a constant. This is a Hooke's law in the thermodynamic field. It is one of the subjects of a future study to unify the deformation, the electromagnetic and the thermodynamic fields in terms of the differential forms.

11 Conclusion

The two basic operators in the differential forms lead the corresponding two basic equations in the continuum mechanics. By the combination of the basic operators, we can obtain the useful equations such as Navier equation, Laplace equation and the incompatibility equations. The Hodge star operator allows us to (i) stipulate the the phenomenological properties of materials through the constitutive equation, and (ii) link the deformation filed with other physical fields such as the electromagnetic field and the thermodynamic field.

References

[1] Eringen, A.C.; Maugin, G.A. Electrodynamics of Continua; Springer: New York, 1990; Vol.1, pp 436.

[2] Nakahara, M. Geometry, Topology and Physics; A. Hilger: Bristol, 1990, pp 505.

[3] Bachman, D.A. Geometric Approach to Differential Forms; Birkhauser: Boston, 2006, pp 133.

[4] Baldomir, D.; Hammond, P. Geometry of Electromagnetic Systems; Clarendon Press: New York, 1996, pp 239.

[5] Hehl, F.W.; Obukhov, Y.N. Foundations of Classical Electrodynamics: Charge, Flux, and Metric; Birkhauser: Boston, 2003, pp 410.

[6] Edelen, D.G.B. Applied Exterior Calculus; Dover Publications: New York, 2005, pp 505.

[7] Yamasaki, K.; Nagahama, H. J. Phys. A: Math. Gen. 1999, 32, L475-481.

[8] Yamasaki, K.; Nagahama, H. J. Phys. A: Math. Gen. 2002, 35, 3767-3778.

[9] Minagawa, S. Int. J. Theor. Phys. 1990, 29, 1271-1276.

[10] Amari, S. Int. J. Eng. Sci. 1981, 19, 1581-1594.

[11] Oden, J.T. Reddy, J.N. Int. J. Eng. Sci. 1974, 12, 1-29.

[12] Tonti, E. In Systemics of Emergence: Research and Development; Minati, G.; Pessa, E.; Abram, M. Ed.; Springer: New York, 2005, pp 695-706

[13] Kadić, A.; Edelen, D.G.B. A Gauge Theory of Dislocations and Disclinations; Springer: New York, 1983, pp 290.

[14] Edelen, D.G.B.; Lagoudas, D.C. Gauge Theory and Defects in Solids; Elsevier: Amsterdam, 1988, pp 427.

[15] Eringen, A.C. Microcontinuum Field Theories; Springer: New York, 1999; Vol. 1, pp 325.

[16] Rubin, M.B. Cosserat Theories: Shells, Rods and Points; Kluwer Academic Publishers: Dordrecht, 2000, pp 480.

[17] Schaefer, V.H. ZAMP 1969, 20, 891-899.

[18] Kondo, K. Proc. 2nd Japan National Congr. on Applied Mechanics (Tokyo) 1952, 41-47.

[19] Kröner, E. In Physics of Defects; Balian, R. Ed.; Elsevier: Amsterdam,1981, pp 214-315.

[20] Kleinert, H. Gauge Fields in Condensed Matter; World Scientific: Singapore, 1989; Vol. 2, pp. 1456.

[21] DeWitt, R. Int. J. Eng. Sci. 1981, 12, 1475-1506.

[22] Eshelby, J.D. Philos. Trans. R. Soc. Lond. Ser. A 1951, 244, 87-112.

[23] Rice, J.R. J. Appl. Mech. 1968, 35, 379-386.

[24] Yamasaki, K.; Nagahama, H. ZAMM 2008, 88, 515-520.

[25] Mühlhaus, H.B.; Pasternak, E. Int. J. Fracture 2002, 113, L21-L26.

[26] Ohnami, M. Plasticity and High Temperature Strength of Materials: Combined Micro- and Macro-Mechanical Approaches; Elsevier: London, 1988, pp. 525.

[27] Katanaev, M.O. PHYS-USP 2005, 48, 675-701.

[28] Miklashevich, I.A. Micromechanics of Fracture in Generalized Spaces; Academic Press: London, 2008, pp. 280.

[29] Lazar, M.; Anastassiadis, C. Philos. Mag. 2008, 88, 1673-1699.

[30] Yamasaki, K. Forma 2007, 22, 191-197.

[31] Nishiyama, Y.; Nanjo, K.Z.; Yamasaki, K. Physica A 2008, 387, 6252-6262.

[32] Clark, A.E.; Teter, J.P.; Wun-Fogie, M. J. Appl. Phys. 1990, 67, 5007-5009.

In: Continuum Mechanics ISBN: 978-1-60741-585-5
Editors: Andrus Koppel and Jaak Oja, pp. 223-244

Chapter 8

STRESS DEPENDENT MORPHOGENESIS: CONTINUUM MECHANICS AND SYSTEM OF TRUSSES

***J.J. Muñoz*[1], *V. Conte*[2] *and M. Miodownik*[2]**
[1] Laboratori de Càlcul Numèric (LaCàN), Dept. Applied. Mathematics III,
Universitat Polit. Catalunya, Barcelona, Spain
[2] Materials Research Group, Div. of Engineering,
King's College, London, UK

ABSTRACT

The mechanical analysis of soft tissues in biomechanics has experienced increasing progress during the last decade. Part of this success is due to the development and application of some techniques of continuum mechanics, in particular, the decomposition of the deformation gradient, and the introduction of mass, density or volume changes in the reference configuration. Resorting to the common terminology employed in the literature, the changes in biomechanical processes may be classified as growth (change of mass), remodelling (change of density or other material properties such as fibre orientation) or morphogenesis (change of shape). Although the use of those concepts in bone and cardiovascular analysis is well extended, their use in morphogenesis during embryo development has been far less studied. The reasons for this fact may be found in the large shape changes encountered during this process, or the complexity of the material changes involved.

In this chapter we develop a general framework for the modelling of morphogenesis by introducing a growth process in the structural elements of the cell, which in turn depends on the stress state of the tissue. Some experimental observations suggest this feedback mechanism during embryo development, and only very recently this behaviour has started to be simulated.

We here derive the necessary equilibrium equations of a stress controlled growth mechanism in the context of continuum mechanics. In these derivations we assume a free energy source which is responsible for the active forces during the elongation process, and a passive hyperelastic response of the material. In addition, we write the necessary conditions that the active elongation law must satisfy in order to be thermodynamically consistent. We particularise these equations and conditions for the relevant elements of the cytoskeleton, namely, microfilaments and microtubules. We apply the model to simulate the shape changes observed during embryo morphogenesis in truss element. As a salient result, the model reveals that by imposing boundary stress conditions, unbounded elongation would be obtained. Therefore, either prescribed displacements or cross-links between fibres are necessary to reach a homeostatic state.

1 Introduction

Resorting to the common terminology employed biomechanics [Tab95], the changes in biomechanical processes may be classified as growth (change of mass), remodelling (change of density or other material properties such as fibre orientation) or morphogenesis (change of shape). In this chapter we introduce the basics to model the reorganisation process of the structural elements in the cytoskeleton (namely microfilaments and microtubules) during morphogenesis of embryo development. Here we assume that the mass of the cell remains constant but that some microtubules and microfilaments grow while others shrink, a process involving active elongation of the structural elements of the cell, which causes the cell to change shape. We resort to the usual decomposition of the total deformation into an active and a passive component. In our model, the former is due to the mechanotransduction of the chemical bonds in the cytoskeleton, while the latter corresponds to the passive elastic response of the deformation. It is not our aim to analyse the source of the cytoskeleton reorganisation, which is a widely debated topic [MTSM07].

We derive the spatial and material equilibrium equations using a variational approach, and deduce an Eshelby stress tensor for the continuum case and particularise the derivations to truss-like structures. We apply special attention to the derivation of a thermodynamically consistent constitutive law for the growth process. The effects of this law on the balance equations is also studied, and a particular form of a simple elastic constitutive law is considered for the solution of a truss element.

Some constitutive laws for growth (mass change) and remodelling (density change) can be found in [AG07, KS04, Lub04, HKMS05]. In these references the evolution laws are derived from the dependence of the internal energy on variables involved in the growth process (density, active deformation). On the other hand, the non-linear elastic behaviour of adaptive isotropic chain networks have been studied in [BA00, MGL04], and the modelling of oriented chains, commonly found in biology, can be found in [KGAG05]. Biological structures have been also modelled resorting to tensegrity concepts, and although this approach has been fruitful in the modelling of self equilibrated systems [Ing97], its application to morphogenesis and growth has not been pursued.

In our case, the constitutive laws of the growth process is a function of the elastic stresses, in a similar manner to other models of blood vessels [Hum01, Tab98], or more recent works in morphogenesis [AG07, Tab08]. The former are based on experimental results, where it is observed that the tissues tend to a homeostatic state, where no further growth/resorption occurs. We here assume that this situation is attained when the stresses achieve a target stress. This idea is also applied to morphogenesis (shape changes) of continua, with a constant [RT08] or variable [Tab08] target stress. The latter works are based on Beloussov's hyper-restoration (HR) hypotheses [BSNN94]: whenever a change is produced in the amount of local stress applied to a cell, it will actively generate forces in order to restore the initial stress value. This is the idea we will mimic in this chapter, which in the development context, is based on experimental observations that indicate the presence of stress controlled morphogenesis [BSF04, Far03, Shr05]. We remark that, this stress-dependent evolution law, may be derived by assuming a thermodynamically consistent form of the remodelling/growth force [AG07, DQ02]. In fact, we will use these ideas to connect the two approaches: Beloussov hyper-restoration hypothesis and the constitutive laws deduced by a thermodynamical reasoning. The particular form of the stress controlled law has been motivated in our case to the stress profiles obtained in our earlier models where the active kinematic response of the cells was imposed externally [MBM07, CMM08a].

2 Continuum Mechanics

2.1 Kinematics

With the aim of developing the theory of the reorganisation process in a truss system, we introduce here the theory for the more familiar case of continuum mechanics. Our aim is to study the deformation of a body $\mathcal{B}$ with *reference configuration* $\Omega^0 \in R^3$ and material coordinates X^0 into a *deformed configuration* $\Omega_t \in R^3$, where the material coordinates are now located in $x = \chi(X^0, t)$ where $\chi(X^0, t) : R^3 \times R \to R^3$ is the motion map of the whole deformation. For clarity in the exposition we will remove the dependence on the time variable t and the position X^0, and simply denote by x the map χ.

As it is customary since the seminal work of [RHM94], we use a multiplicative decomposition of the deformation gradient $F = \frac{\partial x}{\partial X^0}$ as $F = F_e F_a$. In the present case, the deformation gradient $F_a = I + \frac{\partial u_a}{\partial X^0}$ is due to the active deformation u_a of the cell, that is, the mechanotransduction of the chemical reactions that take place in the cytoskeleton, mainly in the actin filaments and microtubules. On the other hand, the tensor $F_e = I + \frac{\partial u_e}{\partial X}$ represents the passive elastic deformation due to the elastic response of either the cytoskeleton and the cytoplasm. The intermediate configuration Ω, which will be henceforth called the *relaxed deformation*, is the one resulting from removing the elastic deformation by also allowing material incompatibilities (overlappings). Therefore, the two tensors F_a and F_e may be discontinuous, while F is continuous (see Figure 1).

From the decomposition of the deformation gradient we have $J = J_a J_e$ with $J = detF$, $J_a = F_a$ and $J_e = detF_e$.

2.2 Balance equations

2.2.1 Balance of mass

We denote by ρ^0, ρ and ρ_t the densities in the reference configuration, the relaxed configuration and in the deformed configuration, respectively. Analogously, we denote by (dM^0, dV^0), (dM, dV) and (dm, dv) the pairs of differentials of mass and volume in the same configurations, which are related by $dM^0 = \rho^0 dV^0$, $dM = \rho J_a$ and $dm = \rho dv$. From the density-preserving hypothesis of the active deformation ($\rho^0 = \rho$), the mass-preserving hypothesis of the elastic deformation ($dM = dm$), and relation $J = J_a J_e$ it can be deduced that $\rho^0 = \rho_t J_e$, and therefore,

$$(\rho_t J_e)^{\cdot} = \dot{\rho}^0 = 0, \tag{1}$$

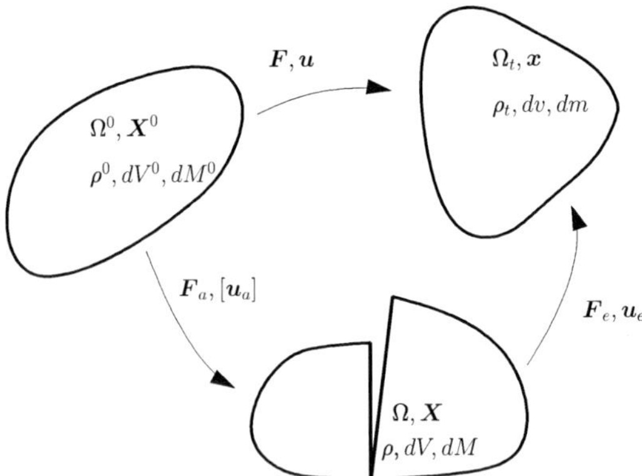

Figure 1: Decomposition of the deformation gradient as $F = F_a F_e$. The reference, relaxed and deformed configurations are indicated by Ω^0, Ω and Ω_t, respectively. Other variables associated to each configuration and map are indicated in the Figure and described in the text.

where a superimposed dot denotes material time differentiation, material time derivative, i.e. $(\dot{\bullet}) = \frac{\partial}{\partial t}|_{X=cnst}$. In order to express the balance of mass, we hypothise that the final growth of mass is directly proportional to the current mass, i.e.

$$(\rho_t J)^{\cdot} = \Gamma \rho_t J \tag{2}$$

with Γ is the *mass production rate*. After making use of mb1 and the relation $\dot{J}_a = J_a F_a^{-\mathrm{T}} : F_a$, the time differentiation of equation mb0 yields,

$$\Gamma = \frac{\dot{J}_a}{J_a} = trace(L_a) = \nabla_X \cdot V, \tag{3}$$

where $V = \dot{X}$ and $L_a = \dot{F}_a F_a^{-1}$. The operator ∇_X is the gradient with respect to the coordinates in the relaxed configuration X. On the other hand, denoting by ∇_x the gradient operator with respect to the coordinates in the deformed configuration Ω_t, using the relation $(\dot{\bullet}) = \frac{\partial \bullet}{\partial t} + v \cdot \nabla(\bullet)$, and equation mb0, we deduce the following equation for the balance of mass:

$$\frac{\partial \rho_t}{\partial t} + \nabla_x \cdot (\rho_t v) = \Gamma \rho_t$$

with $v = \dot{} \; x$.

2.2.2 Balance of linear momentum

No growth phenomena will be considered in this section, and therefore, no distinction will be made between configurations Ω^0 and Ω. By assuming a field of external body loads per unit of mass $\bar{b}$, the balance of linear momentum read in Lagrangian form as follows (see for instance [GS08, MB08]): $\rho^0 \frac{dv}{dt} = \nabla_0 \cdot P + \rho^0 \bar{b}$,

where we have introduced the first Piola stress tensor P. The operator ∇_0 denotes the gradient with respect to the reference coordinates X^0.

2.3 Thermodynamics

The motivation of this section is the construction of proper constitutive laws for the elastic and active parts of the deformation which are thermodynamically consistent. We first present some general well known facts that we will particularise to our needs in the model.

2.3.1 Balance equations in the absence of growth

Let us define the total internal energy E and the total kinetic energy K of a solid as,
E=$\int_{\Omega^0} \rho^0 \phi^0 \, dV^0$
$K = \int_{\Omega^0} \frac{1}{2} \rho^0 v \cdot v \, dV^0$,
where ϕ^0 is the internal energy per unit of mass in the reference configuration. Variations of ϕ^0 may be due to mechanical (elastic) effects or to thermal effects. The *balance of thermal and mechanical energy* demands that the time variations of $K + E$ are due to thermal fluxes and external loads. In symbols, in the absence of any external body heating, $\dot{E} + \dot{K} = \int_{\partial\Omega^0} (\bar{p}^0 \cdot v - q \cdot N^0) \, dV^0 + \int_{\Omega^0} \rho^0 \bar{b} \cdot v \, dV^0$, where $\bar{p}^0$ and q^0 are the material representations of the external surface loads and thermal fluxes, respectively, and the vector N^0 is the outward normal of the solid in the reference configuration. By using the divergence theorem, the boundary condition $PN^0 = \bar{p}^0$, and the standard localisation argument [GS08], equation eq:btm may be expressed as, $\rho^0 v \cdot \frac{dv}{dt} + \rho^0 \dot{\phi}^0 = \nabla_0 \cdot (P^{\mathrm{T}} v) - \nabla_0 \cdot q + \rho^0 \bar{b}$.

On the other hand, after premultiplying teq1 by v, and using relation $\nabla_0 \cdot (P^T v) = v \cdot (\nabla_0 \cdot P) + \nabla_0 v : P$ together with the definition $\dot{F} = \nabla_0 v$, we can derive the so-called *theorem of kinetic energy*:

$$\rho^0 v \cdot \frac{dv}{dt} + P : \dot{F} = \nabla_0 \cdot (P^{\mathrm{T}} v) + \rho^0 \bar{b} \cdot v, \tag{4}$$

which subtracted from teq2, yields the *First Law of Thermodynamics* (also called theorem of internal energy),

$$\rho^0 \dot{\phi}^0 = P : \dot{F} - \nabla_0 \cdot q. \tag{5}$$

On the other hand, the *Second Law of thermodynamics* can be stated as [GS08, MB08],

$$\dot{S} + \nabla_0 \cdot \left(\frac{q}{\theta} \right) \geq 0, \tag{6}$$

where S in the entropy density per unit of reference volume, and $\theta > 0$ is the absolute temperature. Relations 1lt and 2lt may be rewritten by using the Helmholtz free energy function per unit of mass ψ^0, such that $\rho^0 \psi^0(F, \theta) = \rho^0 \phi^0(F, S) - S\theta$, which is the Legendre transformation of $\phi^0(F, S)$ with respect to S, i.e., we define $\theta = \frac{\partial \rho^0 \psi^0(F,S)}{\partial S}$ and can obtain the entropy as $S = \frac{\partial \rho^0 \psi^0}{\partial \theta}$. Introducing the internal heat source as

$$h^{int} = P : \dot{F} - \rho^0 \dot{\psi}^0, \tag{7}$$

inserting relation $\rho^0 \dot{\psi}^0 = \rho^0 \dot{\phi} - (S\theta)^{\cdot}$ in 1lt, and premultiplying 2lt by θ, together with equation 1lt yields: (Sθ)$^{\cdot}$ + $\nabla_0 \cdot q = h^{int}$
$S\dot{\theta} + \frac{q}{\theta} \cdot \nabla_0 \theta \leq h^{int}$.

Equation cd is the Clausius-Duhem (C-D) inequality, whereas relation 1st2 states that the internal heat sources contribute to either a variation of thermal energy $S\theta$ or to the heat flux q. When considering purely (non-dissipative) hyperelastic solids, the free energy is solely due to elastic effects, i.e. $\psi^0 = \hat{\psi}^0(F)$. In this case, we have that $\rho^0\dot{\psi}^0 = \frac{\partial \rho^0\psi^0}{\partial F} : \dot{F}$, and therefore $P = \frac{\partial \rho_t \psi_t}{\partial F}$, which also implies $h^{int} = 0$, as expected. Furthermore, if no heat flux is present ($q = 0$), the C-D inequality states that the stored thermal energy can only decrease. Alternatively, in isothermal processes, such as those that will be modelled here, the inequality in cd reads $h^{int} \geq 0$, that is, D:=P:$\dot{F} - \rho^0\dot{\psi}^0 \geq 0$, where $\mathcal{D}$ is the dissipated energy. This is the so-called *reduced dissipation inequality*, which in words, states that:

- $P : \dot{F} \geq 0$: when the mechanical power increases, $\rho^0\psi^0$ will *increase at most* the same amount (some energy may be lost in a dissipative process).
- $P : \dot{F} \leq 0$: when the mechanical power decreases, $\rho^0\psi^0$ must *decrease at least* the same amount (again, some energy may be lost in a dissipative process).

Equation 2nds, which should be satisfied for all admissible values of $\dot{F}$ and P, furnishes the necessary conditions that the constitutive law relating these quantities should satisfy. Similarly, equation cd yields the condition required by constitutive laws relating the heat flux q and the temperature θ, such as for instance Fourier's law: $q = -\mathbf{K}\nabla_0\theta$, where K is a semi-positive definite second-order tensor of conductivity coefficients.

2.3.2 Thermodynamics of active elongation

Since in our model we will neglect thermal and inertial effects, our point of departure will be equation 2nds. However, in the presence of a growing process, we will though rewrite this equation as a function of the Helmholtz free energy per unit of mass at the relaxed configuration $\psi(F_a, F_e)$, which we *a priori* assume a function of the active and elastic tensors. It is illustrating to split additively function ψ into a chemical and a mechanical component: $\psi(F_a, F_e) = \psi_e(F_e) + \psi_a(F_a)$. While ψ_e accounts for the elastic energy stored in the solid, ψ_a represents the energy stored in the chemical bonds of the structural elements of the cell and thus its variation will be associated to changes in F_a. Similar decompositions may be found elsewhere, for instance in [GON+06], where the active (remodelling) deformations are due to a particular dependence of ψ on the fibre reorentiation. A discussion on the interpretation of the active deformations in the present case of morphogenesis will be given in Section 3.

The term $P : \dot{F}$, which corresponds to the elastic internal energy in equation 2nds, is expressed in the presence of the active elongation as the sum of an elastic internal energy and a chemically induced internal energy. The two terms are represented by the products $P_e : \dot{F}F_a^{-1}$ and $P_a : \dot{F}_aF_a^{-1}$, respectively, where P_e and P_a are the elastic and active internal stress tensors. Inserting these terms into equation 2nds, the Second Law of Thermodynamics is expressed as,

P$_e : L_e + P_a : L_a - \rho^0\dot{\psi} \geq 0$. where we have introduced the elastic velocity gradient $L_e = \dot{F}F_a^{-1}$. [1]

[1]We note that, for any vector $a = F_aA$, the tensors L_a and L_e are such that,

Resorting to the relations $\psi\, dV = \psi J_a\, dV^0$, $\dot{J}_a = J_a F_a^{-\mathrm{T}} : \dot{F}_a$ and $\dot{F}_e = \dot{F} F_a^{-1} - F_e \dot{F}_a F_a^{-1}$, it follows that the last term in 2ndsg is given by: $\rho^0 \dot{\psi} = \rho^0 \frac{\partial \psi}{\partial F_a} : \dot{F}_a + \rho^0 \frac{\partial \psi}{\partial F_e} : \dot{F}_e - \rho^0 J_a \psi I : L_a$
$= \rho^0 \left(I + \frac{\partial \psi}{\partial F_a} F_a^{\mathrm{T}} - F_e^T \frac{\partial \psi}{\partial F_e} \right) : L_a + \rho^0 F_e^{\mathrm{T}} \frac{\partial \psi}{\partial F_e} : L_e$.

Inserting this relation into equation 2ndsg, and using the fact that the inequality must be satisfied for all admissible deformations F_a and F, the following conditions on the stresses P_a and P_e are deduced: $\mathrm{P}_a = \rho^0 \left(\psi I + \frac{\partial \psi}{\partial F_a} F_a^T - F_e^{\mathrm{T}} \frac{\partial \psi}{\partial F_e} \right) + P_a^+$,
$P_e = \frac{\partial \psi}{\partial F_e} + P_e^+$, where P_a^+ and P_e^+ must satisfy the following inequality,

$$P_a^+ : L_a + P_e^+ : L_e \geq 0. \tag{8}$$

We will henceforth assume that $P_e^+ = 0$ and $P_a^+ = c\dot{F}_a F_a^{-1}$, with $c \geq 0$, which allows us to express the condition in eq:claw as $\mathrm{P}_a = \rho^0 \left(\psi I + \frac{\partial \psi}{\partial F_a} F_a^T \right) - F_e^{\mathrm{T}} P_e + c\dot{F}_a F_a^{-1}$,
$P_e = \frac{\partial \psi}{\partial F_e}$,

where the tensor $F_e^T P_e$ is sometimes called in the literature the Mandel stress tensor [EM00, HKMS05]. We note that the parameter c determines the amount of dissipated energy $\mathcal{D} = c\|\dot{F}_a F_a^{-1}\|$ in the system, which is positive whenever $c > 0$.

2.4 Equilibrium equations for unconstrained deformed configuration

Three approaches may be pursued when deriving the equilibrium equations in the presence of truss active elongation: (i) resorting to Noether's theorem, (ii) using the Virtual Power Principle (VPP) [DQ02, AG07], or (iii) using the minimisation of an energy functional (variational approach). Method (i) will be omitted here, but the reader may find the necessary steps in [KH00]. Method (ii) becomes useful when, due to the incompatibility of the relaxed configuration Ω, the maps $X(X^0, t) : \Omega^0 \to \Omega$ are not invertible, not even piecewise, and therefore $u_a = X - X^0$ and $u_e = x - X$ may not be integrable. In this case, method (iii) may not be applicable, and the tensor F_a and the deformation gradient $F = \frac{\partial u}{\partial X^0}$ are taken as the primary kinematic variables. If we assume u_a and u_e are defined, method (iii) consists on interpreting the tensors F_a and F_e as displacement gradients, respectively given by $F_a = \frac{\partial u_a}{\partial X^0}$ and $F_e = \frac{\partial u_e}{\partial X}$. In this case, we can write the free energy function as $\psi(u_a, u_e)$, and find the equilibrium process as the minimisation of a functional.

Although the equilibrium equations using methods (ii) and (iii) are deduced next for illustrative purposes, in the remaining sections we will only use the latter. This is motivated by the fact that in our actual method, described in Section 3, the elongation growth process is embedded in a one dimensional model. In this situation, the piecewise integrability of u_a and u_e is guaranteed.

2.4.1 Variational approach

We assume in this section that the displacement fields u_a and u_e exist and are integrable (or at least piecewise integrable, in which case a similar deduction may be pursued). We

---

d
$dta\Big|_{A=cnst.} = L_a a, \quad \frac{d}{dt} F_e a\Big|_{A=cnst.} = L_e a.$

consider a body subjected to the spatial force $\bar{p}^0$ on the boundary $\partial\Omega^0$, and an external active stress tensor $\bar{P}_a^0$ acting on volume Ω^0, but we neglect any external body loads. After reminding the reader that $\rho^0 = \rho$ (see Figure 1), and using a free energy density function per unit of relaxed mass $\psi(u_a, u_e)$, the spatial equilibrium equations are obtained by minimising the following energy functional:

$$\Pi(u_a, u_e) = \int_\Omega \rho\psi(u_a, u_e)dV - \int_{\partial\Omega^0} \bar{p}^0 \cdot u dS^0 - \int_{\Omega^0} \bar{P}_a^0 : \frac{\partial u_a}{\partial X^0} - dV^0,$$

or equivalently, solving the following variational equation:

$$\delta\Pi = \frac{d}{d\varepsilon}\Pi(u_a + \varepsilon\delta u_a, u_e + \varepsilon\delta u_e)\Big|_{\varepsilon=0} = 0. \tag{9}$$

Note that the functional does not depend on the time derivatives of u_a or u_e, and we are therefore neglecting any inertial terms and the kinetic energy. In the present case we assume that the virtual displacements $\delta u_a \neq 0$ and $\delta u_e \neq 0$ are independent. The minimisation of the total virtual work is performed in a similar manner to the time differentiation of ψ given in Section s:ThGrowth. By assuming that no external body loads exists, and that a material stress tensor $\bar{P}_a$ is being applied, the following equilibrium equation is obtained:

$$\begin{aligned}&\textstyle\int_\Omega \rho \left(\frac{\partial\psi}{\partial F_a} : \delta F_a + \frac{\partial\psi}{\partial F_e} : \delta F_e + \psi I : \delta F_a F_a^{-1}\right) dV\\ &- \textstyle\int_{\partial\Omega^0} \bar{p}^0 \cdot \delta u \, dS^0 - \int_{\Omega^0} \bar{P}_a^0 : \frac{\partial \delta u_a}{\partial X^0} \, dV^0 = 0.\end{aligned}$$

In view of the following relations: $\delta(F_e) = (\delta F)F_a^{-1} - F_e(\delta F_a)F_a^{-1}$,
$(\delta F)F_a^{-1} = \frac{\partial \delta u}{\partial X^0}\frac{\partial X^0}{\partial X} = \frac{\partial \delta u}{\partial X}$,
$(\delta F_a)F_a^{-1} = \frac{\partial \delta u_a}{\partial X^0}\frac{\partial X^0}{\partial X} = \frac{\partial \delta u_a}{\partial X}$,
the first integral can be rewritten in terms of the virtual displacements δu and δu_a as,

$$\begin{aligned}&\textstyle\int_\Omega \rho \left(\frac{\partial\psi}{\partial F_a} F_a^{\mathrm{T}} - F_e^T \frac{\partial\psi}{\partial F_e} + \psi I\right) : \frac{\partial \delta u_a}{\partial X} dV + \int_\Omega \frac{\partial\psi}{\partial F_e} : \frac{\partial \delta u}{\partial X} dV\\ &- \textstyle\int_{\partial\Omega^0} \bar{p}^0 \cdot \delta u \, dS^0 - \int_{\Omega^0} \bar{P}_a^0 : \frac{\partial \delta u_a}{\partial X^0} \, dV^0 = 0.\end{aligned}$$

Finally, from the arbitrariness of the virtual displacements, we obtain, after integrating by parts and using relation $P_e = \frac{\partial\psi}{\partial F_e}$, the following set of differential equations:

$$\nabla_X \cdot \left(\rho^0 \frac{\partial\psi}{\partial F_a} F_a^{\mathrm{T}} + \rho^0 \psi I - F_e^{\mathrm{T}} P_e\right) = \nabla_X \cdot \bar{P}_a, \forall\, X \in \Omega$$

$$\nabla_X \cdot P_e = 0, \forall\, X \in \Omega$$

$$\rho^0 \left(\frac{\partial\psi}{\partial F_a} F_a^{\mathrm{T}} N + \psi N\right) = F_e^{\mathrm{T}} P_e N + \bar{P}_a N, \forall\, X \in \partial\Omega$$

$J_a P_e F_a^{-\mathrm{T}} N^0 = \bar{p}^0, \forall\, x \in \partial\Omega_t$, with $\bar{P}_a = J_a^{-1} \bar{P}_a^0 F_a^{\mathrm{T}}$, that is, $\bar{P}_a$ is the inverse Piola transform of $\bar{P}_a^0$. In the last equation, we have used Nanson's formula $NdS = J_a F_a^{-\mathrm{T}} N^0 dS^0$, with N and N^0 the normal vectors in the relaxed and reference configuration, respectively. The operator $\nabla_X \cdot$ denotes the divergence with respect to the coordinates in the relaxed configuration.

We remark that equations eq:smcb and eq:smcd are the standard *spatial equilibrium equations*, whereas eq:smca and eq:smcc are the *material equilibrium equations*. In the absence of active elongation, only the former are recovered. Instead, the latter are those that govern the deformations during active elongation.

Alternatively, the equations in eq:smc may be expressed in the reference or current configuration. For instance, if the virtual variation $\delta\Pi$ is expressed as a function of the free energy per unit of mass in the reference configuration $\psi^0 = J_a\psi$, it gives rise to, $\delta\Pi = \int_{\Omega^0} \rho^0 \left(\frac{\partial\psi^0}{\partial F_a} : \frac{\partial\delta u_a}{\partial X^0} + \frac{\partial\psi^0}{\partial F_e} : \frac{\partial\delta u}{\partial X^0} F_a^{-1} - F_e^{\mathrm{T}} \frac{\partial\psi^0}{\partial F_e} : \frac{\partial\delta u_a}{\partial X^0} F_a^{-1} \right) dV^0$
$- \int_{\partial\Omega^0} \bar{p}^0 \delta u \, dS^0 - \int_{\Omega^0} \bar{P}_a^0 : \frac{\partial\delta u_a}{\partial X^0} dV^0 = 0$, which yields the following set of differential equations: $\nabla_0 \cdot \rho^0 \left(\frac{\partial\psi^0}{\partial F_a} - F_e^{\mathrm{T}} \frac{\partial\psi^0}{\partial F_e} F_a^{-\mathrm{T}} \right) = \nabla_0 \cdot \bar{P}_a^0, \forall\, X^0 \in \Omega^0$
$\nabla_0 \cdot \rho^0 \frac{\partial\psi^0}{\partial F_e} F_a^{-\mathrm{T}} = 0, \forall\, X^0 \in \Omega^0$
$\rho^0 \left(\frac{\partial\psi^0}{\partial F_a} - F_e^{\mathrm{T}} \frac{\partial\psi^0}{\partial F_e} F_a^{-\mathrm{T}} \right) N^0 = \bar{P}_a^0 N^0, \forall\, X^0 \in \partial\Omega^0$
$\rho^0 \frac{\partial\psi^0}{\partial F_e} N^0 = \bar{p}^0, \forall\, X^0 \in \partial\Omega^0.$

These are the same balance equations in eq:smc but expressed in the reference domain. Similar manipulations permit in turn to transport equations in eq:smc onto the current configuration.

2.4.2 Non-variational approach

We will assume here that the incompatibility of the relaxed configuration does not allow us to assume the existence of the displacements fields u_e and u_a. In this case, the tensors F_a and F_e are not gradients. However, we will show that similar equations to those in eq:smc are recovered. Instead of deriving them from a functional minimisation, the point of departure will be the virtual power principle. In the present context, the set of admissible velocities includes a set of vector field v and a tensor field V, which allows us to express the total internal and external virtual power as,

$\dot{\mathcal{W}}^{int} = \int_\Omega \left(P_a : V + P_e : \frac{\partial v}{\partial X} \right) dV$
$\dot{\mathcal{W}}^{ext} = \int_\Omega \bar{P}_a : V + \int_{\Omega_t} \bar{p} \cdot v ds$

The identity

$$\dot{\mathcal{W}}^{int} = \dot{\mathcal{W}}^{int}, \quad \forall\, v, V$$

gives rise to the following set of equations: $\mathrm{P}_a = \bar{P}_a, \forall\, X \in \Omega$,
$\nabla_X \cdot P_e = 0, \forall\, X \in \Omega$ which are complemented with the corresponding boundary conditions: $\mathrm{P}_a N = \bar{P}_a N, \forall\, X \in \partial\Omega$
$J_e^{-1} P_e F_e^{\mathrm{T}} n = \bar{p}, \forall\, x \in \partial\Omega_t.$

The reader will recognise that while equation eq:NAp is identical to eq:smcb, some differences exists between equation eq:NAa and the one in eq:smca. We here point out two of them: the former does not include the divergence operator, and more importantly, after using the constitutive condition in clawa, it can be seen that eq:NAa is a dynamical equilbirum equation, due to the presence of the term P_a^+ (as yet undefined) in the expression of P_a in clawa. In this regard, equation eq:smca may be turned into an equation of motion if the energy functional depends on $\dot{F}_a$, in which case the (dynamical) equations of motion, equivalent to those in eq:NAa, are obtained using Hamilton's principle.

Despite the fact that the non-variational approach may represent more general situations, we omit here any dependence of ψ on $\dot{F}_a$, and we will use the (static) equilibrium equations in eq:smc. The use of the virtual power principle for growing processes has been already employed in [DQ02, AG07]. In these works, the tensor of active forces P_a is interpreted as an *accretive* force, responsible of the active deformation. In the present chapter, this force is induced by the dependence of the minimised function ψ on F_a. In other works dealing with growth and remodelling processes (see for instance [EM00, HKMS05]), the internal energy $\rho\psi$ is considered as a function of the coordinates X, and with a non-constant density ρ^0, or a function of other parameters such as fibre reorentiation. In this case, the variation of these parameters is the factor determining the shape or structural changes.

2.5 Constrained deformed configuration

We will here derive the equilibrium equations for a body whose whole deformation is fixed, i.e. $\delta u = 0$, but with variable active displacements and elastic displacements. The equilibrium equations will be obtained resorting to the methodology followed in the variational approach.

We first note that the condition $\delta u = 0$ implies $\delta F = \frac{\partial \delta u}{\partial X^0} = 0$, which in turn yields the following relation between δF_e and δF_a: $\delta F_e = -F_e \delta F_a F_a^{-1} = -F_e \frac{\partial \delta u_a}{\partial X}$.

Inserting the second equality into equation eq:EqRelCont, and using $\delta F = 0$ in eq:VarRefCont, we arrive to the following expressions of $\delta\Pi$, solely as a function of δu_a:
$\delta\Pi = \int_\Omega \rho^0 \left(\psi I + \frac{\partial \psi}{\partial F_a} F_a^{\mathrm{T}} - F_e^{\mathrm{T}} \frac{\partial \psi}{\partial F_e}\right) : \frac{\partial \delta u_a}{\partial X} dV - \int_{\Omega^0} \bar{P}_a^0 : \frac{\partial \delta u_a}{\partial X^0}\, dV^0 = 0,$
$\delta\Pi = \int_{\Omega^0} \rho^0 \left(\frac{\partial \psi^0}{\partial F_a} - F_e^{\mathrm{T}} \frac{\partial \psi^0}{\partial F_e} F_a^{-\mathrm{T}}\right) : \frac{\partial \delta u_a}{\partial X^0} dV^0 - \int_{\Omega^0} \bar{P}_a^0 : \frac{\partial \delta u_a}{\partial X^0}\, dV^0 = 0.$

It is not difficult to see that, after integrating by parts the terms in the parenthesis and recalling that $P_e = \rho^0 \frac{\partial \psi}{\partial F_e}$, the two expressions above yield the material equilibrium equations in eq:EqRelCont and eq:EqRefCont, respectively. Therefore, as a result, it turns out that the material equilibrium equations are the balance conditions that any motion must satisfy when the current configuration is fully constrained ($\delta u = 0$), but the relaxed configuration is changing.

2.6 Stress dependent active deformations

Motivated by experimental observations, we here postulate that the active displacements u_a depend on the elastic displacements u_e via the corresponding deformation gradients, F_a and F_e. We write this dependence by introducing a *control function* that relates the active and elastic deformations as follows:

$$\dot{F}_a = \alpha(F_e). \tag{10}$$

The function α is as yet undefined, and some explicit expressions will be given in Section 3. This equation couples both components of the decomposition of the deformation gradient, in a similar manner as the virtual displacements were coupled in the derivations of the material equilibrium equations in Section 2.5. This coupling has the following two main implications:

1. The evolution law for F_a induces an expression of the gradient of the active velocity L_a, which inserted into condition clawa, will in turn determine the amount of dissipated energy $\mathcal{D} = c\|\alpha F_a^{-1}\|$, and the constitutive law of the internal active stresses P_a.

2. Spatial and material equilibrium cannot be treated independently. Due to the coupling introduced in aclaw, both material and spatial equilibrium must be combined into a

single equilibrium equation. This dependence is treated in a similar manner as the constraint $\delta u = 0$ introduced in the previous section.

Regarding point 1 above, we note that due to equation aclaw, the velocity gradient is now given by $L_a = \dot{F}_a F_a^{-1} = \alpha(F_e)F_a^{-1}$. By inserting this expression into the mass production rate Γ in mb0 and the the internal stress tensor P_e in clawa, we may write them as, $\Gamma = trace(\alpha F_a^{-1})$,
$P_a = \rho^0 \left(\psi I + \frac{\partial \psi}{\partial F_a} F_a^T\right) - F_e^{\mathrm{T}} P_e + c\alpha F_a^{-1}$.
With regard to point 2 stated above, it is convenient to discretise in time equation aclaw. We perform this task by introducing the following time-stepping,

$$F_{a,n+1} = F_{a,n} + \Delta t \alpha(F_{e,n+\theta}),$$

where $0 \geq \theta \geq 1$, $(\bullet)_{n+\theta} = (1-\theta)(\bullet)_n + \theta(\bullet)_{n+1}$, and $\Delta t = t_{n+1} - t_n$. The values $\theta = 0, 1/2, 1$ correspond to the standard forward Euler, mid-point rule, or backward Euler, respectively. For the general time-stepping, we obtain the following relationship between the virtual active displacements δu_a and δu_e: $\delta F_a = \theta \Delta t \nabla \alpha(u_{e,n+\theta}) : ((\delta F)F_a^{-1} - F_e(\delta F_a)F_a^{-1})$, where $[\nabla \alpha]_{ijkl} = \frac{\partial [\alpha]_{ij}}{\partial [F_e]_{kl}}$. This relation may be written as,

$$\delta F_a = (\mathcal{I} + F_e^{\mathrm{T}} \mathcal{K})^{-1} \mathcal{K} : \delta F = \mathcal{L} : \delta F, \tag{11}$$

whenever the fourth-order tensor $(\mathcal{I} + F_e^{\mathrm{T}} \mathcal{K})$ is invertible. $\mathcal{I}_{ijkl} = \delta_{ij}\delta_{kl}$ is the fourth-order identity tensor, and $\mathcal{K} = \theta \Delta t \nabla \alpha F_a^{-\mathrm{T}}$ is given, in indicial notation, as $[\mathcal{K}]_{ijkl} = [\nabla \alpha]_{ijkm} [F_a]_{lm}$, whereas the product $F_e^{\mathrm{T}} \mathcal{K}$ denotes $[F_e^{\mathrm{T}} \mathcal{K}]_{ijkl} = [F_e]_{mk} [\nabla \alpha]_{ijml}$. The relationship in ts may be interpreted as a constriction on the motion, in a similar manner to equation eq:Constr in the constrained case in Section 2.5. Inserting relation ts into equation eq:EqRelCont, the following equilibrium equation can be derived, $\nabla_X \cdot \left(P_e + \left(\rho^0 \frac{\partial \psi}{\partial F_a} F_a^{\mathrm{T}} + \rho^0 \psi I - F_e^{\mathrm{T}} P_e\right) F_a^{-\mathrm{T}} : \mathcal{L} F_a^{\mathrm{T}}\right) = \nabla_X \cdot \bar{P}_a, \forall\, X \in \Omega$, which should be complemented with the corresponding boundary conditions. In the next Section we will apply the same ideas described so far to a system of unidimensional trusses. It will be seen that in this case, some of the terms in eq:SDep can be simplified.

3 Mechanics of Trusses

We will here particularise the growing process described in the previous sections to a system of trusses. Each truss can undergo an active elongation process and an elastic deformation. The active elongation process is motivated by the sliding of the myosin heads onto the actin helix, whereas the elastic deformation is the result of applying the external forces $\bar{q}$, as has been depicted in Figure 2.

3.1 Truss kinematics

The truss elements employed in our model are defined by the following kinematical assumptions:

- Each truss is a body much longer in one direction that in the other 2 perpendicular directions.

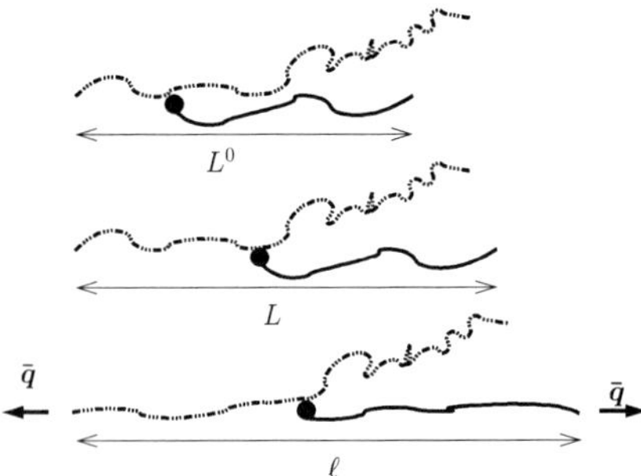

Figure 2: Representation of the three configurations for an actin-myosin complex. The reference, relaxed and current lengths are denoted by L^0, L and ℓ.

- The truss will remain as a straight body in all the configurations, and with constant area.
- The cross section has a constant area A, which remains perpendicular to the centroid axis.
- The truss in the reference configuration is oriented in such a way that its long axis is parallel to the vector E_1 of the reference triad.
- The active deformations deform the truss only in the longitudinal direction, that is in E_1. Also, and in agreement with our hypothesis in the continuum case, no density changes occur during this active deformation.
- The elastic deformations correspond to a change in the longitudinal axis, that is also in E_1, plus a rotation R, constant for each trus.

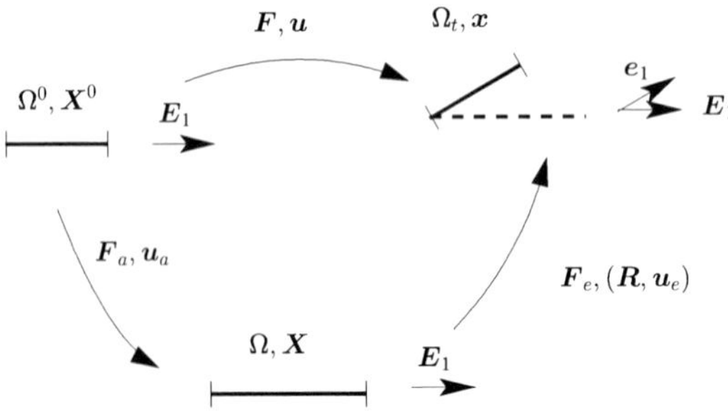

Figure 3: Maps of each truss.

Accordingly, in the present case, the motion of the truss can be described with the maps indicated in Figure 3. The positions in each configuration are given by, $\mathrm{X}^0 = X_i^0 E_i$
$X = X_i E_i$
$x = x_i e_i = x_i R E_i$ where matrix $R \in SO(3)$ represents a rotation that transforms the vector E_i into e_i, i.e. $R = e_i \otimes E_i$ and $e_i = RE_i$. From the kinematic assumptions, the three configurations are related through the following relations: $\mathrm{X}=\mathrm{X}^0 + u_a(X_1^0)E_1 = (X_1^0 + u_a(X_1^0))E_1 + X_2^0 E_2 + X_3^0 E_3$

$x = R(X + u_e(X_1)E_1) = (X_1 + u_e(X_1))e_1 + X_2e_2 + X_3e_3$
$= (X_1^0 + u_a(X_1^0) + u_e(X_1))e_1 + X_2^0e_2 + X_3^0e_3$, where the scalar functions $u_a(X_1^0)$ and $u_e(X_1)$ determine the deformations in the relaxed and current configurations, respectively. We will also use the notation $u = uE_1$, $u_e = u_eE_1$ and $u_a = u_aE_1$.

We denote by $[0, L^0]$, $[0, L]$ and $[0, \ell]$ the domains of X_1^0, X_1 and x_1, respectively. The values L^0, L and ℓ correspond to the lengths of the trusses in the reference, relaxed and deformed configurations, respectively. The expressions of the deformation gradients then are expressible as:

$\mathrm{F}_a = \frac{\partial X}{\partial X^0} = (1 + u_a')E_1 \otimes E_1 + E_2 \otimes E_2 + E_3 \otimes E_3$
$F_e = \frac{\partial x}{\partial X} = (1 + u_e')e_1 \otimes E_1 + e_2 \otimes E_2 + e_3 \otimes E_3$
$F = \frac{\partial x}{\partial X^0} = (1 + u')e_1 \otimes E_1 + e_2 \otimes E_2 + e_3 \otimes E_3.$

In the last equation we have introduced a scalar function $u(X_1^0)$ that, after comparing the last equations in $t_kineandfet, isdefinedbythefollowingequality$:$1 + u' = 1 + u_a' + u_e' \frac{\partial X_1}{\partial X_1^0} = 1 + u_a' + u_e'u_a' = (1 + u_a')(1 + u_a')$.(12)

In view of this result and the expression in fet, it is easy to verify that the multiplicative decomposition $F = F_eF_a$ holds. Note that $u_a' = \frac{\partial u_a}{\partial X_1^0}$ and $u' = \frac{\partial u}{\partial X_1^0}$, but $u_e' = \frac{\partial u_e}{\partial X_1}$.

The associated strain and stress measures of the truss are obtained by particularising the expression of the elastic power $\dot{W} = \int_{L^0} P : \dot{F} dX_1^0$, to the kinematic assumptions given above. It is in fact demonstrated in Appendix A that $\dot{W}$ turns into the following expression:

$$\dot{W} = \int_L (1 + u_a')^{-1} q_e \cdot \dot{\gamma} dX_1 \tag{13}$$

where $\gamma = (1 + u')E_1$, $q_e = \int_{A^0} P_1 dA^0$, $q_e = q_e \cdot E_1$ and $\gamma = \gamma \cdot E_1$. Here, the vector P_1 is the tension in the deformed configuration per unit of area A^0 perpendicular to vector E_1 of the reference configuration, and consequently q_e correspond to the axial force in the truss. We note that the stress and strain measures obtained for the trusses are those of the geometrically exact beam theory [Sim85], but with only axial stiffness. We also remark that the obtained strain measure $\dot{\gamma}(1+u_a')^{-1}$ mimics the tensor product $\dot{F}F_a^{-1}$ in the continuum case.

In general, we assume that the free energy of the truss ψ depends on the elastic strain measure $\gamma_e = 1 + u_e'$, and an active deformation $\gamma_a = 1 + u_a'$. The relation between those measures and the deformation $\gamma = 1 + u'$ in wdot is obtained from the relations of the displacement derivatives in tgrel, i.e.

$$\gamma = \gamma_a \gamma_e, \tag{14}$$

which, as mentioned above, is the equivalent to the deformation gradient decomposition $F = F_eF_a$ in the continuum case.

3.2 Unconstrained equilibrium equations

It has been shown in Section 2.4.1, that when allowing independent virtual displacements δu_a and δu_e, the equilibrium equations in eq:EqRelCont are obtained. Using the same methodology, the variation of the energy functional

$$\Pi = \int_L \rho^0 \psi dX_1 - q^0 \cdot u \Big|_{X_1^0=0}^{X_1^0=L^0} - \int_{L^0} \bar{Q}_a^0 \cdot u_a' dX_1^0$$

yields the following equilibrium equations:

$$\partial \overline{\partial X_1\left(\rho^0 \frac{\partial \psi}{\partial \gamma_a}(1+u'_a)+\rho^0\psi E_1-\gamma_e q_e\right)=\frac{\partial}{\partial X_1}\bar{Q}_a,\forall\ X_1\in[0,L]\frac{\partial}{\partial X_1}q_e=0,\forall\ X_1\in[0,L]\rho^0\left(\frac{\partial\psi}{\partial\gamma_a}\gamma_a+\psi E_1-\gamma_e q_e\right)\cdot N=}$$
$\bar{Q}_a \cdot N$, at $X_1 = \{0, L\} q_e \cdot N = \bar{q}^0 \cdot N$, at $X_1 = \{0, L\}$

where in this case $N|_{X_1=L} = N|_{X_1^0=L^0} = E_1 = -N|_{X_1=0} = -N^0|_{X_1^0=0}$. Equations eq:tmatb and eq:tmatd are the standard spatial equations due to the variations δu, which state that the axial load q_e is constant along the truss and equal to the values of the external load applied at the truss ends. We note that in the present uni-dimensional case, the equivalent version of the inverse Piola transformation yields $\bar{Q}_a = (1+u'_a)^{-1}\bar{Q}_a^0\gamma_a = \bar{Q}_a^0$.

When dealing with the system of I trusses and (unloaded) J joints, the local equilibrium equations eq:tmata and eq:tmatb are applied to each truss $i = 1, \ldots, I$, whereas the boundary condition eq:tmatd must be replaced by the static equilibrium at each joint j, $j = 1, \ldots, J$, that is

$$\sum_{i\in j} R^i q_e^{ij} \cdot N^i = \sum_{i\in j} R^i \bar{q}^{0,ij},\ j = 1, \ldots, J \tag{15}$$

where $\bar{q}^{0,ij}$ is the boundary load at joint j of truss i connected to j. Note that this equilibrium equation is a consequence of the fact that all the spatial virtual displacements δu^{ij} of the truss ends connected to a joint j are equal. Instead, the active virtual displacement $\delta u_a ij$ are internal and not shared among different trusses.

Analogously to the continuum case, we can recast equations eq:tmat in the reference configuration as,

$$\partial \overline{\partial X_1^0 \rho^0\left(\frac{\partial\psi^0}{\partial\gamma_a}-\frac{\partial\psi^0}{\partial\gamma_e}\right)=\frac{\partial}{\partial X_1^0}\bar{Q}_a,\forall X_1^0\in[0,L^0]\frac{\partial}{\partial X_1^0}\rho^0\frac{\partial\psi^0}{\partial\gamma_e}(1+u'_a)^{-1}=0,\forall X_1^0\in[0,L^0]\rho^0\left(\frac{\partial\psi^0}{\partial\gamma_e}-\frac{\partial\psi^0}{\partial\gamma_e}\right)\cdot N=}$$
at $X_1^0 = \{0, L^0\}\rho^0\frac{\partial\psi^0}{\partial\gamma_e} \cdot N = \bar{q}^0 \cdot N$, at $X_1^0 = \{0, L^0\}$.

3.3 Stress dependent active displacements

In parallel with equation aclaw, we here hypothesise that the elastic and active displacements are related through,

$$\dot{u}'_a = \alpha(u'_e). \tag{16}$$

where the control function $\alpha : R \to R$ will be assumed to be linear. In parallel to the continuum case, we insert the evolution law in dept into equations eq:Gamma and clawa, which, after noting that $\gamma_a = 1 + u'_a$ is equivalent to J_a, furnishes the following conditions,
$\Gamma = \frac{\dot{u}'_a}{\gamma_a} = \frac{\alpha}{\gamma_a}$
$q_a = \rho^0\left(\psi E_1 + \frac{\partial\psi}{\partial\gamma_a}\gamma_a\right) - \gamma_e q_e + c\frac{\alpha}{\gamma_a}E_1$, where here, and in the following expression the argument of function α is removed. These are the same equations given in eq:CGPa particularised to the truss elements considered here. Also, similarly to the formulation in continuum with stress dependent active deformations, we discretise in time the previous equation, which for a general single step algorithm leads to the following relation between the virtual displacements: $\delta u'_a = \theta\Delta t\alpha'_{n+\theta}(\delta u' - \gamma_{e,n+1}\delta u'_a)\gamma_{a,n+1}{}^{-1}$ or, rearranging

terms,

$$\delta u_a' = \frac{\theta \Delta t \alpha_{n+\theta}'}{\gamma_{a,n+1} + \theta \Delta t \alpha_{n+\theta}' \gamma_{e,n+1}} \delta u' = \frac{\theta \Delta t \alpha_{n+\theta}'}{\gamma_{a,n+1} + \theta \Delta t \alpha_{n+\theta}' \gamma_{e,n+1}} \gamma_a \frac{\partial \delta u}{\partial X_1} \tag{17}$$

where $\alpha_{n+\theta}' = \frac{\partial}{\partial u_e'} \alpha(u_e')|_{t=t_n+\theta\Delta t}$.

From these expressions, the minimisation of Π yields equivalent equations to those in eq:SDep:

$$\partial \frac{}{\partial X_1 \left(q_e + \left(\rho^0 \frac{\partial \psi}{\partial \gamma_a} \gamma_a + \rho^0 \psi E_1 - \gamma_e q_e \right) \frac{\theta \Delta t \alpha_{n+\theta}' \gamma_a}{\gamma_a + \gamma_e \theta \Delta t \alpha_{n+\theta}'} \right)} = \frac{\partial}{\partial X_1} \bar{Q}_a, \forall X_1 \in [0, L]$$

4 Modelling of Cytoskeletal Elements

4.1 Active deformations

While a common agreement in the constitutive law of the elastic deformation exists, different approaches have been pursued for the choice of a plausible constitutive law of the active deformations or the associated growth process. The latter has been formalised differently depending on the physical phenomena being modelled. In the works of [LH02, HKMS05], the growth rate is a function of the second Piola-Kirchoff [LH02] or Kirchoff stresses [HKMS05]. Alternatively, in [AG07, DQ02], the growth rate is induced by accretive forces, which equilibrate the externally supplied forces. In both cases, though, the evolution laws are such that the growth rate of he accretive stresses tend to achieve a stable (homeostatic) value, at which no further active deformation occurs. A similar idea can be found in [RT08, Tab08], based on Beloussov hyperrestoration hypothesis [BSNN94]. This hypothesis states that whenever a change in the elastic stresses is detected, the tissue tends to deform in order to establish the previous stress state. This stress state may be represented by a constant [RT08] or stress dependent [Tab08] target stress.

Regarding the use of evolutionary laws for the active deformations, it is also worth mentioning the works in [RS04] and [KGAG05]. In the first paper, the form of the relaxed deformation is obtained by maximising the mechanical dissipation. In the second reference [KGAG05], the forces produced by a network chain embedded in a cell arise due to the particular form of the Helmholtz free energy associated to the chain. In our case, we will neglect any interaction between the active-myosin chains, other than their connectivity at the chain ends. In addition, we will assume that the active deformation is at constant density (purely an active elongation process) but, in agreement with Figure 2, with a local mass increase due to the introduction of new actin filaments from other parts of the cell. This additional mass increases the elastic energy of the system, but the free energy does not depend explicitly on the active elongation.

4.2 Passive deformations

The passive deformation of the cytoskeleton is taken as purely non-linear elastic with an energy function that mimics the behaviour of a spring. We will henceforth consider the following stored energy function ψ:

$$\psi = \frac{1}{2} k (tr(F_e) - 3)^2 = \frac{1}{2} k (u_e')^2 \tag{18}$$

with k a constant material parameter. We note that the elastic deformation of the trusses is not assumed incompressible, since a variation of the cross-section area of the microfilament as it deforms would seem unrealistic.

4.3 Control function

We will here consider a simple linear control function, with a constant target stress $q_T = q_T E_1$. More specifically, equation dept will be explicitly expressed by,

$$\dot{u}'_a = \beta(q_{e,n+\theta} - q_T) = \beta(\rho^0 k u'_{e,n+\theta} - q_T), \tag{19}$$

where β is a material parameter. The physical interpretation of this law is simple: when $q_e > q_T$, that is the truss is under tension, the truss will elongate in order to diminish the actual elastic stress, and vice-versa. Such a law is represented in Figure 4.

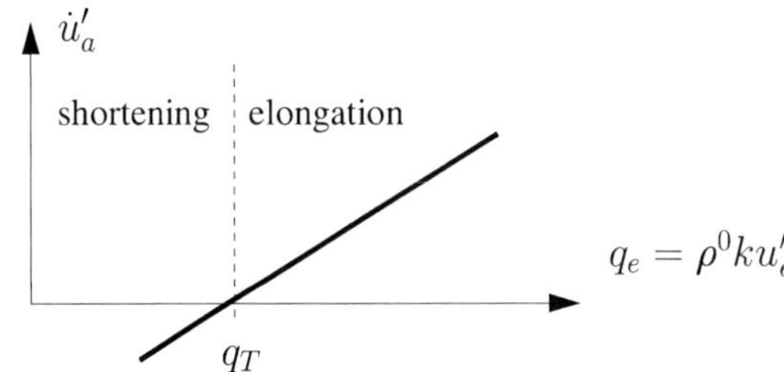

Figure 4: Evolution law for the active deformations.

As a consequence of the linear evolution law in eq:ControlF, we have that $\alpha'_{n+\theta} = \beta\rho^0 k$, which is constant. Moreover, by introducing the following definitions: $\tau = \theta\Delta t\beta\rho^0 k$
$G(u'_e) = \frac{\tau}{\gamma_{a,n+1}+\tau\gamma_{e,n+1}}$, the relation between $\delta u'_a$ and $\delta u'$ in tst, and $\delta u'_e$ and δu, reads in the present case, $\delta u'_a = G\delta u'$
$\delta u'_e = (\delta u' - \gamma_e \delta u'_a)\gamma_a^{-1} = \gamma_a^{-1}(1 - \gamma_e G(u'_e))\delta u'$.

We emphasise that $G(u'_e)$ only depends on the elastic displacements, since the active displacements are related to those through the evolution law eq:ControlF, which after time-discretisation reads, $u'_{a,n+1} = u'_{a,n} + \Delta t\beta(\rho^0 k u'_{e,n+1} - q_T)$.

We will henceforth denote by G the function $G(u'_{e,n+1})$, and omit the arguments for clarity. Resorting to the relations in eq:tstl, the variation of $G(u'_e)$ is obtained as, $\delta G = -\frac{G^2}{\tau}(\tau\delta u'_e + \delta u'_a) = G^2\left(\gamma_e G - 1 - \frac{\gamma_a G}{\tau}\right)\gamma_a^{-1}\delta u'$

4.4 Equilibrium equations

As a particular case, we will consider the energy density functions $\psi + \psi_{inc}$ described above, and a situation where no external active stresses exists, that is $\frac{\partial\psi}{\partial\gamma_a} = 0$ and $\bar{Q}_a = 0$. In this case, after noting that $C' = \beta\rho^0 k$, the equation in eq:tmdX turns into,

$\frac{\partial}{\partial X_1\left(ku'_{e,n+1}-(u'_{e,n+1}+\frac{1}{2}k(u'_{e,n+1})^2)G_{n+1}\gamma_{a,n+1}\right)=0,\forall X_1\in[0,L]\left(ku'_{e,n+1}-(u'_{e,n+1}+\frac{1}{2}k(u'_{e,n+1})^2)G_{n+1}\right)=}$
$\bar{q}\cdot E_1, X_1 = \{0, L\}$.

The local equation in eq:EqTra differs substantially from the usual spatial equilibrium for trusses, that is,

$$ku_e'' = 0.$$

After comparing this equation and eq:EqTr, it can be concluded that the active elongation process, together with the evolution law in eq:ControlF, are equivalent to an additional body force given by the expression underlined in eq:EqTra.

4.5 Implementation

The equilibrium equations in eq:EqTr is solved at each time t resorting to the finite element method. We therefore multiply the local equation in eq:EqTra along direction e_1 for each truss i by a test function (or virtual displacement) $w = we_1$, and integrate over the domain of each truss, which yields $\int_L w \cdot e_1 \frac{\partial}{\partial X_1}\left(ku'_{e,n+1} - \left(u_{e,n+1} + \frac{1}{2}k(u'_{e,n+1})^2\right) G_{n+1}\gamma_{a,n+1}\right) dL = 0, \forall w \in \mathcal{H}^1$, where $\mathcal{H}^1$ is the Hilbert space of functions w whose L_2 norm, and the L_2 norm of the first derivatives are bounded. Due to the use of a Newton-Raphson iterative procedure, which requires the linearisation of the resulting equations, it will be convenient to modify the previous expression as an integral along the reference length L^0. After integrating by parts, making use of eq:EqTrb, and using the finite element interpolation $w \approx w_i N_i(X_1^0)$, with $N_i(X_1^0)$ a set of complete functions and w_i the set of arbitrary nodal values of the test functions $w(X_1^0)$, we obtain the following system of non-linear equations:

$$g = \bar{g}. \tag{20}$$

The component i, associated to node i in the vectors g and $\bar{g}$ are respectively given by, $g_i = \int_{L^0} N_i' \left(ku_e' - \left(u_e' + \frac{1}{2}k(u_e')^2\right) G\gamma_a\right) e_1 dL^0$
$\bar{g}_i = \bar{q}$, where here and in the remaining derivations we omit the subscript $n+1$. In the present implementation, we used linear interpolating functions $N_i(X_1^0)$, which are also used to interpolate the unknown current positions $x(X_1^0) = N_i(X_1^0)x^i$. As a consequence, the vectors u' and $\delta u'$ are elementwise constant and equal to: 1+u' =x' $\cdot e_1 = \frac{x^2 - x^1}{L^0} \cdot e_1 = \frac{\ell e_1}{L^0} \cdot e_1 = \frac{\ell}{L^0}$,
$\delta u' e_1 = \frac{\delta \ell}{L^0} e_1 = e_1 \otimes e_1 \delta x'$ where x^1 and x^2 are the nodal current positions. The non-linear system of equations is computed resorting to a Newton-Raphson iterative process, which requires the computation of the Jacobian matrix A. It can be verified that, by using relations eq:tstl and eq:dG, the component A_{ij}, corresponding to the contribution of nodes i and j, is given by, $A_{ij} = \int_{L^0} N_i' N_j' (c_1 + c_2) e_1 \otimes e_1 \, dL^0 + \int_{L^0} N_i' N_j' c_3 \frac{L^0}{\ell} M \, dL^0$ with $M = I - e_1 \otimes e_1$, and $c_0 = u_e' + \frac{1}{2}k(u_e')^2$, $c_1 = (k\gamma_a^{-1} - (1 + ku_e')G)(1 - \gamma_e G)$,
$c_2 = c_0 G^3 \left(\gamma_e - \frac{\gamma_a}{\tau}\right)$, $c_3 = ku_e' - c_0 G\gamma_a$.

5 Results

We will model the behaviour of a single truss in two situations: (i) with one end constrained and one free, and (ii) with the two ends constrained. We remark that the first situation corresponds to the application of Neumann conditions, that is to apply a prescribed stress, zero in this case, at one end. The second, corresponds to a full Dirichlet condition, that is to prescribe the displacements. The material parameters, and those of the evolution law and the time-stepping are indicated in Table 5. According to the value of q_T, the active elongation process will stop when the internal stress $q_e = -1$, that is under a compression state. While this situation may be reached for the fully constrained truss, the stress of the free truss is always zero, and thus the truss elongates continuously, and a constant growth rate. Figure 5 shows the current length ℓ and the relaxed length L for the two situations. As expected, the constrained truss converges towards a homoeostatic state ($\dot{u}_a' = 0$), while the free truss elongates indefinitely. Consequently, for the evolution law considered here, it is necessary to use displacement constraints give rise to stable static configurations. In systems with multiple chains, this constraints may be replaced by the existence of some cross-links.

Table 1: Material and time-integration parameters, with T_{end} the final time.

k	ρ^0	q_T	θ	Δt	T_{end}
1	1	-1	0.5	0.05	2

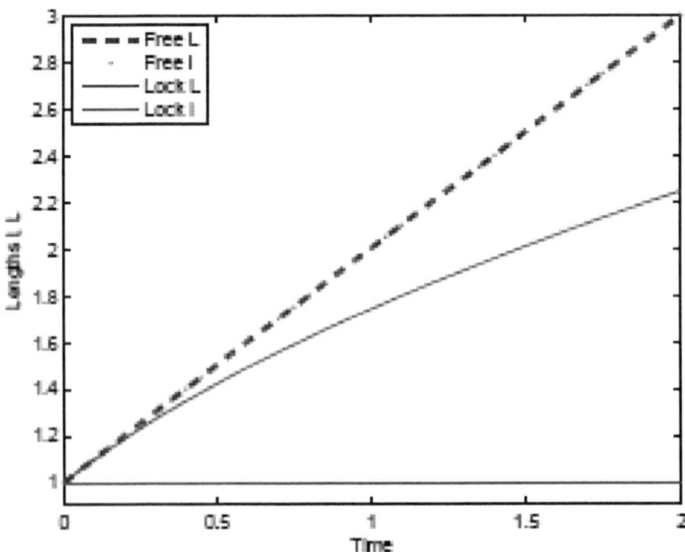

Figure 5: Current length ℓ and relaxed lengths L for a truss with two ends constrained (Lock) and one end constrained (Free).

We point out that the trend of the active and elastic deformation resembles a simplified model previously developed by the authors [CMM08b]. In the present case, though, the behaviour of the truss is supported by a thermomechanically consistent model.

6 Conclusion

This paper has described the necessary ingredients to develop an elastic dependent active elongation process, which is able to reproduce active movements observed in morphogenesis. The theory has been described in continua, and applied to truss elements. The reason for this is twofold: the equilibrium and evolution equations are significantly simplified, and it is easier to retrieve experimental data from the unidimensional elements of the cytoskeleton than from the general cell.

Although the theory presented here has been applied to a simple truss, other approaches pursued by the authors [CMM08a], but with a similar stress dependent evolutionary law, have shown that the response of the trusses that form an embryo cross-section can reproduce the patterns observed during morphogenesis.

A Strain and Stress Measures for Trusses

Let us express the associated first Piola-Kirchoff stress as $P = P_i \otimes E_i$, where $P_i = P_i e_i$ (no summation on i) is the tension in the deformed configuration per unit of area perpendicular to the E_i direction of the reference configuration. The time differentiation of the deformation gradient F is, according to relations in $t_k ine, equalto, \dot{F} = \hat{\omega} e_i \otimes E_i + \dot{u}' e_1 \otimes E_1 + u' \hat{\omega} e_1 \otimes E_1 (21) where$ $\omega^{\mathrm{T}} = (\omega_1, \omega_2, \omega_3)$ is the angular velocity and the symbol $\hat{\omega}$ denotes the skew symmetric matrix $\hat{\omega} = \begin{bmatrix} 0 & -\omega_3 & \omega 2 \\ \omega_3 & 0 & -\omega_1 \\ -\omega_2 & \omega_1 & 0 \end{bmatrix}$, such that $\hat{u} v = -\hat{v} u = u \times v$. With this notation, and the expression in fdot, the tensor product $P : \dot{F}$ results in,

$$P : \dot{F} = -tr((P_i \otimes e_i)\hat{\omega}) - u' tr((P_1 \otimes e_1)\hat{\omega}) + \dot{u}' P_1 \cdot e_1 = -\omega \cdot (\hat{P}_i e_i + u' \hat{P}_1 e_1) + \dot{u}' P_1 \cdot e_1. \tag{22}$$

To proceed further, we note that the ij component of $PF^{\mathrm{T}} - FP^{\mathrm{T}}$ is given by,

$$[PF^{\mathrm{T}} - FP^{\mathrm{T}}]_{ij} = \sum_k \left([P_k]_i \frac{\partial x_j}{\partial X_k} - \frac{\partial x_i}{\partial X_k} [P_k]_j \right) = \sum_k \epsilon_{ijl} \left[\hat{P}_k \frac{\partial x}{\partial X_k} \right]_l$$

with ϵ_{ijl} the permutation index. Since $PF^T - FP^T = 0$, it follows that $\sum_k \hat{P}_k \frac{\partial x}{\partial X_k} = 0$, which for the kinematics of the truss described here yields,

$$\hat{P}_i e_i + u' \hat{P}_1 e_1 = 0.$$

Using this relation in equation pfdot, and denoting by $q_q = \int_{A^0} P_1 dA^0$, we have that the stress power $\dot{W} = \int_{V^0} P : \dot{F}\, dV^0$ reduces to,

$$\dot{W} = \int_{L^0} \dot{u}' e_1 \cdot q_e dX_1^0 = \int_{L^0} \dot{u}' E_1 \cdot q_e dX_1^0,$$

where $q_e = R^T q_e$, with R the rotation matrix introduced in Section 3.1. Since $u'|_{t=0} = 0$, it follows from this equation that we can then define $\gamma = (1 + u')E_1$ as the strain measure of the truss, conjugate to the stress measure q_e, and then rewrite the stress power as:

$$\dot{W} = \int_{L^0} \dot{\gamma} \cdot q_e dX_1^0 = \int_L (1 + u_a')^{-1} \dot{\gamma} \cdot q_e dX_1, \tag{23}$$

where, after assuming $u'_a + 1 > 0$, we have made use of the identity $dX_1 = |1 + u'_a| dX_1^0 = (1 + u'_a) dX_1$.

References

[AG07] D Ambrosi and F Guana. Stress-Modulated Growth. *Mathem. Mech. Solids*, 12:319–342, 2007.

[BA00] M C Boyce and E M Arruda. Constitutive models of rubber elasticity: a review. *Rubber Chem. Technol.*, 73:504–523, 2000.

[BSF04] E Brouzés, W Supatto, and E Farge. Is mechano-sensitive expression of twist involved in mesoderm formation? *Biol. Cell*, 96:471–477, 2004.

[BSNN94] L V Beloussov, S V Saveliev, II Naumidi, and V V Novoselov. Mechanical stresses in embryonic tissues: patterns, morphogenetic role, and involvement in regulatory feedback. . *Int. Rev. Cytol.*, 150:1–34, 1994.

[CMM08a] V Conte, J J Muñoz, and M Miodownik. 3D finite element model of ventral furrow invagination in the Drosophila melanogaster embryo. *J. Mech. Behav. Biomed. Mater.*, 2:188–198, 2008.

[CMM08b] V Conte, J J Munoz, and M Miodownik. Stress Controlled Analysis of Morphogenesis. In *8th. World Congress on Computational Mechanics (WCCM8)*, Venice, Italy, June30-July 5 2008.

[DQ02] A DiCarlo and S Quiligotti. Growth and balance. *Mech. Res. Comm.*, 29:449–456, 2002.

[EM00] M Epstein and G A Maugin. Thermomechanics of volumetric growth in uniform bodies . *Int. J. Plast.*, 16:951–978, 2000.

[Far03] E Farge. Mechanical Induction of Twist in the Drosophila Foregut/Stomodeal Primordium . *Current Biol.*, 13:1365–1377, 2003.

[GON+06] K Garikipati, J E Olberding, H Narayanan, E M Arruda, K Grosh, and S Calve. Biological remodelling: Stationary energy, configurational change, internal variables and dissipation. *J. Mech. Phys. Solids.*, 54:1493–1515, 2006. arXiv:q-bio/0506023v2.

[GS08] O Gonzalez and A M Stuart. *A First Course in Continuum Mechanics*. Cambridge Univ. Press, 2008.

[HKMS05] G Himpel, E Kuhl, A Menzel, and P Steinmann. Computational modelling of isotropic multiplicative growth. *Comp. Mod. Eng. Sci.*, 8:119–134, 2005.

[Hum01] J D Humphrey. *Cardiovascular Solid Mechanics*. Springer, Berlin, 2001.

[Ing97] D E Ingber. Tensegrity: the architectural basis of cellular mechanotransduction. *Annu. Rev. Physiol.*, 59:575–599, 1997.

[KGAG05] E Kuhl, K Garikipati, E M Arruda, and K Grosh. Remodeling of biological tissue: Mechanically induced reorientation of a transversely isotropic chain network. *J. Mech. Phys. Solids.*, 53:1552–1573, 2005.

[KH00] R Kienzler and G Herrmann. *Mechanics in Material Space.* Springer, 2000.

[KS04] E Kuhl and P Steinmann. Material forces in open systems mechanics. *Comp. Meth. Appl. Mech. Engng.*, 193:2357–2381, 2004.

[LH02] V A Lubarda and A Hoger. On the mechanics of solids with a growing mass. *Int. J. Solids Struct.*, 39:4627–4664, 2002.

[Lub04] V A Lubarda. Constitutive theories based on the multiplicative decomposition of deformation gradient: Thermoelasticity, elastoplasticity, and biomechanics. *Appl. Mech. Rev.*, 57:95–108, 2004.

[MB08] G A Maugin and A Berezovski. Introduction to the thermodynamics of configurational forces. *Atti Accad. Pelorit. Peric.*, LXXXVI(Supl. 1), 2008. DOI:10.1478/C1S0801016.

[MBM07] J J Muñoz, K Barrett, and M Miodownik. A deformation gradient decomposition method for the analysis of the mechanics of morphogenesis. *J. Biomechanics*, pages 1372–1380, 2007.

[MGL04] C Miehe, S Göktepe, and F Lulei. Progressive delamination using interface elements. *J. Mech. Phys. Solids.*, 52:2617–2660, 2004.

[MTSM07] D Mizuno, C Tardin, C F Schmidt, and F C MacKintosh. Nonequilibrium Mechanics of Active Cytoskeletal Networks. *Science*, 315:370–373, 2007.

[RHM94] E K Rodriguez, A Hoger, and A D McCulloch. Stress-dependent finite growth in soft elastic tissues. *J. Biomechanics*, 27:455–467, 1994.

[RS04] K R Rajagopal and A R Srinivasa. On the thermomechanics of materials that have multiple natural configurations. Part I: Viscoelasticity and classical plasticity. *J. Appl. Math. Phys. (ZAMP)*, 55:861–893, 2004.

[RT08] A Ramasubramanian and L A Taber. Computational modeling of morphogenesis regulated by mechanical feedback. *Biomech. Model. Mechanobiol.*, 7:77–91, 2008.

[Shr05] B I Shraiman. Mechanical feedback as a possible regulator of tissue growth. *Proc. Nat. Acad. Sci. USA*, 102(9):3318–23, 2005.

[Sim85] J C Simo. A finite strain beam formulation. The three dimensional dynamic problem. Part I. *Comp. Meth. Appl. Mech. Engng.*, 49:55–70, 1985.

[Tab95] L A Taber. Biomechanics of growth, remodeling, and morphogenesis. *Appl. Mech. Rev.*, 48(8):487–545, 1995.

In: Continuum Mechanics ISBN: 978-1-60741-585-5
Editor: Andrus Koppel and Jaak Oja, pp. 245-272

Chapter 9

NEAREST-NODES FINITE ELEMENT METHOD

Yunhua Luo*
Department of Mechanical & Manufacturing Engineering,
University of Manitoba, Winnipeg, Canada

Abstract

In the nearest-nodes finite element method (NN-FEM) [1, 2, 3], finite elements are mainly used for numerical integration; for each quadrature point, shape functions are constructed from a set of nodes that are the nearest to the quadrature point, nodes from neighbour elements may be involved in the construction. Based on this strategy, there are several techniques available for constructing shape functions. In this paper, the moving local polynomial interpolation method is adopted. Benefiting from the above strategy, NN-FEM has several attractive features. High-order shape functions can be constructed from simplex finite element meshes; Analysis accuracy of NN-FEM is not influenced by element distortion [4]; NN-FEM can deal with extremely large deformation [2], etc. Furthermore, NN-FEM provides a favourable environment for implementing an adaptive algorithm.

Keywords: Nearest-Nodes Finite Element Method, Meshless Methods, Finite Element Method, Moving Local Polynomial Interpolation

1 Introduction

The classical finite element method (FEM), since its appearance around the middle of last century [5, 6, 7], has been a great success and the prevalent numerical tool for solving scientific and engineering problems. The FEM has lots of advantages compared with other numerical methods. Compared with the newly emerged meshless or meshfree methods, one big advantage of the FEM is that it does not need extra time to construct shape functions, as the shape functions for a specific element type are pre-defined. Therefore, the FEM is computationally more efficient. Although there are still issues related to finite element

*E-mail address: luoy@cc.umanitoba.ca

meshes and their generation, it is with a mesh that the FEM can give a better approximation to a complex problem domain than any other numerical methods. It also is crucial to improve accuracy of numerical solutions. Now even for a very complex geometric domain, the geometry-based adaptive mesh generation algorithm [8, 9] can generate high quality unstructured finite element meshes. A finite element mesh can approximate a complex geometric domain to any desired accuracy, cf Fig. 1. The dilemma is that although high quality

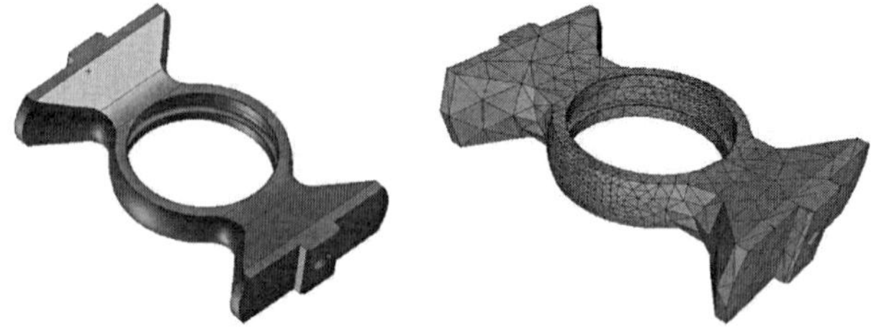

Figure 1: The geometric model of a bearing and its finite element mesh

meshes consisting of triangular elements or tetrahedral elements can be generated, the low order elements based on a triangle or a tetrahedron have a slow convergence. On the other hand, higher order elements with edge or interior nodes have a faster convergence, but the corresponding meshes are hard to handle, especially if mesh adaptation is involved. In the FEM, we already have a very large library of elements, linear elements, 2-D elements and 3-D elements; triangular/tetrahedral elements, quadrilateral/hexahedral elements; various high order elements and hybrid meshes; and new elements are being continuously added in. No single mesh generator can efficiently

handle all these element types in generating high quality meshes. For more and more stringent accuracy requirement, and more and more complex geometry, the solution of a complex problem may need millions and even billions of elements [10]. The existence of a single invalid or ill-conditioned element may ruin the whole solution, or at best compromise the accuracy of the solution. As reviewed in [11], although generating a valid mesh is not a big deal, tuning the quality of a mesh to make all elements in ideal shapes is very time consuming, as it usually involves an optimization process. In certain applications like shape optimization, crash analysis, metal forming, fluid flow analysis, and large deformation analysis, in spite of an ideal starting mesh, the quality of elements could deteriorate, causing severely distorted elements. In extreme cases, some elements become degenerate and further progress of analysis is meaningless or even impossible without taking care of those distorted or degenerate elements. There are basically two ways to eliminate unwanted elements: one is by local mesh modification [10], the other is remeshing. The former needs a robust algorithm workable for all types of elements, which is still not available; The latter, as well known, is very time-consuming. Element distortion has been a major obstacle for the classical FEM to efficiently solve the mentioned engineering problems.

Aiming at resolving issues arising from using a finite element mesh, a new category of methods, collectively called meshless or meshfree methods, e. g. [12, 13, 14, 15, 16, 17, 18, 19, 20, 21, 22, 23, 24, 25, 26, 27] among many others, have been developed in recent years. These methods are gaining more and more attention for some of their advantages over the classical FEM. One big advantage of meshless methods is their flexibility in constructing shape functions. Most of the developed meshless methods differ from each other mainly by their ways of constructing shape functions. The following is an incomplete list of meshless methods appearing in the literature, more new methods are being reported:

- Element-Free Galerkin [14, 25, 17];
- Partition of unit [12, 26];
- Reproducing kernel [23, 18, 24, 22];
- Natural element [28];
- *hp*-meshless cloud method [19];
- Smooth particle hydrodynamics [15];
- Finite point method [29].

Compared with the classical FEM, in a meshless method, cells for numerical integration and nodes for constructing shape functions are completely independent. Connectivity between nodes is not defined or loosely defined. Meshless methods are thus more attractive in dealing with geometric discontinuities, moderate deformation, etc. Nevertheless, meshless methods also have their own disadvantages, and most of these disadvantages are related to 'meshless' — no connectivity between nodes. In a meshless method, adjacency information between nodes is needed in constructing shape functions and in adaptation, but it is not defined and stored. Although greater flexibility in constructing shape functions is obtained, it is at the price of non-trivial extra computational time spent in looking for neighbor nodes. The amount of extra computational time is even not bounded within a linear order of total nodal number. To alleviate the above problem, in some implementation of meshless methods, connection between nodes is established with the aid of a data organizer such as quadtree or octree. With this implementation, space is needed to store information of nodal adjacency, and the implemented meshless method is actually degenerated and retreats to the classical FEM. There are some other issues in meshless methods to be resolved. For example, how to optimally locate nodes onto the boundary of a complex geometric domain to reduce error from geometry approximation is still an open question. As the nodes and cells are independent of each other, another issue is how to make the size of integration cells consistent with the density of node distribution. In simulating large deformation such as metal forming, two material particles that are near each other at one time instant may be far apart from each other at the next time instant. Therefore, meshless methods have the same issue as in the FEM, they need to update adjacency information between nodes from time to time, to construct high-quality shape functions. One typical scenario is given in Fig. 2. Obviously, after experiencing a large deformation, the nodes in the deformed influence domain are no longer the best ones for constructing high quality shape functions.

Therefore, both of the FEM and meshless methods have their pros and cons. One current trend is to develop new numerical methods, e.g. [30, 31, 16, 32, 33, 34, 35, 36, 37, 38, 39, 40] among others, that can combine the advantages from the FEM and meshless methods, and at the same time avoid their disadvantages. Indeed, as long as a numerical method

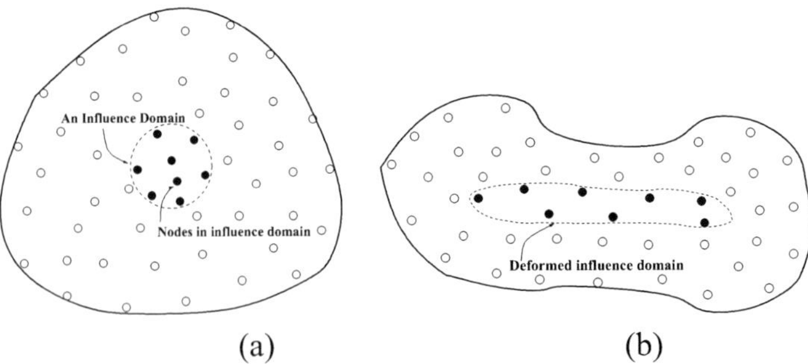

Figure 2: Effect of deformation of influence domain. (a) initial nodal configuration and an influence domain; (b) deformed configuration

is based on an integral formulation or a weak form, numerical integration is inevitable and some kind of integration cell is needed. One great success of meshless methods is the great flexibility of constructing shape functions. In the classical FEM, a finite element is used both for constructing shape functions and for numerical integration. Actually this is not necessary. In this paper, the classical FEM and newly emerged meshless methods are combined in such a way: finite elements are only used for numerical integration; while shape functions are constructed in a similar way as in meshless methods, i.e. by using a number of nodes that are the nearest to a concerned quadrature point, and some of the nodes may be from adjacent elements. The proposed Finite Element Method is thus called Nearest-Nodes Finite Element Method (NN-FEM). As will be demonstrated in the following sections by derivation and numerical examples, with the above strategy, NN-FEM inherits the major advantages of the classical FEM and meshless methods.

The layout of this paper is given in the following: in Section 2, the proposed Nearest-Nodes Finite Element Method is described; Moving local polynomial interpolation (MLPI) method is described in Section 3; Results of numerical investigations are reported in Section 4, followed by concluding remarks in Section 5.

2 Nearest-Nodes Finite Element Method (NN-FEM)

The issues related to element distortion in the classical FEM arise from the restriction that only nodes belonging to an element are used for constructing the shape functions for that element, Fig. 3. One big advantage of this restriction is that with the aid of a local coordinate system and a set of normalized element coordinates, the shape functions of a specific element type can be pre-defined. Therefore, there is no need to solve a local problem. Nevertheless there are also disadvantages arising from this restriction. Two of them are obvious. One is that the number of nodes available from a single element for constructing higher order shape functions is limited. Furthermore the use of edge or interior nodes is inconvenient for mesh generation, element implementation and mesh adaptation. The other even worse disadvantage is that if the shape of an element is distorted, the nodes of that element are also in unfavorable positions for constructing high quality shape functions. To obtain high quality shape functions, all elements in a mesh must be in good shapes.

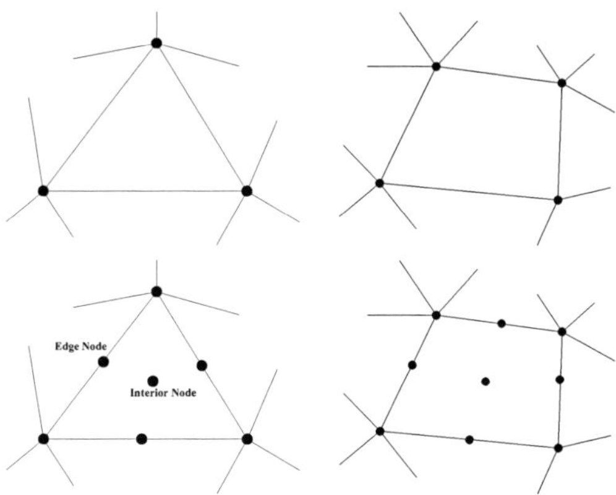

Figure 3: Triangle and quadrilateral elements and their higher-order versions

In the NN-FEM proposed in the following, a finite element is used only for numerical integration, while shape functions are constructed in a similar way as in meshless methods, i. e. by using a set of number of nodes that are the nearest to a concerned quadrature point. In principle, the proposed NN-FEM is applicable to all fields where the classical FEM is workable. In the following, NN-FEM is described in the content of solid mechanics and based on the principle of minimum potential energy.

Let Ω be an open bounded domain, and $\{\Omega_i\ (i=1,2,\cdots,N)\}$ be a subdivision of Ω, representing N finite elements; Ω_i is the i-th subdomain or element; $\Omega_i \cap \Omega_j = \emptyset$ (if $i \neq j$) and $\Omega = \cup_{i=1}^{N}\Omega_i$. The total strain energy stored in Ω is obtained by summing up contributions from all the subdomains,

$$\Pi = \sum_{i=1}^{N} \pi_i \tag{1}$$

where π_i is the strain energy stored in subdomain Ω_i

$$\pi_i = \int_{\Omega_i} e_i(\boldsymbol{x}) d\Omega_i \qquad (i=1,2,\cdots,N) \tag{2}$$

In Equation (2), $e_i(\boldsymbol{x})$ is the strain energy density and it is a function of spatial position $\boldsymbol{x}$.

The integration in Equation (2) is usually conducted numerically, i. e.

$$\pi_i = A_i \sum_{p=1}^{P_i} w_p\, e_i(\boldsymbol{x}_p) \tag{3}$$

where A_i is the area of element i; P_i the number of quadrature points over the element; $e_i(\boldsymbol{x}_p)$ the value of e_i at p-th quadrature point and w_p is the corresponding weight.

The strain energy density, $e_i(\boldsymbol{x})$, is a scalar product of the stress tensor and the strain tensor; it contains the first derivative of displacements. The displacements are the primary unknown fields to be found. After introducing shape functions, the displacements are represented by nodal unknowns, i. e. nodel displacements in this case, and so is the strain energy density. After numerical integration is conducted, the strain energy in the element is also represented by nodal unknowns. The Hessian matrix of the strain energy with respect to nodal unknowns is the element stiffness matrix. Therefore, in calculating an element stiffness matrix, what we essentially need are the estimations of displacements and their derivatives at quadrature points over the element. Based on interpolation theory, with a given set of nodes and corresponding function values at the nodes, the best approximation of the function at a concerned point is obtained by interpolation using a subset of nodes that are the nearest to that point.

2.1 Strategy for constructing shape functions

With the above deduction, we take a different strategy for constructing shape functions in NN-FEM. Unlike in the classical FEM, where shape functions are constructed once for a whole element only using nodes belonging to that element, in the proposed NN-FEM, shape functions are constructed for each quadrature point, using a number of nodes that are the nearest to a concerned quadrature point. Some of those nodes may not belong to the element where the concerned quadrature point is located. A typical scenario is shown in Fig.4, where element stiffness matrix for the shaded element is being calculated. For

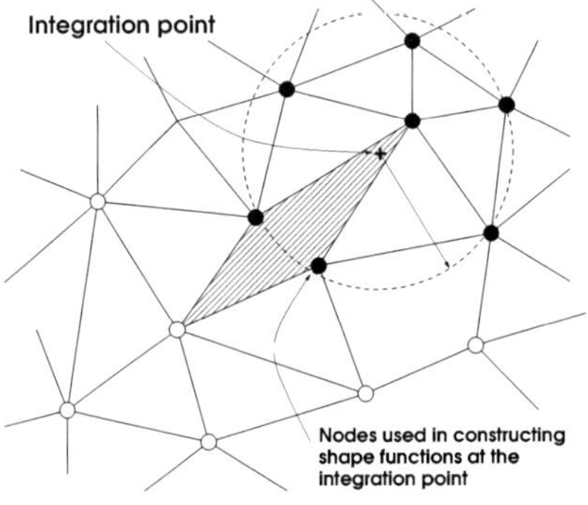

Figure 4: Nearest-Nodes FEM

the quadrature point denoted '+', a number of nearest nodes, marked as dark circles in the figure, are selected for constructing shape functions. Among those nodes, not all the nodes belong to the shaded element; and one node of the shaded element is not included as it is not near enough to the concerned quadrature point. For the next quadrature point, the same procedure is repeated and the selected nodes may or may not be the same as the previous ones. Even before conducting any numerical investigation, it can be expected that the quality of shape functions obtained in the above way is mainly determined by the locations of selected nodes; It has very little to do with the shape of an element. The shape change of an element may shift quadrature points around, but it does not obstruct selecting the nearest nodes for a concerned quadrature point.

As the number of nodes used for constructing shape functions is specified, the order of the resulting shape functions is either known, e. g. in the Moving Local Polynomial Interpolation introduced later, or can be estimated, e. g. for the Moving Least Square (MLS) method [14, 41]. A proper numerical quadrature order can thus be determined. One necessary requirement in NN-FEM is that only one numerical quadrature rule is allowed for each individual element. Therefore, for quadrature points in the same element, the same number of nearest nodes must be selected for constructing shape functions.

2.2 Assembly of global stiffness matrix

One step in the FEM is to assemble element stiffness matrices to the global stiffness matrix. In the classical FEM, the assembly is actually done in two steps. First, contributions from all quadrature points over an element are put together in an element stiffness matrix; and then the element stiffness matrix is assembled into the global stiffness matrix. For the NN-FEM, the same procedure can be applied, but the dimension of an element stiffness matrix is not so obvious as in the classical FEM, as different nodes may be used for constructing shape functions for different quadrature points over an element. To follow the same procedure as in the classical FEM, a common set of nodes used for constructing shape functions for quadrature points in an element must be first found. A more efficient way is to assemble the contribution from a quadrature point directly into the global stiffness matrix. This is explained in the following. The stain energy contribution from the p-th quadrature point in i-th element is denoted π_{ip}, and the corresponding contribution to the global stiffness matrix is

$$\boldsymbol{K}_{ip} = \frac{\partial^2 \pi_{ip}}{\partial \bar{\boldsymbol{u}}_{ip}^2} \tag{4}$$

where $\bar{\boldsymbol{u}}_{ip}$ is a vector consisting of nodal unknowns from nodes involved in constructing shape functions at quadrature point p of element i. The entry indices of $\boldsymbol{K}_{ip}$ in the global stiffness matrix are given by the corresponding degree-of-freedom assigned to the involved nodes.

2.3 Searching nearest nodes by element adjacency

With nodal connectivity defined, searching the nearest nodes for a concerned quadrature point in NN-FEM is much easier than in a meshless method. Nevertheless in a typical implementation of the classical FEM, only an element-node relation is defined, that is, if provided with an element label, nodes belonging to that element can be easily extracted. However, in NN-FEM, to construct shape functions for a concerned quadrature point, some of the nearest nodes may be from adjacent elements, cf Fig. 4. Certainly, the nodes from adjacent elements can be found by traversing over all elements, but that is obviously not efficient. A more efficient way is to expand the definition of nodal connectivity to include element adjacency information. An advanced mesh database such as the Algorithm Oriented Mesh Database (AOMD) [42] can provide any wanted information about element adjacency and is definitely workable. But for NN-FEM, only a simplified version is needed. What is additionally needed in NN-FEM is a node-element relation, that is, if a node is picked up, all elements connected to this node can be extracted. If the node-element relation is defined, any number of nodes from adjacent elements can be quickly found by alternately applying the element-node and node-element relations. This procedure is illustrated in Fig. 5 and described in the following.

1) Starting for element $\hat{E}_i$, where the currently concerned quadrature point, $\boldsymbol{x}_p$, is located, $\boldsymbol{x}_p \in \hat{E}_i$. The quadrature point is labeled as 'x' in Fig. 5.

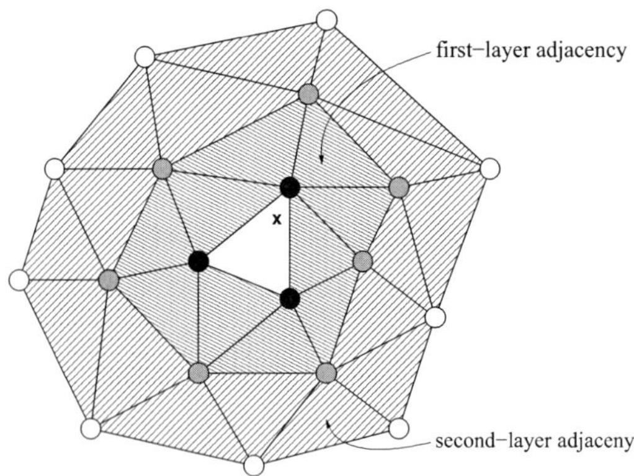

Figure 5: Element adjacency

2) Apply element-node relation to extract all nodes belonging to element $\hat{E}_i$. The node set is denoted $\hat{N}_i = \{n_j\,(j = 1, 2, \cdots, N_i); n_j \in \hat{E}_i\}$, where n_j and N_i are, respectively, the j-th node and the total number of nodes in element i.

3) For each node n_j in $\hat{N}_i$, use node-element relation to find out all elements connected to it; The common set of elements connecting to all nodes in $\hat{N}_i$ is represented by $\hat{A}_i^1 = \{\hat{E}_k(k = 1, 2, \cdots, A_i); \hat{E}_k \leftrightarrow \hat{N}_i\}$. Elements collected in $\hat{A}_i^1$ are in the first layer adjacency to element $\hat{E}_i$.

4) For each element $\hat{E}_k$ in $\hat{A}_i^1$, repeat Step 2); All nodes in the first layer adjacency are thus identified.

5) By repeating Steps 2), 3) and 4), nodes in second and higher layer adjacency can be identified.

2.4 Construction of shape functions for a quadrature point near a crack

In addition to element distortion due to large deformation, simulation of crack propagation is another difficulty for the classical FEM; while it is another merit for meshless methods. The easy treatment of a crack and its propagation with a meshless method is again attributed to the flexibility of constructing shape functions. This merit of meshless methods is inherited by NN-FEM for the similar strategy adopted for constructing shape functions. There are two possible scenarios in simulating crack propagation, an existing crack and a new crack. For an existing crack, element nodes are already aligned with the crack edges in mesh generation, Fig. 6(a), and all elements are complete. For a newly developed crack, some elements may be cut through or partially by a crack path, Fig. 6(b). On a quadrature point-by-point base, the two scenarios can be treated in the same way: in selection of nodes for construction shape functions at a quadrature point, in addition to the nearest criterion, the visibility criterion[43] is also applied. The visibility criterion requires that the line segment connecting a quadrature point and a candidate node should not intersect with any crack path. The two sets of selected nodes for quadrature point '+' and for quadrature point 'x' are marked, respectively, by empty circles and dark circles in Fig. 6. They are on

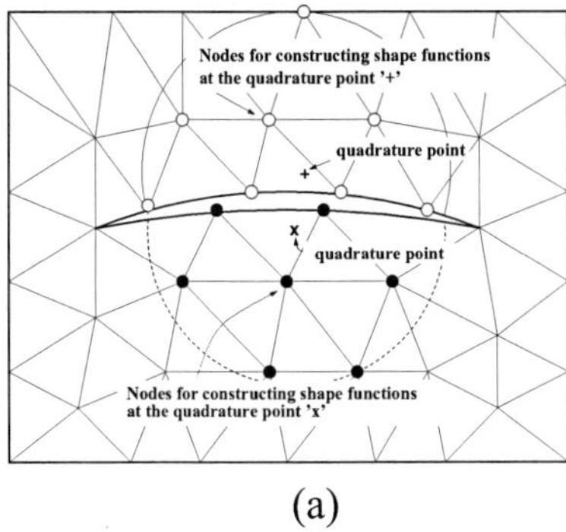

(a)

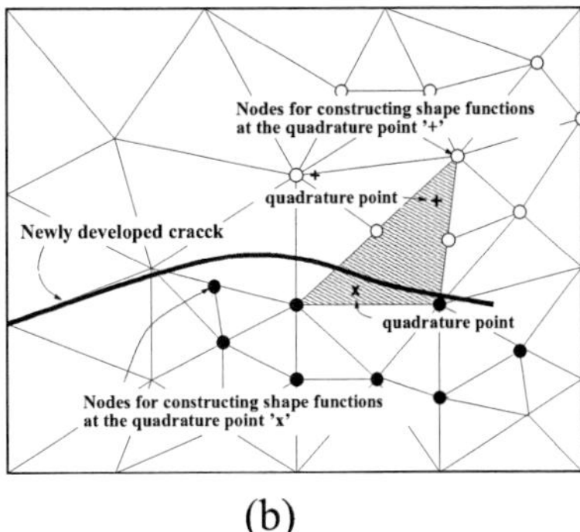

(b)

Figure 6: Selection of nearest nodes for a quadrature point near a crack. (a) an existing crack; (b) a new crack

different sides of the crack path. For a newly developed crack, a more efficient way is to align element nodes with crack edges that can be handled by mesh modification. But even without doing that, a following step of analysis can still be continued.

3 Moving Local Polynomial Interpolation

In principle, most techniques used in meshless methods for constructing shape functions can be modified and applied in NN-FEM. In a meshless method, to construct shape functions at a concerned point, a common practice is that, first an influence domain with a given size is specified; The influence domain can be in different shapes, e. g. in circle or square for a 2-D problem. Then, nodes covered by the influence domain are identified. This practice has several drawbacks. First, if a consistent shape function order is desired over the problem domain, the sizes of influence domains should be related to node distribution, meaning that a smaller influence domain is used in a region where node distribution is dense, and a larger influence domain is needed in a region where node distribution is sparse. To that end, information about node distribution is needed, which takes substantial computational time to figure out. Therefore in some implementation of meshless method, the size of an influence domain is given simply based on personal experience. Second, as the number of nodes covered by an influence domain is not exactly known, the order of constructed shape functions is also unknown. Therefore, although theoretically automatic p-version adaptation is a merit of meshless methods, it is difficult to be implemented in an efficient way. Third, if the order of obtained shape functions is not exactly known, it brings difficulty in selecting a proper numerical quadrature order.

To eliminate the above drawbacks, a slightly different practice is adopted in NN-FEM. To construct shape functions for a concerned quadrature point, first the desired number of nodes, which is related to the order of resulting shape functions, is given; Then, that number of nodes are selected based on the nearest criterion, the visibility criterion, and any other criterion; The size of the influence domain is determined by calculating the distance from the quadrature point to the outmost nodes. Another drawback for most techniques used in meshless methods for constructing shape functions is that the resulting shape functions do not satisfy the so-called Kronecker-delta condition, and thus brings in inconvenience on applying essential boundary conditions. For this reason, a moving local polynomial interpolation method is adopted in NN-FEM for constructing shape functions.

The polynomial interpolation method has been widely used in numerical analysis and approximation [44, 45, 46]. Nevertheless Runge's phenomenon [44] shows that if a large number of data points are involved, polynomial interpolation may oscillate wildly between data points. In this regard local or piecewise polynomial interpolation is more efficient. Piecewise polynomial interpolation has been adopted in various finite element formulations [47]. A reformulated version of local polynomial interpolation known as Point Interpolation Method (PIM) is introduced in [21] and applied in a meshless method. One attractive feature of local polynomial interpolation is that resulting shape functions satisfy the Kronecker delta condition. But one big concern for the method is that the Vandermonde matrix or moment matrix may be singular if the selected nodes are located on the same supersurface[45]. Several remedies are proposed in [21]. A simpler, more reliable and easier to implement remedy is proposed in the following. The strategy described in Sub-section 2.1 is adopted here for selecting nearest nodes. The improved local polynomial interpolation is called Moving Local Polynomial Interpolation (MLPI), as the center of the selected nodes is shifted from one concerned point to another. In the following we strictly distinguish a node from a point. By a node we mean that the function value there is known and the value is taken as accurate; By a point we mean that the function value is to be calculated and it is approximate.

Consider the approximation of a function $f(\boldsymbol{x})$ with the following provided conditions and data. The function is defined over a domain Ω bounded by Γ_Ω. Its variable $\boldsymbol{x}$ is a spatial position vector. A set of nodes, $\{\boldsymbol{x}_i, (i = 1, 2, \cdots, N)\}$, are representatively scattered over the domain and on its boundary. The distribution of the nodes can be regular or irregular. Function values, $\{f_i(i = 1, 2, \cdots, N)\}$, at the nodes are known. As discussed in Sub-section 2.3, with the use of a finite element mesh, nodal connectivity and element adjacency information are available and time spent on searching needed nearest nodes can be significantly reduced. A finite element mesh can be structured, Fig. 7(a), or unstructured, Fig. 7(b). The procedure described in the following is valid for both of them. To evaluate

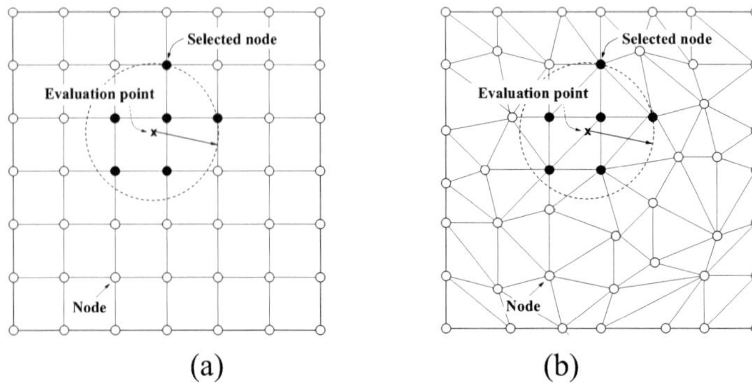

Figure 7: Nodal connectivity. (a) Structured; (b) Unstructured

function value at an arbitrary point, $\boldsymbol{x}$, a group of n nearest nodes around the point are selected from the total N nodes ($n < N$), Fig. 7, based on a set of criteria. The number of nearest nodes n is determined based on the desired order of shape functions or local polynomials. The included criteria are dependent on the problem to be solved, e. g. for crack propagation, visibility criterion is included. It will be seen later, to guarantee that Vandermonde matrix is non-singular, the so-called non-singularity criterion need be added in. With the n nearest nodes, the function value at $\boldsymbol{x}$ can be approximated as

$$\tilde{f}(\boldsymbol{x}) = \sum_{i=1}^{n} a_i p_i(\boldsymbol{x}) = \boldsymbol{p}^T(\boldsymbol{x}) \cdot \boldsymbol{a} \tag{5}$$

where $\boldsymbol{p}(\boldsymbol{x}) = \{p_1(\boldsymbol{x}),\ p_2(\boldsymbol{x}),\ \cdots,\ p_n(\boldsymbol{x})\}$ is a base vector consisting of monomials; $\boldsymbol{a} = \{a_1,\ a_2,\ \cdots,\ a_n\}$ is a coefficient vector. The monomials included in the base vector should satisfy the completeness and symmetry requirement [47]. The number of nearest nodes needed to construct a complete polynomial for different dimension problems is listed in Table 1.

Table 1: Number of needed nearest nodes for complete polynomials

	Polynomial Order				
220mm[18pt]Dimension	1	2	3	4	5
1-D	2	3	4	5	6
2-D	3	6	10	15	21
3-D	4	10	20	35	56

The coefficient vector $\boldsymbol{a}$ is determined by enforcing the approximation in Equation (5) at the n selected nodes, i. e.

$$\boldsymbol{p}^T(\boldsymbol{x}_i) \cdot \boldsymbol{a} = f_i \qquad (i = 1, 2, \cdots, n) \tag{6}$$

The above equations can be collected in a matrix form

$$\boldsymbol{P}\boldsymbol{a} = \bar{\boldsymbol{f}} \tag{7}$$

where $\bar{\boldsymbol{f}} = \{f_1,\ f_2,\ \cdots,\ f_n\}$ is a vector consisting of function values at the n nodes. $\boldsymbol{P}$ is the Vandermonde matrix with dimensions $n \times n$. The expressions of entries in matrix $\boldsymbol{P}$ depend on problem dimension and polynomial order. For example, for a two-dimensional second-order polynomial, matrix $\boldsymbol{P}$ has the following form

$$\boldsymbol{P} = \begin{bmatrix} 1 & x_1 & y_1 & x_1^2 & x_1 y_1 & y_1^2 \\ 1 & x_2 & y_2 & x_2^2 & x_2 y_2 & y_2^2 \\ & \cdots & & \cdots & & \cdots \\ 1 & x_6 & y_6 & x_6^2 & x_6 y_6 & y_6^2 \end{bmatrix} \tag{8}$$

If it is assumed that $\boldsymbol{P}$ is non-singular and its inverse exists, the coefficient vector $\boldsymbol{a}$ is determined by

$$\boldsymbol{a} = \boldsymbol{P}^{-1}\bar{\boldsymbol{f}} \tag{9}$$

and the approximation is now expressed as

$$\tilde{f}(\boldsymbol{x}) = \boldsymbol{\phi}^T(\boldsymbol{x})\bar{\boldsymbol{f}} \tag{10}$$

where

$$\boldsymbol{\phi}^T(\boldsymbol{x}) = \boldsymbol{p}^T(\boldsymbol{x})\boldsymbol{P}^{-1} \tag{11}$$

contains shape functions or local polynomials.

Approximate first-derivatives at point $\boldsymbol{x}$ are calculated as

$$\frac{\partial \tilde{f}}{\partial \alpha} = \boldsymbol{\phi}_{,\alpha}^T \bar{\boldsymbol{f}}, \qquad (\alpha = x, y, z) \tag{12}$$

with

$$\boldsymbol{\phi}_{,\alpha} = \frac{\partial \boldsymbol{p}^T(\boldsymbol{x})}{\partial \alpha}\boldsymbol{P}^{-1}, \qquad (\alpha = x, y, z) \tag{13}$$

Any higher order derivatives can be calculated in a similar way, as long as sufficient high order monomials are included in the base vector in Equation (5). To guarantee that the Vandermonde matrix $\boldsymbol{P}$ is non-singular, in filling $\boldsymbol{P}$ matrix with nearest nodes, the so-called non-singularity criterion is applied. The criterion is based on the following theorem.

Theorem 1 (Rank in terms of determinants) [48]
An $m \times n$ matrix $\boldsymbol{A} = [a_{jk}]$ has rank r $(r \geq 1)$ if and only if $\boldsymbol{A}$ has an $r \times r$ submatrix with nonzero determinant, whereas the determinant of every square submatrix with $r + 1$ or more rows that $\boldsymbol{A}$ has (or does not have!) is zero. In particular, if $\boldsymbol{A}$ is square of $n \times n$, it has rank n if and only if $\det(\boldsymbol{A}) \neq 0$.

Based on Theorem 1, to construct a d-dimensional local polynomial, the non-singularity criterion is stated as: in filling $\boldsymbol{P}$ matrix with nearest nodes, for each added new row, say, now the i-th row is added, $i > d$, the determinant of master submatrix, $\boldsymbol{M}_P^i$, is examined, i. e.

$$\text{if} \quad \det(\boldsymbol{M}_P^i) \begin{cases} \geq \delta, & i\text{-th row is kept.} \\ < \delta, & i\text{-th row is discarded.} \end{cases} \tag{14}$$

where δ is a small positive real number, $\delta = 10^{-10} \sim 10^{-20}$. If a larger value is taken for δ, the quality of resultant $\boldsymbol{P}$ matrix is higher, but the kept nodes are scattered apart from the evaluation point. On the other hand, if a smaller value is taken for δ, the kept nodes are more compact around the evaluation point, but the quality of the resultant $\boldsymbol{P}$ matrix may be lower, meaning that it has a larger condition number.

A master submatrix is a square submatrix including all the main entries of current $\boldsymbol{P}$ matrix. The current $\boldsymbol{P}$ matrix after filling i-th row is denoted $\boldsymbol{P}^i$. For example, for constructing a 2-D second-order polynomial, suppose that the forth row is added, the current

$\boldsymbol{P}$ matrix and its master submatrix are given in the following:

$$\boldsymbol{P}^4 = \begin{bmatrix} 1 & x_1 & y_1 & x_1^2 & x_1y_1 & y_1^2 \\ 1 & x_2 & y_2 & x_2^2 & x_2y_2 & y_2^2 \\ 1 & x_3 & y_3 & x_3^2 & x_3y_3 & y_3^2 \\ 1 & x_3 & y_3 & x_3^2 & x_3y_3 & y_3^2 \\ \cdots & \cdots & \cdots & \cdots & \cdots & \cdots \end{bmatrix},$$

$$\boldsymbol{M}_P^4 = \begin{bmatrix} 1 & x_1 & y_1 & x_1^2 \\ 1 & x_2 & y_2 & x_2^2 \\ 1 & x_3 & y_3 & x_3^2 \\ 1 & x_3 & y_3 & x_3^2 \end{bmatrix} \tag{15}$$

Non-singularity criterion (14) is a sufficient condition to guarantee that $\boldsymbol{P}$ matrix is non-singular and its inverse exists. Nevertheless, as the criterion is based on determinants of submatrices rather than on the final $\boldsymbol{P}$ matrix, some of the discarded nodes, which are nearer to the evaluation point, may be still workable. Therefore a re-checking procedure is devised to pick up back workable ones from the discarded nodes. The procedure is simple and it is described as follows. A discarded node and the corresponding row number of $\boldsymbol{P}$ matrix where the node is discarded are stored in a temporary container. For example, if the i-th node is discarded when it is used to fill j-th row of $\boldsymbol{P}$ matrix, then the numbers i and j are stored. After $\boldsymbol{P}$ matrix is full, for each discarded node, use it to repeatedly replace a row in $\boldsymbol{P}$ matrix. If $\det(\boldsymbol{P}) \geq \delta$, the replacement is kept; otherwise, the replacement is moved to the next row. The replacement is not necessarily started from the first row, but from the row where a node was discarded. For example, the i-th node was discarded when the j-th row was filled, therefore the replacement starts from j-th row. With this re-checking procedure, the kept nodes are more compact around the evaluation point. An illustrative case is given in Fig. 8, where a 2-D second-order local polynomial is being constructed for evaluation point

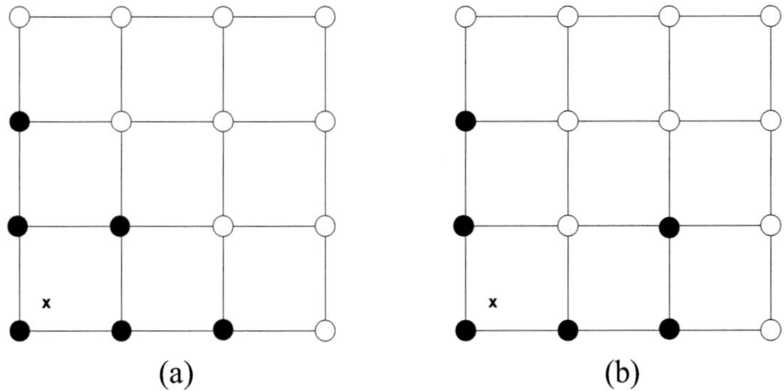

Figure 8: Nearest Nodes. (a) with re-checking; (b) without re-checking

'x', six nearest nodes are needed according to Table 1. Configurations of final nearest nodes with and without re-checking procedure are shown, respectively, in Fig. 8(a) and Fig. 8(b). It is obvious that the nearest nodes selected with re-checking procedure is more compact with respect with the evaluation point.

4 Numerical Investigations and Results

In this section, results of numerical investigations on the proposed NN-FEM are reported. The conducted numerical investigations include:

(1) Function approximation with the moving local polynomial interpolation.

(2) Convergence of NN-FEM.

(3) Three-dimensional problem.

(4) Effect of element distortion.

(5) Crack analysis.

4.1 Function approximation with moving local polynomial interpolation

As well-known, the quality of constructed shape functions has a significant effect on the performance of a numerical method such as the FEM and meshless methods. Therefore, the performance of the moving local polynomial interpolation was first studied by applying it in function approximation. The Peaks function is a complex function and it was used in the study. The Peaks function has two variables, x and y, and defined over domain $\Omega = \{-3 \leq x \leq 3, -3 \leq y \leq 3\}$. The function has the following expression:

$$\begin{aligned} f(x,y) &= 3(1-x)^2\,e^{-x^2-(y+1)^2} \\ &- 10(\frac{x}{5} - x^3 - y^5)\,e^{-x^2-y^2} \\ &- \frac{1}{3}\,e^{-(x+1)^2-y^2} \end{aligned} \tag{16}$$

The exact function surface and its first order derivatives are plotted in Fig. 9. To study

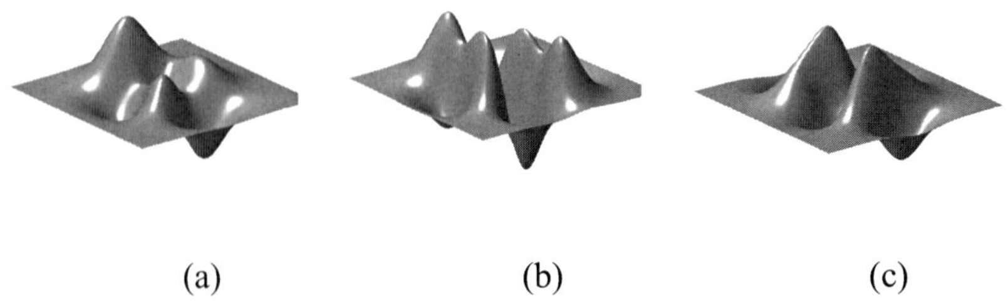

(a) (b) (c)

Figure 9: Peaks Function and its first-order derivatives. (a) Function surface; (b) $\dfrac{\partial f}{\partial x}$; (c) $\dfrac{\partial f}{\partial y}$

the performance of MLPI, the definition domain is first discretized by a set of uniformly distributed nodes. Then, function values at nodes are calculated. Evaluation points are intentionally selected different from nodes. A typical configuration of nodes and evaluation points is given in Fig. 10. For clarity, connections between nodes are not displayed in the figure. The moving local polynomial interpolation is used to calculate approximate

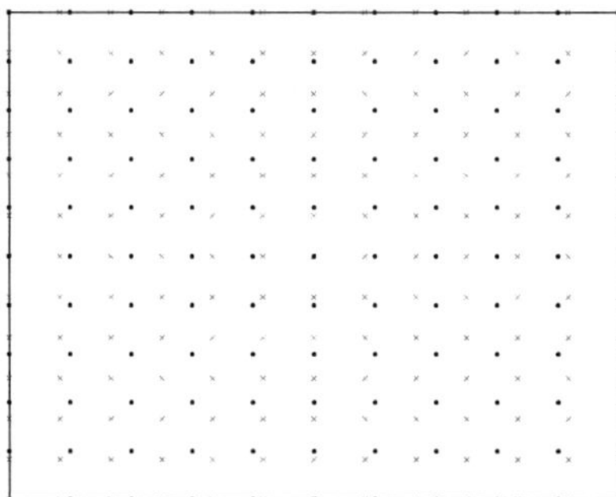

Figure 10: A typical configuration of nodes (marked by '.') and evaluation points (marked by 'x')

function values and derivatives at evaluation points. h- and p-convergence are studied. By h-convergence it means that the order of local polynomials is fixed and the density of nodes is gradually increased; while p-convergence implies that for a fixed set of nodes, the order of local polynomials is gradually increased. The obtained approximate function and derivative surfaces are shown in Figs. 11, 12 and 13. By comparing the approximate surfaces in Figs. 11, 12 and 13 with the exact surfaces in Fig. 9, the following conclusions can be drawn. Approximate function converges faster than approximate derivatives; With a fixed grid and an increasing polynomial order, approximate function surfaces, especially approximate derivative surfaces, become even rougher. This is due to the Runge's phenomenon. It suggests that, to improve accuracy, a pure p-version adaptation is not efficient and a more efficient way is to combine h- and p- adaptation. Therefore, the fastest convergence is observed along the diagonal from upper-left to lower-right in Figs. 11, 12 and 13.

h- and p-convergence were also studied with unstructured nodes. A set of 163 nodes and a set of 632 nodes were used. The configurations of nodal distributions and evaluation points are shown in Fig. 14. The obtained approximate function and derivative surfaces are displayed in Figs. 15, 16 and 17. It can be concluded from the studies that the proposed MLPI with non-singularity and re-checking procedure is more robust and reliable. After nodes for constructing shape functions are collected, the amount of time spent by MLPI is approximately the same as that needed by the Moving Least Square (MLS) method. Both of them need to check the existence of inverse matrix and then to inverse the matrix. It is evident that shape functions yielded by MLPI satisfy the Kronecker delta condition, as the number of undetermined coefficients in approximation (5) is equal to the number of nodes where the approximation is enforced. This is exactly what has been done in the FEM with Lagrange interpolations [47].

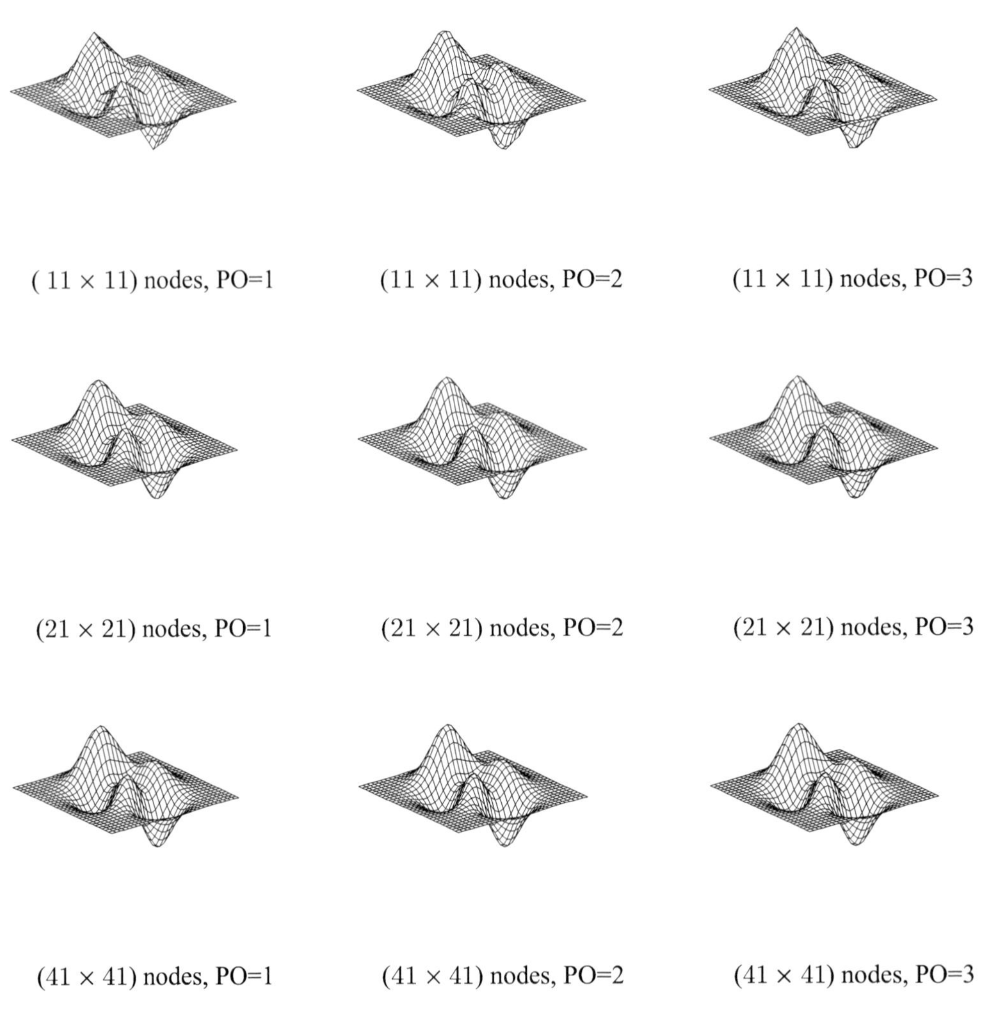

Figure 11: h- and p-convergence of function surface (PO—Polynomial Order)

4.2 Convergence of NN-FEM

Convergence of the proposed NN-FEM with moving local polynomial interpolation was studied by a benchmark problem. A cantilever beam under an end shear force, Fig. 18(a), was analyzed as a plane stress problem. The beam has the following geometric and material parameters: length L=10, height $2b$=2, thickness t=1, elasticity modulus E=1000.0, Poisson's ratio ν=0.3. The parameters have consistent units. The initial uniform mesh, Fig. 18(b), was progressively refined, and two of them are displayed in Figs. 18(c) and (d).

The obtained results are plotted in Fig. 19. For comparison purpose, results from the FEM and from the Element-Free Galerkin (EFG) method are also displayed.

Finite element solutions were produced using the commercial software ANSYS. Element PLANE2 from the ANSYS element library was used for the calculation. PLANE2

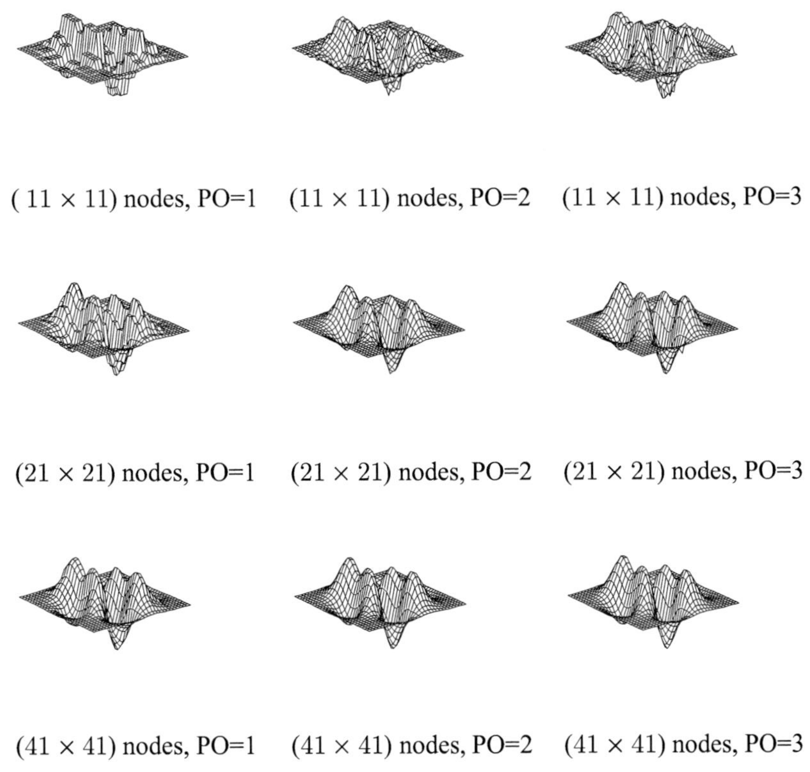

Figure 12: *h*- and *p*-convergence of derivative $\dfrac{\partial f}{\partial x}$ (PO—Polynomial Order)

is a second order plane stress element with six element nodes. For the EFG method, the radius of a circular domain of influence was determined in such a way that approximately six nodes are covered by the domain of influence, and a complete set of second order monomials were included in the base. A fair base was tried for the comparison, that is, for all the compared methods they have the same order of shape functions, a similar number of total nodes, etc. Nevertheless, as a weight function is usually used in the MLS method, shape functions resulted from the MLS method with a second-order base actually have a higher order. From the obtained results, it can be observed that compared with the FEM and EFG method, the proposed NN-FEM with a second order base has a similar convergence rate. NN-FEM with a first order of base is equivalent to the constant strain element, therefore it has a slower convergence. It was also found that although by increasing the order of local polynomials, NN-FEM can have a gain in convergence rate, the bandwidth of global stiffness matrix also becomes larger. Therefore more storage and memory are demanded and more solution time is needed.

Based on a comprehensive consideration of balance between convergence rate, computational time and memory consumption, second- or third-order local polynomials would be the most efficient for NN-FEM.

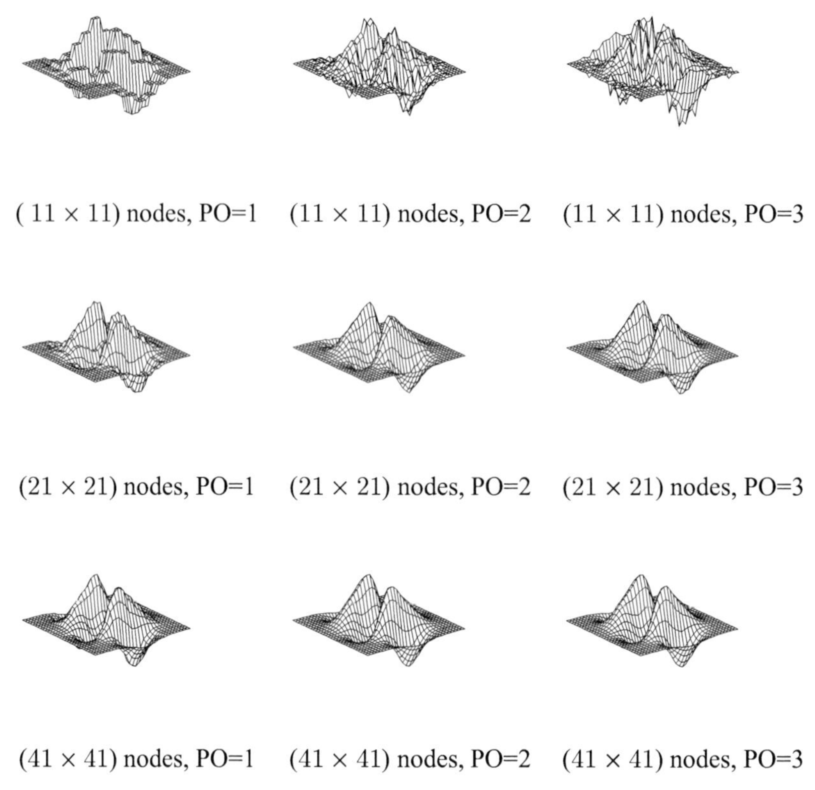

Figure 13: h- and p-convergence of derivative $\dfrac{\partial f}{\partial y}$ (PO—Polynomial Order)

4.3 Three-dimensional problem

A cantilever beam under an end shear force shown in Fig. 20(a) was analyzed using a three-dimensional continuum model. The beam has the following geometric and material parameters: length L=6, width $b = 1$, height $h = 1$, elasticity modulus E=1000.0, Poisson's ratio ν=0. Unit consistence is again assumed. The beam cross-section at the right end is loaded with a uniformly distributed shear force. Convergence was studied by increasing mesh density. One representative mesh is displayed in Fig. 20(b).

With $\nu = 0.$, three-dimensional solutions should be close to the beam solutions. The obtained displacements of the beam cross-section center at the loaded end are normalized by the analytical beam solution and plotted in Fig. 21. A deformed configuration of the beam and the corresponding effective stress distribution are given in Fig. 22. For comparison purpose, results from the FEM and from the Element-Free Galerkin (EFG) method [14] are also displayed. The FEM results were produced by using SOLID187 in ANSYS. SOLID187 is a second order tetrahedral element with ten nodes. For the EFG method, the radius of a circular domain of influence was so determined that approximately 10 nodes are covered by the domain of influence, and a complete set of second order monomials were included in the base. Similar observations can be made from the numerical results as in the

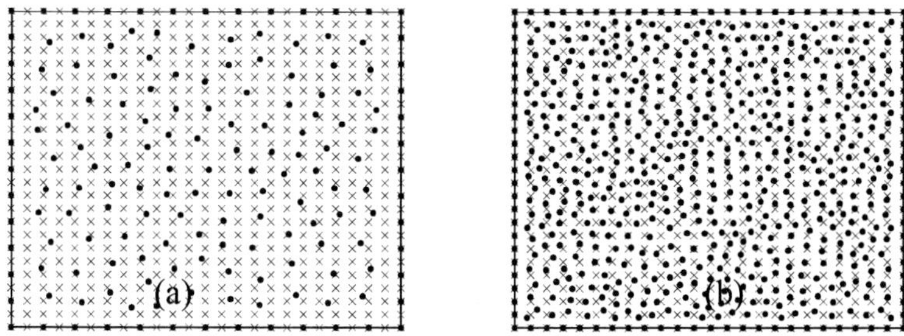

Figure 14: Unstructured nodes (marked as '.') and evaluation points (marked as 'x'). (a) 163 nodes; (b) 632 nodes

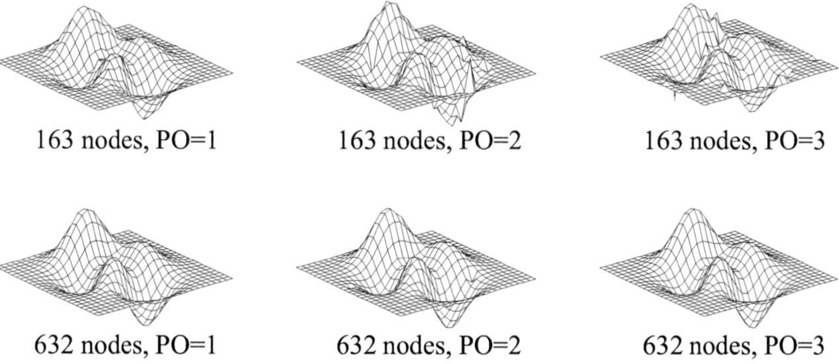

Figure 15: Convergence of approximate function with unstructured nodes (PO—Polynomial Order)

two-dimensional problem.

4.3.1 Effects of mesh distortion

Insensitiveness to element distortion is one big advantage of the proposed NN-FEM. This advantage was verified by specially devised meshes. The cantilever beam in Fig. 18(a) was re-analyzed with the two meshes shown in Figs. 23(a) and (b). The two meshes have the same number of nodes, and these nodes have exactly the same coordinates in the two meshes. The mesh in Fig. 23(a) is a regular undistorted mesh, all elements have a good shape; while the mesh in Fig. 23(b) is severely distorted. For some of the elements in the mesh shown in Fig. 23(b), their shape is extremely flat and the three element nodes are nearly on the same straight line. The obtained results are given in Table 2. Generally speaking, differences between results yielded by mesh (a) and mesh (b) are small. The difference between results obtained by first order local polynomials is relatively more significant. The reason is that, with first order local polynomials, only three nodes are selected for constructing shape functions; while the distortion of an element will change the locations of quadrature points. The effect of element distortion can nearly not be seen with second order local polynomials, as more nodes are used in shape function construction.

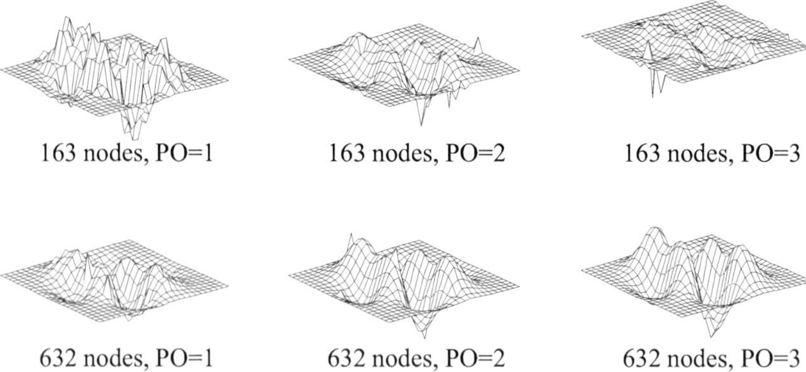

Figure 16: Convergence of approximate $\frac{\partial f}{\partial x}$ with unstructured nodes (PO—Polynomial Order)

Table 2: Transverse displacement at the free end of cantilever

Local Polynomial Order	1	2
Mesh (a)	0.325793	0.516522
Mesh (b)	0.303255	0.516534
Analytical solution	0.516540	

4.4 Crack analysis

Ability in dealing with geometric discontinuities and their development without any constraint is another big advantage of the proposed NN-FEM. As described in Sub-section 2.4, the treatment of an existing crack and that of a new crack are in principle the same. The only difference is that for a new crack some nodes may not be aligned with the edges of the crack. In this test, a plate with an existing edge crack was analyzed, Fig. 24(a). The aim of this test is mainly to check the applicability of the visibility criterion in NN-FEM. The undeformed mesh is shown in Fig. 24(b). Applicability of the visibility criterion in NN-FEM is confirmed by the deformed configuration and the plot of effective stress given in Fig. 25. The displacements across the crack are obviously discontinuous and the predicted location of stress concentration is as expected.

5 Discussions and Concluding Remarks

In the proposed NN-FEM, the classical FEM and newly emerged meshless methods are combined in the following way: the finite elements are only used for numerical integration; while shape functions are constructed in a similar way as in meshless methods, i.e. by using a number of nodes that are the nearest to a concerned quadrature point. Some of the nodes may be from adjacent elements. Although most techniques of constructing shape

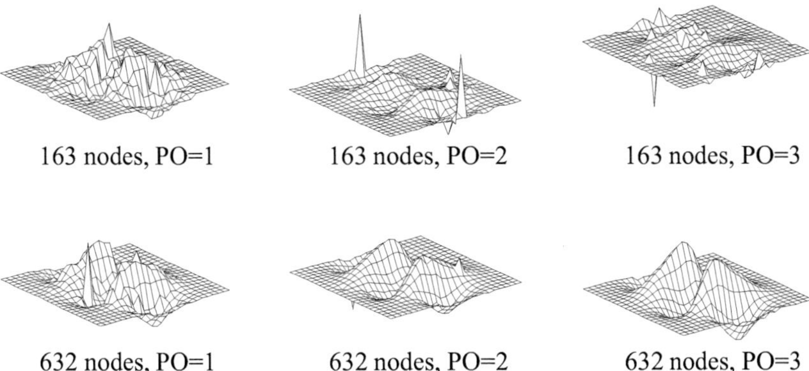

Figure 17: Convergence of approximate $\frac{\partial f}{\partial y}$ with unstructured nodes (PO—Polynomial Order)

functions in meshless methods can be modified and used in NN-FEM, a local polynomial interpolation method was adopted, as the resulting shape functions satisfy the Kronecker delta condition. In the above way, the proposed NN-FEM inherits most of the advantages from both the FEM and meshless methods. More specifically, NN-FEM has the following attractive advantages:

- The shape of finite elements can be arbitrary, e.g. triangles, quadrilaterals, arbitrary polygons, or any hybrid mesh for a 2-D case. Finite elements can be extremely distorted. The minimum requirement on a mesh in NN-FEM is that elements are not overlapped to each other and not self-penetrated, to avoid difficulty in numerical integration. The quality of shape functions is determined by the locations of element nodes. Therefore NN-FEM, especially if higher order shape functions are used, is almost not affected by element distortion as demonstrated by numerical results.

- Compared with meshless methods, NN-FEM can take the advantages of advanced mesh generation techniques. Problem domains with very complex geometry can be well represented by finite element meshes, which is not so easily handled in other numerical methods.

- The order of shape functions can be arbitrarily high, as long as enough element nodes are available. Therefore, p-version of adaptation can be easily implemented in NN-FEM.

- A crack can develop in an arbitrary path in NN-FEM, as in the selection of nodes for constructing shape functions, the visibility criterion can be applied.

- As the number of used nodes is specified, the order of constructed shape functions is either known or can be easily estimated; therefore a consistent numerical integration order can be determined.

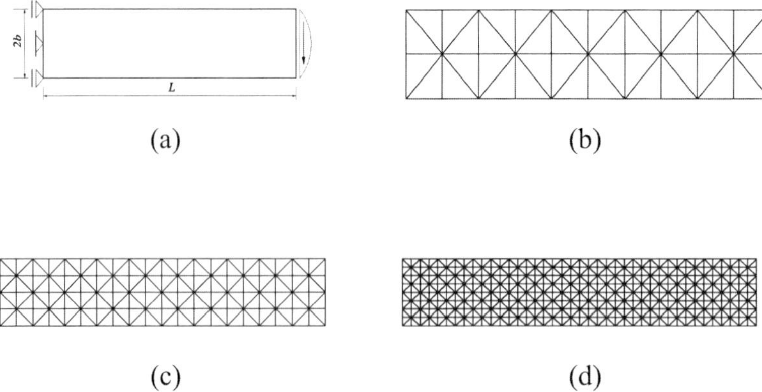

Figure 18: Cantilever beam and meshes

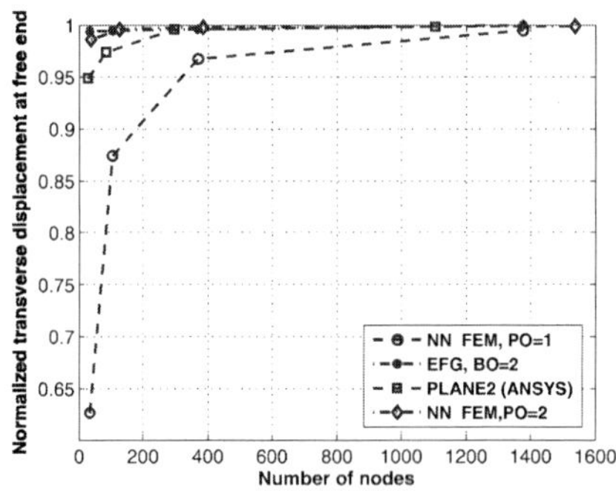

Figure 19: h-Convergence in two-dimensional problem. (PO—Polynomial Order; BO—Base order)

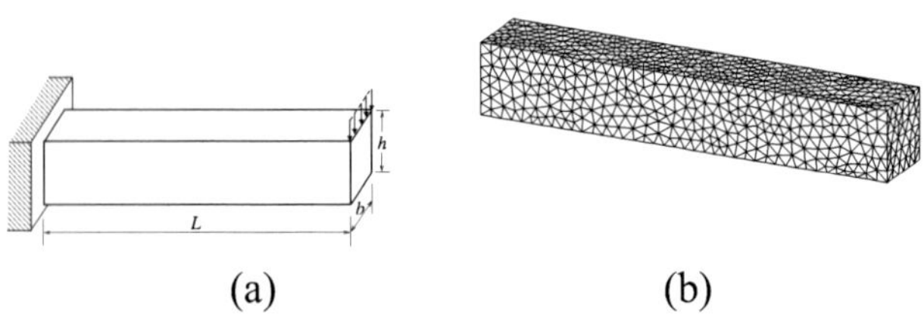

Figure 20: A cantilever beam (a) physical model; (b) a mesh

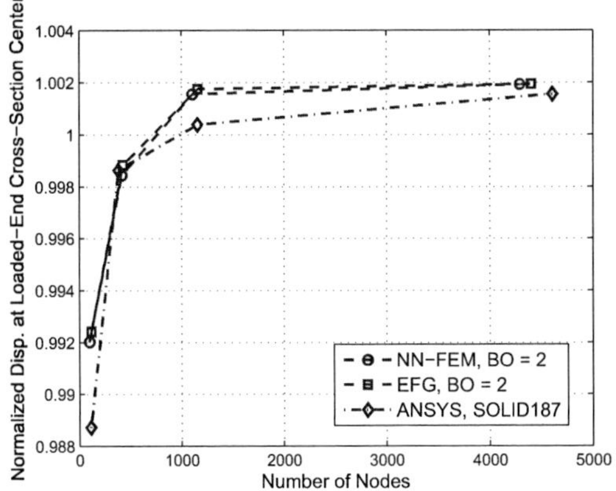

Figure 21: *h*-Convergence in three-dimensional problem. (BO—Base order)

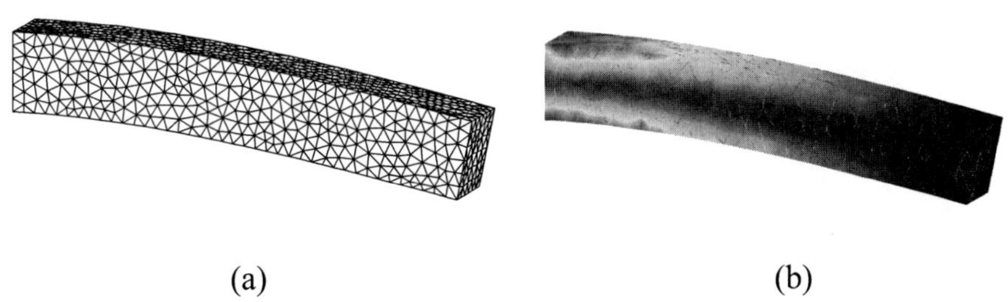

(a) (b)

Figure 22: (a) Deformed configuration; (b) Effective stress distribution

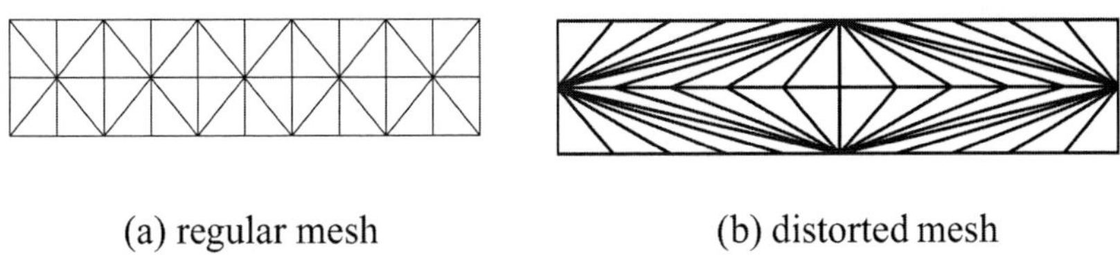

(a) regular mesh (b) distorted mesh

Figure 23: Regular and distorted meshes

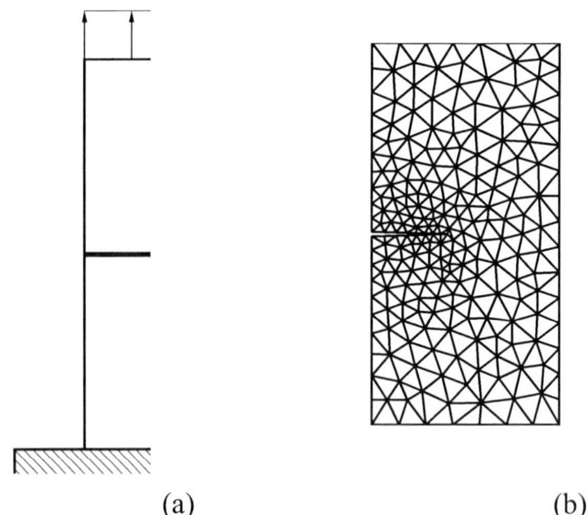

Figure 24: A plate with an edge crack. (a) boundary and loading conditions; (b) undeformed configuration

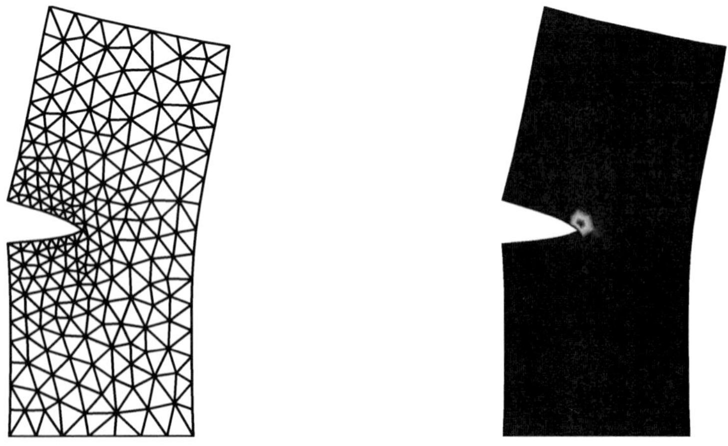

Figure 25: Deformed configuration and effective stress

- With the moving local polynomial interpolation, essential boundary conditions can be applied in the same way as in the classical FEM.
- With element adjacency information, time spent in search of nearest nodes for constructing shape functions is much shorter than that in meshless methods used for identifying nodes covered under a domain of influence.

References

[1] Y. Luo. A nearest-nodes finite element method with local multivariate lagrange interpolation. *Finite Elements in Analysis and Design*, 44:797–803, 2008.

[2] Y. Luo. Dealing with extremely large deformation by nearest-nodes fem with algorithm for updating element connectivity. *International Journal of Solids and Structures*, 45:5074–5087, 2008.

[3] Y. Luo. 3D Nearest-Nodes Finite Element Method for solid continuum analysis. *Advances in Theoretical and Applied Mechanics*, 1:131–139, 2008.

[4] Y. Luo. An effective way for dealing with element distortion by nearest-nodes fem. *Communications in Numerical Methods in Engineering*, (in press, DOI: 10.1002/cnm.1184), 2008.

[5] J.H. Argyris, H. Kelsey, and H. Kamel. Matrix methods of structural analysis: a précis of recent development. In B. Fraeijs de Veubeke, editor, *Matrix Methods of Structural Analysis*, pages 1–164. Pergamon Press, Oxford, 1964.

[6] Fraeijs de Veubeke B. Displacement and equilibrium models in the finite element method. In O. C. Zienkiewicz and G. S. Holister, editors, *Stress Analysisin*, pages 145–197, London, 1965. John Wiley & Sons.

[7] O.C. Zienkiewicz and Y.K. Cheung. *The Finite Element Method in Continuum and Structural Mechanics*. McGraw Hill, New York, 1967.

[8] M.S. Shephard. Meshing environment for geometry-based analysis. *Int. J. Numer. Meth. Engng.*, 47:169–190, 2000.

[9] M.S. Shephard, J.E. Flaherty, K.E. Jansen, X. Li, X. Luo, N. Chevaugeon, J.-F. Remacle, M.W. Bell, and R.M. O'Bara. Adaptive mesh generation for curved domains. *Applied Numerical Mathematics*, 52:251–271, 2005.

[10] Y. Luo, I. Malik, Z. Li, M.S. Shephard, and K. Ko. Adaptive mesh refinement accelerator modeling with omega3p. In *7th U.S. National Congress on Computational Mechanics*. Sandia National Laboratory, July 28 - 30, 2003.

[11] N.P. Weatherill. A review of mesh generation. In *Advances in Finite Element Technology*, pages 1–10, Budapest, Hungary, 21-23 Aug. 1996.

[12] I. Babuška and J.M. Melenk. Partition of unity method. *Int. J. Numer. Meth. Engng.*, 40:727–758, 1997.

[13] T. Belytschko, Y. Krongauz, D. Organ, M. Fleming, and P. Krysl. Meshless methods: an overview and recent developments. *Comput. Methods Appl. Mech. Engrg.*, 139:3–47, 1996.

[14] T. Belytschko, Y.Y. Lu, and L. Gu. Element-free galerkin methods. *Int. J. Numer. Meth. Engng.*, 37:229–256, 1994.

[15] R.A. Gingold and J.J. Monaghan. Smoothed particle hydrodynamics: Theory and application to non-spherical stars. *Mon. Not. R. Astr. Soc.*, 181:375–389, 1977.

[16] S. Hao, H.S. Park, and W.K. Liu. Moving particle finite element method. *Int. J. Numer. Meth. Engng.*, 53:1937–1958, 2002.

[17] Y. Krongauz and T. Belytschko. Enforcement of essential boundary conditions in meshless approximation using finite elements. *Comput. Methods Appl. Mech. Engrg.*, 131:133–145, 1996.

[18] S. Li and W.K. Liu. Moving least squares reproducing kernel particle method, part ii: Fourier analysis. *Comput. Methods Appl. Mech. Engrg.*, 139:159–194, 1996.

[19] T.J. Liszka, C.A.M. Duarte, and W.W. Tworzydlo. *hp*-meshless cloud method. *Comput. Methods Appl. Mech. Engrg.*, 139:263–288, 1996.

[20] G.R. Liu. *Meshless Methods*. CRC Press, 2002.

[21] G.R. Liu. *Mesh Free Methods: Moving beyond the Finite Element Method*. CRC Press, 2003.

[22] W.K. Liu, M. Han, H. Lu, S. Li, and J. Cao. Reproducing kernel element method. part i: Theoretical formulation. *Comput. Methods Appl. Mech. Engrg.*, 193:933–951, 2004.

[23] W.K. Liu, S. Jun, and Y.F. Zhang. Reproducing kernel particle methods. *Int. J. Numer. Meth. Engng.*, 20:1081–1106, 1995.

[24] W.K. Liu, S. Li, and T. Belytschko. Moving least square reproducing kernel particle method: Methodology and convergence. *Comput. Methods Appl. Mech. Engrg.*, 143:422–453, 1997.

[25] Y.Y. Lu, T. Belytschko, and L. Gu. A new implementation of the element free galerkin method. *Comput. Methods Appl. Mech. Engrg.*, 113:397–414, 1994.

[26] J.M. Melenk and I. Babuška. The partition of unity finite element method: Theory and application. *Comput. Methods Appl. Mech. Engrg.*, 139:289–314, 1996.

[27] B. Nayroles, G. Touzot, and P. Villon. Generalizing the fem: Diffuse approximation and diffuse elements. *Comput. Mech.*, 10:307–318, 1992.

[28] N. Sukumar, B. Moran, and T. Belytschko. Natural element method in solid mechanics. *Int. J. Numer. Meth. Engng.*, 43:839–887, 1998.

[29] E. Onãte, F. Perazzo, and J. Miquel. A finite point method for elasticity problems. *Computers & Structures*, 79:2151–2163, 2001.

[30] I. Babuška, U. Banerjee, and J. Osborn. Meshless and generalized finite element methods: A survey of major results. In M. Griebel and M. A. Schweitzer, editors, *Meshfree Methods for Partial Differential Equations* . Springer, 2002.

[31] C.A.M. Duarte, I. Babuška, and J.T.Oden. Generalized finite element method for three dimensional structural problems. *Computers and Structures*, 77:219–232, 2000.

[32] S.R. Idelsohn and E. Oñate. To mesh or not to mesh. that is the question ... *Comput. Methods Appl. Mech. Engrg.*, 195:4681–4696, 2006.

[33] A.M. Matache, I. Babuška, and C. Schwab. Generalized p-fem in homogenization. *Numer. Math.*, 86:319–375, 2000.

[34] A. Moes, J. Dolbow, and T. Belytschko. A finite element method for crack growth without remeshing. *Int. J. Numer. Meth. Engng.*, 46:131–150, 1999.

[35] N. Moes, A. Gravouil, and T. Belytschko. Non-planar 3d crack growth by extended finite element and level sets, part 1: Mechanical model. *Int. J. Numer. Meth. Engng.*, 53:2549–2568, 2002.

[36] J.T. Oden, C.A. Duarte, and O.C. Zienkiewicz. A new cloud based *hp* finite element method. *Comput. Methods Appl. Mech. Engrg.*, 153:117–126, 1998.

[37] T. Strouboulis and I. Babuška. The design and analysis of the generalized finite element method. *Comput. Methods Appl. Mech. Engrg.*, 181:43–69, 2000.

[38] T. Strouboulis, K. Copps K, and I. Babuška. The generalized finite element method. *Comput. Methods Appl. Mech. Engrg.*, 190:4081–4193, 2001.

[39] N. Sukumar, N. Moes, B. Moran, and T. Belytschko. Extended finite element method for three dimensional crack modelling. *Int. J. Numer. Meth. Engng.*, 48:1549–1570, 2000.

[40] R.L. Taylor, O.C. Zienkiewicz, and E. Oñate. A hierarchical finite element method based on partition of unity. *Comput. Methods Appl. Mech. Engrg.*, 152:73–84, 1998.

[41] P. Lancaster and K. Salkauskas. Surfaces generated by moving least squares method. *Math. Comp.*, 37:141–158, 1981.

[42] J.-F. Remacle and M. Shephard. An algorithm oriented mesh database. *Int. J. Numer. Meth. Engng.*, 58:349–374, 2003.

[43] P. Krysl and T. Belytschko. Element free galerkin method for dynamic propagation of arbitrary 3-d cracks. *Int. J. Numer. Meth. Engng.*, 44:767–800, 1999.

[44] K.A. Atkinson. *An Introduction to Numerical Analysis*. John Wiley and Sons, 2nd edition, 1988.

[45] M. Gasca and T. Sauer. On the history of multivariate polynomial interpolation. *Journal of Computational and Applied Mathematics*, 122:23–35, 2000.

[46] M.J.D. Powell. *Approximation Theory and Method*. Cambridge University Press, 1981.

[47] O.C. Zienkiewicz and R.L. Taylor. *The Finite Element Method*. Butterworth-Heinemann, Linacre House, Jordan Hill, 5 edition, 2000.

[48] E. Kreyszig. *Advanced Engineering Mathematics*. Wiley, 8 edition, 1998.

In: Continuum Mechanics
Editor: A. Koppel and J. Oja, pp. 273-294
ISBN:978-1-60741-585-5

Chapter 10

MESH ADAPTATION ALGORITHM BASED ON GRADIENT OF STRAIN ENERGY DENSITY

Yunhua Luo*
Department of Mechanical & Manufacturing Engineering,
University of Manitoba, Winnipeg, Canada

Abstract

In this chapter, an adaptive finite element method is formulated based on the newly developed nearest-nodes finite element method (NN-FEM). In the adaptive NN-FEM, mesh modification is guided by the gradient of strain energy density, i. e., a larger gradient requires a denser mesh and vice versa. A finite element mesh is iteratively modified by a set of operators, including mesh refinement, mesh coarsening and mesh smoothing, to make its density conform with the gradient of strain energy density. The selection of a proper operator for a specific mesh region is determined by a set of criteria that are based on mesh intensity. The iteration loop of mesh modification is stopped when the relative error in the total potential energy is less than a prescribed accuracy. Numerical examples are presented to demonstrate the performance of the proposed adaptive NN-FEM.

Keywords: nearest-nodes finite element method, gradient of strain energy density, mesh intensity, mesh modification operator

1 Introduction

With powerful modern computers, the Finite Element Method (FEM) is now able to solve many complicated engineering problems. Nevertheless, it is still inefficient in dealing with extremely large scale problems that may require billions or even trillions of degrees of freedom (DOF) to achieve a specified accuracy. It seems that advances in computer capacity can never catch up with human demand for solving more complicated problems. Therefore, a lot of research effort has been devoted to developing more efficient adaptive finite element

*E-mail address:luoy@cc.umanitoba.ca

methods, e. g. [1–45] among others, so that more complicated engineering problems can be solved with available existing computer capacity. A rational way to achieve the maximal efficiency of finite element analysis, and thus to most efficiently make use of existing computer capacity, is to take three steps. First, for uniform meshes a finite element formulation should have the optimal convergence; Based on that, an adaptive algorithm is implemented to further improve the efficiency; The third step is to parallelize the above finite element procedure. A so-called nearest-nodes finite element method (NN-FEM) was recently developed in [46–49]. In NN-FEM, elements are mainly used for numerical integration; shape functions are constructed for each quadrature point by selecting a set of nodes that are the nearest to the quadrature point. With this strategy, NN-FEM is nearly not affected by element distortion, and it has a superior convergence compared to the conventional FEM. In implementing the above strategy, there are several ways available for constructing shape functions. In [47], shape functions are constructed by local moving polynomials, while in [46, 48] they are built using local multivariate Lagrange interpolation method [50]; inter-dependent shape functions are adopted in [49]. Among the above options, NN-FEM with inter-dependent shape functions has the optimal convergence rate. Although further improvement is still possible, NN-FEM with inter-dependent shape functions is used here as a base for implementing an adaptive algorithm.

There are basically three types of error estimators or error indicators for guiding mesh modification: smoothing-based [37, 41–44], residual-based [16, 51, 52] and gradient-based [53, 54]. Although in principle any of the above error estimators can be adopted in NN-FEM, the gradient-based error indicator is selected here. The main idea of the gradient-based error indicator is that for regions where there is a large variation of strain energy density, a denser mesh is needed and vice versa. Strain energy density contains all relevant information needed for mesh adaptation, e. g. material distribution, deformation, etc. In other finite element formulations, a counterpart equivalent to the strain energy density can always be found. Therefore, the gradient-based error indicator is generally applicable. The gradient-based error indicator has been successfully applied in an adaptive meshless method [54]. One attractive feature of NN-FEM is that higher-order shape functions can be constructed from simplex finite element meshes, i. e. a 2D mesh consisting of only three-node triangles or a 3D mesh consisting of only four-node tetrahedrons. High-order shape functions are required for calculating the strain energy density and its gradients; while finite element meshes consisting of only simplexes are easy to operate in mesh adaptation. Therefore, NN-FEM provides a favourable environment for implementing the gradient-based error indicator.

In this chapter, the gradient of strain energy density is used as a guide for mesh modification in NN-FEM. The layout of this chapter is as follows. In Section 2, NN-FEM with inter-dependent shape functions is briefly described for completeness; The gradient-based error indicator for NN-FEM is formulated in Section 3; Three mesh modification operators are defined in Section 4; Numerical examples are presented in Section 5 and concluding remarks are provided in Section 6.

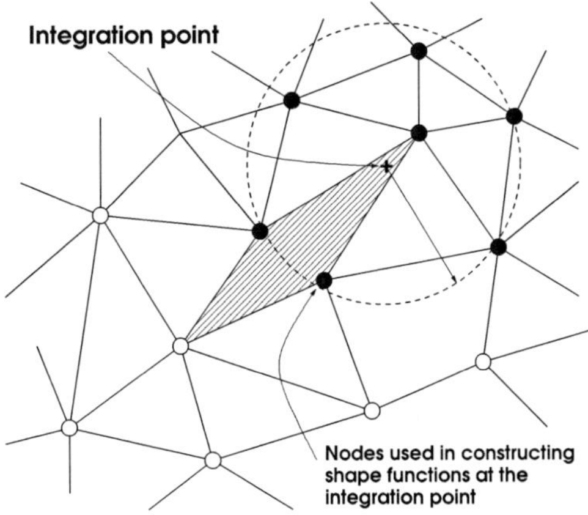

Figure 1: Nearest-Nodes FEM

2 Nearest-Nodes Finite Element Method with Inter-dependent Shape Functions

The major difference between NN-FEM and the conventional FEM is the way of constructing shape functions. In the conventional Finite Element Method, a set of shape functions are defined for a whole element, only nodes belonging to the element are involved in the definition of the shape functions. In NN-FEM, the above restriction is removed and shape functions are constructed for each quadrature point rather than for a whole element. Nodes from neighbour elements may be involved in the construction of shape functions of an element. This idea can be illustrated using a 2-D scenario as shown in Fig. 2. Where stiffness contribution from the quadrature point marked by '+' in the shaded element is being evaluated. To conduct the evaluation, a set of element nodes that are the nearest to the quadrature point are selected and shape functions are constructed using these element nodes. With this strategy, high-order shape functions can be constructed for finite element meshes consisting of only simplexes, which is a very attractive feature for implementing an adaptive algorithm. For a set of irregularly distributed nodes, there are several ways available for constructing shape functions, e. g. the moving least square (MLS) method [55], the local moving polynomial method [47], the local multivariate Lagrange interpolation method [50], the method of constructing inter-dependent shape functions [49], etc. Among them, inter-dependent shape functions leads to the optimal convergence.

Inter-dependent shape functions are constructed from the general solutions of the homogeneous Euler-Lagrange equations. Major steps of the construction are described in the following, detailed derivations can be found in [49]. Although in principle the procedure described in the following is applicable to three dimensional and large deformation problems, for simplicity, a plane stress model with small deformation is adopted here. The homogeneous Euler-Lagrange equations are obtained by, first, plugging the linear strain-displacement relations into the stress-strain relations, and then, the stress-strain relations are substituted into the equilibrium equations,

$$\left\{\begin{array}{l} D_1\dfrac{\partial^2 u}{\partial x^2}+D_3\dfrac{\partial^2 u}{\partial y^2}+(D_2+D_3)\dfrac{\partial^2 v}{\partial x\partial y}=0 \\ D_1\dfrac{\partial^2 v}{\partial y^2}+D_3\dfrac{\partial^2 v}{\partial x^2}+(D_2+D_3)\dfrac{\partial^2 u}{\partial x\partial y}=0 \end{array}\right. \tag{1}$$

in Eq. (1), u and v are the two displacements; D_1, D_2 and D_3 are three material parameters defined by Young's modulus (E) and Poisson's ratio (ν)

$$D_1=\frac{E}{1-\nu^2},\qquad D_2=\frac{\nu E}{1-\nu^2},\qquad D_3=\frac{E}{2(1+\nu)} \tag{2}$$

It is difficult to find the real general solutions of the partial differential equations in (1). Therefore, approximate general solutions in polynomial form are assumed using unknown coefficients, a_0,a_1, $\cdots$, a_5; b_0, b_1, $\cdots$, b_5,

$$\left\{\begin{array}{l} u=a_0+a_1\,x+a_2\,y+a_3\,x^2+a_4\,x\,y+a_5\,y^2 \\ v=b_0+b_1\,x+b_2\,y+b_3\,x^2+b_4\,x\,y+b_5\,y^2 \end{array}\right. \tag{3}$$

The assumed polynomials in Eq. (3) are of second-order. Higher or lower order polynomials can also be adopted. In principle the assumed polynomials should be complete and symmetrical with respect to the involved coordinates. The technique described in [56] is used to 'replace' the unknown coefficients by element nodal displacements. The assumed polynomials in Eq. (3) are substituted into Eq. (1); Two algebraic equations in the unknown coefficients are obtained. By solving the algebraic equations, two of the unknown coefficients are expressed by the rest. Here, a_4 and b_4 are solved to keep the symmetry of the polynomials.

$$\left\{\begin{array}{l} a_4=-\lambda_3 b_3-\lambda_1 b_5 \\ b_4=-\lambda_1 a_3-\lambda_3 a_5 \end{array}\right. \tag{4}$$

where the two non-dimensional material parameters are defined as

$$\lambda_1=\frac{2D_1}{D_2+D_3},\qquad \lambda_3=\frac{2D_3}{D_2+D_3} \tag{5}$$

With the relations in Eq. (4), the assumed polynomials become

$$\left\{\begin{array}{l} u=a_0+a_1\,x+a_2\,y+a_3\,x^2-b_3\lambda_3 x\,y+a_5 y^2-b_5\lambda_1 x\,y \\ v=b_0+b_1\,x+b_2\,y-a_3\lambda_1 x\,y+b_3\,x^2-a_5\lambda_3 x\,y+b_5 y^2 \end{array}\right. \tag{6}$$

Put in a matrix form, Eq. (6) is transformed into

$$\boldsymbol{u}=\boldsymbol{p}(x,y)\boldsymbol{a}^T \tag{7}$$

where

$$\begin{aligned} &\boldsymbol{u}=\left[\begin{array}{ll} u & v \end{array}\right]^T \\ &\boldsymbol{p}(x,y)=\left[\begin{array}{cccccccccc} 1 & 0 & x & 0 & y & 0 & x^2 & -\lambda_3 xy & y^2 & -\lambda_1 xy \\ 0 & 1 & 0 & x & 0 & y & -\lambda_1 xy & x^2 & -\lambda_3 xy & y^2 \end{array}\right] \\ &\boldsymbol{a}=\left[\begin{array}{cccccccccc} a_0 & b_0 & a_1 & b_1 & a_2 & b_2 & a_3 & b_3 & a_5 & b_5 \end{array}\right]^T \end{aligned} \tag{8}$$

The polynomials are now inter-dependent and contain ten unknown coefficients. The unknown coefficients need to be expressed by nodal displacements. To that end, five nearest nodes to a concerned quadrature point are selected, each node has two displacements that can be used as two conditions. The polynomials in Eq. (6) are forced to satisfy the ten conditions at the five nearest nodes, i. e.,

$$\boldsymbol{P}\boldsymbol{a} = \bar{\boldsymbol{u}} \tag{9}$$

where

$$\boldsymbol{P} = \begin{bmatrix} 1 & 0 & x_1 & 0 & y_1 & 0 & x_1^2 & -\lambda_3 x_1 y_1 & y_1^2 & -\lambda_1 x_1 y_1 \\ 0 & 1 & 0 & x_1 & 0 & y_1 & -\lambda_1 x_1 y_1 & x_1^2 & -\lambda_3 x_1 y_1 & y_1^2 \\ 1 & 0 & x_2 & 0 & y_2 & 0 & x_2^2 & -\lambda_3 x_2 y_2 & y_2^2 & -\lambda_1 x_2 y_2 \\ 0 & 1 & 0 & x_2 & 0 & y_2 & -\lambda_1 x_2 y_2 & x_2^2 & -\lambda_3 x_2 y_2 & y_2^2 \\ \cdots & & \cdots & & \cdots & & \cdots & \cdots & \cdots & \cdots \\ 1 & 0 & x_5 & 0 & y_5 & 0 & x_5^2 & -\lambda_3 x_5 y_5 & y_5^2 & -\lambda_1 x_5 y_5 \\ 0 & 1 & 0 & x_5 & 0 & y_5 & -\lambda_1 x_5 y_5 & x_5^2 & -\lambda_3 x_5 y_5 & y_5^2 \end{bmatrix}$$
$$\bar{\boldsymbol{u}} = \begin{bmatrix} u_1 & v_1 & u_2 & v_2 & u_3 & v_3 & u_4 & v_4 & u_5 & v_5 \end{bmatrix}^T \tag{10}$$

In the above expressions, $x_1 \quad y_1 \quad \cdots \quad x_5 \quad y_5$ and $u_1 \quad v_1 \quad \cdots \quad u_5 \quad v_5$ are, respectively, the coordinates and displacements of the five selected nearest nodes. To guarantee the existence of the inverse of matrix $\boldsymbol{P}$, singularity is checked in selecting the nearest nodes as described in [49].

By solving the unknown coefficients from Eq. (9), the displacements now have the following expression,

$$\boldsymbol{u} = \boldsymbol{p}(x, y)\boldsymbol{P}^{-1}\bar{\boldsymbol{u}} = \boldsymbol{\phi}(x, y)\bar{\boldsymbol{u}} \tag{11}$$

where $\boldsymbol{\phi}(x, y)$ contains the shape functions

$$\boldsymbol{\phi}(x, y) = \boldsymbol{p}(x, y)\boldsymbol{P}^{-1} \tag{12}$$

and $\boldsymbol{\phi}(x, y)$ is a (2×5) matrix function,

$$\boldsymbol{\phi}(x, y) = \begin{bmatrix} \varphi_1^u & \psi_1^u & \cdots & \varphi_5^u & \psi_5^u \\ \varphi_1^v & \psi_1^v & \cdots & \varphi_5^v & \psi_5^v \end{bmatrix} \tag{13}$$

The shape functions are different from the conventional ones in the sense that they are inter-dependent. With the obtained shape functions, the principle of minimum potential energy or the principle of virtual work can be used as a base to establish the finite element equations. As the inter-dependent shape functions satisfy the Kronecker delta condition, application of essential boundary conditions can be done exactly in the same way as in the conventional Finite Element Method.

3 Gradient of Strain Energy Density as a Guide for Mesh Modification

A mesh adaptation procedure consists of two major components: an error estimator (or error indicator) and a set of operators for modifying the mesh. The error estimator or indicator

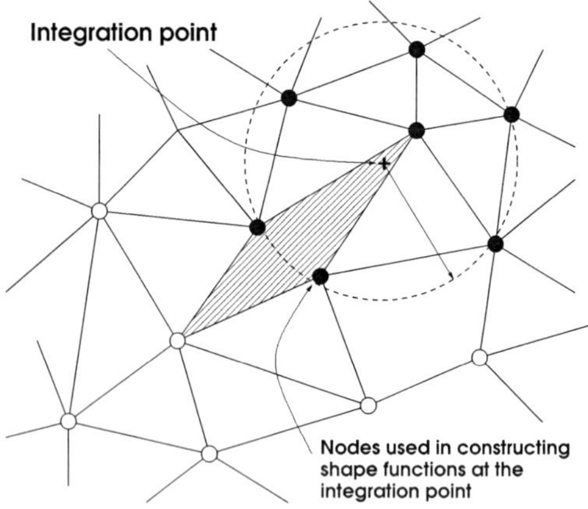

Figure 2: Nearest-Nodes FEM

tells where the mesh need to be refined or coarsened. A corresponding operator is selected to do the job. In this section, how the gradient of strain energy density is used as a guide for mesh modification is described. A set of mesh modification operators will be defined in the following section.

In solid mechanics, the strain energy density is defined as the scalar product of the stress and the strain vectors,

$$e = e(\boldsymbol{x}) = \frac{1}{2}\boldsymbol{\sigma}^T \cdot \boldsymbol{\varepsilon} \tag{14}$$

Theoretically, the strain energy density is a continuous or piece-wise continuous function of spatial coordinates, $\boldsymbol{x}$. In nearest-nodes finite element method, the strain energy density can be approximated by nodal displacements as

$$e \approx \tilde{e}(\boldsymbol{x}) = \frac{1}{2}\bar{\boldsymbol{u}}_q^T \tilde{\boldsymbol{K}}_q \bar{\boldsymbol{u}}_q = \frac{1}{2}\boldsymbol{u}_q^T \boldsymbol{B}_q^T \boldsymbol{D} \boldsymbol{B}_q \boldsymbol{u} \tag{15}$$

where $\tilde{\boldsymbol{K}}_q = \tilde{\boldsymbol{K}}_q(\boldsymbol{x}) = \boldsymbol{B}_q^T \boldsymbol{D} \boldsymbol{B}_q$ is a matrix function; $\boldsymbol{u}_q$ is a vector consisting of nodal displacements from the involved finite element nodes; $\boldsymbol{B}_q$ is the B-matrix relating the strains and the displacements.

After numerical integration, the matrix $\tilde{\boldsymbol{K}}_q$ in Eq. (15) will lead to the element stiffness matrix. Therefore, the quality of the approximate strain energy density in Eq. (15) is closely related to the quality of the element stiffness matrix that in turn affects the solution accuracy. On the other hand, the quality of the approximate strain energy density is related to shape functions. In nearest-nodes finite element method [46–48], the quality of shape functions is directly determined by the locations of the selected nearest nodes. Based on the above reasoning and the calculus theory, the distribution of element nodes should conform with the gradient of strain energy density to reduce errors in finite element solutions, i. e. a larger variation of strain energy density needs a denser mesh and vice versa. Now the question is how to quantitatively correlate the gradient of strain energy density to mesh density. For example, if the mesh in a region needs to be refined, then how fine should the mesh be. For this purpose, mesh intensity is introduced. Mesh intensity is defined as the ratio of a gradient of strain energy density to mesh density.

After solving the finite element equations, values of strain energy density at element nodes are obtained by Eq. (15). The gradients of the strain energy density at element nodes can be calculated by the moving least square (MLS) method [55] or the local multivariate Lagrange interpolation [50].

$$\nabla e = [\ e_x \quad e_y \quad e_z\ ]^T = [\ \frac{\partial e}{\partial x} \quad \frac{\partial e}{\partial y} \quad \frac{\partial e}{\partial z}\ ]^T \tag{16}$$

where x, y and z represent, respectively, the three coordinate axes in a rectangular coordinate system.

Mesh intensity is defined as

$$\mathcal{M}_I = \frac{\sqrt{e_x^2 + e_y^2 + e_z^2}}{\mathcal{M}_D} \tag{17}$$

where $\mathcal{M}_D$ is mesh density that is defined as the number of element nodes (N) per unit volume (S) of the problem domain,

$$\mathcal{M}_D = \frac{N}{S} \tag{18}$$

An alternative definition of mesh intensity is based on variation of strain energy density along element edges, i. e.,

$$\mathcal{M}_I^{(k)} = \frac{\left| e_{(k_1)} - e_{(k_2)} \right|}{l_{(k)}^2} \tag{19}$$

where sub- or superscript (k) represents the k-th element edge; Subscripts k_1 and k_2 indicate the two element nodes connected by the edge. $l_{(k)}$ is the length of the element edge. e_{k_1} and e_{k_2} are the strain energy density at the two nodes. Although the mesh intensity in Eq. (19) is an approximate version of the one defined in Eqs. (17) and (18), it is much easier to implement and computationally more efficient. Therefore, it was used in calculating the results presented in Section 5.

Based on the idea that a larger gradient (or variation) requires a denser mesh and vice versa, for regions with higher gradients, more elements and element nodes will be added and the mesh density there will be increased; for regions with lower gradients, redundant elements and element nodes will be removed and the mesh density there will be reduced. As a result, the ratio of the gradient of strain energy density to the mesh density, i. e. the mesh intensity in Eq. (17), over the problem domain will become more and more uniform. Theoretically, if the adaptation process is allowed to continue for a sufficient number of loops, the mesh intensity over the domain will reach a constant, i. e.

$$\mathcal{M}_I = c \qquad \text{or} \qquad \mathcal{M}_I^{(k)} = c \quad (k = 1, 2, \cdots, N_{\text{edge}}) \tag{20}$$

where c is a constant pertinent to the accuracy of finite element solutions; N_{edge} is the total number of element edges in the finite element mesh.

For a given field, its gradients are also given. In finite element analysis, increasing mesh density would lead to improvement of solution accuracy. Based on Eqs. (17) and (20), increasing mesh intensity would also lead to smaller mesh intensity, i. e. a smaller constant c. Therefore, the smaller the mesh intensity, the higher the solution accuracy is. Nevertheless, it is very difficult, if not impossible, to establish an explicit function relation between a specified solution accuracy and the required mesh intensity. An iteration procedure has to be adopted to gradually tune the mesh intensity to achieve a specified accuracy. The tuning job is done by mesh modification using a set of mesh modification operators, i. e., mesh refinement, mesh coarsening and mesh smoothing. The selection of a specific operator is based on the following criterion

$$\mathcal{M}_I^{(k)} \quad (k = 1, 2, \cdots, N_{\text{edge}}) \begin{cases} > c_{\{i\}} & \text{refine} \\ = c_{\{i\}} & \text{unchanged} \\ < c_{\{i\}} & \text{coarsen} \end{cases} \tag{21}$$

where the subscript $\{i\}$ indicates the i-th iteration loop. If the specified accuracy is not satisfied, the mesh intensity will be reduced and a new iteration loop will be started. The adjustment of the mesh intensity should be done automatically based on convergence information.

There are different convergence criteria adopted in the literature, for example, the displacement at a control point, the stress at a critical location, etc. Most of them are based on local rather than global information. According to the principle of minimum potential energy, among all the admissible solutions for a structure under static external forces, the real solution would make the system have the minimum potential energy. Accordingly, the convergence criterion should be global and it should be based on the total potential energy. Therefore, the relative error in total potential energy is adopted as a convergence criterion for the adaptive NN-FEM,

$$\eta_{\{i\}} = \frac{\left|\tilde{\Pi}_{\{i\}} - \tilde{\Pi}_{\{i-1\}}\right|}{\left|\tilde{\Pi}_{\{i\}}\right|} < \delta \tag{22}$$

where δ is a prescribed positive small number; $\tilde{\Pi}_{\{i\}}$ is the potential energy calculated at the end of the i-th mesh modification loop,

$$\tilde{\Pi}_{\{i\}} = \frac{1}{2}\boldsymbol{u}_{\{i\}}^T \boldsymbol{K}_{\{i\}} \boldsymbol{u}_{\{i\}} - \boldsymbol{u}_{\{i\}}^T \boldsymbol{P}_{\{i\}} \tag{23}$$

Subscript $\{i\}$ indicates that the quantities are related to the i-th iteration loop.

With convergence information from the i-th iteration loop, i. e. the relative error in potential energy, $\eta_{\{i\}}$, the desired mesh intensity for the $(i+1)$-th iteration loop can be adjusted as

$$c_{\{i+1\}} = \left|1 - \eta_{\{i\}}\right| c_{\{i\}} \tag{24}$$

Depending on the solved problem, the relative error (η) from the first few iterations may be larger than unit. Therefore, an absolute value is adopted in Eq. (24) to prevent a negative mesh intensity to occur. When the iteration process approaches the convergence point, the relative error (η) will be smaller and smaller and the adjustment on mesh intensity should be accordingly reduced. As indicated by Eq. (24), if the relative error from previous step is zero, there should be no adjustment on the mesh intensity, and the mesh intensity is the one required to achieve the specified accuracy.

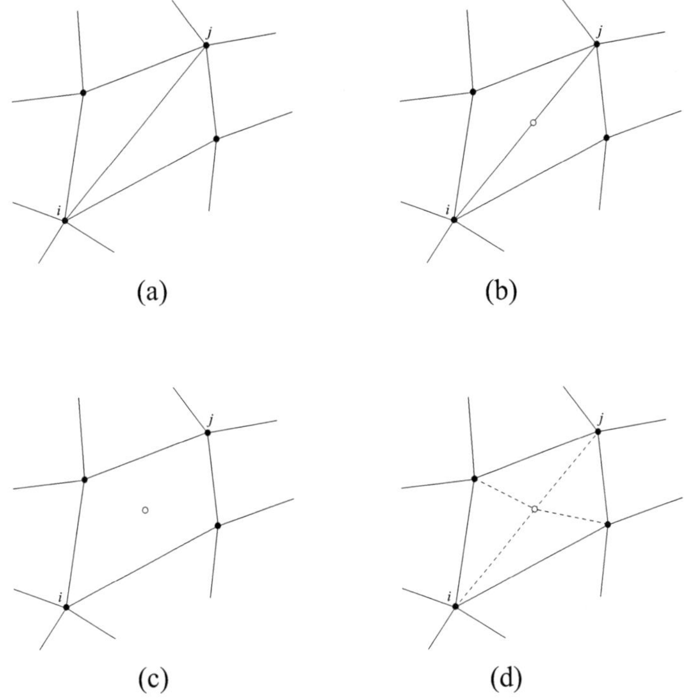

Figure 3: Local mesh refinement

4 Mesh Modification Ooperators

A set of three mesh modification operators, mesh refinement, mesh coarsening and mesh smoothing, are defined in this section. Each of them is a unique tool in modifying a finite element mesh.

4.1 Mesh refinement

Mesh refinement is a basic operation for mesh adaptation. Mesh refinement can be conducted in an element-by-element way or in an edge-by-edge way. The later is adopted in this paper. If an element edge is to be refined based on the refinement criterion stated in Eq. (21), a new node will be added at the middle point of the edge. After addition of new nodes, there are two ways to update element connectivities, one is local triangulation that is more efficient if only a small region of the mesh is to be modified, the other is global triangulation. Both of them are based on Delaunay algorithm [57]. The major difference is that for a local triangulation, a local cavity must be formed before the Delaunay algorithm is applied. The procedure is depicted in Fig. 3 and described in the following. In Fig. 3(a), the element edge i-j is to be refined. Therefore, a new node is added at the middle point of the edge, Fig. 3(b); the old edge i-j is removed and a cavity is formed, Fig. 3(c); the cavity is then triangulated, Fig. 3(d). An alternative way for local mesh refinement is to use templates to split elements, e. g. [58, 59] among others.

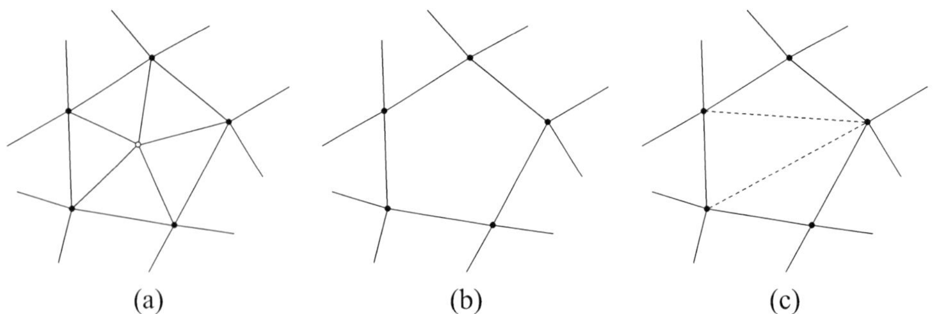

Figure 4: Mesh coarsening operations

4.2 Mesh coarsening

Mesh coarsening is a crucial operator to improve mesh efficiency for solving time-dependent problems. Mesh coarsening can also be conducted in different ways. Here, mesh coarsening is carried out in a node-by-node way. Each node in the finite element mesh is checked if it is allowed to be removed or not. First, all geometric key points and control points can not be removed. For the rest, only those nodes satisfying both of the following conditions can be deleted. First, all the edges connected to the node, see Fig. 4(a), must satisfy the coarsening criterion stated in Eq. (21); Second, if the node and the edges connecting to it were removed, a cavity would be formed, Fig. 4(b). After updating element connectivity, new element edges, as indicated by the dashed line in Fig. 4(c), would be generated. To remove the node, any of the new edges must not satisfy the refinement criterion in Eq. (21). The second condition is important to prevent a dead loop of refinement-coarsening iteration. After marking all the nodes that can be removed, there are again two options to update element connectivity: local and global triangulation. The local triangulation is depicted in Fig. 4.

4.3 Mesh smoothing

After applying refinement and coarsening, the nodes in the finite element mesh may not be at their optimal locations for reducing error in finite element solutions. The nodes can be repositioned to further reduce solution error. There are different mesh smoothing techniques available from the literature, e. g. [60] and the references therein, to relocate the nodes in a finite element mesh. In this paper the r-adaptation algorithm proposed in [61] is adopted. The algorithm is also guided by the gradient of strain energy density in Eq. (16). Details of the algorithm can be found in [61], major formulas are provided in the following. Although theoretically the gradient components in Eq. (16) are continuous functions of spatial coordinates, their approximate expressions obtained from finite element solutions may be discontinuous or even in wild fluctuation due to local errors in finite element solutions. This may result in rough or even ill-conditioned finite element meshes. A function smoothing or averaging technique can be used to resolve the above problem.

In weighted Laplacian smoothing, to re-position an element node (p), all element nodes that share an edge with the node (p) are found using element connectivity, cf. Fig. 5. All the found nodes, assume in total $(n+1)$ nodes including node p, form a star around node p, and the involved elements form a polygon (or a polyhedron for 3D problems). A set of

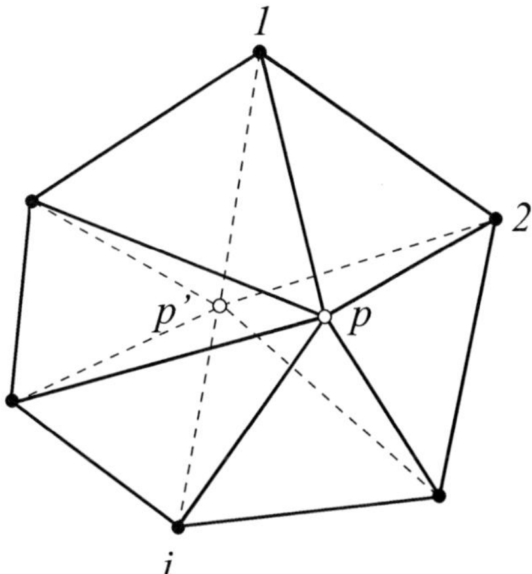

Figure 5: Re-position of a typical interior node

weight functions are defined at the $(n+1)$ nodes as

$$w_i^{\alpha} = \begin{cases} 1, & \text{if } \max\limits_i(|e_{\alpha}^{(i)}|) = 0 \\ \dfrac{|e_{\alpha}^{(i)}|}{\sum\limits_i^{n+1} |e_{\alpha}^{(i)}|}, & \text{if } \max\limits_i(|e_{\alpha}^{(i)}|) > 0 \end{cases} \qquad (\alpha = x,\, y,\, z) \tag{25}$$

where $e_{\alpha}^{(i)}$ is the gradient of strain energy density at node (i) in α-axis.

The new position $(x_{p'},\, y_{p'},\, z_{p'})$ of node (p) is given by

$$x_{p'} = \frac{\sum\limits_i^{n+1} w_i^x x_i}{\sum\limits_i^{n+1} w_i^x}, \quad y_{p'} = \frac{\sum\limits_i^{n+1} w_i^y y_i}{\sum\limits_i^{n+1} w_i^y}, \quad z_{p'} = \frac{\sum\limits_i^{n+1} w_i^z z_i}{\sum\limits_i^{n+1} w_i^z} \tag{26}$$

where $(x_i,\, y_i,\, z_i)$ are the old coordinates of node i of the $(n+1)$ nodes.

For nodes close to or on the problem boundary, some special treatments are needed. For the node (p) shown in Fig. 6(a), the polygon surrounding the node is a concave. The new position (p') obtained by Eqs. (25) and (26) is outside the polygon. The shaded part in the figure may be another element or a concave angle of the problem domain boundary. In either of the cases, the new node (p') need be pulled back into the polygon, to avoid forming overlapped elements or to maintain the shape of the problem boundary. The node (p) in Fig. 6(b) is on the problem boundary. To maintain the shape of the problem boundary, nodes on the boundary are only allowed move on the boundary.

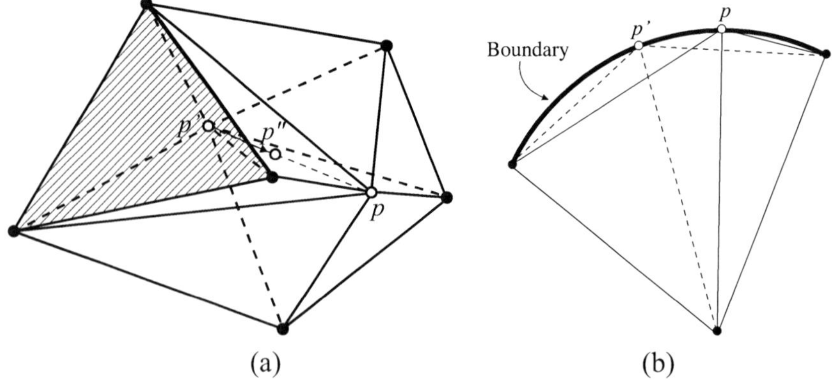

Figure 6: Re-position of special and boundary nodes

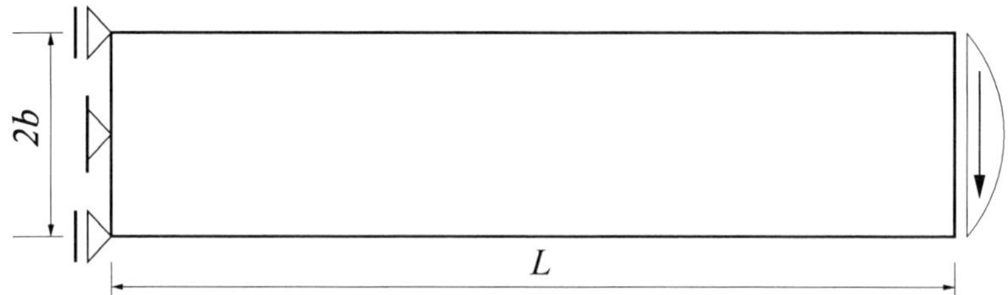

Figure 7: Cantilever beam under shear force

5 Numerical Examples

The performance of the proposed adaptive NN-FEM was examined by numerical examples. The first example is a cantilever beam as shown in Fig. 7. The beam has a length $L = 10$ and a width $2b = 2$. The material is linear elastic with Young's modulus $E = 1000.0$ and Poisson's ratio $\nu = 0.0$. The convergence criterion, i. e. the relative error in the total potential energy, was specified as $\eta \leq 1\%$.

The mesh adaptation was started with a coarse structured mesh, Fig. 8(a). The adapted meshes are displayed in Figs. 8(b), (c) and (d). Although all the three mesh operators were involved in modifying the mesh, refinement and smoothing were playing the dominant role in this example. After three loops of iteration, the convergence criterion was reached. The convergence in the total potential energy and in the displacement (at the cross-section center of the loaded end) are shown in Figs. 9(a) and (b), where the obtained displacements are normalized using the theoretical solution [54]. Relative errors in the total potential energy and in the displacement are plotted in Fig. 10, where the relative error in the displacement was calculated in a similar way as that of the total potential energy. It can be observed from Fig. 9 that the convergence trends in the total potential energy and in the displacement are similar, if the sign of the total potential energy in Fig. 9(a) is reversed. However, it can also be observed from Fig. 10 that relative error in displacement is smaller than that in the total potential energy. Therefore, the above observations confirm that a convergence in the total

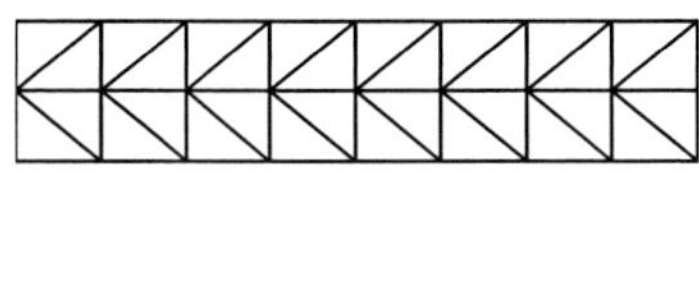

(a) 27 nodes (Iteration 0)

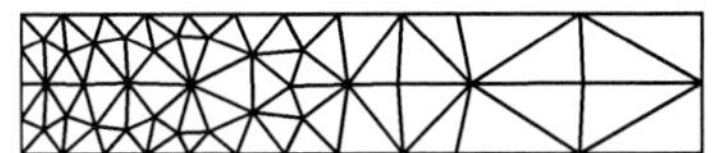

(b) 55 nodes (Iteration 1)

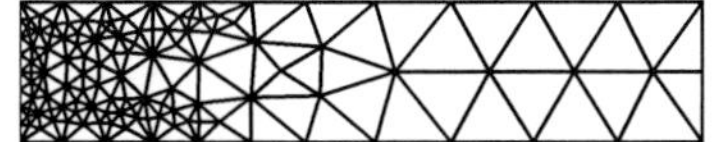

(c) 126 nodes (Iteration 2)

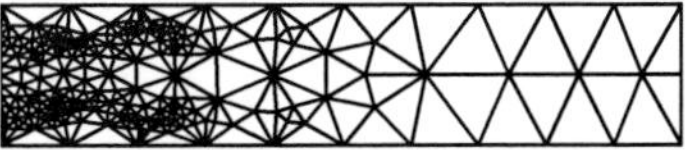

(d) 241 nodes (Iteration 3)

Figure 8: Mesh refinement

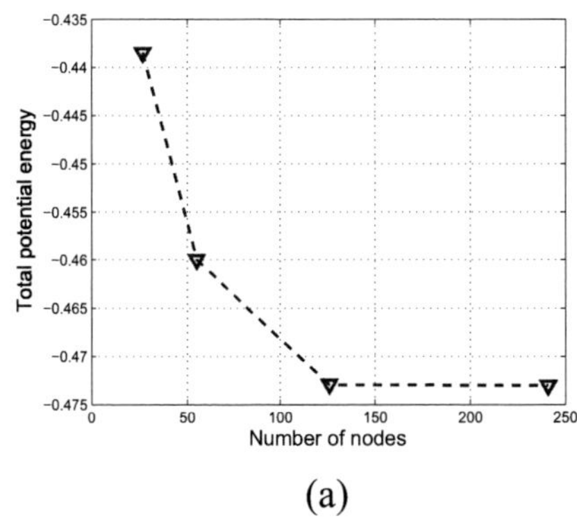

(a)

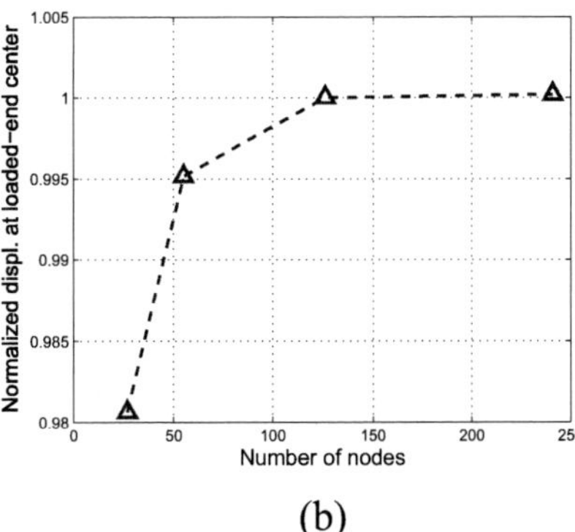

(b)

Figure 9: Convergence (a) Total potential energy; (b) Normalized displacement at cross-section center of loaded end

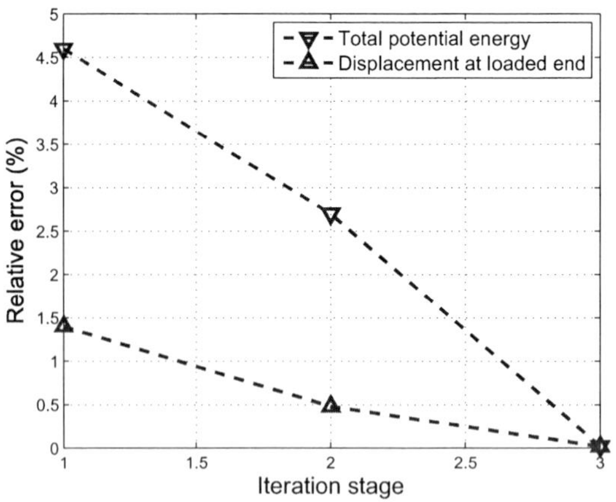

Figure 10: Relative error

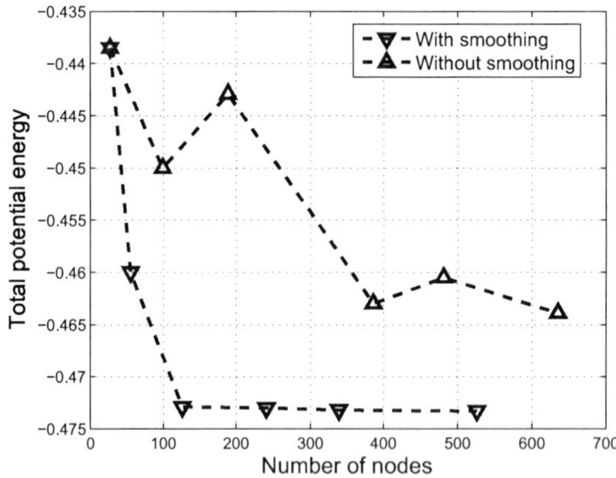

Figure 11: Comparison of convergence in total potential energy with and without mesh smoothing

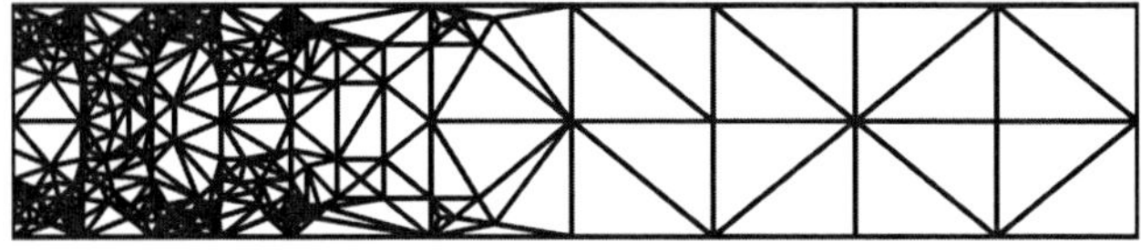

Figure 12: Adapted finite element mesh without smoothing

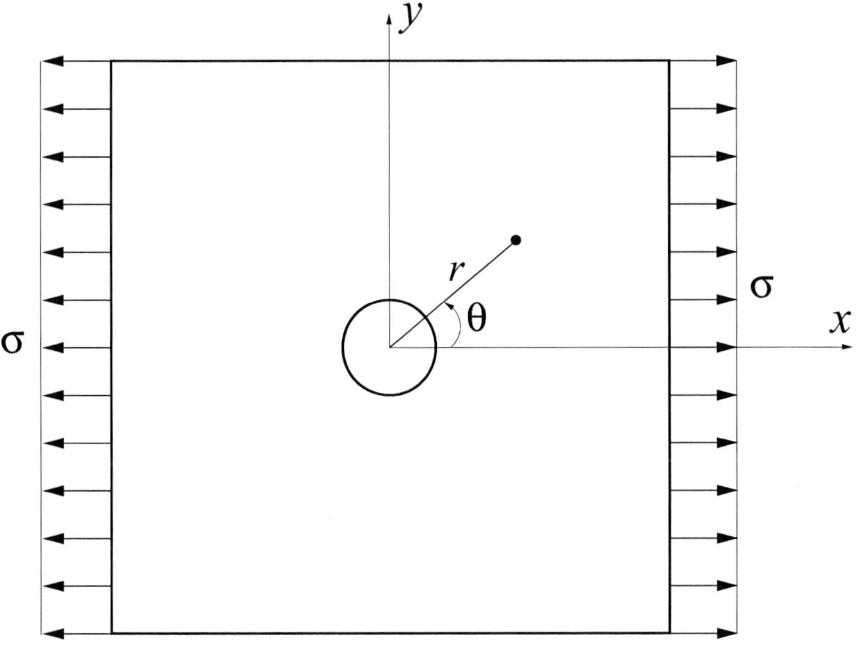

Figure 13: Plate with a small hole

potential energy is an adequate condition for finite element fields such as displacements to converge. Nevertheless, mesh smoothing operator has a significant effect on the convergence process as shown in Fig. 11. If mesh smoothing is used, the convergence process is steady and smooth; otherwise, the convergence process will be oscillating. Furthermore, compared with a process with mesh smoothing, a process without mesh smoothing needs more element nodes to achieve a similar accuracy. One adapted mesh obtained without using mesh smoothing is displayed in Fig. 12, which is obviously not as smooth as those shown in Fig. 8.

The second example is a plate with a hole, Fig. 13. Both of the length and the width of the plate are 10; the material is linear elastic described by Young's modulus $E = 1000.0$ and Poisson's ratio $\nu = 0.25$. The plate has a unit thickness. The radius of the hole is 2. The intensity of the distributed load is 10. All the parameters have consistent units. For the symmetry in both geometry and loading, one quarter of the plate was analyzed. A dense initial mesh was selected, see Fig. 14(a). The mesh is dense enough to produce very accurate results. The purpose of this example is to examine the performance of mesh coarsening operator. The distribution of strain energy density obtained by the initial mesh is shown in Fig. 14(b).

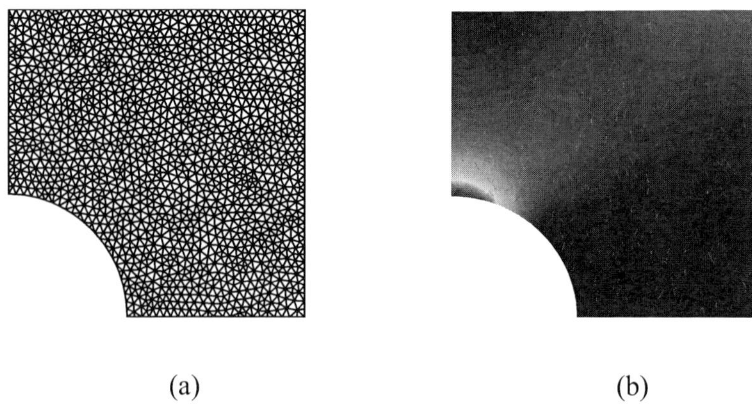

(a) (b)

Figure 14: (a) Initial mesh (1427 nodes); (b) Distribution of strain energy density

The results obtained with the initial and adapted meshes are listed in Table 1. The adapted meshes are plotted in Fig. 15. It can be seen from Table 1, while keeping solution

Table 1: Calculation results from initial and adapted meshes

Iteration Loop	Number of Nodes	Vertical Displacement hole apex	von Mises Stress at at hole apex	Total Potential Energy
0	1427	-0.05485	47.1513	-50.4981
1	263	-0.05490	47.1617	-50.5131
2	192	-0.05492	47.1620	-50.5202

accuracy at approximately the same level, the number of nodes is greatly reduced by mesh adaptation, 86.5% of the nodes in the initial mesh were removed.

In the third example, a square-shaped material body contains another ring-shaped material body, Fig. 16(a). The ratio of their elasticity modulus is 100. Uniform stretch is applied on all the four edges of the square, so that a uniform strain field is generated. The example is used to examine if the proposed adaptive finite element method is able to capture the variation in material property. A uniform unstructured mesh is used for starting mesh adaptation, Fig. 16(b). The resulting adapted mesh is shown in Fig. 17.

6 Concluding Remarks

An adaptive finite element method is formulated in this paper. The adaptive finite element method is based on the newly developed nearest-nodes finite element method (NN-FEM). Mesh modification is guided by the gradient of strain energy density, i. e., a larger gradient requires a denser mesh and vice versa. To relate the gradient of strain energy density to mesh density, mesh intensity is introduced. A finite element mesh is iteratively modified by

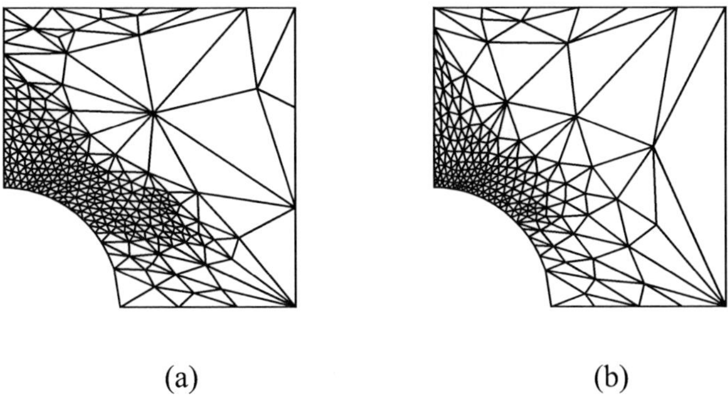

(a) (b)

Figure 15: Adapted meshes (a) 263 nodes; (b) 192 nodes

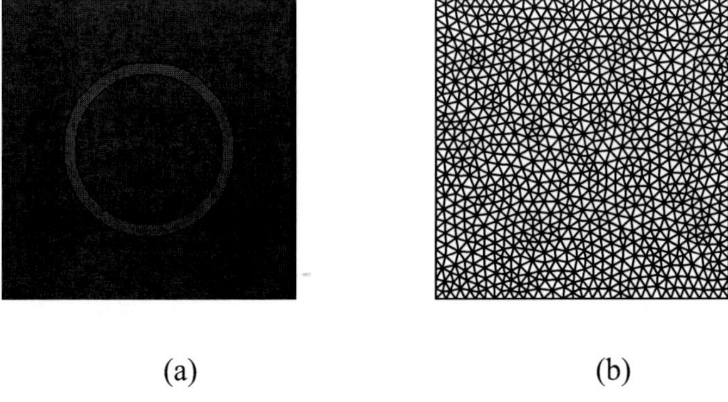

(a) (b)

Figure 16: (a) Heterogeneous materials; (b) Initial finite element mesh

Figure 17: Adapted finite element mesh

a set of operators to make its density conform to the gradient of strain energy density. The operators include mesh refinement, mesh coarsening and mesh smoothing. The selection of a proper operator for a specific mesh region is determined by a set of criteria based on mesh intensity. The loop of mesh modification is stopped when the relative error in the total potential energy is less than a prescribed value. Numerical results demonstrate that the proposed adaptive NN-FEM is sensitive, robust and effective.

References

[1] M. Ainsworth, J.Z. Zhu, A.W. Craig, and O.C. Zienkiewicz. Analysis of the zienkiewicz-zhu a-posteriori error estimator in the finite element method. *International Journal for Numerical Methods in Engineering* , 28(9):2161 – 2174, 1989.

[2] I. Babuška and A. Miller. Post-processing approach in the finite element method - Part 3: a posteriori error estimates and adaptive mesh selection. *International Journal for Numerical Methods in Engineering*, 20:2311 – 2324, 1984.

[3] B. Boroomand and O.C. Zienkiewicz. Recovery procedures in error estimation and adaptivity. ii. adaptivity in nonlinear problems of elasto-plasticity behaviour. *Computer Methods in Applied Mechanics and Engineering* , 176:127 – 146, 1999.

[4] Mark E. Botkin and Hui-Ping Wang. An adaptive mesh refinement of quadrilateral finite element meshes based upon a posteriori error estimation of quantities of interest: Linear static response. *Engineering with Computers*, 20:31 – 37, 2004.

[5] C. Carstensen and B. Faermann. Mathematical foundation of a posteriori error estimates and adaptive mesh-refining algorithms for boundary integral equations of the first kind. *Engineering Analysis with Boundary Elements* , 25:497 – 509, 2001.

[6] K.C. Chellamuthu and Nathan Ida. 'a posteriori' error indicator and error estimators for adaptive mesh refinement. *COMPEL - The International Journal for Computation and Mathematics in Electrical and Electronic Engineering* , 14:139 – 156, 1995.

[7] R.D. Cook and J. Avrashi. Error estimation and adaptive meshing for vibration problems. *Computers and Structures*, 44:619 – 626, 1992.

[8] D. Cook, Robert, Jacob Avrashi, and Chin-Hsu Lin. Buckling analysis: Alternative formulations, error estimation, and adaptive meshing. *Finite Elements in Analysis and Design*, 11:55 – 65, 1992.

[9] A.W. Craig, M. Ainsworth, J.Z. Zhu, and O.C. Zienkiewicz. H and h-p version error estimation and adaptive procedures from theory to practice. *Engineering with Computers (New York)*, 5:221 – 234, 1989.

[10] P. Fernandes, P. Girdinio, G. Molinari, and M. Repetto. Local error estimation procedures as refinement indicators in adaptive meshing. *IEEE Transactions on Magnetics*, 27:4189 – 4192, 1991.

[11] J. P. De S. R. Gago, D. W. Kelly, and O. C. Zienkiewicz. *A posteriori* error analysis and adaptive processes in the finite element method: Part ii — adaptive mesh refinement. *Int. J. Numer. Meth. Engng.*, 19:1621–1656, 1983.

[12] R. Heinzl, M. Spevak, P. Schwaha, and T. Grasser. A novel technique for coupling three dimensional mesh adaptation with an a posteriori error estimator. In *2005 PhD Research in Microelectronics and Electronics (IEEE Cat. No.05EX1148)*, volume 1, pages 201 – 204, Lausanne, Switzerland, 2005.

[13] Weizhang Huang and Weiwei Sun. Variational mesh adaptation. ii. error estimates and monitor functions. *Journal of Computational Physics*, 184:619 – 48, 2003/01/20.

[14] Christopher R. Johnson and Robert S. MacLeod. Nonuniform spatial mesh adaptation using a posteriori error estimates: applications to forward and inverse problems. *Applied Numerical Mathematics*, 14:311 – 326, 1994.

[15] P. Katragadda and I.R. Grosse. Posteriori error estimation and adaptive mesh refinement for combined thermal-stress finite element analysis. *Computers and Structures*, 59:1149 – 1163, 1996.

[16] D. W. Kelly, J. P. De S. R. Gago, O. C. Zienkiewicz, and I. Babuška. *A posteriori* error analysis and adaptive processes in the finite element method: Part i — error analysis. *Int. J. Numer. Meth. Engng.*, 19:1593–1619, 1983.

[17] E. Kita and N. Kamiya. Error estimation and adaptive mesh refinement in boundary element method, an overview. *Engineering Analysis with Boundary Elements*, 25(7):479 – 495, 2001.

[18] N. Mahomed and M. Kekana. Error estimator for adaptive mesh refinement analysis based on strain energy equalization. *Computational Mechanics*, 22:355 – 366, 1998.

[19] Eugenio Onate, Joaquin Arteaga, Julio Garcia, and Roberto Flores. Error estimation and mesh adaptivity in incompressible viscous flows using a residual power approach. *Computer Methods in Applied Mechanics and Engineering*, 195(4-6):339 – 362, 2006.

[20] G.S. Palani, Nagesh R. Iyer, and B. Dattaguru. A new posteriori error estimator and adaptive mesh refinement strategy for 2-d crack problems. *Engineering Fracture Mechanics*, 73:802 – 819, 2006.

[21] R. Radovitzky and M. Ortiz. Error estimation and adaptive meshing in strongly nonlinear dynamic problems. *Computer Methods in Applied Mechanics and Engineering*, 172(1-4):203 – 240, 1999.

[22] Maharavo Randrianarivony. Anisotropic finite elements for the stokes problem: A posteriori error estimator and adaptive mesh. *Journal of Computational and Applied Mathematics*, 169:255 – 275, 2004.

[23] E. Rank and O. C. Zienkiewicz. Simple error estimator in the finite element method. *Communications in Applied Numerical Methods*, 3:243 – 249, 1987.

[24] Rolf Rannacher and Franz-Theo Suttmeier. A posteriori error estimation and mesh adaptation for finite element models in elasto-plasticity. *Computer Methods in Applied Mechanics and Engineering*, 176:333 – 361, 1999.

[25] A. Samuelsson, N.-E. Wiberg, and L.F. Zeng. Effectivity of the zienkiewicz-zhu error estimate and two 2d adaptive mesh generators. *Communications in Numerical Methods in Engineering*, 9:687 – 699, 1993.

[26] T. Strouboulis, K.A. Haque, and M.F. Mahmoud. Recent experiences with adaptive mesh optimization and a-posteriori error estimation. In *American Society of Mechanical Engineers (Paper)*, pages 11 –, Dallas, TX, USA, 1990.

[27] A. Tabarraei and N. Sukumar. Adaptive computations using material forces and residual-based error estimators on quadtree meshes. *Computer Methods in Applied Mechanics and Engineering*, 196:2657 – 2680, 2007.

[28] R. Verfurth. Posteriori error estimation and adaptive mesh-refinement techniques. *Journal of Computational and Applied Mathematics* , 50(1-3):67 – 83, 1994.

[29] Yoshitaka Wada, Mamtimin Geni, Masanao Matsumoto, and Masanori Kikuchi. Effective adaptation of hexahedral mesh using local refinement and error estimation. In *Key Engineering Materials*, volume 243-244, pages 27 – 32, Dunhuang, China, 2003.

[30] H. Wu and I.G. Currie. Adaptive mesh refinement using a-posteriori finite element error estimation. *Transactions of the Canadian Society for Mechanical Engineering* , 24:247 – 261, 2000.

[31] Ryoji Yuuki, Gou-Qiang Cao, and Masatoshi Tamaki. Efficient error estimation and adaptive meshing method for boundary element analysis. *Advances in Engineering Software*, 15:279 – 287, 1992.

[32] Chun-Hua Zhou. Mesh adaptation technique via a posterior error estimate for incompressible navier-stokes equations. *Chinese Journal of Computational Mechanics* , 22:705 – 710, 2005.

[33] J. Z. Zhu and O. C. Zienkiewicz. Adaptive techniques in the finite element method. *Communications in Applied Numerical Methods* , 4:197 – 204, 1988.

[34] J.Z. Zhu and O.C. Zienkiewicz. Superconvergence recovery technique and a posteriori error estimators. *International Journal for Numerical Methods in Engineering* , 30:1321 – 1339, 1990.

[35] O. C. Zienkiewicz, A. W. Craig, J. Z. Zhu, and R. H. Gallagher. Adaptive analysis refinement and shape optimization - some new possibilities. In *General Motors Research Laboratories Symposia Series*, pages 3 – 27, Warren, MI, USA, 1986.

[36] O. C. Zienkiewicz, K. Morgan, J. Peraire, and J. Z. Zhu. Some expanding horizons for computational mechanics - error estimates, mesh generation, and hyperbolic problems. volume 75, pages 281 – 297, Anaheim, CA, USA, 1986.

[37] O. C. Zienkiewicz and J. Z. Zhu. A simple error estimator and adaptive procedure for practical engineering analysis. *Int. J. Numer. Meth. Engng.*, 24:337–357, 1987.

[38] O. C. Zienkiewicz and R. L. Taylor. *The Finite Element Method. Vol. 1: Basic Formulation and Linear Problems*. McGraw-Hill, London, 1989.

[39] O.C. Zienkiewicz, G.C. Huang, and Y.C. Liu. Adaptive fem computation of forming processes. application to porous and non-porous materials. *International Journal for Numerical Methods in Engineering*, 30:1527 – 1553, 1990.

[40] O.C. Zienkiewicz and J.Z. Zhu. Three r's of engineering analysis and error estimation and adaptivity. *Computer Methods in Applied Mechanics and Engineering*, 82:95 – 113, 1990.

[41] O. C. Zienkiewicz and J. Z. Zhu. The superconvergent patch recovery (spr) and finite element refinement. *Comput. Methods Appl. Mech. Engrg.*, 101:207–224, 1992.

[42] O. C. Zienkiewicz and J. Z. Zhu. The superconvergent patch recovery and *a posteriori* error estimates. part 1: the recovery technique. *Int. J. Numer. Meth. Engng.*, 33:1331–1364, 1992.

[43] O.C. Zienkiewicz and J.Z. Zhu. Superconvergence and the superconvergent patch recovery. *Finite Elements in Analysis and Design*, 19:11 – 23, 1995.

[44] O.C. Zienkiewicz, B. Boroomand, and J.Z. Zhu. Recovery procedures in error estimation and adaptivity. i. adaptivity in linear problems. *Computer Methods in Applied Mechanics and Engineering*, 176(1-4):111 – 125, 1999.

[45] Olgierd C. Zienkiewicz. The background of error estimation and adaptivity in finite element computations. *Computer Methods in Applied Mechanics and Engineering*, 195(4-6):207 – 213, 2006.

[46] Y. Luo. A nearest-nodes finite element method with local multivariate lagrange interpolation. *Finite Elements in Analysis and Design*, 44:797–803, 2008.

[47] Y. Luo. 3D Nearest-Nodes Finite Element Method for solid continuum analysis. *Advances in Theoretical and Applied Mechanics*, 1:131–139, 2008.

[48] Y. Luo. Dealing with extremely large deformation by nearest-nodes fem with algorithm for updating element connectivity. *International Journal of Solids and Structures*, 45:5074–5087, 2008.

[49] Y. Luo. Nearest-nodes finite element method with inter-dependent shape functions. *Computational Mechanics*, (submitted), 2008.

[50] Y. Luo. A local multivariate lagrange interpolation method for constructing shape functions. *Communications in Numerical Methods in Engineering*, (in press, DOI: 10.1002/cnm.1149), 2008.

[51] I. Babuška and W. C. Rheinboldt. Error estimates for adaptive finite element computations. *SIAM J. Numer. Anal.*, 15:736–754, 1978.

[52] I. Babuška and W. C. Rheinboldt. On the reliability & optimality of finite element method. *Computers and Structures*, 10:87–94, 1979.

[53] E. Stein and W. Rust. Mesh adaptations for linear 2d finite-element discretizations in structural mechanics, especially in thin shell analysis. *J. Comput. Appl. Math.*, 36:107–129, 1991.

[54] Y. Luo and U. Häussler-Combe. A gradient-based adaptation procedure and its implementation in element-free galerkin method. *International Journal for Numerical Methods in Engineering*, 56:1335–1354, 2003.

[55] P. Lancaster and K. Salkauskas. Surfaces generated by moving least squares method. *Math. Comp.*, 37:141–158, 1981.

[56] Y. Luo and A. Eriksson. Extension of field consistence approach into developing plane stress elements. *Computer Methods in Applied Mechanics and Engineeirng*, 173:111–134, 1999.

[57] Q. Du and D. Wang. Recent progress in robust and quality delaunay mesh generation. *Journal of Computational and Applied Mathematics*, 195:8–23, 2006.

[58] E. Bansch. Local mesh refinement in 2 and 3 dimensions. *Impact of Computing in Science and Engineering*, 3:181191, 1991.

[59] M. T. Jones and P. E. Plassmann. Adaptive refinement of unstructured finite-element meshes. *Finite Element in Analysis and Design*, 40:4160, 1997.

[60] D. Scott McRae. r-Refinement grid adaptation algorithms and issues. *Computer Methods in Applied Mechanics and Engineering*, 189:1161–1182, 2000.

[61] Y. Luo. r-adaptation algorithm guided by gradient of strain energy density. *Communications in Numerical Methods in Engineering*, (in press, DOI:10.1002/cnm.1209), 2008.

In: Continuum Mechanics
Editors: A. Koppel and J. Oja, pp. 295-323

ISBN: 978-1-60741-585-5

Chapter 11

THE NATURAL APPROACH – AN APPRAISAL IN THE CONTEXT OF CONTINUUM MECHANICS

Ioannis St. Doltsinis*
Faculty of Aerospace Engineering and Geodesy,
University of Stuttgart,
Pfaffenwaldring 27, D-70569 Stuttgart, Germany

Abstract

The natural finite element approach introduced by John Argyris in the early sixties is characterized by the distinction between rigid body motion and deformation, on the one hand, and by the description of the latter in compliance with the element purpose and geometry, on the other hand. For triangular and tetrahedral elements the concept suggests strain and stress measures defined along the sides or the edges respectively as homogeneous normal quantities, free of shear. In the mechanics of continua the corresponding infinitesimal elements represent minimum configurations to define local deformation in two- and three dimensions.

This treatise concerns utilization of the natural approach on the continuum level within a consistent theoretical framework. It is proposed to begin with a reference system of supernumerary coordinates associated with the elementary tetrahedron in the space or with the triangle in the plane. Vectorial quantities are defined, the operations of gradient and divergence are interpreted in this system. The natural deformation rate is deduced from the velocity field, the stress is introduced as work conjugate measure. The condition for local equilibrium is presented in natural quantities as well as the stress definition in association with the resultant forces. The set up of material constitutive relations is exemplified for the elastic solid and for viscous media. Beyond the description of the momentary kinematics as from the velocity field, the appearance of finite deformation is considered basing on displacements. Illustration of the methodology for a plane elastic case terminates the part regarding the mechanics of solids. Extension to fluid motion and to thermal phenomena is appended.

Key Words: Deformable continua, elasticity, viscous solids, finite deformation, fluid motion, thermal phenomena.

*E-mail address: doltsinis@ica.uni-stuttgart.de

Professor John Argyris: 1913–2004.
Founding computational mechanics at Imperial College, London,
and at Stuttgart University. He introduced the Natural Finite Element approach.

1. Introduction

1.1. The natural finite element method in retrospective

The idea of a natural stiffness matrix inherent to the structural element under study not depending on its position and orientation in space was conceived by John Argyris while dealing with the stress analysis of elastic aircraft structures undergoing large displacements. This was after the establishment of the energy methods in structural analysis along with the introduction of matrix systematics as published in 1960 [1]. The inspiring natural concept of the late fifties/early sixties has been incorporated in his book published in 1964 [2] reporting on advances in matrix methods in structural analysis – the name of the finite element method at that time. The initial work on flanges, beams and membrane elements of triangular shape has been consistently extended to three-dimensional elastic media by making available the theory for the tetrahedral finite element [3]. The fruits of a first period of pioneering research, development and application of the natural approach undertaken predominantly at the Institute for Statics and Dynamics of Aerospace Structures, University of Stuttgart, under his direction have been highlighted in a remarkable address to the International Conference on Matrix Methods of Structural Mechanics in Dayton, Ohio, 1965 [4]. A second, according to Argyris even more significant acme of the natural method was presented at the first Conference on Finite Elements in Nonlinear Mechanics in Stuttgart, 1978 (Fenomech'78) [10]. The work is a comprehensive survey of the natural mode technique, which had been already extended considerably to the nonlinear regime beyond the large displacement analysis of structures to finite elasticity and finite plastic deformations. The paper comprises a complete bibliography on the subject by that time.

The natural concept exhibits two essential characteristics. These are: the analysis of the finite element kinematics into rigid body motion and deformation modes, and the definition of stress and strain measures pertinent to the purpose and the geometry of the element. Decisive for an investigation of large displacements is the distinction between rigid body motion and deformation modes in the element kinematics; aspects of the continuum level are given by Biot in [5]. Such a separation of the finite element kinematics allows the definition of a natural stiffness appertaining solely to deformation. It is invariant to rigid body motion which plays its role, however, at the stage of assembling elements and establishing the equilibrium at the actual geometry of the deformed structure. The approach favours the modular organization of theory and computational algorithm for small and large displacements. Also, the arithmetics of strain and stress is free of digits that qu antify rigid body motion.

In order to illustrate the argument let the column matrix, resp. the vector, $\mathbf{U}_\mathrm{e}$ comprise the displacements of the element nodal points with reference to the global Cartesian system for the structure. The definition of a rigid body part $\mathbf{U}_0$ and the natural part $\mathbf{U}_\mathrm{N}$ responsible for deformation entails the transformation

$$\begin{bmatrix} \mathbf{U}_0 \\ \mathbf{U}_\mathrm{N} \end{bmatrix} = \begin{bmatrix} \mathbf{a}_0 \\ \mathbf{a}_\mathrm{N} \end{bmatrix} \mathbf{U}_\mathrm{e} \tag{1}$$

Corresponding parts of the stress resultants $\mathbf{S}_\mathrm{e}$ at the element nodal points are established by the virtual work principle as,

$$\mathbf{S}_\mathrm{e} = \begin{bmatrix} \mathbf{a}_0 \\ \mathbf{a}_\mathrm{N} \end{bmatrix}^\mathrm{t} \begin{bmatrix} \mathbf{S}_0 \\ \mathbf{S}_\mathrm{N} \end{bmatrix} = \mathbf{a}_0^\mathrm{t}\mathbf{S}_0 + \mathbf{a}_\mathrm{N}^\mathrm{t}\mathbf{S}_\mathrm{N}. \tag{2}$$

Only the deformation modes contribute to the strain in the element. In natural terms

$$\boldsymbol{\varepsilon}_\mathrm{N} = \boldsymbol{\alpha}_\mathrm{N}\mathbf{U}_\mathrm{N} = [\mathbf{0}\ \boldsymbol{\alpha}_\mathrm{N}] \begin{bmatrix} \mathbf{U}_0 \\ \mathbf{U}_\mathrm{N} \end{bmatrix}, \tag{3}$$

where $\boldsymbol{\varepsilon}_\mathrm{N}$ denotes the arrangement of the natural strains as a vector, and the matrix $\boldsymbol{\alpha}_\mathrm{N}$ is the appertaining small strain operator. Defining the vector $\boldsymbol{\sigma}_\mathrm{N}$ as to comprise the corresponding, that is the work conjugate stresses, the stress resultants at the element nodal points are obtained to

$$\begin{bmatrix} \mathbf{S}_0 \\ \mathbf{S}_\mathrm{N} \end{bmatrix} = \begin{bmatrix} \mathbf{0} \\ \int_{V_\mathrm{e}} \boldsymbol{\alpha}_\mathrm{N}^\mathrm{t}\boldsymbol{\sigma}_\mathrm{N}\mathrm{d}V \end{bmatrix}, \tag{4}$$

the integration extending over the element volume V_e. In elasticity $\boldsymbol{\sigma}_\mathrm{N} = \boldsymbol{\kappa}_\mathrm{N}\boldsymbol{\varepsilon}_\mathrm{N}$, where the material stiffness matrix $\boldsymbol{\kappa}_\mathrm{N}$ may describe anisotropic material properties as well. Then, the expression for the stress resultants becomes

$$\mathbf{S}_\mathrm{N} = (\int_{V_\mathrm{e}} \boldsymbol{\alpha}_\mathrm{N}^\mathrm{t}\boldsymbol{\kappa}_\mathrm{N}\boldsymbol{\alpha}_\mathrm{N}\mathrm{d}V)\mathbf{U}_\mathrm{N} = \mathbf{k}_\mathrm{N}\mathbf{U}_\mathrm{N}, \tag{5}$$

and defines the natural stiffness matrix $\mathbf{k}_\mathrm{N}$ of the elastic element in the small strain regime. The natural stiffness matrix refers to the individual geometry of the element and does not depend on its position and orientation in space. The Cartesian components of the stress resultants in the global reference system are

$$\mathbf{S}_\mathrm{e} = \mathbf{a}_\mathrm{N}^\mathrm{t}\mathbf{S}_\mathrm{N} = \mathbf{k}_\mathrm{e}\mathbf{U}_\mathrm{e}, \tag{6}$$

the associated element stiffness matrix being obtained from the natural one by the transformation

$$\mathbf{k}_\mathrm{e} = \mathbf{a}_\mathrm{N}^\mathrm{t}\mathbf{k}_\mathrm{N}\mathbf{a}_\mathrm{N}. \tag{7}$$

The stress resultants $\mathbf{S}_\mathrm{e}$ and consequently the stiffness matrix $\mathbf{k}_\mathrm{e}$ depend on the motion of the element since equilibrium is established at the displaced state of the element and $\mathbf{a}_\mathrm{N} = \mathbf{a}_\mathrm{N}({}^\circ\mathbf{X}_\mathrm{e}+\mathbf{U}_\mathrm{e})$. As long as the displacement remains negligible, evaluation is performed with the initial coordinates ${}^\circ\mathbf{X}_\mathrm{e}$, however. Applications involving large deformation encounter kinematic nonlinearity and entail incremental solution procedures for the equilibrium condition of the finite element representation of the structure. To this end differentiation of Eqn (6) yields the change of the stress resultants as

$$\mathrm{d}\mathbf{S}_\mathrm{e} = \mathbf{a}_\mathrm{N}^\mathrm{t}\mathrm{d}\mathbf{S}_\mathrm{N} + \mathrm{d}\mathbf{a}_\mathrm{N}^\mathrm{t}\mathbf{S}_\mathrm{N} = \mathbf{k}_\mathrm{e}\mathrm{d}\mathbf{U}_\mathrm{e} + \mathrm{d}\mathbf{a}_\mathrm{N}^\mathrm{t}\mathbf{S}_\mathrm{N}. \tag{8}$$

The first term on the right hand side of the above equation employs the incremental form of the natural Eqn (5) when element deformation is negligible. The transition to the Cartesian one is performed with the matrix $\mathbf{a}_\mathrm{N}$ at the actually displaced position. The second term accounts for the

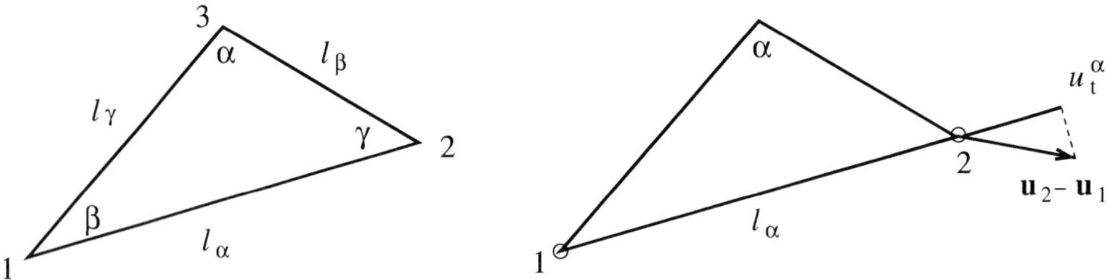

Figure 1. Triangular element and natural deformation.

impact of the incremental displacement on the existing state of stress as specified by the natural resultant forces $\mathbf{S}_\mathrm{N}$; it gives rise to the notion of the geometrical stiffness.

The specification of rigid body motion and of deformation is simple for elements with a linearly varying displacement field, associated with constant strain. These are the two-node flange, the three-node triangle and the four-node tetrahedron also known as simplex elements. At the same time, if these basic elements are interpreted as subelements up to infinitesimal size they can be assembled by integration to define the properties of higher order elements of the respective spatial dimensionality. The subelement technique essentially introduced in [6] has been elaborated for beam elements in [7] and more generally exposed in [10]; for a remote view see [13].

The description of geometrical and mechanical properties in a manner natural to the purpose of the element under consideration has been developed for several applications on the continuum and the structural level also comprising plates and shells [9]. A most transparent elucidation of the natural background for the present account is given by the original constant strain triangular element [2]. The elongation of the sides $l_\alpha, l_\beta, l_\gamma$ of the triangle in Fig 1 are taken as the pertinent natural deformations. Given the element displacements as those of the three nodal points with respect to the global Cartesian system of reference

$$\mathbf{U}_\mathrm{e} = \{\mathbf{u}_1\ \mathbf{u}_2\ \mathbf{u}_3\}, \tag{9}$$

the changes of the sidelengths are obtained by projecting the vectorial differences $\mathbf{u}_2 - \mathbf{u}_1$ etc. onto the direction of the sides:

$$U_\mathrm{N}^\alpha = \mathbf{c}_\alpha^\mathrm{t}\,(\mathbf{u}_2 - \mathbf{u}_1) = u_\mathrm{t}^\alpha. \tag{10}$$

The unit vector $\mathbf{c}_\alpha$ fixes the direction of the side in the global system by the cosines of the respective inclination angles; it is specified in Section 2. The orthogonal projections of vectors onto the natural directions will be denoted total natural quantities marked by the index " t". The set up of the collective kinematic relation

$$\mathbf{U}_\mathrm{N} \Leftarrow \mathbf{u}_\mathrm{t} = \left\{u_\mathrm{t}^\alpha\ u_\mathrm{t}^\beta\ u_\mathrm{t}^\gamma\right\} = \mathbf{a}_\mathrm{N}\mathbf{U}_\mathrm{e}, \tag{11}$$

is straightforward. The associated natural strains are obtained as the unit elongations

$$\boldsymbol{\varepsilon}_\mathrm{N} \Leftarrow \boldsymbol{\varepsilon}_\mathrm{t} = \left\{\varepsilon_\mathrm{t}^\alpha\ \varepsilon_\mathrm{t}^\beta\ \varepsilon_\mathrm{t}^\gamma\right\} = \mathbf{l}^{-1}\mathbf{u}_\mathrm{t}. \tag{12}$$

The braces { } conventionally indicate the horizontal presentation of a column vector,

$$\boldsymbol{\varepsilon}_\mathrm{t} = \begin{bmatrix} \varepsilon_\mathrm{t}^\alpha \\ \varepsilon_\mathrm{t}^\beta \\ \varepsilon_\mathrm{t}^\gamma \end{bmatrix} = \left\{\varepsilon_\mathrm{t}^\alpha\ \varepsilon_\mathrm{t}^\beta\ \varepsilon_\mathrm{t}^\gamma\right\},$$

and

$$\mathbf{l} = \lceil l_\alpha\ l_\beta\ l_\gamma \rfloor$$

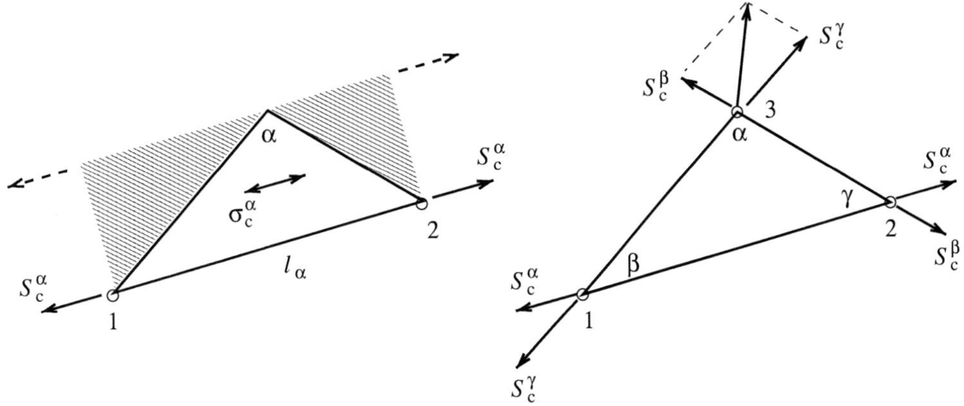

Figure 2. Natural stress and resulting forces.

is the arrangement of the side lengths as a diagonal matrix.

At this point it is instructive considering the definition of natural stress. The forces associated with the natural deformations are directed along the sides of the triangular element. At the vertices they represent vectorial constituents of the resulting nodal force vectors, and are therefore termed component natural forces. The respective magnitudes marked by the index "c" are the pertinent natural quantities collected in the array

$$\mathbf{S}_\mathrm{N} \Leftarrow \mathbf{S}_\mathrm{c} = \left\{S_\mathrm{c}^\alpha \; S_\mathrm{c}^\beta \; S_\mathrm{c}^\gamma\right\}. \tag{13}$$

The appertaining stress is

$$\boldsymbol{\sigma}_\mathrm{N} \Leftarrow \boldsymbol{\sigma}_\mathrm{c} = \left\{\sigma_\mathrm{c}^\alpha \; \sigma_\mathrm{c}^\beta \; \sigma_\mathrm{c}^\gamma\right\}. \tag{14}$$

It is defined corresponding to the natural strain by the virtual work equality for the triangle of area Ω and thickness t,

$$\Omega t \boldsymbol{\varepsilon}_\mathrm{t}^\mathrm{t} \boldsymbol{\sigma}_\mathrm{c} = \mathbf{u}_\mathrm{t}^\mathrm{t} \mathbf{S}_\mathrm{c}. \tag{15}$$

This gives

$$\Omega t \mathbf{l}^{-1} \boldsymbol{\sigma}_\mathrm{c} = \mathbf{S}_\mathrm{c}. \tag{16}$$

The single natural stress reads

$$\sigma_\mathrm{c}^\alpha = \frac{l_\alpha}{\Omega t} S_\mathrm{c}^\alpha, \tag{17}$$

and is realized to be the force $2S_\mathrm{c}^\alpha$ acting on the strip divided by the cross area $2\Omega t/l_\alpha$ (Fig 2). Since forces that act along the other two sides also contribute to the stress in the considered direction, the denomination *component* is justified.

Establishing the stiffness matrix of the element with respect to the natural deformations requires the relation between stress and strain. For elasticity specified in the plane of the element in terms of Cartesian quantities:

$$\boldsymbol{\sigma} = \boldsymbol{\kappa} \boldsymbol{\varepsilon}. \tag{18}$$

The transition to natural terms is via the standard transformation for strain and stress

$$\boldsymbol{\varepsilon}_\mathrm{t} = \mathbf{C}^\mathrm{t} \boldsymbol{\varepsilon}, \qquad \boldsymbol{\sigma} = \mathbf{C} \boldsymbol{\sigma}_\mathrm{c}. \tag{19}$$

The matrix $\mathbf{C}$ will be detailed later in the text, strain and stress transforms comply with the virtual work principle. There follows for the component stress

$$\boldsymbol{\sigma}_\mathrm{c} = \mathbf{C}^{-1} \boldsymbol{\kappa} \mathbf{C}^{-\mathrm{t}} \boldsymbol{\varepsilon}_\mathrm{t} = \boldsymbol{\kappa}_\mathrm{N} \boldsymbol{\varepsilon}_\mathrm{t}, \tag{20}$$

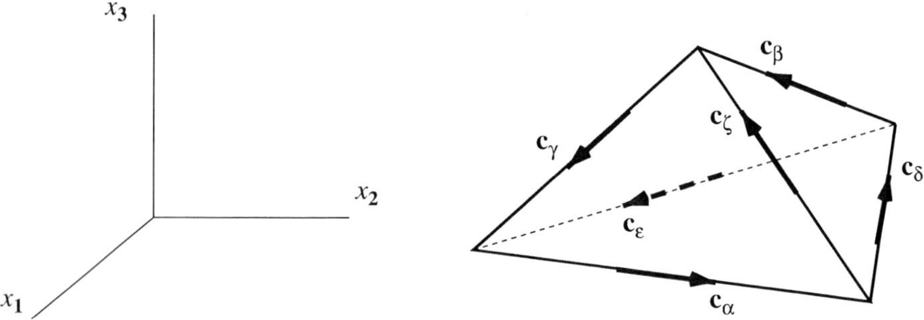

Figure 3. Global Cartesian axes and natural tetrahedron element.

which determines the elastic material stiffness $\boldsymbol{\kappa}_{\mathrm{N}}$ simplifying in case of isotropy. Substitution in Eqn (16) for the resultant forces and use of Eqn (12) for the strain gives

$$\mathbf{S}_{\mathrm{c}} = \Omega t \mathbf{l}^{-1} \boldsymbol{\kappa}_{\mathrm{N}} \mathbf{l}^{-1} \mathbf{u}_{\mathrm{t}} = \mathbf{k}_{\mathrm{N}} \mathbf{u}_{\mathrm{t}}. \tag{21}$$

The expression for the natural stiffness matrix for the triangular element is

$$\mathbf{k}_{\mathrm{N}} = \Omega t \mathbf{l}^{-1} \boldsymbol{\kappa}_{\mathrm{N}} \mathbf{l}^{-1}. \tag{22}$$

1.2. Application to problems of continua

Apart from the original scope of their employment as finite elements, triangle and tetrahedron are appealing as the minimal elements to define deformation in two- and three-dimensional material space, respectively. The appertaining natural strain measures and stress are homogeneous. Steps in this direction have been undertaken by the author in the context of elastoplasticity at finite strain in [11]. The incorporation of thermal phenomena was elaborated in [12] apropos of thermomechanically coupled problems, and in [14]. Utilization of the natural formalism in the description of fluid flow initiated in [15] has been exposed within a wider framework in [17]. Beyond the above natural description of the continuum as a starting point for developing the finite element formalism in various areas, the précis [18] examines the utility of a consistent natural methodology in solid mechanics. This aspect is extended here to encompass the description of fluid media and thermal phenomena as well.

The point of departure for a continuum theory based on the natural approach is the statement of the elementary tetrahedron as the object for local considerations in the material and the definition of a system of six coordinate axes parallel to the edges of the tetrahedral element. The two-dimensional analogon is the triangular element, which defines three coordinate axes parallel to the sides. In such a supernumerary system of reference, vectors can be interpreted as the vectorial sum of components along the axes - the component natural quantities. The orthogonal projections of the vector onto the axes - the total natural quantities - are of equal importance. The above definitions form the basis for the deduction of natural expressions for gradients, the divergence of a vector field and the introduction of strain at infinitesimal as well as finite deformation.

2. Representation of vectors

In continuum mechanics, natural quantities are proposed in conformity with the infinitesimal tetrahedron element which is introduced in place of the classical parallelopiped considered in the Cartesian formulation (Fig 3). A reference system that adheres to the elementary tetrahedron is defined by the six unit vectors $\mathbf{c}_\alpha, \cdots, \mathbf{c}_\zeta$ directed along the edges of the element in accordance with the convention of Fig 3. Each unit vector $\mathbf{c}_\vartheta$ is specified by its components in the Cartesian system of reference $O - x_1x_2x_3$:

$$\mathbf{c}_\vartheta = \{c_{\vartheta 1}\, c_{\vartheta 2}\, c_{\vartheta 3}\}, \quad c_{\vartheta i} = \cos(x_\vartheta, x_i) \quad \vartheta = \alpha, \cdots, \zeta, \quad i = 1, 2, 3. \tag{23}$$

Consider a vector represented in the Cartesian system by its components as

$$\mathbf{a} = \{a_1\, a_2\, a_3\}. \tag{24}$$

In the natural approach, vectorial constituents directed along the edges of the tetrahedron define the vector $\mathbf{a}$ as a resultant vector with the respective *component* natural quantities collected in the 6×1 array

$$\mathbf{a}_\mathrm{c} = \{a_\mathrm{c}^\alpha\, a_\mathrm{c}^\beta\, \cdots\, a_\mathrm{c}^\zeta\}. \tag{25}$$

It is evident, on the other hand, that a given vector can not be decomposed uniquely in the redundant system of six basis vectors. The concept is demonstrated by Fig 4 and Fig 5 for the two-dimensional case.

The orthogonal projections of the vector $\mathbf{a}$ onto the natural axes define what is called the *total* entities of a vector in the natural terminology. They are arranged in the 6×1 array

$$\mathbf{a}_\mathrm{t} = \{a_\mathrm{t}^\alpha\, a_\mathrm{t}^\beta\, \cdots\, a_\mathrm{t}^\zeta\}. \tag{26}$$

The typical element in Eqn (26) is obtained as

$$a_\mathrm{t}^\vartheta = c_{\vartheta\alpha} a_\mathrm{c}^\alpha + c_{\vartheta\beta} a_\mathrm{c}^\beta + \cdots + c_{\vartheta\zeta} a_\mathrm{c}^\zeta, \quad c_{\vartheta\varphi} = \cos(x_\vartheta, x_\varphi). \tag{27}$$

The collective operation for the total quantities reads

$$\mathbf{a}_\mathrm{t} = \begin{bmatrix} a_\mathrm{t}^\alpha \\ a_\mathrm{t}^\beta \\ \vdots \\ a_\mathrm{t}^\zeta \end{bmatrix} = \begin{bmatrix} c_{\alpha\alpha} & c_{\alpha\beta} & \cdots & c_{\alpha\zeta} \\ c_{\beta\alpha} & c_{\beta\beta} & \cdots & c_{\beta\zeta} \\ \vdots & & & \\ c_{\zeta\alpha} & c_{\zeta\beta} & \cdots & c_{\zeta\zeta} \end{bmatrix} \begin{bmatrix} a_\mathrm{c}^\alpha \\ a_\mathrm{c}^\beta \\ \vdots \\ a_\mathrm{c}^\zeta \end{bmatrix} = \mathbf{B}_\mathrm{tc}\mathbf{a}_\mathrm{c}. \tag{28}$$

The matrix $\mathbf{B}_\mathrm{tc}$ comprises the direction cosines $c_{\vartheta\varphi}$ of each natural axis with respect to the others.

With the above definitions the scalar product of two arbitrary vectors $\mathbf{a}$ and $\mathbf{b}$ may be expressed in either of the alternative forms

$$\mathbf{a}^\mathrm{t}\mathbf{b} = \mathbf{a}_\mathrm{c}^\mathrm{t}\mathbf{b}_\mathrm{t} = \mathbf{b}_\mathrm{c}^\mathrm{t}\mathbf{a}_\mathrm{t}. \tag{29}$$

Given the natural constituents $\mathbf{a}_\mathrm{c}$ of the vector, its Cartesian components are obtained as

$$\mathbf{a} = [\mathbf{c}_\alpha \mathbf{c}_\beta\, \cdots\, \mathbf{c}_\zeta]\, \mathbf{a}_\mathrm{c} = \mathbf{B}_\mathrm{ND}^\mathrm{t}\mathbf{a}_\mathrm{c}. \tag{30}$$

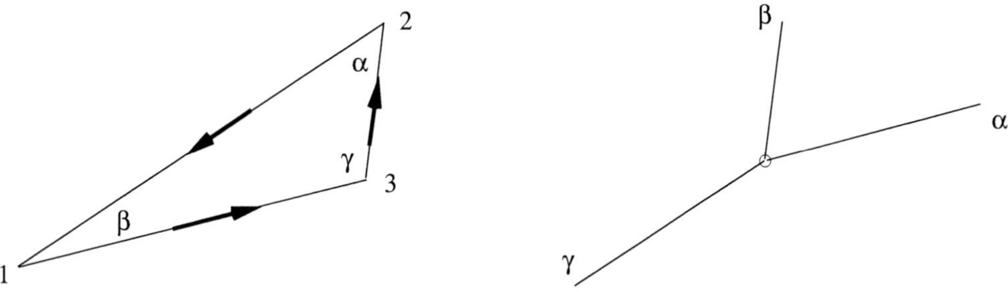

Figure 4. Natural axes pertaining to elementary triangle.

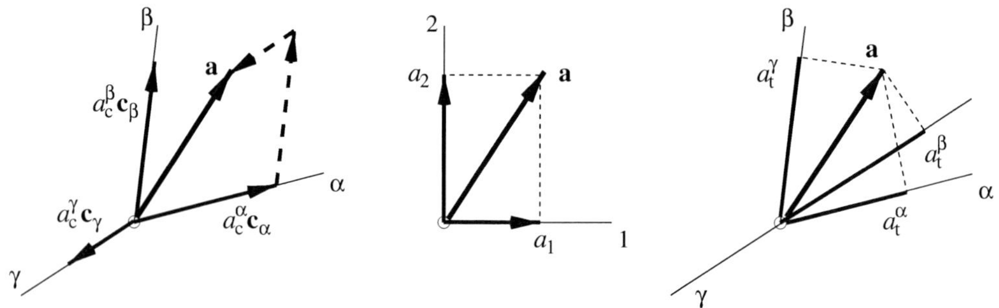

Figure 5. Component (left) and total natural entities of vector in the plane (right).

The matrix

$$\mathbf{B}_{\mathrm{ND}}^{\mathrm{t}} = [\mathbf{c}_\alpha \mathbf{c}_\beta \cdots \mathbf{c}_\zeta] = \begin{bmatrix} c_{1\alpha} & c_{1\beta} & \cdots & c_{1\zeta} \\ c_{2\alpha} & c_{2\beta} & \cdots & c_{2\zeta} \\ c_{3\alpha} & c_{3\beta} & \cdots & c_{3\zeta} \end{bmatrix}, \tag{31}$$

comprises the direction cosines $c_{i\vartheta}$ between the Cartesian axes x_i and the unit vectors $\mathbf{c}_\vartheta$ resp. the x_ϑ–axis.

The projections $a_{\mathrm{t}}^\vartheta = \mathbf{c}_\vartheta^{\mathrm{t}} \mathbf{a}$ of the Cartesian components of the vector furnish the total quantities alternatively to Eqn (28), as

$$\mathbf{a}_{\mathrm{t}} = [\mathbf{c}_\alpha \mathbf{c}_\beta \cdots \mathbf{c}_\zeta]^{\mathrm{t}} \, \mathbf{a} = \mathbf{B}_{\mathrm{ND}} \mathbf{a}. \tag{32}$$

With Eqns (30) and (28) one verifies the sequence

$$\mathbf{a}_{\mathrm{t}} = \mathbf{B}_{\mathrm{ND}} \mathbf{a} = \mathbf{B}_{\mathrm{ND}} \mathbf{B}_{\mathrm{ND}}^{\mathrm{t}} \mathbf{a}_{\mathrm{c}} = \mathbf{B}_{\mathrm{tc}} \mathbf{a}_{\mathrm{c}}, \tag{33}$$

which defines the symmetric operator $\mathbf{B}_{\mathrm{tc}}$ by

$$\mathbf{B}_{\mathrm{tc}} = \mathbf{B}_{\mathrm{ND}} \mathbf{B}_{\mathrm{ND}}^{\mathrm{t}}. \tag{34}$$

It is pointed out that the relationships in Eqn (28), Eqn (30) and consequently in Eqn (33) are not invertible.

The symbolism used for differential changes of a scalar field function $w(\mathbf{x})$ is $\mathrm{d}w = (\mathrm{d}w/\mathrm{d}\mathbf{x})\mathrm{d}\mathbf{x}$. The gradient vector of the scalar field is represented by the column matrix

$$\mathbf{g}(w) = \left\{ \frac{\partial w}{\partial x_1} \, \frac{\partial w}{\partial x_2} \, \frac{\partial w}{\partial x_3} \right\} = \left(\frac{\mathrm{d}w}{\mathrm{d}\mathbf{x}} \right)^{\mathrm{t}}. \tag{35}$$

Its orthogonal projections onto the natural axes, the total entities $g_{\mathrm{t}}^\vartheta(w) = \mathbf{c}_\vartheta^{\mathrm{t}} \mathbf{g}(w)$, give the rate of change of the function w along the respective direction:

$$\frac{\partial w}{\partial x_\vartheta} = \frac{\mathrm{d}w}{\mathrm{d}\mathbf{x}} \mathbf{c}_\vartheta = \mathbf{c}_\vartheta^{\mathrm{t}} \mathbf{g}(w). \tag{36}$$

Collectively, in agreement with Eqn (32),

$$\mathbf{g}_{\mathrm{t}}(w) = \left\{ \frac{\partial w}{\partial x_\alpha} \, \frac{\partial w}{\partial x_\beta} \cdots \frac{\partial w}{\partial x_\zeta} \right\} = \mathbf{B}_{\mathrm{ND}} \mathbf{g}(w). \tag{37}$$

The Cartesian components of distances along the natural directions follow the transformation by Eqn (30). Thus $x_\vartheta = x_{\mathrm{c}}^\vartheta$, and for Eqn (37):

$$\mathbf{g}_{\mathrm{t}}(w) = \left(\frac{\mathrm{d}w}{\mathrm{d}\mathbf{x}_{\mathrm{c}}} \right)^{\mathrm{t}} = \left(\frac{\mathrm{d}w}{\mathrm{d}\mathbf{x}} \frac{\mathrm{d}\mathbf{x}}{\mathrm{d}\mathbf{x}_{\mathrm{c}}} \right)^{\mathrm{t}} = \mathbf{B}_{\mathrm{ND}} \mathbf{g}(w). \tag{38}$$

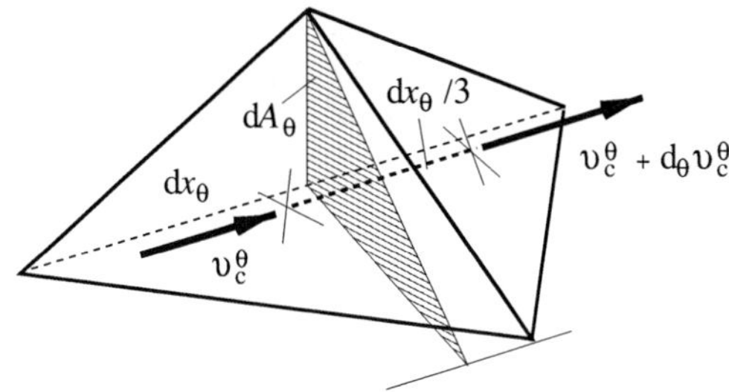

Figure 6. On the divergence of a vector in natural terms.

Regarding a vector valued function $\mathbf{v}(\mathbf{x})$, differential changes $\mathrm{d}\mathbf{v} = (\mathrm{d}\mathbf{v}/\mathrm{d}\mathbf{x})\mathrm{d}\mathbf{x}$ are determined by the tangent matrix $\mathrm{d}\mathbf{v}/\mathrm{d}\mathbf{x} = [\partial v_i/\partial x_j]\ i, j = 1, 2, 3$ which comprises the partial derivatives of the vector components along the spatial coordinates. The total natural entities of $\mathbf{v}$ are in the vector $\mathbf{v}_\mathrm{t} = \mathbf{B}_\mathrm{ND}\mathbf{v}$. Their rate of change in space along the natural directions assembled in the matrix

$$\frac{\mathrm{d}\mathbf{v}_\mathrm{t}}{\mathrm{d}\mathbf{x}_\mathrm{c}} = \left[\frac{\partial v_\mathrm{t}^\varphi}{\partial x_\vartheta}\right] \qquad \varphi, \vartheta = \alpha, \cdots, \zeta, \tag{39}$$

is obtained in analogy to Eqn (38) as

$$\frac{\mathrm{d}\mathbf{v}_\mathrm{t}}{\mathrm{d}\mathbf{x}_\mathrm{c}} = \mathbf{B}_\mathrm{ND}\frac{\mathrm{d}\mathbf{v}}{\mathrm{d}\mathbf{x}}\frac{\mathrm{d}\mathbf{x}}{\mathrm{d}\mathbf{x}_\mathrm{c}} = \mathbf{B}_\mathrm{ND}\frac{\mathrm{d}\mathbf{v}}{\mathrm{d}\mathbf{x}}\mathbf{B}_\mathrm{ND}^\mathrm{t}. \tag{40}$$

The operation for the divergence of the vector field is expressed in matrix form by

$$\mathrm{div}\,\mathbf{v} = \boldsymbol{\nabla}^\mathrm{t}\mathbf{v}. \tag{41}$$

The Hamilton operator $\boldsymbol{\nabla}$ is interpreted as a column vector

$$\boldsymbol{\nabla} = \left\{\frac{\partial}{\partial x_1}\,\frac{\partial}{\partial x_2}\,\frac{\partial}{\partial x_3}\right\}, \tag{42}$$

The homologous natural expression can be obtained by considering the elementary tetrahedron (Fig. 6). Denoting the lengths of the edges $\mathrm{d}x_\alpha, \mathrm{d}x_\beta, \cdots, \mathrm{d}x_\zeta$, and the projection areas of the tetrahedron on the planes normal to the edges $\mathrm{d}A_\alpha$ and so on, the element volume is given by $\mathrm{d}V = \mathrm{d}A_\vartheta \mathrm{d}x_\vartheta/3$ invariant to the respective direction. A single vector along the ϑ–axis of magnitude v_c^ϑ passes through the tetrahedron with input $v_\mathrm{c}^\vartheta\mathrm{d}A_\vartheta$ and output $(v_\mathrm{c}^\vartheta + \mathrm{d}_\vartheta v_\mathrm{c}^\vartheta)\mathrm{d}A_\vartheta$ at a distance $\mathrm{d}x_\vartheta/3$. Simple application of the Gauß-Ostrogradski theorem of integral calculus yields

$$(\mathrm{div}\,\mathbf{v}_\vartheta)\,\mathrm{d}V = \frac{\partial v_\mathrm{c}^\vartheta}{\partial x_\vartheta}\frac{\mathrm{d}x_\vartheta}{3}\mathrm{d}A_\vartheta = \frac{\partial v_\mathrm{c}^\vartheta}{\partial x_\vartheta}\mathrm{d}V. \tag{43}$$

More generally, accounting for several contributions on each side of the tetrahedron

$$\mathrm{div}\,\mathbf{v} = \frac{1}{\mathrm{d}V}\sum_{j=1}^{4}(\mathbf{v}_\mathrm{c}^\mathrm{t}\mathrm{d}\mathbf{A}_\mathrm{t})_j = \frac{\partial v_\mathrm{c}^\alpha}{\partial x_\alpha} + \frac{\partial v_\mathrm{c}^\beta}{\partial x_\beta} + \cdots + \frac{\partial v_\mathrm{c}^\zeta}{\partial x_\zeta} = \boldsymbol{\nabla}_\mathrm{t}^\mathrm{t}\mathbf{v}_\mathrm{c}, \tag{44}$$

where $\mathbf{v}_{\mathrm{c}j}$ comprises the constituents along the natural directions passing through the jth plane (positive if pointing out of the tetrahedron), and $\mathrm{d}\mathbf{A}_{\mathrm{t}j}$ comprises the respective orthogonal projections of the surface element. The appertaining differential operator is

$$\boldsymbol{\nabla}_\mathrm{t} = \left\{\frac{\partial}{\partial x_\alpha}\,\frac{\partial}{\partial x_\beta}\cdots\frac{\partial}{\partial x_\zeta}\right\}. \tag{45}$$

A formal statement of Eqn (44) makes use of Eqn (30) in order to express the Cartesian components $\mathbf{v}$ by the natural constituents $\mathbf{v}_c$ in the divergence Eqn (41):

$$\operatorname{div} \mathbf{v} = \boldsymbol{\nabla}^t \mathbf{v} = (\mathbf{B}_{\mathrm{ND}} \boldsymbol{\nabla})^t \mathbf{v}_c. \tag{46}$$

In the above, the operator in the second expression may be detailed as

$$\mathbf{B}_{\mathrm{ND}} \boldsymbol{\nabla} = \begin{bmatrix} \mathbf{c}_\alpha^t \\ \mathbf{c}_\beta^t \\ \vdots \\ \mathbf{c}_\zeta^t \end{bmatrix} \begin{bmatrix} \dfrac{\partial}{\partial x_1} \\ \dfrac{\partial}{\partial x_2} \\ \dfrac{\partial}{\partial x_3} \end{bmatrix} = \begin{bmatrix} \dfrac{\partial}{\partial x_\alpha} \\ \dfrac{\partial}{\partial x_\beta} \\ \vdots \\ \dfrac{\partial}{\partial x_\zeta} \end{bmatrix} = \boldsymbol{\nabla}_t, \tag{47}$$

the individual entities determining the rate of change of a quantity along the respective natural direction. The relationship to the Cartesian operator is as for total quantities.

The differential operators in Eqns (41), (45) allow representation of the gradients of Eqns (35), (37) as

$$\mathbf{g}(w) = \boldsymbol{\nabla} w, \quad \mathbf{g}_t(w) = \boldsymbol{\nabla}_t w. \tag{48}$$

3. Kinematics of momentary deformation

At the considered instant t particles of the solid are identified by their position $\mathbf{x} = \{x_1 \, x_2 \, x_3\}$ with respect to the Cartesian coordinate system $O - x_1 \, x_2 \, x_3$ fixed in space. The momentary state of motion in the solid is specified by the velocity field $\mathbf{v}(\mathbf{x})$ with the velocity vector

$$\mathbf{v} = \frac{\mathrm{d}\mathbf{x}}{\mathrm{d}t} = \{v_1 \, v_2 \, v_3\}. \tag{49}$$

The relative particle velocity is given by

$$\mathrm{d}\mathbf{v} = \frac{\mathrm{d}\mathbf{v}}{\mathrm{d}\mathbf{x}} \mathrm{d}\mathbf{x} = \mathbf{L} \mathrm{d}\mathbf{x}, \tag{50}$$

where the 3×3 matrix

$$\mathbf{L} = \frac{\mathrm{d}\mathbf{v}}{\mathrm{d}\mathbf{x}}, \tag{51}$$

is known as the velocity gradient.

Information about the deformation of the material in the vicinity of a point while moving will be gained by considering the temporal change of the geometry of the elementary tetrahedron at this place. Natural deformation measures refer to the change in length of the edges of the tetrahedron (Fig 7). To this end the orthogonal projection of the velocity along the edge with direction $\mathbf{c}_\vartheta$ is obtained to

$$v_t^\vartheta = \mathbf{c}_\vartheta^t \mathbf{v}, \tag{52}$$

and its change along the edge defines the unit elongation rate there as

$$\delta_t^\vartheta = \frac{\partial v_t^\vartheta}{\partial x_\vartheta} = \mathbf{c}_\vartheta^t \mathbf{L} \mathbf{c}_\vartheta. \tag{53}$$

Note that merely the symmetric part of the velocity gradient $\mathbf{L}$ is effective in the operation in Eqn (53). The state of the momentary deformation rate at the point is defined by the unit elongation rates of the six edges. They are assembled to the 6×1 vector

$$\boldsymbol{\delta}_t = \left\{ \delta_t^\alpha \; \delta_t^\beta \; \delta_t^\gamma \; \delta_t^\delta \; \delta_t^\varepsilon \; \delta_t^\zeta \right\}, \tag{54}$$

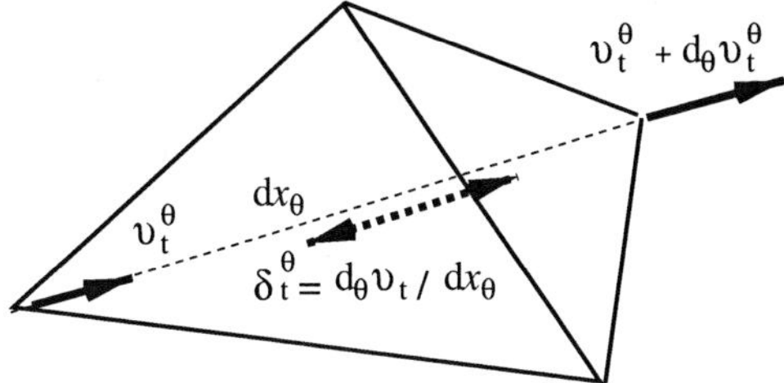

Figure 7. Definition of total deformation rate.

which constitutes the total natural deformation rate. The six homogeneous entities are defined by Eqn (53). Collectively,

$$\boldsymbol{\delta}_{\mathrm{t}} = \boldsymbol{\partial}_{\mathrm{N}} \mathbf{v}_{\mathrm{t}}, \tag{55}$$

with the diagonal matrix operator

$$\boldsymbol{\partial}_{\mathrm{N}} = \begin{bmatrix} \dfrac{\partial}{\partial x_\alpha} & & & \\ & \dfrac{\partial}{\partial x_\beta} & & \\ & & \ddots & \\ & & & \dfrac{\partial}{\partial x_\zeta} \end{bmatrix}. \tag{56}$$

It is common to define the deformation rate by describing the temporal change of the scalar product of two material line elements $\mathrm{d}\mathbf{x}_1$ and $\mathrm{d}\mathbf{x}_2$ emanating from the point under observation in the solid. Making use of the velocity gradient, Eqn (50):

$$\frac{\mathrm{d}}{\mathrm{d}t}\left(\mathrm{d}\mathbf{x}_1^{\mathrm{t}} \mathrm{d}\mathbf{x}_2\right) = \mathrm{d}\mathbf{x}_1^{\mathrm{t}}\left(\mathbf{L} + \mathbf{L}^{\mathrm{t}}\right)\mathrm{d}\mathbf{x}_2 = 2\mathrm{d}\mathbf{x}_1^{\mathrm{t}} \mathbf{D} \mathrm{d}\mathbf{x}_2. \tag{57}$$

The symmetric part of the velocity gradient defines the deformation rate

$$\mathbf{D} = \frac{1}{2}\left(\mathbf{L} + \mathbf{L}^{\mathrm{t}}\right) = \mathbf{D}^{\mathrm{t}}, \tag{58}$$

and determines the temporal change of the product $\mathrm{d}\mathbf{x}_1^{\mathrm{t}} \mathrm{d}\mathbf{x}_2$. The antisymmetric part of the velocity gradient defines the momentary rotation rate or spin

$$\mathbf{W} = \frac{1}{2}\left(\mathbf{L} - \mathbf{L}^{\mathrm{t}}\right) = -\mathbf{W}^{\mathrm{t}}, \tag{59}$$

which does not induce deformation; it leaves the product $\mathrm{d}\mathbf{x}_1^{\mathrm{t}} \mathrm{d}\mathbf{x}_2$ unchanged.

From Eqn (57), there follows the rate of change of line elements $\mathrm{d}\mathbf{x}_\vartheta = (\mathrm{d}x_\vartheta)\mathbf{c}_\vartheta$ along the natural directions:

$$\frac{\mathrm{d}\dot{x}_\vartheta}{\mathrm{d}x_\vartheta} = \mathbf{c}_\vartheta^{\mathrm{t}} \mathbf{D} \mathbf{c}_\vartheta = \delta_{\mathrm{t}}^{\vartheta}. \tag{60}$$

The superposed dot denotes the time rate, and the transition from the Cartesian quantities in $\mathbf{D} = [\delta_{ij}]\ i, j = 1, 2, 3$ reads in detail

$$\delta_{\mathrm{t}}^{\vartheta} = c_{\vartheta 1}^2 \delta_{11} + c_{\vartheta 2}^2 \delta_{22} + c_{\vartheta 3}^2 \delta_{33} + 2\left(c_{\vartheta 1} c_{\vartheta 2} \delta_{12} + c_{\vartheta 2} c_{\vartheta 3} \delta_{23} + c_{\vartheta 1} c_{\vartheta 3} \delta_{13}\right), \tag{61}$$

which is the familiar strain transformation identical to Eqn (53).

Arranging the Cartesian deformation rate as the 6×1 vector

$$\boldsymbol{\delta} = \left\{ \delta_{11} \; \delta_{22} \; \delta_{33} \; \sqrt{2}\delta_{12} \; \sqrt{2}\delta_{23} \; \sqrt{2}\delta_{13} \right\}, \tag{62}$$

Eqn (61) transfers to the collective relation

$$\boldsymbol{\delta}_{\mathrm{t}} = \mathbf{C}^{\mathrm{t}} \boldsymbol{\delta}. \tag{63}$$

The 6×6 matrix

$$\mathbf{C}^{\mathrm{t}} = \begin{bmatrix} c_{\alpha 1}^2 & c_{\alpha 2}^2 & c_{\alpha 3}^2 & \sqrt{2}c_{\alpha 1}c_{\alpha 2} & \sqrt{2}c_{\alpha 2}c_{\alpha 3} & \sqrt{2}c_{\alpha 1}c_{\alpha 3} \\ c_{\beta 1}^2 & c_{\beta 2}^2 & c_{\beta 3}^2 & \sqrt{2}c_{\beta 1}c_{\beta 2} & \sqrt{2}c_{\beta 2}c_{\beta 3} & \sqrt{2}c_{\beta 1}c_{\beta 3} \\ \vdots & & & & & \\ c_{\zeta 1}^2 & c_{\zeta 2}^2 & c_{\zeta 3}^2 & \sqrt{2}c_{\zeta 1}c_{\zeta 2} & \sqrt{2}c_{\zeta 2}c_{\zeta 3} & \sqrt{2}c_{\zeta 1}c_{\zeta 3} \end{bmatrix} \tag{64}$$

reflects the strain transformation by Eqn (61), which furnishes the homogeneous natural measures. Given the natural $\boldsymbol{\delta}_{\mathrm{t}}$, Eqn (63) can be solved for the Cartesian $\boldsymbol{\delta}$. Hence, the state of momentary deformation is completely defined.

4. Statics, natural stress

Stress is conveniently defined corresponding, i.e. work conjugate, to the deformation rate. In this context recall the virtual work principle [1]; in Cartesian space:

$$\int_A \mathbf{v}^{\mathrm{t}}\mathbf{t}\mathrm{d}A + \int_V \mathbf{v}^{\mathrm{t}}\mathbf{f}\mathrm{d}V = \int_V \boldsymbol{\delta}^{\mathrm{t}}\boldsymbol{\sigma}\mathrm{d}V. \tag{65}$$

The surface force per unit area is denoted $\mathbf{t} = \{t_1 \; t_2 \; t_3\}$, the body force per unit volume is $\mathbf{f} = \{f_1 \; f_2 \; f_3\}$. The integration extends over the volume V and the bounding surface A of the solid. The deformation rate $\boldsymbol{\delta}$, Eqn (62), is derived from the velocity $\mathbf{v}$ by

$$\boldsymbol{\delta} = \boldsymbol{\partial}\mathbf{v}, \tag{66}$$

with the differential operator

$$\boldsymbol{\partial}^{\mathrm{t}} = \begin{bmatrix} \frac{\partial}{\partial x_1} & 0 & 0 & \frac{1}{\sqrt{2}}\frac{\partial}{\partial x_2} & 0 & \frac{1}{\sqrt{2}}\frac{\partial}{\partial x_3} \\ 0 & \frac{\partial}{\partial x_2} & 0 & \frac{1}{\sqrt{2}}\frac{\partial}{\partial x_1} & \frac{1}{\sqrt{2}}\frac{\partial}{\partial x_3} & 0 \\ 0 & 0 & \frac{\partial}{\partial x_3} & 0 & \frac{1}{\sqrt{2}}\frac{\partial}{\partial x_2} & \frac{1}{\sqrt{2}}\frac{\partial}{\partial x_1} \end{bmatrix}. \tag{67}$$

The transposed form to which preference has been given here for typographical brevity will be encountered below in connection with the stress [19]. The latter is specified by the 6×1 vector that comprises the Cartesian components of Cauchy stress

$$\boldsymbol{\sigma} = \left\{ \sigma_{11} \; \sigma_{22} \; \sigma_{33} \; \sqrt{2}\sigma_{12} \; \sqrt{2}\sigma_{23} \; \sqrt{2}\sigma_{13} \right\}. \tag{68}$$

The statement of Eqn (65) bases as usual on the transformation of the surface integral to one over the volume. In the present notation,

$$\int_A \mathbf{v}^{\mathrm{t}}\mathbf{t}\mathrm{d}A = \int_V \left[\mathbf{v}^{\mathrm{t}}(\boldsymbol{\partial}^{\mathrm{t}}\boldsymbol{\sigma}) + (\boldsymbol{\partial}\mathbf{v})^{\mathrm{t}}\boldsymbol{\sigma} \right] \mathrm{d}V. \tag{69}$$

Developing with the above a right-hand side for the work expression in Eqn (65) allows the transition to the condition of local equilibrium for the stress field

$$\boldsymbol{\partial}^{\mathrm{t}}\boldsymbol{\sigma} + \mathbf{f} = \mathbf{0}, \tag{70}$$

and leaves us with the virtual work equation as stated; Eqn (66) giving the deformation rate is accounted for.

For completeness, the static boundary conditions are

$$\mathbf{N}^{\mathrm{t}}\boldsymbol{\sigma} = \mathbf{t}. \tag{71}$$

The matrix

$$\mathbf{N}^{\mathrm{t}} = \begin{bmatrix} l & 0 & 0 & \frac{1}{\sqrt{2}}m & 0 & \frac{1}{\sqrt{2}}n \\ 0 & m & 0 & \frac{1}{\sqrt{2}}l & \frac{1}{\sqrt{2}}n & 0 \\ 0 & 0 & n & 0 & \frac{1}{\sqrt{2}}m & \frac{1}{\sqrt{2}}l \end{bmatrix}, \tag{72}$$

is composed of the direction cosines l, m, n of the surface element with respect to the x_1, x_2 and x_3-axes, respectively.

In order to state the virtual work principle in natural terms, Eqn (65) is written as

$$\int_A \mathbf{v}_{\mathrm{t}}^{\mathrm{t}}\mathbf{t}_{\mathrm{c}}\mathrm{d}A + \int_V \mathbf{v}_{\mathrm{t}}^{\mathrm{t}}\mathbf{f}_{\mathrm{c}}\mathrm{d}V = \int_V \boldsymbol{\delta}_{\mathrm{t}}^{\mathrm{t}}\boldsymbol{\sigma}_{\mathrm{c}}\mathrm{d}V, \tag{73}$$

where the 6×1 vectors

$$\mathbf{t}_{\mathrm{c}} = \left\{t_{\mathrm{c}}^{\alpha}\ t_{\mathrm{c}}^{\beta}\ \cdots\ t_{\mathrm{c}}^{\zeta}\right\}, \qquad \mathbf{f}_{\mathrm{c}} = \left\{f_{\mathrm{c}}^{\alpha}\ f_{\mathrm{c}}^{\beta}\ \cdots\ f_{\mathrm{c}}^{\gamma}\right\}, \tag{74}$$

collect the component natural constituents of specific surface and body forces, respectively. Equation (73) implies that the component natural stress

$$\boldsymbol{\sigma}_{\mathrm{c}} = \left\{\sigma_{\mathrm{c}}^{\alpha}\ \sigma_{\mathrm{c}}^{\beta}\ \sigma_{\mathrm{c}}^{\gamma}\ \sigma_{\mathrm{c}}^{\delta}\ \sigma_{\mathrm{c}}^{\varepsilon}\ \sigma_{\mathrm{c}}^{\zeta}\right\}, \tag{75}$$

satisfies the work equality

$$\boldsymbol{\delta}_{\mathrm{t}}^{\mathrm{t}}\boldsymbol{\sigma}_{\mathrm{c}} = \boldsymbol{\delta}^{\mathrm{t}}\boldsymbol{\sigma}. \tag{76}$$

With Eqn (63) between the virtual deformation rates, the dual relation for the stresses then follows to

$$\boldsymbol{\sigma} = \mathbf{C}\boldsymbol{\sigma}_{\mathrm{c}}. \tag{77}$$

The equilibrium condition for the stress field inherent to Eqn (73) is obtained in terms of natural stresses as

$$\mathbf{B}_{\mathrm{ND}}^{\mathrm{t}}(\boldsymbol{\partial}_{\mathrm{N}}\boldsymbol{\sigma}_{\mathrm{c}} + \mathbf{f}_{\mathrm{c}}) = \mathbf{0}. \tag{78}$$

The above makes use in Eqn (70) of the relation

$$\boldsymbol{\partial}^{\mathrm{t}}\boldsymbol{\sigma} = \mathbf{B}_{\mathrm{ND}}^{\mathrm{t}}(\boldsymbol{\partial}_{\mathrm{N}}\boldsymbol{\sigma}_{\mathrm{c}}), \tag{79}$$

which is easily confirmed for the stress transformation by Eqn (77). The utility of Eqns (73) and (78) is limited to the domain of applicability of the reference system.

The static boundary conditions, Eqn (71), may be expressed in natural terms:

$$(\mathbf{N}^{\mathrm{t}}\mathbf{C})\boldsymbol{\sigma}_{\mathrm{c}} = \mathbf{t} = \mathbf{B}_{\mathrm{ND}}^{\mathrm{t}}\mathbf{t}_{\mathrm{c}}. \tag{80}$$

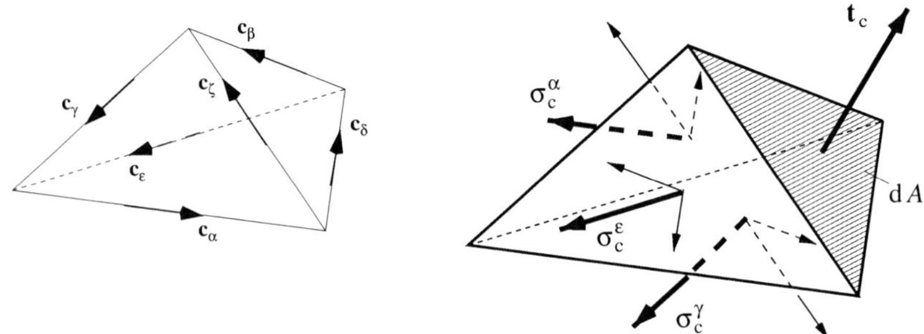

Figure 8. Boundary condition in natural system.

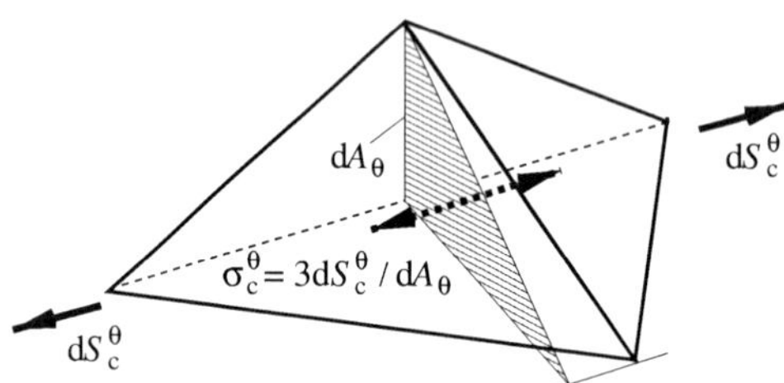

Figure 9. Component stress and resulting force.

It is meaningful paying attention to a surface element that makes one of the sides of the tetrahedron, for instance the side marked by the unit vectors $\mathbf{c}_\delta$, $\mathbf{c}_\beta$, $\mathbf{c}_\zeta$ in Fig 8. The entities in the matrix $\mathbf{N}$, the Cartesian components of the exterior surface normal $\mathbf{n} = \{l \ m \ n\}$, are determined by the vectorial product of any two of the three vectors:

$$\mathbf{n} = \begin{bmatrix} l \\ m \\ n \end{bmatrix} = \frac{1}{s_{\delta\beta}} \begin{bmatrix} 0 & -c_{\delta 3} & c_{\delta 2} \\ c_{\delta 3} & 0 & -c_{\delta 1} \\ -c_{\delta 2} & c_{\delta 1} & 0 \end{bmatrix} \begin{bmatrix} c_{\beta 1} \\ c_{\beta 2} \\ c_{\beta 3} \end{bmatrix}, \tag{81}$$

with $s_{\delta\beta} = \sin(x_\delta, x_\beta)$, and the antisymmetric matrix operation with $\mathbf{c}_\delta$ performed on $\mathbf{c}_\beta$. The condition of Eqn (80) can then be stated with the applied traction given either as $\mathbf{t}$ or $\mathbf{t}_c$.

A direct statement of the boundary condition resolves the traction vector along the three out of plane directions $\mathbf{c}_\alpha, \mathbf{c}_\gamma, \mathbf{c}_\varepsilon$ such that $\mathbf{t}_c = \{t_c^\alpha \ 0 \ t_c^\gamma \ 0 \ t_c^\varepsilon \ 0\}$. From Fig 8, the traction is equilibrated by the stresses $\sigma_c^\alpha, \sigma_c^\gamma, \sigma_c^\varepsilon$ along the distinct directions, all other component stresses being in mutual equilibrium. Let $\mathrm{d}A$ denote the surface area relevant to $\mathbf{t}_c$, $\mathrm{d}A_\vartheta$ the projection onto the plane perpendicular to the stress σ_c^ϑ, and $\mathrm{d}S_c^\vartheta$ the resultant force along the associated tetrahedron edge with length $\mathrm{d}x_\vartheta$. Then, in a condensed form,

$$\mathbf{t}_c = \begin{bmatrix} t_c^\alpha \\ t_c^\gamma \\ t_c^\varepsilon \end{bmatrix} = \begin{bmatrix} \mathrm{d}A_\alpha/\mathrm{d}A & & \\ & \mathrm{d}A_\gamma/\mathrm{d}A & \\ & & \mathrm{d}A_\varepsilon/\mathrm{d}A \end{bmatrix} \begin{bmatrix} \sigma_c^\alpha \\ \sigma_c^\gamma \\ \sigma_c^\varepsilon \end{bmatrix} = \frac{3}{\mathrm{d}A} \begin{bmatrix} \mathrm{d}S_c^\alpha \\ \mathrm{d}S_c^\gamma \\ \mathrm{d}S_c^\varepsilon \end{bmatrix}. \tag{82}$$

The forces along the edges of the elementary tetrahedron equivalent to the stress $\boldsymbol{\sigma}_c$ are indicated in Fig 9. Their magnitudes form the 6×1 vector

$$\mathrm{d}\mathbf{S}_c = \left\{\mathrm{d}S_c^\alpha \ \mathrm{d}S_c^\beta \ \cdots \ \mathrm{d}S_c^\zeta\right\}. \tag{83}$$

The virtual work equality for the individual constituent $\mathrm{d}S_{\mathrm{c}}^{\vartheta}$ and the stress $\sigma_{\mathrm{c}}^{\vartheta}$ in the volume $\mathrm{d}V$ can be written in the form

$$\left(\mathrm{d}_{\vartheta} v_{\mathrm{t}}^{\vartheta}\right) \mathrm{d}S_{\mathrm{c}}^{\vartheta} = \delta_{\mathrm{t}}^{\vartheta} \sigma_{\mathrm{c}}^{\vartheta} \mathrm{d}V. \tag{84}$$

Expressing the elongation velocity along the edge $\mathrm{d}x_{\vartheta}$ by

$$\mathrm{d}_{\vartheta} v_{\mathrm{t}}^{\vartheta} = \frac{\partial v_{\mathrm{t}}^{\vartheta}}{\partial x_{\vartheta}} \mathrm{d}x_{\vartheta} = \delta_{\mathrm{t}}^{\vartheta} \mathrm{d}x_{\vartheta}, \tag{85}$$

the force follows to

$$\mathrm{d}S_{\mathrm{c}}^{\vartheta} = \sigma_{\mathrm{c}}^{\vartheta} \frac{\mathrm{d}V}{\mathrm{d}x_{\vartheta}}. \tag{86}$$

With the elementary volume $\mathrm{d}V = \mathrm{d}A_{\vartheta} \mathrm{d}x_{\vartheta}/3$ the stress is seen to be defined as

$$\sigma_{\mathrm{c}}^{\vartheta} = 3 \frac{\mathrm{d}S_{\mathrm{c}}^{\vartheta}}{\mathrm{d}A_{\vartheta}}. \tag{87}$$

The matrix form of Eqn (86) for all six edges of the element is

$$\mathrm{d}\mathbf{S}_{\mathrm{c}} = \mathrm{d}V \left\lceil \mathrm{d}x_{\vartheta} \right\rfloor^{-1} \boldsymbol{\sigma}_{\mathrm{c}} \qquad \vartheta = \alpha, \cdots, \zeta. \tag{88}$$

The Cartesian components of the stress resultants are obtained by the transformation of Eqn (30):

$$\mathrm{d}\mathbf{S} = \mathbf{B}_{\mathrm{ND}}^{\mathrm{t}} \mathrm{d}\mathbf{S}_{\mathrm{c}}. \tag{89}$$

5. Material constitutive description

The stress-strain relations for small strain elasticity were presented in [2], [3] as indicated in the Introduction, plasticity has been the subject of the work in [8]. The aforementioned publications have set the fundaments for subsequent extensions of the natural formalism to the large strain regime. In what follows the description is exemplified for the viscous solid with reference to [16], on the background of the equivalence to the familiar Cartesian formulation.

It is worth noticing that the normal stresses obtained along each natural direction from the Cartesian stress components analogously to the deformation rate, Eqns (61), (63), emanate from the entire stress state and are therefore denoted as the natural total stresses listed in the 6×1 vector $\boldsymbol{\sigma}_{\mathrm{t}}$:

$$\boldsymbol{\sigma}_{\mathrm{t}} = \mathbf{C}^{\mathrm{t}} \boldsymbol{\sigma} = \mathcal{A} \boldsymbol{\sigma}_{\mathrm{c}}. \tag{90}$$

The second equation takes account of Eqn (77) with the matrix

$$\mathcal{A} = \mathbf{C}^{\mathrm{t}} \mathbf{C} = \begin{bmatrix} 1 & c_{\alpha\beta}^2 & \cdots & c_{\alpha\zeta}^2 \\ & 1 & \cdots & c_{\beta\zeta}^2 \\ & & & \vdots \\ & \text{symmetric} & & 1 \end{bmatrix}, \tag{91}$$

effecting the transformation from the component stress definition. The work conjugate deformation rate to $\boldsymbol{\sigma}_{\mathrm{t}}$ is the component one

$$\boldsymbol{\delta}_{\mathrm{c}} = \boldsymbol{\partial}_{\mathrm{N}} \mathbf{v}_{\mathrm{c}}. \tag{92}$$

With the relationship between stresses in Eqn (90) there follows,

$$\boldsymbol{\delta}_{\mathrm{c}} = \mathcal{A}^{-1} \boldsymbol{\delta}_{\mathrm{t}} = \mathbf{C}^{-1} \boldsymbol{\delta}. \tag{93}$$

Material constitutive relations will be stated between component stress and total strain. The latter, resolved from the displacements in the global Cartesian frame, determines the natural stress $\boldsymbol{\sigma}_{\mathrm{c}}$

which ultimately composes the forces for establishing the equilibrium in the global system.

Isotropy suggests a distinction between hydrostatic/volumetric and deviatoric conditions. The hydrostatic part of the stress $\boldsymbol{\sigma} = \boldsymbol{\sigma}_\mathrm{D} + \boldsymbol{\sigma}_\mathrm{H}$ is specified by the invariant sum of the normal components: $\sigma_{11} + \sigma_{22} + \sigma_{33}$. Using matrix notation and utilizing the transformation by Eqn (77),

$$\sigma_\mathrm{H} = \frac{1}{3}\mathbf{e}_{3,3}^\mathrm{t}\boldsymbol{\sigma} = \frac{1}{3}\mathbf{e}_6^\mathrm{t}\boldsymbol{\sigma}_\mathrm{c}. \tag{94}$$

The vectors

$$\mathbf{e}_{3,3} = \{1\ 1\ 1\ 0\ 0\ 0\}\,, \qquad \mathbf{e}_6 = \{1\ 1\ 1\ 1\ 1\ 1\}\,,$$

are employed as summation/distribution operators in the Cartesian and the natural forms, respectively. Also, $\mathbf{e}_{3,3}^\mathrm{t}\mathbf{C} = \mathbf{e}_6^\mathrm{t}$ and $\mathbf{e}_6^\mathrm{t}\mathcal{A}^{-1}\mathbf{e}_6 = \mathbf{e}_{3,3}^\mathrm{t}\mathbf{e}_{3,3} = 3$ is an interesting consequence of the former. The hydrostatic stress state is defined by $\boldsymbol{\sigma}_\mathrm{H} = \sigma_\mathrm{H}\mathbf{e}_{3,3}$ resp. $\boldsymbol{\sigma}_\mathrm{tH} = \mathbf{C}^\mathrm{t}\boldsymbol{\sigma}_\mathrm{H} = \sigma_\mathrm{H}\mathbf{e}_6$. Hence, with Eqn (94),

$$\boldsymbol{\sigma}_\mathrm{tH} = \sigma_\mathrm{H}\mathbf{e}_6 = \frac{1}{3}\mathbf{e}_6\mathbf{e}_6^\mathrm{t}\boldsymbol{\sigma}_\mathrm{c}. \tag{95}$$

The deviatoric part becomes

$$\boldsymbol{\sigma}_\mathrm{tD} = \boldsymbol{\sigma}_\mathrm{t} - \boldsymbol{\sigma}_\mathrm{tH} = \left(\mathcal{A} - \frac{1}{3}\mathbf{e}_6\mathbf{e}_6^\mathrm{t}\right)\boldsymbol{\sigma}_\mathrm{c}. \tag{96}$$

The scalar equivalent deviatoric stress is determined by the expression

$$\bar{\sigma}^2 = \frac{3}{2}\boldsymbol{\sigma}_\mathrm{D}^\mathrm{t}\boldsymbol{\sigma}_\mathrm{D} = \frac{3}{2}\boldsymbol{\sigma}_\mathrm{D}^\mathrm{t}\boldsymbol{\sigma} = \frac{3}{2}\boldsymbol{\sigma}_\mathrm{tD}^\mathrm{t}\boldsymbol{\sigma}_\mathrm{c}, \tag{97}$$

which utilizes the transformation by Eqn (90).

Volumetric and deviatoric parts of the deformation rate $\boldsymbol{\delta} = \boldsymbol{\delta}_\mathrm{D} + \boldsymbol{\delta}_\mathrm{V}$ are defined in analogy to the stress quantities. The volumetric rate reads

$$\delta_\mathrm{V} = \frac{1}{3}\mathbf{e}_6^\mathrm{t}\boldsymbol{\delta}_\mathrm{c} = \frac{1}{3}\mathbf{e}_6^\mathrm{t}\mathcal{A}^{-1}\boldsymbol{\delta}_\mathrm{t}, \tag{98}$$

and induces the isotropic state

$$\boldsymbol{\delta}_\mathrm{tV} = \delta_\mathrm{V}\mathbf{e}_6. \tag{99}$$

The deviatoric deformation rate is

$$\boldsymbol{\delta}_\mathrm{tD} = \boldsymbol{\delta}_\mathrm{t} - \boldsymbol{\delta}_\mathrm{tV} = \left(\mathcal{A} - \frac{1}{3}\mathbf{e}_6\mathbf{e}_6^\mathrm{t}\right)\boldsymbol{\delta}_\mathrm{c} = \left(\mathbf{I} - \frac{1}{3}\mathbf{e}_6\mathbf{e}_6^\mathrm{t}\mathcal{A}^{-1}\right)\boldsymbol{\delta}_\mathrm{t}. \tag{100}$$

The equivalent deviatoric deformation rate is given on account of Eqn (93) as

$$\bar{\delta}^2 = \frac{2}{3}\boldsymbol{\delta}_\mathrm{D}^\mathrm{t}\boldsymbol{\delta}_\mathrm{D} = \frac{2}{3}\boldsymbol{\delta}_\mathrm{tD}^\mathrm{t}\boldsymbol{\delta}_\mathrm{c} = \frac{2}{3}\boldsymbol{\delta}_\mathrm{tD}^\mathrm{t}\mathcal{A}^{-1}\boldsymbol{\delta}_\mathrm{t}. \tag{101}$$

Isotropic viscosity relates the deviatoric quantities by $\boldsymbol{\sigma}_\mathrm{D} = 2\mu\boldsymbol{\delta}_\mathrm{D}$ and hydrostatic/volumetric measures by $\sigma_\mathrm{H} = 3\kappa\delta_\mathrm{V}$, with μ, κ denoting here the appertaining viscosity coefficients not necessarily constant, but possibly depending on the deformation rate. In natural terms this becomes for the deviatoric stress

$$\boldsymbol{\sigma}_\mathrm{cD} = 2\mu\boldsymbol{\delta}_\mathrm{cD} = 2\mu\mathcal{A}^{-1}\boldsymbol{\delta}_\mathrm{tD}, \tag{102}$$

and for the hydrostatic stress

$$\boldsymbol{\sigma}_\mathrm{cH} = 3\kappa\boldsymbol{\delta}_\mathrm{cV} = 3\kappa\mathcal{A}^{-1}\boldsymbol{\delta}_\mathrm{tV}. \tag{103}$$

Superposition of Eqns (102) and (103) relates component stress and total deformation rate by

$$\boldsymbol{\sigma}_\mathrm{c} = 2\mu\left(\mathcal{A}^{-1} + \frac{3\kappa - 2\mu}{6\mu}\mathcal{A}^{-1}\mathbf{e}_6\mathbf{e}_6^\mathrm{t}\mathcal{A}^{-1}\right)\boldsymbol{\delta}_\mathrm{t} = \boldsymbol{\mu}_\mathrm{N}\boldsymbol{\delta}_\mathrm{t}, \tag{104}$$

which specifies the viscosity matrix $\boldsymbol{\mu}_\mathrm{N}$ for the isotropic material. In anisotropy, if the material viscosity matrix has been specified with respect to Cartesian axes such that $\boldsymbol{\sigma} = \boldsymbol{\mu}\boldsymbol{\delta}$, the natural one follows by the transformation:

$$\boldsymbol{\sigma}_\mathrm{c} = \mathbf{C}^{-1}\boldsymbol{\mu}\mathbf{C}^{-\mathrm{t}}\boldsymbol{\delta}_\mathrm{t} = \boldsymbol{\mu}_\mathrm{N}\boldsymbol{\delta}_\mathrm{t}, \qquad \boldsymbol{\mu}_\mathrm{N} = \mathbf{C}^{-1}\boldsymbol{\mu}\mathbf{C}^{-\mathrm{t}}. \tag{105}$$

Viscous deformation is associated with the dissipation of mechanical work which in the unit volume occurs at the rate

$$\boldsymbol{\sigma}^\mathrm{t}\boldsymbol{\delta} = \boldsymbol{\sigma}_\mathrm{c}^\mathrm{t}\boldsymbol{\delta}_\mathrm{t} = \boldsymbol{\delta}_\mathrm{t}^\mathrm{t}\boldsymbol{\mu}_\mathrm{N}\boldsymbol{\delta}_\mathrm{t} \geq 0. \tag{106}$$

The inequality is required by the second law of thermodynamics for the entropy production [20]. If the quadratic viscosity matrix $\boldsymbol{\mu}_\mathrm{N}$ is otherwise arbitrary, only its symmetric part is effective in dissipation and this is requested to be positive definite. But the forces $\boldsymbol{\sigma}_\mathrm{c}$ and the fluxes $\boldsymbol{\delta}_\mathrm{t}$ – to use the terminology of irreversible thermodynamics – are related by a symmetric coefficient matrix [21].

The natural fashion of the stress-strain relations for isotropic elasticity in the small strain regime are set up in analogy to the viscous case. Instead of Eqn (102), the relation between deviatoric stress $\boldsymbol{\sigma}_\mathrm{cD}$ and elastic strain $\boldsymbol{\varepsilon}_\mathrm{cD}$, $\boldsymbol{\varepsilon}_\mathrm{tD}$ becomes

$$\boldsymbol{\sigma}_\mathrm{cD} = 2G\boldsymbol{\varepsilon}_\mathrm{cD} = 2G\mathcal{A}^{-1}\boldsymbol{\varepsilon}_\mathrm{tD}, \tag{107}$$

and Eqn (103) here relates the hydrostatic stress $\boldsymbol{\sigma}_\mathrm{cH}$ to the elastic volumetric strain $\boldsymbol{\varepsilon}_\mathrm{cV}$, $\boldsymbol{\varepsilon}_\mathrm{tV}$:

$$\boldsymbol{\sigma}_\mathrm{cH} = 3k\boldsymbol{\varepsilon}_\mathrm{cV} = 3k\mathcal{A}^{-1}\boldsymbol{\varepsilon}_\mathrm{tV}. \tag{108}$$

Small strains are derived from the displacement field by the same operation as the deformation rate derives from the velocity field; the modulus of volume expansion is $k = E/3(1-2\nu)$, the shear modulus $G = E/2(1+\nu)$, with E, ν the modulus of elasticity in tension and Poisson's ratio, respectively. Superposition of Eqns (107) and (108) yields the relation between component stress and total elastic strain as

$$\boldsymbol{\sigma}_\mathrm{c} = 2G\left(\mathcal{A}^{-1} + \frac{\nu}{1-2\nu}\mathcal{A}^{-1}\mathbf{e}_6\mathbf{e}_6^\mathrm{t}\mathcal{A}^{-1}\right)\boldsymbol{\varepsilon}_\mathrm{t} = \boldsymbol{\kappa}_\mathrm{N}\boldsymbol{\varepsilon}_\mathrm{t}, \tag{109}$$

and specifies the elasticity matrix $\boldsymbol{\kappa}_\mathrm{N}$ for the isotropic material subjected to small strains.

6. Finite deformation

While the solid deforms, particles originally at ${}^\mathrm{o}\mathbf{x}$ move to positions $\mathbf{x} = {}^\mathrm{o}\mathbf{x} + \mathbf{u}({}^\mathrm{o}\mathbf{x})$ where $\mathbf{u}({}^\mathrm{o}\mathbf{x})$ denotes the displacement. Line elements $\mathrm{d}\mathbf{x}$ in the material emanating from the particle transform as

$$\mathrm{d}\mathbf{x} = \mathbf{F}\mathrm{d}^\mathrm{o}\mathbf{x}, \tag{110}$$

where the matrix

$$\mathbf{F} = \frac{\mathrm{d}\mathbf{x}}{\mathrm{d}^\mathrm{o}\mathbf{x}} = \mathbf{I} + \frac{\mathrm{d}\mathbf{u}}{\mathrm{d}^\mathrm{o}\mathbf{x}} \tag{111}$$

is known as the deformation gradient. The definition of strain refers to the change of the scalar product of two line elements $\mathrm{d}^\mathrm{o}\mathbf{x}_1$ and $\mathrm{d}^\mathrm{o}\mathbf{x}_2$ emanating from the particle:

$$\mathrm{d}\mathbf{x}_1^\mathrm{t}\mathrm{d}\mathbf{x}_2 - \mathrm{d}^\mathrm{o}\mathbf{x}_1^\mathrm{t}\mathrm{d}^\mathrm{o}\mathbf{x}_2 = \mathrm{d}^\mathrm{o}\mathbf{x}_1^\mathrm{t}\left(\mathbf{F}^\mathrm{t}\mathbf{F} - \mathbf{I}\right)\mathrm{d}^\mathrm{o}\mathbf{x}_2^\mathrm{t} = 2\mathrm{d}^\mathrm{o}\mathbf{x}_1^\mathrm{t}\boldsymbol{\Gamma}\mathrm{d}^\mathrm{o}\mathbf{x}_2. \tag{112}$$

The symmetric 3×3 matrix

$$\boldsymbol{\Gamma} = \frac{1}{2}\left(\mathbf{F}^\mathrm{t}\mathbf{F} - \mathbf{I}\right) = \boldsymbol{\Gamma}^\mathrm{t}, \tag{113}$$

corresponds to the definition of Green strain. The entities in $\mathbf{\Gamma} = [\gamma_{ij}]\ i,j = 1,2,3$ specify the 6×1 strain vector

$$\boldsymbol{\gamma} = \left\{ \gamma_{11}\ \gamma_{22}\ \gamma_{33}\ \sqrt{2}\gamma_{12}\ \sqrt{2}\gamma_{23}\ \sqrt{2}\gamma_{13} \right\}. \tag{114}$$

For material line elements $\mathrm{d}^{\mathrm{o}}\mathbf{x}_{\vartheta} = (\mathrm{d}^{\mathrm{o}}x_{\vartheta})^{\mathrm{o}}\mathbf{c}_{\vartheta}$ along the natural directions at the original state before deformation Eqn (112) gives the total strains

$$\frac{1}{2}\frac{(\mathrm{d}x_{\vartheta})^2 - (\mathrm{d}^{\mathrm{o}}x_{\vartheta})^2}{(\mathrm{d}^{\mathrm{o}}x_{\vartheta})^2} = {}^{\mathrm{o}}\mathbf{c}_{\vartheta}^{\mathrm{t}}\mathbf{\Gamma}\,{}^{\mathrm{o}}\mathbf{c}_{\vartheta} = \gamma_{\mathrm{t}}^{\vartheta}, \tag{115}$$

the relation to the Cartesian measures being as in Eqns (60), (61) for the deformation rate. The strain vector

$$\boldsymbol{\gamma}_{\mathrm{t}} = \left\{ \gamma_{\mathrm{t}}^{\alpha}\ \gamma_{\mathrm{t}}^{\beta}\ \gamma_{\mathrm{t}}^{\gamma}\ \gamma_{\mathrm{t}}^{\delta}\ \gamma_{\mathrm{t}}^{\varepsilon}\ \gamma_{\mathrm{t}}^{\zeta} \right\}, \tag{116}$$

follows the transformation as by Eqn (63),

$$\boldsymbol{\gamma}_{\mathrm{t}} = {}^{\mathrm{o}}\mathbf{C}^{\mathrm{t}}\boldsymbol{\gamma}. \tag{117}$$

Alternatively, the natural strains are obtained directly by interpreting the second part of Eqn (115) as

$$\gamma_{\mathrm{t}}^{\vartheta} = \frac{1}{2}\left[(\mathbf{F}\,{}^{\mathrm{o}}\mathbf{c}_{\vartheta})^{\mathrm{t}}(\mathbf{F}\,{}^{\mathrm{o}}\mathbf{c}_{\vartheta}) - 1 \right]. \tag{118}$$

In the above, the vector

$$\mathbf{F}\,{}^{\mathrm{o}}\mathbf{c}_{\vartheta} = \frac{\mathrm{d}\mathbf{x}}{\mathrm{d}^{\mathrm{o}}\mathbf{x}}{}^{\mathrm{o}}\mathbf{c}_{\vartheta} = \frac{\partial \mathbf{x}}{\partial^{\mathrm{o}}x_{\vartheta}} = {}^{\mathrm{o}}\mathbf{c}_{\vartheta} + \frac{\partial \mathbf{u}}{\partial^{\mathrm{o}}x_{\vartheta}} \tag{119}$$

specifies the line element along the original ϑ-direction after deformation .

Observing Eqn (111) for the deformation gradient, the strain of Eqn (118) can be presented as a function of the displacement $\mathbf{u} = \mathbf{x} - {}^{\mathrm{o}}\mathbf{x}$ which reads,

$$\begin{aligned} \gamma_{\mathrm{t}}^{\vartheta} &= {}^{\mathrm{o}}\mathbf{c}_{\vartheta}^{\mathrm{t}}\frac{\mathrm{d}\mathbf{u}}{\mathrm{d}^{\mathrm{o}}\mathbf{x}}{}^{\mathrm{o}}\mathbf{c}_{\vartheta} + \frac{1}{2}\left(\frac{\mathrm{d}\mathbf{u}}{\mathrm{d}^{\mathrm{o}}\mathbf{x}}{}^{\mathrm{o}}\mathbf{c}_{\vartheta}\right)^{\mathrm{t}}\left(\frac{\mathrm{d}\mathbf{u}}{\mathrm{d}^{\mathrm{o}}\mathbf{x}}{}^{\mathrm{o}}\mathbf{c}_{\vartheta}\right) \\ &= {}^{\mathrm{o}}\mathbf{c}_{\vartheta}^{\mathrm{t}}\frac{\mathrm{d}\mathbf{u}}{\partial^{\mathrm{o}}x_{\vartheta}} + \frac{1}{2}\left(\frac{\partial\mathbf{u}}{\partial^{\mathrm{o}}x_{\vartheta}}\right)^{\mathrm{t}}\left(\frac{\partial\mathbf{u}}{\partial^{\mathrm{o}}x_{\vartheta}}\right). \end{aligned} \tag{120}$$

The first, linear term in the above expression

$${}^{\mathrm{o}}\mathbf{c}_{\vartheta}^{\mathrm{t}}\frac{\mathrm{d}\mathbf{u}}{\partial^{\mathrm{o}}x_{\vartheta}} = \frac{\partial u_{\mathrm{t}}^{\vartheta}}{\partial^{\mathrm{o}}x_{\vartheta}}, \tag{121}$$

with the orthogonal projection $u_{\mathrm{t}}^{\vartheta}$ of the displacement onto the ϑ–axis, is analogous to the deformation rate (or velocity strain, Eqn (53)). This term appertains to infinitesimal deformation; the second, quadratic term completes the determination of finite strain according to Eqn (115).

Since the line elements are specified in the material before deformation occurs, reference is entirely to the original configuration. The natural axes follow the deformation of the material, however, and they are taken at the actual configuration if the line elements are specified in it as in the definition of Almansi strain [20].

7. A plane elastic case

The parallelogram panel in Fig 10 is considered in two-dimensional space under the condition of plane stress. The natural reference system $O - x_{\alpha}x_{\beta}x_{\gamma}$ is defined as shown in the figure. With

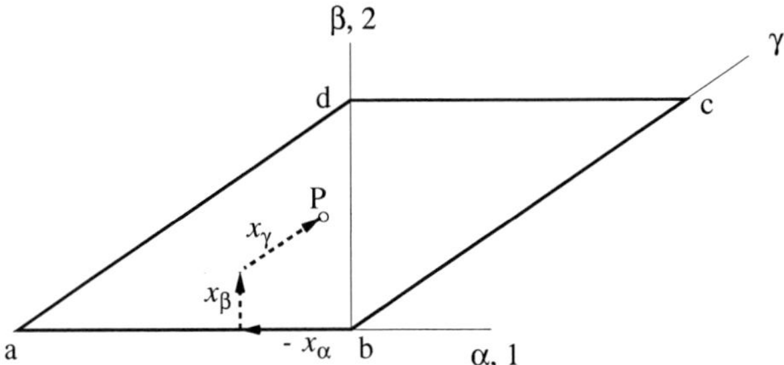

Figure 10. Parallelogram panel. Selected path to point P along the natural directions α, β, γ, defines the component coordinates.

the notation of Section 2, the component coordinates $\mathbf{x}_\mathrm{c}$ of a point in the plane transform to the Cartesian $\mathbf{x}$ according to

$$\mathbf{x} = \begin{bmatrix} x_1 \\ x_2 \end{bmatrix} = \begin{bmatrix} 1 & 0 & c_{1\gamma} \\ 0 & 1 & c_{2\gamma} \end{bmatrix} \begin{bmatrix} x_\alpha \\ x_\beta \\ x_\gamma \end{bmatrix} = \mathbf{B}_{\mathrm{ND}}^{\mathrm{t}} \mathbf{x}_\mathrm{c}. \tag{122}$$

This specifies the matrix $\mathbf{B}_{\mathrm{ND}}^{\mathrm{t}}$:

$$\mathbf{B}_{\mathrm{ND}}^{\mathrm{t}} = \begin{bmatrix} 1 & 0 & c_{1\gamma} \\ 0 & 1 & c_{2\gamma} \end{bmatrix}, \tag{123}$$

with $c_{1\gamma} = c_{\gamma 1}, c_{2\gamma} = c_{\gamma 2}$ being the direction cosines of the oblique x_γ–axis to the axes $x_1 = x_\alpha$ and $x_2 = x_\beta$, respectively.

Apropos of a functional transformation $\hat{\phi}(\mathbf{x}) = \hat{\phi}(\mathbf{B}_{\mathrm{ND}}^{\mathrm{t}} \mathbf{x}_\mathrm{c}) = \phi(\mathbf{x}_\mathrm{c})$ it is noticed that for fixed location $\mathbf{x}(\mathrm{P})$ of point P the condition

$$\mathbf{B}_{\mathrm{ND}}^{\mathrm{t}} \mathbf{x}_\mathrm{c}(\mathrm{P}) = \mathbf{x}(\mathrm{P}), \;\; \mathbf{B}_{\mathrm{ND}}^{\mathrm{t}} \mathrm{d}\mathbf{x}_\mathrm{c}(\mathrm{P}) = \mathbf{0}, \tag{124}$$

imposes an interdependence on the constituents of $\mathbf{x}_\mathrm{c}(\mathrm{P})$, while non uniqueness is maintained (Fig 10). Derivatives are related as

$$\left(\frac{\mathrm{d}\phi}{\mathrm{d}\mathbf{x}_\mathrm{c}}\right)^{\mathrm{t}} = \begin{bmatrix} \dfrac{\partial \phi}{\partial x_\alpha} \\ \dfrac{\partial \phi}{\partial x_\beta} \\ \dfrac{\partial \phi}{\partial x_\gamma} \end{bmatrix} = \begin{bmatrix} 1 & 0 \\ 0 & 1 \\ c_{1\gamma} & c_{2\gamma} \end{bmatrix} \begin{bmatrix} \dfrac{\partial \hat{\phi}}{\partial x_1} \\ \dfrac{\partial \hat{\phi}}{\partial x_2} \end{bmatrix} = \mathbf{B}_{\mathrm{ND}} \left(\frac{\mathrm{d}\hat{\phi}}{\mathrm{d}\mathbf{x}}\right)^{\mathrm{t}}. \tag{125}$$

In detail,

$$\frac{\partial \phi}{\partial x_\alpha} = \frac{\partial \hat{\phi}}{\partial x_1}, \;\; \frac{\partial \phi}{\partial x_\beta} = \frac{\partial \hat{\phi}}{\partial x_2}, \;\; \frac{\partial \phi}{\partial x_\gamma} = c_{1\gamma} \frac{\partial \hat{\phi}}{\partial x_1} + c_{2\gamma} \frac{\partial \hat{\phi}}{\partial x_2}.$$

The formalism of Section 3 applies to the infinitesimal elastic strain $\boldsymbol{\varepsilon}$ dealt with here. The transformation for the total strain $\boldsymbol{\varepsilon}_\mathrm{t}$ in the plane becomes

$$\boldsymbol{\varepsilon}_\mathrm{t} = \begin{bmatrix} \varepsilon_\mathrm{t}^\alpha \\ \varepsilon_\mathrm{t}^\beta \\ \varepsilon_\mathrm{t}^\gamma \end{bmatrix} = \begin{bmatrix} 1 & 0 & 0 \\ 0 & 1 & 0 \\ c_{\gamma 1}^2 & c_{\gamma 2}^2 & \sqrt{2} c_{\gamma 1} c_{\gamma 2} \end{bmatrix} \begin{bmatrix} \varepsilon_{11} \\ \varepsilon_{22} \\ \sqrt{2}\varepsilon_{12} \end{bmatrix} = \mathbf{C}^{\mathrm{t}} \boldsymbol{\varepsilon}. \tag{126}$$

The above specifies the matrix **C**; setting up the inverse relationship is simple:

$$\mathbf{C} = \begin{bmatrix} 1 & 0 & c_{\gamma1}^2 \\ 0 & 1 & c_{\gamma2}^2 \\ 0 & 0 & \sqrt{2}c_{\gamma1}c_{\gamma2} \end{bmatrix}, \quad \mathbf{C}^{-1} = \begin{bmatrix} 1 & 0 & -c_{\gamma1}/\sqrt{2}c_{\gamma2} \\ 0 & 1 & -c_{\gamma2}/\sqrt{2}c_{\gamma1} \\ 0 & 0 & 1/\sqrt{2}c_{\gamma1}c_{\gamma2} \end{bmatrix}. \tag{127}$$

The matrices enter the stress transformation of Section 4 as well. Alternatively, they may be deduced therefrom.

For completeness, and for subsequent use, recall the kinematic compatibility condition in the familiar Cartesian notation,

$$\frac{\partial^2 \varepsilon_{11}}{\partial x_1^2} + \frac{\partial^2 \varepsilon_{22}}{\partial x_2^2} = 2\frac{\partial^2 \varepsilon_{12}}{\partial x_1 \partial x_2}. \tag{128}$$

Substituting from the inverse Eqn (126) and interpreting the derivatives along the natural directions this becomes,

$$c_{\gamma1}\frac{\partial^2 \varepsilon_{\mathrm{t}}^{\alpha}}{\partial x_\beta \partial x_\gamma} + c_{\gamma2}\frac{\partial^2 \varepsilon_{\mathrm{t}}^{\beta}}{\partial x_\alpha \partial x_\gamma} = \frac{\partial^2 \varepsilon_{\mathrm{t}}^{\gamma}}{\partial x_\alpha \partial x_\beta}. \tag{129}$$

Both forms, the Cartesian and the natural one, ensure that the components of the displacement vector in the plane are continuous functions of the position.

Static quantities have been the subject of Section 4. The transformation of the component natural stress reads in the present case

$$\boldsymbol{\sigma} = \begin{bmatrix} \sigma_{11} \\ \sigma_{22} \\ \sqrt{2}\sigma_{12} \end{bmatrix} = \begin{bmatrix} 1 & 0 & c_{\gamma1}^2 \\ 0 & 1 & c_{\gamma2}^2 \\ 0 & 0 & \sqrt{2}c_{\gamma1}c_{\gamma2} \end{bmatrix} \begin{bmatrix} \sigma_{\mathrm{c}}^{\alpha} \\ \sigma_{\mathrm{c}}^{\beta} \\ \sigma_{\mathrm{c}}^{\gamma} \end{bmatrix} = \mathbf{C}\boldsymbol{\sigma}_{\mathrm{c}}. \tag{130}$$

For the condition of interior equilibrium one deduces the explicit form

$$\begin{aligned} \mathbf{B}_{\mathrm{ND}}^{\mathrm{t}}(\boldsymbol{\partial}_{\mathrm{N}}\boldsymbol{\sigma}_{\mathrm{c}}) &= \begin{bmatrix} 1 & 0 & c_{1\gamma} \\ 0 & 1 & c_{2\gamma} \end{bmatrix} \left\{ \frac{\partial \sigma_{\mathrm{c}}^{\alpha}}{\partial x_\alpha}\ \frac{\partial \sigma_{\mathrm{c}}^{\beta}}{\partial x_\beta}\ \frac{\partial \sigma_{\mathrm{c}}^{\gamma}}{\partial x_\gamma} \right\} = \\ &= \begin{bmatrix} \dfrac{\partial \sigma_{\mathrm{c}}^{\alpha}}{\partial x_\alpha} + c_{1\gamma}\dfrac{\partial \sigma_{\mathrm{c}}^{\gamma}}{\partial x_\gamma} \\ \dfrac{\partial \sigma_{\mathrm{c}}^{\beta}}{\partial \sigma_{\mathrm{c}}^{\beta}} + c_{2\gamma}\dfrac{\partial \sigma_{\mathrm{c}}^{\gamma}}{\partial x_\gamma} \end{bmatrix} = \mathbf{0}. \end{aligned} \tag{131}$$

The static boundary condition for the component natural stress $(\mathbf{N}^{\mathrm{t}}\mathbf{C})\boldsymbol{\sigma}_{\mathrm{c}} = \mathbf{t} = \mathbf{B}_{\mathrm{ND}}^{\mathrm{t}}\mathbf{t}_{\mathrm{c}}$ becomes along the horizontal sides ab and cd, respectively, of the panel:

$$\mp \begin{bmatrix} c_{1\gamma}c_{2\gamma}\sigma_{\mathrm{c}}^{\gamma} \\ \sigma_{\mathrm{c}}^{\beta} + c_{2\gamma}^2\sigma_{\mathrm{c}}^{\gamma} \end{bmatrix} = \begin{bmatrix} t_1 \\ t_2 \end{bmatrix} = \begin{bmatrix} t_{\mathrm{c}}^{\alpha} + c_{1\gamma}t_{\mathrm{c}}^{\gamma} \\ t_{\mathrm{c}}^{\beta} + c_{2\gamma}t_{\mathrm{c}}^{\gamma} \end{bmatrix}, \tag{132}$$

and along the oblique sides bc, da:

$$\pm \begin{bmatrix} c_{\gamma2}\sigma_{\mathrm{c}}^{\alpha} \\ -c_{\gamma1}\sigma_{\mathrm{c}}^{\beta} \end{bmatrix} = \begin{bmatrix} t_1 \\ t_2 \end{bmatrix} = \begin{bmatrix} t_{\mathrm{c}}^{\alpha} + c_{1\gamma}t_{\mathrm{c}}^{\gamma} \\ t_{\mathrm{c}}^{\beta} + c_{2\gamma}t_{\mathrm{c}}^{\gamma} \end{bmatrix}. \tag{133}$$

For the relationship between strain $\boldsymbol{\varepsilon}_{\mathrm{t}} = \{\varepsilon_{\mathrm{t}}^{\alpha}\,\varepsilon_{\mathrm{t}}^{\beta}\,\varepsilon_{\mathrm{t}}^{\gamma}\}$ and stress $\boldsymbol{\sigma}_{\mathrm{c}} = \{\sigma_{\mathrm{c}}^{\alpha}\,\sigma_{\mathrm{c}}^{\beta}\,\sigma_{\mathrm{c}}^{\gamma}\}$, refer to Section 5. Isotropic elasticity is described by separate relations for volumetric/hydrostatic, and deviatoric quantities. Bearing in mind the plane stress condition, Eqns (108) and (107) give:

$$\begin{aligned} \varepsilon_{\mathrm{tV}} &= \frac{1}{3k}\mathcal{A}\boldsymbol{\sigma}_{\mathrm{cH}} = \frac{1}{3k}\frac{1}{3}\mathbf{e}_3\mathbf{e}_3^{\mathrm{t}}\boldsymbol{\sigma}_{\mathrm{c}}, \\ \varepsilon_{\mathrm{tD}} &= \frac{1}{2G}\mathcal{A}\boldsymbol{\sigma}_{\mathrm{cD}} = \frac{1}{2G}\left(\mathcal{A} - \frac{1}{3}\mathbf{e}_3\mathbf{e}_3^{\mathrm{t}}\right)\boldsymbol{\sigma}_{\mathrm{c}}. \end{aligned} \tag{134}$$

The vector $\mathbf{e}_3 = \{1\ 1\ 1\}$ effects the operations of summation and distribution in the reduced space of plane stress. With the modulus of volume expansion $k = E/3(1-2\nu)$ and the shear modulus $G = E/2(1+\nu)$, superposition of the strain parts in Eqn (134) yields

$$\varepsilon_{\rm t} = \frac{1}{2G}\left(\mathcal{A} - \frac{\nu}{1+\nu}\mathbf{e}_3\mathbf{e}_3^{\rm t}\right)\boldsymbol{\sigma}_{\rm c} = \boldsymbol{\kappa}_{\rm N}^{-1}\boldsymbol{\sigma}_{\rm c}, \tag{135}$$

which defines the flexibility matrix $\boldsymbol{\kappa}_{\rm N}^{-1}$ for the isotropic elastic material in plane stress.

In the particular case under investigation, the relevant matrix $\mathcal{A}$ reads

$$\mathcal{A} = \mathbf{C}^{\rm t}\mathbf{C} = \begin{bmatrix} 1 & 0 & c_{1\gamma}^2 \\ & 1 & c_{2\gamma}^2 \\ \text{sym} & & 1 \end{bmatrix}, \tag{136}$$

where the equalities $c_{\alpha\gamma} = c_{1\gamma}, c_{\beta\gamma} = c_{2\gamma}$ have been accounted for. Thus, a specific form of Eqn (135) is

$$\begin{bmatrix} \varepsilon_{\rm t}^{\alpha} \\ \varepsilon_{\rm t}^{\beta} \\ \varepsilon_{\rm t}^{\gamma} \end{bmatrix} = \frac{1}{E}\begin{bmatrix} 1 & -\nu & (1+\nu)c_{1\gamma}^2 - \nu \\ & 1 & (1+\nu)c_{2\gamma}^2 - \nu \\ \text{sym} & & 1 \end{bmatrix}\begin{bmatrix} \sigma_{\rm c}^{\alpha} \\ \sigma_{\rm c}^{\beta} \\ \sigma_{\rm c}^{\gamma} \end{bmatrix}. \tag{137}$$

The stress-strain relations of Eqn (137) are utilized to express the kinematic compatibility condition, Eqn (129), in terms of stress:

$$c_{\gamma 1}\frac{\partial^2 \sigma_{\rm H}}{\partial x_\beta \partial x_\gamma} + c_{\gamma 2}\frac{\partial^2 \sigma_{\rm H}}{\partial x_\alpha \partial x_\gamma} = \frac{\partial^2 \sigma_{\rm H}}{\partial x_\alpha \partial x_\beta}. \tag{138}$$

In order to discuss solutions by Airy's stress function $\phi(x_1, x_2)$ the familiar formalism for the Cartesian stresses

$$\sigma_{11} = \frac{\partial^2 \hat{\phi}}{\partial x_2^2}, \quad \sigma_{22} = \frac{\partial^2 \hat{\phi}}{\partial x_1^2}, \quad \sigma_{12} = -\frac{\partial^2 \hat{\phi}}{\partial x_1 \partial x_2},$$

transfers to the natural component stresses by Eqn (130) and Eqn (125) as

$$\sigma_{\rm c}^{\alpha} = \frac{1}{c_{\gamma 2}}\frac{\partial^2 \phi}{\partial x_\beta \partial x_\gamma}, \quad \sigma_{\rm c}^{\beta} = \frac{1}{c_{\gamma 1}}\frac{\partial^2 \phi}{\partial x_\alpha \partial x_\gamma}, \quad \sigma_{\rm c}^{\gamma} = -\frac{1}{c_{\gamma 1}c_{\gamma 2}}\frac{\partial^2 \phi}{\partial x_\alpha \partial x_\beta}. \tag{139}$$

Satisfaction of the interior equilibrium condition, Eqn (131), by the above expressions for ϕ continuous, is easily verified. The function ϕ is specified by kinematic compatibility and boundary conditions. From Eqn (138) with Eqn (139),

$$\begin{aligned} & c_{\gamma 1}^2\frac{\partial^4 \phi}{\partial x_\beta^2 \partial x_\gamma^2} + c_{\gamma 2}^2\frac{\partial^4 \phi}{\partial x_\alpha^2 \partial x_\gamma^2} + \frac{\partial^4 \phi}{\partial x_\alpha^2 \partial x_\beta^2} = \\ = \; & 2\left(c_{\gamma 1}\frac{\partial^4 \phi}{\partial x_\alpha \partial x_\beta^2 \partial x_\gamma} + c_{\gamma 2}\frac{\partial^4 \phi}{\partial x_\alpha^2 \partial x_\beta \partial x_\gamma} - c_{\gamma 1}c_{\gamma 2}\frac{\partial^4 \phi}{\partial x_\alpha \partial x_\beta \partial x_\gamma^2}\right). \end{aligned} \tag{140}$$

With reference to the treatment of rectangular panels in [22], solutions $\phi(x_\alpha, x_\beta, x_\gamma)$ of Eqn (140) will be considered in the form of polynomials of various degrees. It is observed that here only mixed terms in $x_\alpha, x_\beta, x_\gamma$ are of relevance. Polynomials up to the third degree satisfy Eqn (140) identically, the coefficients adjusted to the actual boundary conditions.

The second degree polynomial

$$\phi = a x_\beta x_\gamma + b x_\alpha x_\gamma + c x_\alpha x_\beta, \tag{141}$$

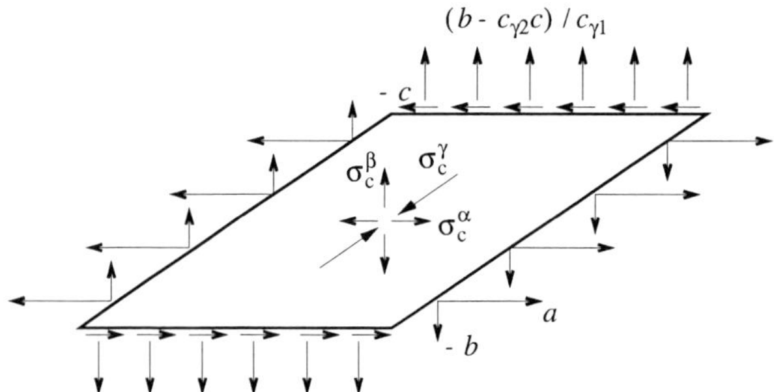

Figure 11. Boundary tractions defined in the Cartesian system.

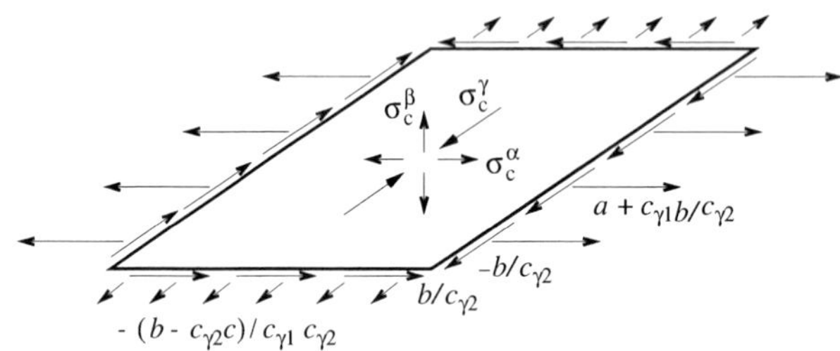

Figure 12. Boundary tractions referring to the natural system ($t_c^\beta = 0$).

is the lowest one associated with stress. From Eqn (139):

$$\sigma_c^\alpha = \frac{a}{c_{\gamma 2}}, \quad \sigma_c^\beta = \frac{b}{c_{\gamma 1}}, \quad \sigma_c^\gamma = -\frac{c}{c_{\gamma 1} c_{\gamma 2}}. \tag{142}$$

The three stresses are constant throughout the panel. Satisfaction of the static boundary conditions along the sides of the parallelogram, Eqn (132) and Eqn (133), determines the coefficients a, b, c as shown in Fig 11 for actions defined along the Cartesian axes.

It is evident, that the same stress state can be induced by a variety of alternative decompositions of the boundary tractions. For instance, if load application is possible in the direction of the α–axis and the γ–axis but not along the β–axis, the associated statement of the boundary conditions in Eqn (132) and Eqn (133) gives in conjunction with the stress expressions from Eqn (142):
for the horizontal sides of the parallelogram

$$t_c^\alpha = \pm \frac{b}{c_{\gamma 2}}, \quad t_c^\gamma = \mp \frac{b - c_{\gamma 2} c}{c_{\gamma 1} c_{\gamma 2}}, \tag{143}$$

and for the oblique sides,

$$t_c^\alpha = \pm \left(a + \frac{c_{\gamma 1}}{c_{\gamma 2}} b \right), \quad t_c^\gamma = \mp \frac{b}{c_{\gamma 2}}. \tag{144}$$

This is demonstrated in Fig 12. Figure 13 shows resolution of the boundary tractions on the sides ab, cd and bc, da along the other directions, respectively. This is an application of the direct statement referring to Eqn (82), which translates for the horizontal sides to

$$t_c^\beta = \mp \sigma_c^\beta = \mp \frac{b}{c_{\gamma 1}}, \quad t_c^\gamma = \mp c_{\gamma 2} \sigma_c^\gamma = \pm \frac{c}{c_{\gamma 1}}, \tag{145}$$

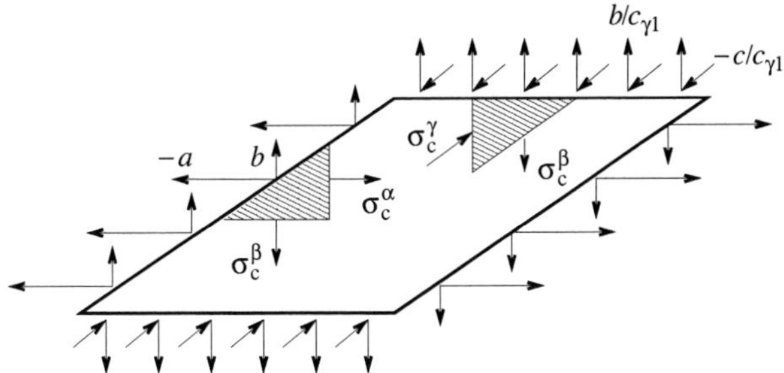

Figure 13. Direct statement of boundary conditions in natural terms.

and for the oblique sides to

$$t_{\rm c}^{\alpha} = \pm c_{\gamma 2}\sigma_{\rm c}^{\alpha} = \pm a, \;\; t_{\rm c}^{\beta} = \mp c_{\gamma 1}\sigma_{\rm c}^{\beta} = \mp b. \tag{146}$$

The stresses are substituted from Eqn (142).

The stress function stated in Eqn (141) may be arrived at from the expression

$$\hat{\phi} = \frac{\hat{a}}{2}x_1^2 + \hat{b}x_1x_2 + \frac{\hat{c}}{2}x_2^2,$$

by the transformation of Eqn (122) between the coordinates $\mathbf{x}$ and $\mathbf{x}_{\rm c}$. Maintainig only mixed terms in the natural coordinates, the stress function of Eqn (141) is reproduced for

$$a = \hat{b}c_{\gamma 1} + \hat{c}c_{\gamma 2}, \;\; b = \hat{a}c_{\gamma 1} + \hat{b}c_{\gamma 2}, \;\; c = \hat{b}.$$

Deducing with the above the Cartesian stress measures and determining then the component natural stresses, confirms those obtained directly in Eqn (142).

The polynomial of the third degree

$$\phi = ax_\alpha x_\beta x_\gamma + \frac{b}{2}x_\beta^2 x_\gamma + \frac{c}{2}x_\gamma^2 x_\beta + \frac{d}{2}x_\alpha^2 x_\gamma + \frac{e}{2}x_\gamma^2 x_\alpha + \frac{f}{2}x_\alpha^2 x_\beta + \frac{g}{2}x_\beta^2 x_\alpha, \tag{147}$$

furnishes stresses that vary linearly along $x_\alpha, x_\beta, x_\gamma$ within the panel:

$$\begin{aligned}
\sigma_{\rm c}^{\alpha} &= \frac{1}{c_{\gamma 2}}(ax_\alpha + bx_\beta + cx_\gamma),\\
\sigma_{\rm c}^{\beta} &= \frac{1}{c_{\gamma 1}}(ax_\beta + dx_\alpha + ex_\gamma),\\
\sigma_{\rm c}^{\gamma} &= -\frac{1}{c_{\gamma 1}c_{\gamma 2}}(ax_\gamma + fx_\alpha + gx_\beta).
\end{aligned} \tag{148}$$

Displaying in matrix form this becomes

$$\begin{bmatrix} c_{\gamma 2}\sigma_{\rm c}^{\alpha} \\ c_{\gamma 1}\sigma_{\rm c}^{\beta} \\ -c_{\gamma 1}c_{\gamma 2}\sigma_{\rm c}^{\gamma} \end{bmatrix} = \begin{bmatrix} a & b & c \\ d & a & e \\ f & g & a \end{bmatrix} \begin{bmatrix} x_\alpha \\ x_\beta \\ x_\gamma \end{bmatrix},$$

and facilitates discussion for various patterns of the coefficient matrix in conjunction with the boundary conditions.

8. Fluid media

8.1. Pressure

Equation (70) may be adapted to the statement of local equilibrium for a fluid at rest. In the absence of viscosity the fluid sustains only hydrostatic pressure p such that

$$\boldsymbol{\sigma} \Leftarrow \boldsymbol{\sigma}_{\mathrm{H}} = -\begin{bmatrix} p \\ p \\ p \\ 0 \\ 0 \\ 0 \end{bmatrix} = -p\,\mathbf{e}_{3,3},$$

$$\boldsymbol{\partial}^{\mathrm{t}}\boldsymbol{\sigma} \Leftarrow \boldsymbol{\partial}^{\mathrm{t}}\boldsymbol{\sigma}_{\mathrm{H}} = -\mathbf{g}(p). \tag{149}$$

Then Eqn (70) becomes

$$-\mathbf{g}(p) + \mathbf{f} = \mathbf{0}. \tag{150}$$

The vector $\mathbf{f}$ comprises volume forces as these arising from gravity, $\mathbf{g}(p)$ denotes the pressure gradient

$$\mathbf{g}(p) = \left(\frac{\mathrm{d}p}{\mathrm{d}\mathbf{x}}\right)^{\mathrm{t}} = \begin{bmatrix} \dfrac{\partial p}{\partial x_1} \\ \dfrac{\partial p}{\partial x_2} \\ \dfrac{\partial p}{\partial x_3} \end{bmatrix}. \tag{151}$$

The relation to natural terms is by

$$\mathrm{d}p = \frac{\mathrm{d}p}{\mathrm{d}\mathbf{x}_{\mathrm{c}}}\mathrm{d}\mathbf{x}_{\mathrm{c}} = \frac{\mathrm{d}p}{\mathrm{d}\mathbf{x}}\mathrm{d}\mathbf{x},$$

where $\mathrm{d}\mathbf{x} = \mathbf{B}_{\mathrm{ND}}^{\mathrm{t}}\mathrm{d}\mathbf{x}_{\mathrm{c}}$, Eqn (30). Thus, for the orthogonal projection of the gradient along the natural directions, collected in the vector

$$\mathbf{g}_{\mathrm{t}}(p) = \left(\frac{\mathrm{d}p}{\mathrm{d}\mathbf{x}_c}\right)^{\mathrm{t}} = \begin{bmatrix} \dfrac{\partial p}{\partial x_\alpha} \\ \dfrac{\partial p}{\partial x_\beta} \\ \vdots \\ \dfrac{\partial p}{\partial x_\zeta} \end{bmatrix}, \tag{152}$$

there follows in agreement with Eqn (30),

$$\mathbf{g}_{\mathrm{t}}(p) = \mathbf{B}_{\mathrm{ND}}\mathbf{g}(p). \tag{153}$$

Application of the above transformation to Eqn (150) establishes equilibrium along the natural directions by

$$-\mathbf{g}_{\mathrm{t}}(p) + \mathbf{f}_{\mathrm{t}} = \mathbf{0}, \tag{154}$$

the vector $\mathbf{f}_{\mathrm{t}}$ comprising the respective projections of the volume force $\mathbf{f}$.

Considering the equilibrium statement in the form of Eqn (78) one has

$$\boldsymbol{\sigma}_{\mathrm{c}} \Leftarrow \boldsymbol{\sigma}_{\mathrm{CH}} = \mathcal{A}^{-1}\boldsymbol{\sigma}_{\mathrm{tH}} = -p\mathcal{A}^{-1}\mathbf{e}_6,$$

$$\boldsymbol{\partial}_{\mathrm{N}}\boldsymbol{\sigma}_{\mathrm{c}} \Leftarrow -\boldsymbol{\partial}_{\mathrm{N}}(p\mathcal{A}^{-1}\mathbf{e}_6) = -(\boldsymbol{\partial}_{\mathrm{N}}p)\mathcal{A}\mathbf{e}_6. \tag{155}$$

The total hydrostatic stress $\boldsymbol{\sigma}_{\mathrm{tH}}$, Eqn (95), has been introduced by the transformation of Eqn (90). The symbol $\boldsymbol{\partial}_{\mathrm{N}}p$ denotes a diagonal matrix comprising the partial derivatives $\partial p/\partial x_\vartheta$ of the pressure along the natural directions, the entities of the vector $\mathbf{g}_{\mathrm{t}}(p)$ in Eqn (151): $\mathbf{g}_{\mathrm{t}}(p) = (\boldsymbol{\partial}_{\mathrm{N}}p)\mathbf{e}_6$. Equation (78) becomes

$$\mathbf{B}^{\mathrm{t}}_{\mathrm{ND}}[-(\boldsymbol{\partial}_{\mathrm{N}}p)\mathcal{A}^{-1}\mathbf{e}_6 + \mathbf{f}_{\mathrm{c}}] = \mathbf{0}. \tag{156}$$

8.2. Velocity field

Non-equilibrium gives rise to accelerated motion. In the spatial reference system the particle velocity $\mathbf{v}(t, \mathbf{x})$ varies with time t and particle position $\mathbf{x}$:

$$\dot{\mathbf{v}} = \frac{\mathrm{d}\mathbf{v}}{\mathrm{d}t} = \frac{\partial \mathbf{v}}{\partial t} + \frac{\partial \mathbf{v}}{\partial \mathbf{x}}\mathbf{v}. \tag{157}$$

Accounting for the associated inertia force in Eqn (150) results to the Euler equations in the form

$$\varrho\dot{\mathbf{v}} = -\mathbf{g}(p) + \mathbf{f}, \tag{158}$$

where ϱ denotes the density of the fluid, and the acceleration $\dot{\mathbf{v}}$ consists of the local part $\partial\mathbf{v}/\partial t$ and the convective term as from Eqn (157); the convective term introduces a nonlinear dependence on the velocity.

Multiplying through by the matrix $\mathbf{B}_{\mathrm{ND}}$ transforms Eqn (158) to one between total natural quantities:

$$\varrho\dot{\mathbf{v}}_{\mathrm{t}} = -\mathbf{g}_{\mathrm{t}}(p) + \mathbf{f}_{\mathrm{t}}. \tag{159}$$

The particle acceleration along the natural directions is

$$\dot{\mathbf{v}}_{\mathrm{t}} = \frac{\mathrm{d}\mathbf{v}_{\mathrm{t}}}{\mathrm{d}t} = \frac{\partial \mathbf{v}_{\mathrm{t}}}{\partial t}\frac{\partial \mathbf{v}_{\mathrm{t}}}{\partial \mathbf{x}_{\mathrm{c}}}\mathbf{v}_{\mathrm{c}}. \tag{160}$$

The above implies the equivalence

$$\frac{\partial(\cdot)}{\partial \mathbf{x}}\frac{\mathrm{d}\mathbf{x}}{\mathrm{d}t} = \frac{\partial(\cdot)}{\partial \mathbf{x}_{\mathrm{c}}}\frac{\mathrm{d}\mathbf{x}_{\mathrm{c}}}{\mathrm{d}t},$$

see also Eqn (40). Also, $\mathrm{d}\mathbf{x}/\mathrm{d}t = \mathbf{v}$, $\mathrm{d}\mathbf{x}_{\mathrm{c}}/\mathrm{d}t = \mathbf{v}_{\mathrm{c}}$.

Alternatively to Eqn (159), substituting from Eqn (153) for the non-equilibrated forces on the right-hand side of Eqn (158) and multiplying both sides by the matrix $\mathbf{B}_{\mathrm{ND}}$ one obtains

$$\varrho\dot{\mathbf{v}}_{\mathrm{t}} = \mathbf{B}_{\mathrm{tc}}[-(\boldsymbol{\partial}_{\mathrm{N}}p)\mathcal{A}^{-1}\mathbf{e}_6 + \mathbf{f}_{\mathrm{c}}]. \tag{161}$$

Thereby, the equilibrium condition of Eqn (156) has been transformed to an equation of motion for the fluid. The matrix $\mathbf{B}_{\mathrm{tc}} = \mathbf{B}_{\mathrm{ND}}\mathbf{B}^{\mathrm{t}}_{\mathrm{ND}}$, Eqn (34), relates total natural quantities to the component ones of a vector.

Viscosity adds deviatoric stresses; instead of Eqn (149),

$$\boldsymbol{\sigma} \Leftarrow 2\mu\boldsymbol{\delta} - p\mathbf{e}_{3,3},$$

$$\boldsymbol{\partial}^{\mathrm{t}}\boldsymbol{\sigma} \Leftarrow \boldsymbol{\partial}^{\mathrm{t}}(2\mu\boldsymbol{\delta} - p\mathbf{e}_{3,3}) = 2\mu\boldsymbol{\partial}^{\mathrm{t}}\boldsymbol{\delta} - \mathbf{g}(p). \tag{162}$$

Here the deformation rate $\boldsymbol{\delta} = \boldsymbol{\partial}\mathbf{v}$, Eqn (66), appertains to the isochoric medium: $\boldsymbol{\delta} = \boldsymbol{\delta}_{\mathrm{D}}$. The transition to the last expression is for constant viscosity coefficient μ in the field. The above modifies Eqn (158) to

$$\varrho\dot{\mathbf{v}} = \boldsymbol{\partial}^{\mathrm{t}}(2\mu\boldsymbol{\delta} - p\mathbf{e}_{3,3}) + \mathbf{f} = 2\mu\boldsymbol{\partial}^{\mathrm{t}}\boldsymbol{\partial}\mathbf{v} - \mathbf{g}(p) + \mathbf{f}, \tag{163}$$

as a collective form of the familiar Navier-Stokes equations governing viscous fluid motion.

In natural terms, Eqn (155) must be supplemented by the deviatoric stress from Eqn (102):

$$\boldsymbol{\sigma}_{\mathrm{c}} \Leftarrow \mathcal{A}^{-1}(2\mu\boldsymbol{\delta}_{\mathrm{t}} - p\mathbf{e}_6),$$

$$\boldsymbol{\partial}_{\mathrm{N}}\boldsymbol{\sigma}_{\mathrm{c}} \Leftarrow \boldsymbol{\partial}_{\mathrm{N}}\mathcal{A}^{-1}(2\mu\boldsymbol{\delta}_{\mathrm{t}} - p\mathbf{e}_6) = 2\mu\boldsymbol{\partial}_{\mathrm{N}}\mathcal{A}^{-1}\boldsymbol{\delta}_{\mathrm{t}} - (\boldsymbol{\partial}_{\mathrm{N}}p)\mathcal{A}^{-1}\mathbf{e}_6. \tag{164}$$

The total deformation rate is $\boldsymbol{\delta}_{\mathrm{t}} = \boldsymbol{\partial}_{\mathrm{N}}\mathbf{v}_{\mathrm{t}}$, Eqn (55), and since the medium is assumed isochoric $\boldsymbol{\delta}_{\mathrm{t}} = \boldsymbol{\delta}_{\mathrm{tD}}$. The consequence for Eqn (161) is

$$\begin{aligned} \varrho\dot{\mathbf{v}}_{\mathrm{t}} &= \mathbf{B}_{\mathrm{tc}}[\boldsymbol{\partial}_{\mathrm{N}}\mathcal{A}^{-1}(2\mu\boldsymbol{\partial}_{\mathrm{N}}\mathbf{v}_{\mathrm{t}} - p\mathbf{e}_6) + \mathbf{f}_{\mathrm{c}}] \\ &= \mathbf{B}_{\mathrm{tc}}[2\mu\boldsymbol{\partial}_{\mathrm{N}}\mathcal{A}^{-1}\boldsymbol{\partial}_{\mathrm{N}}\mathbf{v}_{\mathrm{t}} - (\boldsymbol{\partial}_{\mathrm{N}}p)\mathcal{A}^{-1}\mathbf{e}_6 + \mathbf{f}_{\mathrm{c}}]. \end{aligned} \tag{165}$$

The last expressions in Eqns (159), (165) are for constant viscosity μ in the field.

For the incompressible fluid the continuity condition may be stated in terms of component natural velocity $\mathbf{v}_{\mathrm{c}}$ by means of Eqn (44) derived with reference to Fig 6:

$$\operatorname{div}\mathbf{v} = \frac{\partial v_{\mathrm{c}}^{\alpha}}{\partial x_{\alpha}} + \frac{\partial v_{\mathrm{c}}^{\beta}}{\partial x_{\beta}} + \cdots + \frac{\partial v_{\mathrm{c}}^{\zeta}}{\partial x_{\zeta}} = \mathbf{e}_6^{\mathrm{t}}\boldsymbol{\partial}_{\mathrm{N}}\mathbf{v}_{\mathrm{c}} = \boldsymbol{\nabla}_{\mathrm{t}}^{\mathrm{t}}\mathbf{v}_{\mathrm{c}} = 0. \tag{166}$$

The equivalence $\mathbf{e}_6^{\mathrm{t}}\boldsymbol{\partial}_{\mathrm{N}}(\cdot) = \boldsymbol{\nabla}_{\mathrm{t}}^{\mathrm{t}}(\cdot)$ of the differential operators as applied to scalar or vectorial quantities is noticed at this place.

An alternative form of Eqn (166) for the vanishing volume change is deduced from Eqn (98) in terms of natural total velocity $\mathbf{v}_{\mathrm{t}}$:

$$3\delta_{\mathrm{V}} = \mathbf{e}_6^{\mathrm{t}}\mathcal{A}^{-1}\boldsymbol{\delta}_{\mathrm{t}} = \mathbf{e}_6^{\mathrm{t}}\mathcal{A}^{-1}\boldsymbol{\partial}_{\mathrm{N}}\mathbf{v}_{\mathrm{t}} = 0. \tag{167}$$

In the case of compressiblity Fig 6 transfers to the mass supply $\varrho\mathbf{v}$, and the condition of mass conservation under non stationary conditions simply modifies to

$$\begin{aligned} \frac{\partial\varrho_{\mathrm{t}}}{\partial t} &= -\operatorname{div}(\varrho\mathbf{v}) \\ &= -\mathbf{e}_6^{\mathrm{t}}[(\boldsymbol{\partial}_{\mathrm{N}}\varrho)\mathbf{v}_{\mathrm{c}} + \varrho(\boldsymbol{\partial}_{\mathrm{N}}\mathbf{v}_{\mathrm{c}})] \\ &= -[\mathbf{g}_{\mathrm{t}}^{\mathrm{t}}(\varrho)\mathbf{v}_{\mathrm{c}} + \varrho\boldsymbol{\nabla}_{\mathrm{t}}^{\mathrm{t}}\mathbf{v}_{\mathrm{c}}]. \end{aligned} \tag{168}$$

9. Thermal phenomena

The following demonstrates employment of the natural formalism in the energy balance and entropy in thermodynamics. The statement of the energy balance for the unit mass of the medium in the form

$$\dot{u} = \dot{w} + \dot{q}, \tag{169}$$

involves the time rate of internal energy $\dot{u}$, mechanical work $\dot{w}$ and heat supply $\dot{q}$ per unit mass, respectively.

Without going into details as for particular assumptions [12], [20] the rate of specific internal energy is considered as

$$\dot{u} = c\dot{T}. \tag{170}$$

Here, c is the specific heat capacity of the medium under the actual conditions, and $T(t,\mathbf{x}) = T(t,\mathbf{x}_{\mathrm{c}})$ the absolute temperature. If particles at momentary spatial position $\mathbf{x}, \mathbf{x}_{\mathrm{c}}$ move with velocity $\mathbf{v}, \mathbf{v}_{\mathrm{c}}$:

$$\dot{T} = \frac{\mathrm{d}T}{\mathrm{d}t} = \frac{\partial T}{\partial t} + \mathbf{g}^{\mathrm{t}}(T)\mathbf{v} = \frac{\partial T}{\partial t} + \mathbf{g}_{\mathrm{t}}^{\mathrm{t}}(T)\mathbf{v}_{\mathrm{c}} \tag{171}$$

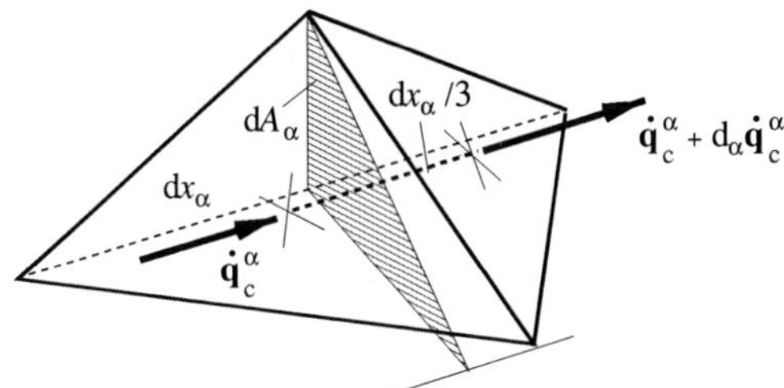

Figure 14. Heat supply to the natural element due to the component heat flux $\dot{\mathbf{q}}_{\mathrm{c}}^{\alpha}$.

The convective contribution to the temperature rate in the medium is a consequence of the velocity $\mathbf{v}, \mathbf{v}_{\mathrm{c}}$ and the respective temperature gradient

$$\mathbf{g}(T) = \left(\frac{\partial T}{\partial \mathbf{x}}\right)^{\mathrm{t}}, \qquad \mathbf{g}_{\mathrm{t}}(T) = \left(\frac{\partial T}{\partial \mathbf{x}_{\mathrm{c}}}\right)^{\mathrm{t}}, \tag{172}$$

defined analogously to Eqns (151) and (152) for the pressure.

The mechanical power per unit mass (see Eqn (76)) is

$$\dot{w} = \frac{1}{\varrho}\boldsymbol{\sigma}^{\mathrm{t}}\boldsymbol{\delta} = \frac{1}{\varrho}\boldsymbol{\sigma}_{\mathrm{t}}^{\mathrm{t}}\boldsymbol{\delta}_{\mathrm{c}}. \tag{173}$$

The rate of deformation $\boldsymbol{\delta}, \boldsymbol{\delta}_{\mathrm{c}}$ derives from the velocity field $\mathbf{v}, \mathbf{v}_{\mathrm{c}}$. While the medium may be isochoric under isothermal conditions, temperature variation causes volumetric changes; hydrostatic pressure performs thereby work.

Heat supplied to the unit mass in an undirected way is not subject of the natural formalism. Directed heat supply may be described by the heat flux vector

$$\dot{\mathbf{q}} = \{\dot{q}_1\ \dot{q}_2\ \dot{q}_3\}, \tag{174}$$

referred to the Cartesian system, or by its component natural counterpart

$$\dot{\boldsymbol{q}}_{\mathrm{c}} = \{\dot{q}_{\mathrm{c}}^{\alpha}\ \dot{q}_{\mathrm{c}}^{\beta} \cdots \dot{q}_{\mathrm{c}}^{\zeta}\}. \tag{175}$$

The two quantities are related by Eqn (30).

The entities in the arrays of Eqns (174), (175) specify the heat flow passing the unit area perpendicular to the respective direction. The heat supply per unit volume may be formally given as by Eqns (44), (46) in either the Cartesian or the natural formulation:

$$\varrho\dot{q} = -\mathrm{div}\dot{\mathbf{q}} = -\boldsymbol{\nabla}^{\mathrm{t}}\dot{\mathbf{q}} = -\boldsymbol{\nabla}_{\mathrm{t}}^{\mathrm{t}}\dot{\mathbf{q}}_{\mathrm{c}}. \tag{176}$$

The natural fashion of Eqn (176) is deduced in detail with reference to Fig 14. Noting the input $\dot{q}_{\mathrm{c}}^{\alpha}\mathrm{d}A_{\alpha}$ and the output $(\dot{q}_{\mathrm{c}}^{\alpha} + \mathrm{d}_{\alpha}\dot{q}_{\mathrm{c}}^{\alpha})\mathrm{d}A_{\alpha}$ of the heat rate emerging at distance $dx_{\alpha}/3$ from the point of input, the rate of heat supply to the element contributed by the component natural heat flux $\dot{q}_{\mathrm{c}}^{\alpha}$ reads

$$-\mathrm{d}_{\alpha}\dot{q}_{\mathrm{c}}^{\alpha}\mathrm{d}A_{\alpha} = -\frac{1}{3}\frac{\partial\dot{q}_{\mathrm{c}}^{\alpha}}{\partial x_{\alpha}}\mathrm{d}A_{\alpha} = -\frac{\partial\dot{q}_{\mathrm{c}}^{\alpha}}{\partial x_{\alpha}}\mathrm{d}V. \tag{177}$$

Summing up contributions along all natural directions,

$$\varrho\dot{q} = -\left(\frac{\partial\dot{q}_{\mathrm{c}}^{\alpha}}{\partial x_{\alpha}} + \frac{\partial\dot{q}_{\mathrm{c}}^{\beta}}{\partial x_{\beta}} + \cdots + \frac{\partial\dot{q}_{\mathrm{c}}^{\zeta}}{\partial x_{\zeta}}\right) = -\mathbf{e}_6^{\mathrm{t}}\boldsymbol{\partial}_{\mathrm{N}}\dot{\mathbf{q}}_{\mathrm{c}} = -\boldsymbol{\nabla}_{\mathrm{t}}^{\mathrm{t}}\dot{\mathbf{q}}_{\mathrm{c}}. \tag{178}$$

The specific entropy s per unit mass of the medium is a state variable which changes at a rate

$$\dot{s} = \dot{s}_{\mathrm{e}} + \dot{s}_{\mathrm{i}}. \tag{179}$$

The part $\dot{s}_{\mathrm{e}}$ refers to reversible entropy supply under thermodynamic equilibrium, the part $\dot{s}_{\mathrm{i}} \geq 0$ is due to irreversible entropy production. Considering the entropy effect of heat flux

$$\begin{aligned} \dot{s}_{\mathrm{e}} = -\frac{1}{\varrho}\mathrm{div}\left(\frac{1}{T}\dot{\mathbf{q}}\right) &= \frac{\dot{q}}{T} + \frac{1}{\varrho}\frac{1}{T^2}\mathbf{g}^{\mathrm{t}}(T)\dot{\mathbf{q}} \\ &= \frac{\dot{q}}{T} + \frac{1}{\varrho}\frac{1}{T^2}\mathbf{g}_{\mathrm{t}}^{\mathrm{t}}(T)\dot{\mathbf{q}}_{\mathrm{c}}, \end{aligned} \tag{180}$$

with $\dot{q}$ from Eqns (176), (178). The irreversible entropy production

$$\dot{s}_{\mathrm{i}} = -\frac{1}{\varrho}\frac{1}{T^2}\mathbf{g}^{\mathrm{t}}(T)\dot{\mathbf{q}} = -\frac{1}{\varrho}\frac{1}{T^2}\mathbf{g}_{\mathrm{t}}^{\mathrm{t}}(T)\dot{\mathbf{q}}_{\mathrm{c}}, \tag{181}$$

is invariant to the expression of the product

$$\mathbf{g}^{\mathrm{t}}(T)\dot{\mathbf{q}} = \mathbf{g}_{\mathrm{t}}^{\mathrm{t}}(T)\dot{\mathbf{q}}_{\mathrm{c}}$$

in either Cartesian or natural terms.

10. Conclusions

The natural finite element approach has been reviewed with regard to applications in the mechanics of deformable continua. For this purpose, the infinitesimal tetrahedron is adopted as the minimum configuration to define material deformation in space as the triangle does in the plane, and specifies a local reference system which to a certain extent is adjustable to the problem under investigation.

The methodology essentially starts with the velocity field as given in the global Cartesian system, which is resolved in the local natural one to obtain the so-called total deformation rate. The material constitutive relations furnish the component natural stress that results to forces entering the statement of the global equilibrium condition. The equivalence between the natural and the Cartesian description has been established, the former implying homogeneous, normal strain and stress measures omitting shear.

Apart from the mechanics of solids, the natural formalism has been considered with respect to its utility for the description of fluid motion and thermal phenomena.

Acknowledgment

The author is thankful to Mrs Grethe Knapp Christiansen for the preparation of the manuscript in LaTex.

References

[1] J. Argyris and S. Kelsey, Energy Theorems and Structural Analysis, Butterworths, London, 1960.

[2] J. Argyris, Recent Advances in Matrix Methods in Structural Analysis, Pergamon Press, Oxford, 1964.

[3] J. Argyris, Three-dimensional anisotropic and inhomogeneous elastic media, matrix analysis for small and large displacements, Ingenieur-Archiv **34** (1965) 33–55.

[4] J. Argyris, Continua and Discontinua, Internat. Conf. on Matrix Methods of Structural Mechanics, Dayton, Ohio, Wright-Patterson USAF Base, 1965. Proceedings 1967.

[5] M.A. Biot, Mechanics of Incremental Deformations, Wiley, New York, 1965.

[6] J. Argyris and D.W. Scharpf, The curved tetrahedronal and triangular elements TEC and TRIC for the matrix displacement method, Parts I and II, Aeron. J. Roy. Aeron. Soc. **73** (1969) 55–65.

[7] J. Argyris and D.W. Scharpf, Some general considerations on the natural mode technique. Part I, Small displacements, Aeron. J. Roy. Aeron. Soc. **73** (1969) 219–226, Part II, Large displacements, ibid, 361–368.

[8] J. Argyris, D.W. Scharpf and J.B. Spooner, Die elastoplastische Berechnung von allgemeinen Tragwerken und Kontinua, Ingenieur Archiv **37** (1969) 326–352.

[9] J. Argyris, M. Haase, G.A. Malegiannakis, Natural geometry of surfaces with specific reference to the matrix displacement analysis of shells. In: Koninklijke Nederlandse Akademie van Wetenschappen, Amsterdam. Proceedings, Series B. **76** (1973) 361–410.

[10] J. Argyris, H. Balmer, I. Doltsinis, P.C. Dunne, M. Haase, M. Kleiber, G. Malejannakis, H.-P. Mlejnek, M. Müller, D.W. Scharpf, Finite element method – the natural approach, Comput. Methods Appl. Mech. Engng. **17/18** (1979) 1–106.

[11] J. Argyris and I. Doltsinis, On the large strain inelastic analysis in natural formulation. Part I: Quasistatic problems, Comput. Methods Appl. Mech. Engng. **20** (1979) 213–251. Part II: Dynamic problems, ibid, **21** (1980) 91–126.

[12] J. Argyris and I. Doltsinis, On the natural formulation and analysis of large deformation coupled thermomechanical problems, Comput. Methods Appl. Mech. Engng. **25** (1981) 195–253.

[13] J. Argyris, M. Haase, H.-P. Mlejnek, Some considerations on the natural approach. In: Computer Meth. Appl. Mech. Engng. **30** (1982) 335–346.

[14] J. Argyris, I. Doltsinis, P.M. Pimenta and H. Wüstenberg, Thermomechanical response of solids at high strains–natural approach, Comput. Methods Appl. Mech. Engng. **32** (1982) 3–57.

[15] J. Argyris, I. Doltsinis, P.M. Pimenta and H. Wüstenberg, Natural finite element techniques for viscous fluid motion, Comput. Methods Appl. Mech. Engng. **45** (1984) 3–55.

[16] J. Argyris and I. Doltsinis, A primer on superplasticity in natural formulation, Comput. Methods. Appl. Mech. Engng. **46** (1984) 83–131.

[17] J. Argyris, I. Doltsinis, H. Fischer and H. Wüstenberg, TA ΠANTA PEI, Comput. Methods Appl. Mech. Engng. **51** (1985) 289–362.

[18] I. Doltsinis, Continuum aspects of the natural approach, Comput. Methods. Appl. Mech. Engng. **195** (2006) 5403–5421.

[19] I. Doltsinis, Elements of Plasticity – Theory and Computation, WIT-Press, Southampton, 1999.

[20] I. Doltsinis, Large Deformation Processes of Solids, WIT-Press, Southampton, 2004.

[21] S.R. de Groot, Thermodynamics of Irreversible Processes, North-Holland, Amsterdam, 1951.

[22] S.P. Timoshenko and J.N. Goodier, Theory of Elasticity, Third Edition, McGraw-Hill Kogakusha, Tokyo, 1970.

INDEX

A

B

C

D

E

F

G

H

I

J

K

L

M

N

O

P

Q

R

S

T

U

V

W

Y

Z